Systematics and Evolution of Fungi

Series on
Progress in Mycological Research

- Fungi from Different Environments
- Systematics and Evolution of Fungi

Systematics and Evolution of Fungi

Editors

J.K. Misra

Formerly Head, Department of Botany
Sri Jai Narain Postgraduate College
Lucknow
India

J.P. Tewari

Department of Agricultural, Food, and Nutritional Science
University of Alberta, Edmonton
Alberta
Canada

S.K. Deshmukh

Piramal Life Sciences Limited
Goregaon (East), Mumbai
India

CRC Press
Taylor & Francis Group
an Informa business
www.crcpress.com

6000 Broken Sound Parkway, NW
Suite 300, Boca Raton, FL 33487
270 Madison Avenue
New York, NY 10016
2 Park Square, Milton Park
Abingdon, Oxon OX14 4RN, UK

Science Publishers
Jersey, British Isles
Enfield, New Hampshire

Published by Science Publishers, an imprint of Edenbridge Ltd

- St. Helier, Jersey, British Channel Islands
- P.O. Box 699, Enfield, NH 03748, USA

E-mail: *info@scipub.net* Website: *www.scipub.net*

Marketed and distributed by:

CRC Press
Taylor & Francis Group
an **informa** business
www.crcpress.com

6000 Broken Sound Parkway, NW
Suite 300, Boca Raton, FL 33487
270 Madison Avenue
New York, NY 10016
2 Park Square, Milton Park
Abingdon, Oxon OX14 4RN, UK

ISBN: 978-1-57808-723-5

Cover illustrations: Reproduced by kind courtesy of Dr. Martha J. Powell, Dr. Peter M. Letcher, Dr. Tapani Yli-Mattila and Dr. Tatiana Gagkaeva.

Library of Congress Cataloging-in-Publication Data

Systematics and evolution of fungi / editors, J.K. Misra, J.P. Tewari, S.K. Deshmukh. -- 1st ed.

p. cm. -- (Progress in mycological research ; 2)

Includes bibliographical references and index.

ISBN 978-1-57808-723-5 (hardcover)

1. Fungi--Classification. 2. Fungi--Phylogeny. I. Misra, J. K. II. Tewari, J. P. III. Deshmukh, S. K. (Sunil K.) IV. Series: Progress in mycological research ; 2.

QK603.2.S97 2011

579.5′5--dc22

2011011246

The views expressed in this book are those of the author(s) and the publisher does not assume responsibility for the authenticity of the findings/conclusions drawn by the author(s). Also no responsibility is assumed by the publishers for any damage to the property or persons as a result of operation or use of this publication and/or the information contained herein.

Printed in the United States of America

Preface

The availability of molecular tools and techniques has revolutionized the thinking of morphosystematists about the relationships of fungi, their evolution and classification since the mid-1990s. Thus, the morphological approach of classifying taxa is getting the support of molecular studies that incorporate the relationships of taxa and their evolutionary trends on more reliable and stable molecular sequence characters. Hence, the current phylogenetic classification is legitimately gaining an edge over the solely morphology based system. But the exclusive use of molecular data in handling classification and identification without incorporating morphological characters causes problems particularly in deducing the correct identification of an organism. In future, however, molecular methods may further help investigators to identify species and their variants or assign a species to the one that has been identified using these methods/characters, though these have to be accepted with caution and hence polyphasic approach is desirable so that molecular databases should be treated as complementary to systematics rather than conflicting. Therefore, a balanced use of all available characters in a consensus way is advocated. Such considerations have favored some taxonomic groups and established some new while abandoning a few older ones.

This book, comprising 13 chapters is, therefore, intended to present the progress and shifts that have taken place towards the understanding of systematics and evolution of fungi in recent years. An attempt has been made to cover almost all the major groups of fungi such as the Chytridiomycota, Zygomycota, Ascomycota, and Basidiomycota but as one can appreciate, the huge mass of information available cannot be accommodated in just one book. Besides the above, advances in morphological and molecular taxonomy of highly toxigenic Fusarium species and understanding the phylogeny of the alternarioid hyphomycetes have also been dealt with in their respective chapters. Methods used in fungal evolutionary biology, their theory, examples and potential applications, and proteomics research for rapid diagnosis of invasive candidiasis have been reviewed in two different chapters. How the molecular biologists and morphosystematists can develop synergy between them has been elaborated in the introductory chapter-1.

We hope that the book will be of interest to systematists and evolutionary mycologists.

The editors are grateful to the contributors who took pains in providing the chapters for the book. One of us (SKD) is also grateful to Dr. Somesh Sharma, Managing Director, Piramal Life Sciences Limited, Mumbai for his help.

J.K. Misra
J.P. Tewari
S.K. Deshmukh

Contents

List of Contributors

Barseghyan Gayane S.

Institute of Evolution and Department of Evolutionary & Environmental Biology, Faculty of Science & Science Education, University of Haifa, Mt Carmel, Haifa 31905, Israel.

E-mail: gbarseghyan@gmail.com

Benny Gerald L.

Department of Plant Pathology, 1453 Fifleld Hall, University of Florida, Gainesville, Florida, USA.

E-mail: gbenny @ulf.edu

Didukh Maryna

Mushroom Biology and Fungal Biotechnology Laboratory, School of Agriculture and Environmental Sciences, North Carolina Agricultural and Technical State University, Greensboro, NC 27411, USA.

E-mail: maryna.didukh@gmail.com

Gannibal Philipp B.

All-Russian Institute of Plant Protection (VIZR), Shosse Podbelskogo 3, Pushkin, St. Petersburg, 196608, Russia.

E-mail: phbgannibal@yandex.ru

Gardner William G.

Northeastern Ohio Universities College of Medicine and Pharmacy, Rootstown, Ohio, U.S.A. 44272.

E-mail: wgardner@neoucom.edu

Hawksworth David L.

Departamento de Biología Vegetal II, Facultad de Farmacia, Universidad Complutense de Madrid, Plaza Ramón y Cajal, Madrid 28040, Spain and Department of Botany, Natural History Museum, Cromwell Road, London, SW7 5BD, UK.

E-mail: d.hawksworth@nhm.ac.uk

Isikhuemhen Omoanghe S.

Mushroom Biology and Fungal Biotechnology Laboratory, School of Agriculture and Environmental Sciences, North Carolina Agricultural and Technical State University, Greensboro, NC 27411, USA.

E-mail: omon@ncat.edu

Kosakyan Anush

Laboratory of Soil Biology, University of Neuchâtel, Rue Emile Argand 11, CH-2009 Neuchâtel, Switzerland.

E-mail: akosakyan@yahoo.com

Letcher Peter M.

Department of Biological Sciences, Box 870344, The University of Alabama, Tuscaloosa, AL 35487 USA.

E-mail: letch006@bama.ua.edu

Lichtwardt Robert W.

Department of Ecology & Evolutionary Biology, University of Kansas, 1200 Sunnyside Avenue, Lawrence, Kansas, 66045-7534 USA.

E-mail: licht@ku.edu

Misra J.K.

Bhartiya Sanskriti Evam Shiksha Samiti (Indian Cultural and Educational Society), Saroj Lalji Mehrotra (SLM) Bhartiya Vidya Bhavan Girls Degree College, 6 Vineet Khand, Gomti Nagar, Lucknow 226 010, India.

E-mail: jitrachravi@gmail.com

Nagy G. László

Department of Microbiology, Faculty of Science and Informatics, University of Szeged, Közép fasor 52., Szeged, H-6726, Hungary.

E-mail: cortinarius2000@yahoo.co.uk

Nevo Eviatar

Department of Evolutionary and Environmental Biology, Institute of Evolution, Faculty of Natural Sciences, University of Haifa, Mt. Carmel, Haifa 31905, Israel.

E-mail: nevo@research.haifa.ac.il

Papp Tamás

Department of Microbiology, Faculty of Science and Informatics, University of Szeged, Közép fasor 52., Szeged, H-6726, Hungary.

E-mail: pappt@bio.u-szeged.hu

Powell Martha J.

Department of Biological Sciences, Box 870344, The University of Alabama, Tuscaloosa, AL 35487 USA.

E-mail: mpowell@bama.us.edu

Raats Dina

Department of Evolutionary and Environmental Biology, Institute of Evolution, Faculty of Natural Sciences, University of Haifa, Mt. Carmel, Haifa 31905, Israel.

E-mail: ddraats@gmail.com

Tewari Jalpa P.

Department of Agricultural, Food, and Nutritional Science, University of Alberta, Edmonton, Alberta, Canada T6G 2P5.

E-mail: JP.Tewari@ualberta.ca; jptewari@shaw.ca

Tewari Sanjiv

Akron General Medical Center, Akron, Ohio USA 44307.

E-mail: stewari@agmc.org

Tripathi S.K.M.

Birbal Sahni Institute of Palaeobotany, 53 University Road, Lucknow 226 007, India.

E-mail: suryatripathi.2009@rediffmail.com

Tura Daniel

Department of Evolutionary and Environmental Biology, Institute of Evolution, Faculty of Natural Sciences, University of Haifa, Mt. Carmel, Haifa 31905, Israel.

E-mail: turadaniel21@yahoo.com

Vágvölgyi Csaba

Department of Microbiology, Faculty of Science and Informatics, University of Szeged, Közép fasor 52., Szeged, H-6726, Hungary.

E-mail: mucor1959@gmail.com

Wasser Solomon P.

Institute of Evolution and Department of Evolutionary and Environmental Biology, Faculty of Science and Science Education, University of Haifa, Mt. Carmel, 31905, Haifa, Israel and N.G. Kholodny Institute of Botany, National Academy of Sciences of Ukraine, 2, Tereshchenkovskaya St., Kiev, 010601, Ukraine.

E-mail: spwasser@research.haifa.ac.il

Yli-Mattila Tapani

Molecular Plant Biology, Department of Biochemistry and Food Chemistry, University of Turku, FI-20014 Turku, Finland.

E-mail: tymat@utu.fi

Zmitrovich Ivan V.

V.L. Komarov Botanical Institute, Russian Academy of Sciences, 2, Popova St., St. Petersburg, 197376, Russia.

E-mail: iv_zmitrovich@mail.ru

Chapter 1

Integrating Morphological and Molecular Data in Fungal Systematics

David L. Hawksworth

Departamento de Biología Vegetal II, Facultad de Farmacia
Universidad Complutense de Madrid, Plaza Ramón y Cajal, Madrid 28040
Spain; and Department of Botany, Natural History Museum Cromwell Road
London, SW7 5BD, UK
E-mail: d.hawksworth@nhm.ac.uk

ABSTRACT

The impacts of molecular approaches on fungal systematics are considered with respect to the elucidation of major evolutionary lineages, convergence and divergence of characters, generic concepts (including pleomorphism), species concepts (including cryptic species), the development of identification methods, and issues arising from the discovery of unattributable environmental sequences. The importance of collaborations between morphosystematists and molecular biologists is emphasized, including in the development of nomenclatural systems. Attention is drawn to caveats meriting consideration when undertaking molecular systematic studies. References to examples from the primary and recent literature are included throughout.

INTRODUCTION

"Molecules *vs.* Morphology" headlined a debate held by the British Mycological Society and British Society for Medical Mycology in Manchester in 2003 (Whalley, 2003). It encapsulates the concept of perceived conflict in approaches to fungal systematics, and reflects an underlying unease of a perceived threat to traditional mycology. This is not, however, a concern unique to mycologists, and similar sentiments have been expressed and assessed by, for example, botanists (Daly et al., 2001).

Many disagreements, inside and outside science, prove to be a consequence of either misunderstandings, or minds that are closed or reluctant to accept new ideas. In the case of systematics, the results of new research are of particularly wide concern as they frequently lead to changes in classification and, most irritatingly to those involved, scientific names in widespread use. Most users of the scientific names of fungi are not, of course, systematists; they are brewers, conservationists, ecologists, environmental health officers, field mycologists, food mycologists, geneticists, industrial mycologists, medical mycologists, pharmacologists, physiologists, plant pathologists, and quarantine officers, amongst others. Understandably, users react almost instinctively against changes that appear to be enforced without any explanations and for which they do not see a necessity – and especially if there are repeated changes in names commended for use within a short period of time.

During the four decades in which I have been involved in systematics, there have been several major innovations to embrace: cytology, numerical taxonomy, scanning electron microscopy, thin-layer chromatography, chemical products, serology, electrophoresis (especially of isozymes), transmission electron microscopy, monoclonal and polyclonal antibodies, cladistic (phylogenetic) analysis, computerized identification systems, and databasing. Now, there is a role in almost every aspect of a systematist's work for the worldwide web – even access from a personal computer or a mobile "telephone". Some of these approaches have occasioned much controversy and acrimony, most heatedly in the case of cladistics (Hull, 1988). However, in general, systematics has absorbed and benefited from a multifaceted and increasingly "polyphasic" approach.

What is different about the molecular revolution, that became energized with the discovery of the PCR reaction by Kary Mullis in 1983 (Bartlett and Stirling, 2003), is that it is pertinent to, and potentially permeates and affects, almost every aspect of systematics today – and that while refinements will continue, it will be with us to stay. The primary objective of this contribution is to draw attention to the ways molecular data are advancing fungal systematics, from illuminating our understanding of evolution and relationships, to the refinement of species concepts, and introduction of novel identification methods. A secondary objective is to reflect on the implications of molecular approaches for the future of the subject, and also to express some cautionary caveats that merit consideration by both morphological and molecular practitioners. I do this from the standpoint of a morphosystematist who now uses molecular approaches in some of his investigations. As this is such a vast topic, the examples chosen to illustrate particular points are necessarily eclectic and are not intended to be exhaustive.

Evolution of major lineages

The fungi, organisms studied by mycologists, have been revealed mostly to comprise a kingdom *Fungi* which has a common ancestor with *Animalia* and is placed with them in *Opisthokonta* (Adl et al., 2005). Some other organisms traditionally studied by mycologists belong elsewhere. The oomycete "fungi", for example, belong to a predominantly algal group (termed *Chromista* or *Straminipila*), and the cellular and plasmodial slime-moulds belong to different protist groups. These diverse organisms have been referred to as "fungal analogues" (Kirk et al., 2008). On the other hand, *Microsporidia* which mainly occur in invertebrates but also humans, prove to be a separate phylum within the kingdom *Fungi* (Adl et al., 2005; Corradi and Keeling, 2009); these comprise 160 genera and 1300 species and were not historically studied by mycologists.

As more fungi are being sequenced, especially representatives of orders and families not previously represented in molecular phylogenetic analyses, a more stable and evolutionarily meaningful classification system is emerging. There is now a broad consensus at the levels of phyla and order (Hibbett et al., 2007). That is not to say all issues are resolved in all groups; for example, there are proposals to divide *Chytridiomycota* into four phyla (M. J. Powell and P. M. Letcher; Ch 3, this volume), and *Zygomycota* into four subphyla, but with their relationships unresolved at the phylum level (G. L. Benny; Ch 4, this volume).

Attention is now also focussing on the age of different fungal lineages through the correlation of fossil remains of fungi: estimates are derived from rates of change in gene sequences, the so-called molecular clocks. There is still much to be done and some uncertainties (S. K. M. Tripathi; Ch 2, this volume), but it now seems clear that *Ascomycota* and *Basidiomycota* could have diverged as early as 1.2 Byr ago (Heckman et al., 2001). Berbee and Taylor (2010) endeavoured to establish the age of fungal groups on the basis of assumptions made as to the systematic position of some well-researched fossils of known geochronology. However, they showed that assumptions as to the systematic position of particular fossils had a dramatic effect on interpretation; for example, different interpretations of *Palaeopyrenomycites devonicus* yielded dates for *Ascomycota* divergence ranging from 452 Myr to 1489 Myr (i.e. 1.49 Byr). A further complication is that the speed of the molecular clock might vary in different fungal groups. Indeed, Wang et al. (2010) found rates in *Ascomycota* to be higher than in *Basidiomycota*, and especially fast in *Sordariomycetes*.

One of the most intriguing possibly fungal fossils is the Silurian and Lower Devonian *Prototaxites* which could reach an amazing 8 m in height and 1.25 m in width, and could have been lichenized (Selosse, 2002; Taylor et al., 2009). Indeed, the view that the earliest filamentous ascomycetes were lichenized has been given new life as the "Protolichenes Hypothesis" (Eriksson, 2005). That view has been challenged by Schoch et al. (2009) who considered that ascomata originated twice, and Lücking et al. (2009) who interpreted the asci of

Palaeopyrenomycites in a different way. However, these views have been critically analyzed by Eriksson (2009), and the debate will continue (L. Nagy et al.; Ch 13, this volume).

CHARACTER CONVERGENCE AND DIVERGENCE

One of the most amazing results of the re-circumscription of orders and families of fungi, based on molecular phylogenetic studies, relates to fungal sporocarps[1]. Features treated as fundamental and diagnostic even for classes, such as the flask-shaped perithecia of *Pyrenomycetes*, or the closed sporocarps of *Gasteromycetes*, have proved to be a consequence of character convergence or divergence and do not reflect evolutionary relationships. It is now recognized that several different types of sporocarp can occur in a single lineage. For example, *Agaricales* includes fungi which have stipitate or sessile caps (pilei); these can be lamellate, porate, have a smooth hymenium, one which is uneven (merulioid), or even resupinate with the hymenium covering teeth-like projections (hydnoid) (Hibbett, 2007; L. Nagy et al.; Ch 13, this volume). Conversely, the ascomycete truffle order *Tuberales* proved to comprise representatives of some 15 different lineages within *Pezizales* where they were mixed with fungi with open apothecia (Læssøe and Hansen, 2007). Similar situations were found to occur in relation to thallus form in *Lecanorales* which may be leprose, microfilamentous, gelatinous, crustose, squamulose, placodioid, foliose, umbilicate, shrubby, or pendent (Grube and Hawksworth, 2007). Re-definitions of taxa at all levels consequently become necessary, as in *Agaricomycetes* (I. V. Zmitrovich and S. P. Wasser; Ch 8, this volume).

A further aspect of character evolution, and especially divergence, is the issue of co-speciation or co-evolution of fungi along with their host organisms. This is a topic which has fascinating mycologists since the mid-1950s, especially with regard to rusts and their hosts (Pirozynski and Hawksworth, 1988), and the case with trichomycetes is considered here (R. W. Lichtwardt; Ch 5, this volume). Molecular tools, however, enable the phylogenies of both hosts and fungi to be compared and evidence for correlations sought. A fine example, also taking into account biogeographical aspects, is the revisiting of the *Cyttaria*/*Nothofagus* case with molecular phylogenetics by Peterson et al. (2010). More

[1]The term "sporocarp" is used throughout this contribution for specialized structures within which, or on which, the spore-producing cells are located, i.e. as a general term for many-celled bodies arising from a sexual act such as the ascomata in *Ascomycota* and the basidiomata in *Basidiomycota*. The alternative and often-used "fruit-body" or "fruiting body" is better avoided as these structures are not homologous with plant fruits. Also, "carpophores" is potentially confusing as it has been restricted to the stipe region of basidiomes by some authors (Snell and Dick 1957) paralleling its usage in plants for fruit stalks (Jackson 1916), including those supporting carpels or ovaries in flowers, and sporing structures in ferns (Beentje 2010). Further, "sporophore" has been employed in conidial fungi (especially hyphomycetes) for spore-producing or supporting structures (Kirk *et al.* 2008) in which no separately differentiated organs are formed.

fascinating insights as a consequence of such phylogenetic comparisons can be expected.

Generic concepts

Re-circumscriptions have also become necessary at the generic level. For example, "*Coprinus*" was found to comprise four genera placed in two different families, *Agaricaceae* and *Psathyrellaceae* (Redhead et al., 2001). On the other hand, molecular data showed that five genera of parmelioid lichens should be subsumed under the single generic name *Xanthoparmelia* (Blanco et al., 2004). In both these examples, there were correlations with characters which had not previously been considered of particular phylogenetic significance. Examples are becoming commonplace, and the cases of *Inonotus* and *Phellinus* (D. Tura et al.; Ch 9, this volume), and of *Alternaria*-like genera (P. B. Gannibal; Ch 11, this volume) are discussed here.

In some instances remarkable levels of convergence have been demonstrated. For instance, *Coniochaeta*, which despite having morphologically uniform ascomata and ascospore types, proved to include species that belonged to two different families placed in different orders, *Coniochaetaceae* (in *Coniochaetales*) and *Xylariaceae* (in *Xylariales*) – a result supported by the production of different conidial states (Garcia et al., 2006).

Molecular data also mean that fungi which produce no spores at all, or which produce only asexual spores (e.g. conidia) can be firmly placed in genera and families characterized by the formation of sexual spores[2]. For example, the placement of the completely sterile filamentous lichenized genera *Cystocoleus* and *Racodium* amongst the *Dothideomycetes* (Muggia et al., 2008). It also follows that there is no longer a scientific reason to retain a dual nomenclatural system for fungi with pleomorphic life-cycles, something already recognized by some mycologists as early as 1991 (Reynolds and Taylor, 1991, 1992). Currently, there is a concerted move to revise or eliminate this provision in fungal nomenclature, and most mycologists favour having one name for one fungal species (Norvell et al., 2010). Although the pertinent rules of nomenclature have not yet been modified fully to address this issue, mycologists are increasingly adopting a pragmatic approach. Examples include the description of a species in *Penicillium* in which a sexual stage is known (Houbraken et al., 2010), the use of *Calonectria* even for species where only the *Cylindrocladium* state is known (Lombard et al., 2010), and retention of the most-familiar names regardless of the state represented when taxa are re-modelled (Aveskamp et al., 2010). While the matter is still under discussion by the mycological community, it has been

[2]I prefer to use the readily understood terms "sexual" and "asexual" rather than "teleomorph" or "anamorph" as the latter pair serve only to obfuscate and hinder communication with non-mycologists.

recommended that mycologists refrain from introducing new scientific names when additional states are discovered (Hawksworth, 2011).

Species concepts

The definition of the term "species" is the subject of an enormous literature, including two recent books (Wilkins, 2009; Richards, 2010), and much pertinent information for fungi is included in Burnett (2003). Different species concepts abound across biology, and 25 are enumerated and discussed by Mayden (1997) – although that is not a complete list. The most important recent development is the emphasis which is now given to the phylogenetic species concept, as opposed to the biological or morphological one; i.e. an irreducible group whose members are descended from a common ancestor and who all possess a combination of certain defining, or individually-derived, traits. This phylogenetic approach to species recognition in fungi has been advocated by Taylor et al. (2000), and is being increasingly adopted. However, this is not simply a matter of automatically recognizing each clade as a separate species, and a degree of subjectivity, i.e. of taxonomic opinion, remains. Richards (2010) offers a conceptual solution of a definitional fixed core and then a more subjectively defined circumscription.

Molecular phylogenetic studies are valuable in clarifying whether already named species really merit recognition as being separate or not, and examples in *Morchella* and *Macrolepiota* are included here (G. S. Barseghyan et al.; Ch 7, this volume). However, the most exciting results arise when large numbers of specimens of morphologically defined fungal "species" are examined by molecular phylogenetic methods. This is especially the case when the material originates from a wide range of geographical regions, ecological situations, or hosts. Here, single morphologically defined species almost invariably prove to consist of several so-called cryptic species.

A cryptic species is one which is not, or is scarcely, distinguishable from its closest relatives by traditional morphological characters or other criteria, but which is reproductively isolated or molecularly distinct. There are numerous cases of cryptic species now documented amongst both macromycetes and micromycetes. For example, O'Donnell et al. (2004) distinguished nine species within the earlier concept of *Fusarium graminearum,* several of which were then discovered to be distinguishable by spore shape and cultural features, and also differed in the host plants they attacked; issues in this genus are discussed further by T. Yli-Mattila (Ch 10, this volume). In the *Aspergillus fumigatus* complex, some nine species can be distinguished molecularly, but there are correlations with conidium ornamentation, as viewed in the scanning electron microscope (SEM) and maximum temperatures for growth (Balajee et al., 2007). Most striking is the recent recognition of 22 new species in the *Cladosporium cladosporioides* complex, where again there were some correlative features found

(Bensch et al., 2010). Cryptic speciation is also now well-documented in lichen-forming fungi, where again, in some cases previously overlooked, morphological characters are found to correlate with the phylogenetic data (Crespo and Pérez-Ortega, 2009; Crespo and Lumbsch, 2010).

While molecular work commonly results in a more critical scrutiny of other characters, and the development of more evolutionarily meaningful species concepts, in some cases either no correlative features can be found, or they are not easy to demonstrate. In such instances, mycologists will have to accept that they will not always be able to identify species, and learn to append notations such as *s. lat.*, aggr., complex, or group, to species binominals. At the same time, the traditional use of special form (f. sp.) notations for biologically or physiologically specialized infraspecific categories of plant pathogenic fungi can be expected gradually to disappear as their status becomes resolved by molecular phylogenetic methods.

Environmental species

While not immediately pertinent to the primary objective of this contribution, it is relevant to note that molecular data have revealed an enormous, and hitherto, unexpected diversity of as yet unidentified fungi as endophytes in plant tissues, within insect guts, and in soils (Hawksworth, 2006). The relevance arises from demonstrating unequivocally that there are large numbers of fungi in nature that are not accommodated into current classifications. Sequence data from environmental samples, and also ones from little-studied habitats, can include lineages that are not represented in sequence databases. With modern technology, particularly 454-pyrosequencing, enormous numbers of sequences can be generated. For instance, of 18 000 sequences obtained from an oak phyllosphere, Jumpponen and Jones (2009) recognized 689 operational taxonomic units using a 95% sequence identity criterion, of which 214 (31%) could not be referred to a named species[3]. Similarly, in a study on anaerobic fungi in animal guts, an amazing 38.3% of *Neocallimastigomycota* sequences obtained did not cluster with previously identified genera and formed eight distinct and previously unknown lineages (Liggenstoffer et al., 2010).

In some cases, the fungi can be isolated into culture, but fail to form any spores, but many are known only from sequences. This is the case with rock-inhabiting fungi, some of which represent unnamed lineages in *Arthoniomycetes* and *Dothideomycetes* (Ruibal et al., 2009).

Molecular data are consequently making a major contribution to answering the question often posed in response to the 1.5 million species estimate of fungal diversity (Hawksworth, 1991): Where are all the undescribed fungi?

[3]In interpreting pyrosequence results, it should be noted that in the case of bacteria, Quince et al. (2009) found that the method over-estimated the number of species known to be present in a sample by a factor of six.

Identification

Now, many regions of the world have declining numbers of mycologists able reliably to identify fungi from macro- and microscopic features. Consequently, those involved in applied mycology and ecology increasingly turn to molecular sequence data for answers. However, there is a problem in that the largest public repository for fungal sequence data, GenBank, has data under the names of just 20 765 species, which represents only about 21% of the 100 000 known fungal species – but most importantly a mere 1.4% of the estimated species (Hawksworth, 1991). Further, there is a problem in that several studies have now indicated that around 20–25% of the fungal sequences in GenBank are based on wrongly identified material or are contaminants (Nilsson et al., 2006). A further complication is that in BLAST searches, the closest matches often come out as species of flowering plants. In many or most cases this is, perhaps, a consequence of botanists working with plant material with internal or external fungal contaminants. There may also be problems because internal amplification controls have not been employed (Paterson, 2006), and the issue of whether all repeat sequences in a single organism are the same, not least in heterokaryotic mycelia, is hardly addressed.

As a part of the international effort to have a molecular BarCode for all known species of organisms, an international group to coordinate this for fungi is now being established, with the ITS region envisaged as the universal BarCode for all fungi (Schoch and Seifert, 2010; 'fungalbarcoding.org'). This initiative has great promise since a key component of the project is the use of correctly-identified fungi; but it has much to accomplish. Further, in some groups of fungi, the ITS sequences fail to differentiate accepted species, as in certain groups of *Aspergillus* (Samson et al., 2007). Molecular methods are of especial value for the detection of the presence of particular species, and here microarrays can be developed, for example, for medically important and plant-pathogenic fungi. A discussion of those is beyond the scope of this contribution, but a synopsis of applications for plant pathogenic fungi is included in Tan et al. (2009). A particular example of what can be achieved is the microarray developed by Spiess et al. (2007) for 14 medically relevant fungi. An alternative approach for *Candida* species identification, using proteonomic biomarkers, is reported elsewhere in this volume (J. P. Tewari et al.; Ch 12).

An exciting extension of the barcoding approach is that of "micocoding" (Summerbell et al., 2005) which allows the identification of a particular genus or species. A fine example of this technology is the development of a microarray PhyloChip for *Leptographium* based on data from 56 species, the application of which not only correctly identified known strains, but led to the discovery of eight species new to science (van Zuydam et al., 2010).

Discussion

Molecular methods have revolutionized approaches to the elucidation of fungal relationships and their classification since the mid-1990s. Fungal systematics has become an exciting area, attracting many new researchers, who now have tools with which to test past and newly formulated hypotheses. The development of hypothesis-driven research has also resulted in aspects of fungal systematics being considered more favourably by funding agencies.

The future holds the promise of more fundamental understanding, especially as more whole genome comparisons become possible. There is also the prospect of future generations having the means to make identifications directly from specimens or environmental samples. But these visions will not be realizable without the close collaboration of mycologists, who really understand the fungi with which they work, and skilled molecular biologists. In formulating hypotheses to be tested and developing improved taxonomies, it is essential to consider as many aspects of an organism as possible, i.e. to adopt what has been termed a polyphasic approach. It is, therefore, no surprise to find that most insightful and far-reaching molecular phylogenetic studies have always involved a combination of experienced whole-organism systematics and skilled molecular biologists. This is a synergy which needs to be maintained and so a pool of skilled morphomycologists needs to be ensured for collaboration in the future. Unfortunately, that pool is becoming severely depleted in many nations at this time - but fortunately not in all. There will always be a need for those able to develop keys, descriptive accounts, and identification aids for use by applied mycologists, ecologists, and citizen scientists. Contributions such as that of J. K. Misra (Ch 6, this volume) consequently will always be of value.

In order to maintain the quality of molecular databases, the importance of utilizing correctly identified material cannot be over-estimated. And, whether material has been examined by a specialist or not, it is critical that voucher cultures and/or specimens are permanently preserved in reference collections so that they can be checked by subsequent workers (Agerer et al., 2000). A prerequisite of scientific work is that it should be verifiable, and without the same material being available, that cannot be guaranteed. The issue of vouchers is particularly critical where sequences are to be used as reference points, for example, as the source of a BarCode sequence. In undertaking molecular studies, it is always prudent to check and re-check when some unexpected result appears, ideally with material from different sources. The possibility of contaminants having been inadvertently sequenced is high when dealing with specimens on natural materials and hosts, as opposed to pure cultures.

It is also important not to rush from phylogenetic analyses directly to the introduction of formal changes in classification and, more importantly, scientific names. A common situation is that a conclusion which is based on a small number of taxa or collections cannot be upheld when many taxa or collections are subsequently investigated. In both basic and applied science, scientific names provide the means of accessing all accumulated knowledge on particular organisms. Consequently, changes to that system should never be made hastily. They should only be considered when the evidence is overwhelming.

Molecular phylogenetics is also posing fresh challenges that must be satisfactorily accommodated into nomenclatural systems. This is not only an issue of resolving the debate over the dual nomenclatural system in pleomorphic fungi (see above), but also developing mechanisms to permit the naming of environmental sequences where no information other than sequence data is available (Hibbett et al., 2009). These are matters in which those using molecular approaches need to become involved so that the established committees can take actions that best suit the needs of mycologists as a whole. In the future, there is the prospect of this dialogue being facilitated by a transfer of some nomenclatural responsibility from International Botanical Congresses to International Mycological Congresses (Hawksworth et al., 2009).

In a previous publication (Hawksworth, 2006), I compared the opening of the molecular phylogenetic toolbox in mycology to the mythical Pandora's Box from which either all manner of miseries or evils, or blessings, flew rapidly all over the Earth with no possibility of escape from their impact. There is no escape from molecular approaches in the mycology of the 21st century, and it is the task of its practitioners to ensure that these enhance rather than confuse.

ACKNOWLEDGEMENTS

This contribution was undertaken while I was in receipt of research grant I+D CGL 2008-01600 from the Ministerio de Educación y Ciencia of Spain. I am also grateful to my wife, Patricia E J Wiltshire-Hawksworth, for her improvements to this presentation.

REFERENCES

Adl, S. M., Simpson, A. G. B., Farmer, M. A., Andersen, R. A., Anderson, O. R., Barta, J. R., Bowser, S. S., Brugerolle, G., Fensome, R. A., Fredericq, S., James, T. Y., Karpov, S., Kugrens, P., Krug, J., Lane, C. E., Lewis, L. A., Lodge, J., Lynn, D. H., Mann, D. G., McCourt, R. M., Mendoza, L., Moestrup, Ø., Mozley-Standridge, S. E., Nerad, T. A., Shearer, C. A., Smirnov, A. V., Spiegel, F. W. and Taylor, M. F. J. R. 2005. The new higher level classification of eukaryotes with emphasis on the taxonomy of protists. Journal of Eukaryotic Microbiology, 52: 399-451.

Agerer, R., Ammirati, J., Blanz, P., Courtecuise, R., Desjardin, D. E., Gams, W., Hallenberg, N., Halling, R., Hawksworth, D. L., Horak, E., Korf, R. P., Mueller, G. M., Oberwinkler, F., Rambold, G., Summerbell, R. C., Triebel, D. and Watling, R. 2000. Always deposit vouchers. Mycological Research, 104: 642-644.

Aveskamp, M., de Gruyter, H., Woudenberg, J., Verkeley, G. and Crous, P. W. 2010. Highlights of the *Didymellaceae*: a polyphasic approach to characterise and related pleosporalean genera. Studies in Mycology, 65: 1-64.

Balajee, S. A., Houbraken, J., Verweij, P. E., Hong, S-B., Yaghuchi, T., Varga, J. and Samson, R. A. 2007. *Aspergillus* species identification in the clinical setting. Studies in Mycology, 59: 39-46.

Bartlett, J. M. S. and Stirling, D. 2003. A short history of the polymerase chain reaction. Methods in Molecular Biology, 226: 3-6.

Beentje, H. 2010. The Kew Plant Glossary: an illustrated dictionary of plant terms. Kew: Kew Publishing.

Bensch, K., Groenewald, J. Z., Dijksterhuis, J., Starink-Willemse, M., Andersen, B., Summerell, B. A., Shin, H-D., Dugan, F. M., Schroers, H-J., Braun, U. and Crous, P. W. 2010. Species and ecological diversity within the *Cladosporium cladosporioides* complex (*Davidiellaceae*, *Capnodiales*). Studies in Mycology, 67: 1-94.

Berbee, M. L., and Taylor, J. W. 2010. Dating the molecular clock in fungi – how close are we? Fungal Biology Reviews, 24: 1-16.

Blanco, O., Crespo, A., Elix, J. A., Hawksworth, D. L. and Lumbsch, H. T. 2004. A molecular phylogeny and a new classification of parmelioid lichens containing *Xanthoparmelia*-type lichenan (*Ascomycota*: *Lecanorales*). Taxon, 53: 959-975.

Burnett, J. [H]. 2003. Fungal Populations and Species. Oxford University Press, Oxford.

Corradi, N. and Keeling, P. J. 2009. *Microsporidia*: a journey through radical taxonomical revisions. Fungal Biology Reviews, 23: 1-8.

Crespo, A. and Lumbsch, H. T. 2010. Cryptic species in lichen-forming fungi. IMA Fungus, 1: 167-170.

Crespo, A. and Peréz-Ortega, S. 2009. Cryptic species and species pairs in lichens: a discussion on the relationship between molecular phylogenies and morphological characters. Anales del Jardin Botánico de Madrid, 66 (S1): 71-81.

Daly, D. C., Cameron, K. M. and Stevenson, D. W. 2001. Plant systematics in the age of genomics. Plant Physiology, 127: 1328-1333.

Eriksson, O. E. 2005. Ascomyceternas ursprung och evolution – Protolichenes-hypotesen. Svensk Mykologisk Tidskrift, 26: 22-33.

Eriksson, O. E. 2009. Ascomyceternas ursprung – argument granskade. [The origin of the ascomycetes - argument scrutinized.] Svensk Mykologisk Tidskrift, 30: 61-64.

García, D., Stchigel, A. M., Cano, J., Calduch, M., Hawksworth, D. L. and Guarro, J. 2006. Molecular phylogeny of *Coniochaetales*. Mycological Research, 110: 1269-1287.

Grube, M. and Hawksworth, D. L. 2007. Trouble with lichen: the re-evaluation and re-interpretation of thallus form and fruit body types in the molecular era. Mycological Research, 111: 1116-1132.

Hawksworth, D. L. 1991. The fungal dimension of biodiversity: magnitude, significance and conservation. Mycological Research, 95: 641-655.

Hawksworth, D. L. 2006. Pandora's mycological box: molecular sequences *vs.* morphology in understanding fungal relationships and biodiversity. Revista Iberoamericana de Micología, 23: 127-133.

Hawksworth, D. L. 2011. Naming *Aspergillus* species: progress towards one name for each species. Medical Mycology, 49 (Suppl. 1): S70-S76.

Hawksworth, D. L., Crous, P. W., Dianese, J. C., Gryzenhout, M., Norvell, L. L. and Seifert, K. A. 2009. Proposals to amend the *Code* to make it clear that it covers the nomenclature of fungi, and to modify the governance with respect to names of organisms treated as fungi. Taxon, 58: 658-659; Mycotaxon, 108: 1-4.

Heckman, D.S., Geiser, D. M., Eudell, B. R., Stauffer, R. L., Kardos, N. L. and Hedges, S.B. 2001. Molecular evidence for the early colonization of land by fungi and plants. Science, 293: 1129-1133.

Hibbett, D. S. 2007. After the gold rush, or before the flood? Evolutionary morphology of mushroom-forming fungi (*Agaricomycetes*) in the early 21st century. Mycological Research, 111: 1001-1018.

Hibbett, D. S., Binder, M., Bischoff, J. F., Blackwell, M., Cannon, P. F., Eriksson, O. E., Huhndorf, S., James, T., Kirk, P. M., Lücking, R., Lumbsch, H.T., Lutzoni, F., Matheny, P. B., McLaughlin, D. J., Powell, M. J., Redhead, S., Schoch, C. L., Spatafora, J. W., Stalpers, J. A., Vilgalys, R., Aime, M. C., Aptroot, A., Bauer, R., Begerow, D., Benny, G. L., Castelbury, L. A., Crous, P. W., Dai, Y. C., Gams, W., Geiser, D. M., Griffith, G. W., Gueidan, C., Hawksworth, D. L., Hestmark, G., Kosaka, K., Humber, R. A., Hyde, K. D., Ironside, J. E., Kõljalg, U., Kurtzman, C. P., Larsson, K.-H., Lichtwardt, R., Longcore, J., Miądlikowska, J., Miller, A., Moncalvo, J.-M., Mozley-Standridge, S., Oberwinkler, F., Parmasto, E., Reeb, V., Rogers, J. D., Roux, C., Ryvarden, L., Sampaio, J. P., Schüßler, A., Sugiyama, J., Thorn, R. G., Tibell, L., Untereiner, W. A., Walker, C., Wang, Z., Weir, A., Weiss, M., White, M. M., Winka, K., Yao and N. Zhang, Y.-J. 2007. A higher-level classification of the fungi. Mycological Research, 111: 509-547.

Hibbett, D. S., Ohman, A. and Kirk, P. M. 2009. Fungal ecology catches fire. New Phytologist, 184: 279-282.

Houbraken, J., Frsivad, J. C. and Samson, R. A. 2010. Sex in *Penicillium* series *Roqueforti.* IMA Fungus, 1: 167-170.

Hull, D. L. 1988. Science as a Process: an evolutionary account of the social and conceptual development of science. University of Chicago Press, Chicago.

Jackson, B. D. 1916. A Glossary of Botanic Terms with their Derivation and Accent. 3rd edn. Duckworth and Co., London.

Jumpponen, A. and Jones, K. L. 2009. Massively parallel 45-sequencing indicates hyperdiverse fungal communities in temperate *Quercus macrocarpa* phyllosphere. New Phytologist, 184: 438-448.

Kirk, P. M., Cannon, P. F., Minter, D. W. and Stalpers, J. A. 2008. Ainsworth & Bisby's Dictionary of the Fungi. 10th edn. CAB International, Wallingford.

Læssøe, T. and Hansen, K. 2007. Truffle trouble: whatever happened to the *Tuberales*? Mycological Research, 111: 1075-1099.

Liggenstoffer, A. S., Youseff, N. H., Couger, M. B. and Elshahed, M. S. 2010. Phylogenetic diversity and community structure of anaerobic gut fungi (phylum *Neocallimastigomycota*) in ruminant and non-ruminant herbivores. The ISME Journal, 4: 1225-1235.

Lombard, L., Crous, P. W., Wingfield, B. D. and Wingfield, M. J. 2010. Systematics of *Calonectria*: a genus of root, shoot and foliar pathogens. Studies in Mycology, 66: 1-71.

Lücking, R. L., Huhndorf, S., Pfister, D. H., Rivas-Plata, E. and Lumbsch, H. T. 2009. Fungi evolved right on track. Mycologia, 101: 810-822.

Mayden, R. L. 1997. A hierarchy of species concepts: the denouement in the saga of the species problem, pp. 381-424. In: Species: the units of biodiversity. M. F. Claridge, H. A. Dawah and M. R. Wilson (eds.). Chapman & Hall, London.

Muggia, L., Hafellner, J., Wirtz, N., Hawksworth, D. L. and Grube, M. 2008. The sterile microfilamentous lichenized fungi *Cystocoleus ebeneus* and *Racodium rupestre* are relatives of plant pathogens and clinically important dothidealean fungi. Mycological Research, 112: 51-57.

Nilsson, R. H., Ryberg, M., Kristiansson, E., Abarenkov, K., Larsson, K-H. and Köljalg, U. 2006. Taxonomic reliability of DNA sequences in public sequence databases: a fungal perspective. PLoS One: e59.

Norvell, L. L., Hawksworth, D. L., Petersen, R. H. and Redhead, S. A. 2010. The IMC9 Edinburgh Nomenclature Sessions. Mycotaxon, 113: 503-511; IMA Fungus, 1: 143-148.

O'Donnell, K., Ward, T. J., Geiser, D. M., Kistler, H. C. and Aoki, T. 2004. Geneological concordance between the mating type locus and seven other nuclear genes supports formal recognition of nine phylogenetically distinct species within the *Fusarium graminearum* clade. Fungal Genetics and Biology, 41: 600-623.

Paterson, R. R. M. 2006. Internal amplification controls have not been employed in fungal PCR hence potential false negative results. Journal of Applied Microbiology, 102: 1070-1080.

Peterson, K. R., Pfister, D. H. and Bell, C. D. 2010. Cophylogeny and biogeography of the fungal parasite *Cyttaria* and its host Nothofagus, southern beech. Mycologia, 102: 1417-1425.

Pirozynski, K. A. and Hawksworth, D. L. 1988. Coevolution of fungi with plants and animals: introduction and overview, pp. 1-29. In: Coevolution of Fungi with Plants and Animals. K. A. Pirozynski and D. L. Hawksworth (eds). Academic Press, London.

Quince, C., Lanzén, A., Curtis, T. P., Davenport. R. J., Hall, N., Head, I. M., Read, L. F. and Sloan, W. T. 2009. Accurate determination of microbial diversity from 454 pyrosequencing data. Nature Methods, 6: 639-641.

Redhead, S. A., Vilgalys, R., Moncalvo, J-M., Johnson, J. and Hoppe, J. S. Jr 2001. *Coprinus* Pers. and the disposition of *Coprinus* species *sensu lato*. Taxon, 50: 203-241.

Reynolds, D. R. and Taylor, J. W. 1991. Nucleic acids and nomenclature: name stability under Article 59. Regnum Vegetabile, 123: 171-177.

Reynolds, D. R. and Taylor, J. W. 1992. Article 59: reinterpretation or revision? Taxon, 41: 91-98.

Richards, R. A. 2010. The Species Problem: a philosophical analysis. Cambridge University Press, Cambridge.

Ruibal, C., Gueidan, C., Selbmann, L., Gorbushina, A. A., Crous, P. W., Groenewald, J. Z., Muggia, L., Grube, M., Isola, D., Schoch, C. L., Staley, J. T., Lutzoni, F., and Hoog, G. S. De. 2009. Phylogeny of rock-inhabiting fungi related to Dothideomycetes. Studies in Mycology, 64: 123–133.

Samson, R. A., Noonim, P., Meijer, M., Houbraken, J., Frisvad, J. C. and Varga, J. 2007. Diagnostic tools to identify black aspergilla. Studies in Mycology, 59: 129-145,

Schoch, C. [L] and Seifert, K. [A] 2010. The home stretch for fungal barcoding. IMA Fungus, 1(2): (2)-(3).

Schoch, C. L., Sung, G.-H., López-Giráldez, F., Townsend, J. P., Miadlikowska, J., Hofstetter, V., Barbara R., Matheny, P. B., Kauff, F, Wang, Z., Gueidan, C., Andrie, R. M., Trippe, K., Ciufetti, L. M., Wynns, A., Fraker, E., Hodkinson, B. P., Bonito, G., Groenewald, J. Z., Arzanlou, M., de Hoog, G. Sybren, C., Pedro W., Hewitt, D., Pfister, D. H., Peterson, K., Gryzenhout, M., Wingfield, M. J., Aptroot, A., Suh, S.-O., Blackwell, M., Hillis, D. M., Griffith, G. W., Castlebury, L. A., Rossman, A. Y., Lumbsch, H. T., Lücking, R., Büdel, B., Rauhut, A., Diederich, P., Ertz, D., Geiser, D. M., Hosaka, K., Inderbitzin, P., Kohlmeyer, J., Volkmann-Kohlmeyer, B., Mostert, L., O'Donnell, K., Sipman, H., Rogers, J. D., Shoemaker, R. A., Sugiyama, J., Summerbell, R. C., Untereiner, W., Johnston, P. R., Stenroos, S., Zuccaro, A., Dyer, P. S., Crittenden, P. D., Cole, M. S., Hansen, K., Trappe, J. M., Yahr, R., Lutzoni, F. and Spatafora, J. W. 2009. The *Ascomycota* tree of life: a phylum-wide phylogeny clarifies the origin and evolution of fundamental reproductive and ecological traits. Systematic Biology, 58: 224-239.

Selosse, M. A. 2002. *Prototaxites*: a 400 myr giant fossil – a homobasidiomycete or a lichen? Mycological Research, 106: 642-644.

Snell, W. H. and Dick, E. A. 1957. A Glossary of Mycology. Harvard University Press, Cambridge, MA.

Spiess, B., Seifarth, W., Hummel, M., Frank, O., Fabarius, A., Zheng, C., Mörz, H., Hehlman, R. and Buchheidt, D. 2007. DNA microarray-based detection and identification of fungal pathogens in clinical samples from neutropenic patients. Journal of Clinical Microbiology, 45: 3743-3753.

Summerbell, R. C., Lévesque, C. A., Siefert, K. A., Bovers, M., Fell, J. W., Diaz, M. R., Boekhoet, T., de Hoog, G. S., Stalpers, J. A. and Crous, P. W. 2005. Microcoding: the second step in DNA barcoding. Philosophical Transactions of the Royal Society of London, B 360: 1897-1903.

Tan, K-C., Ipcho, S. V. S., Trenglove, R. D., Oliver, R. P. and Solomon, P. S. 2009. Assessing the impact of transcriptomics, proteomics and metabolomics on fungal phytopathology. Molecular Plant Pathology, 10: 703-715.

Taylor, J. W., Jacobson, D. J., Kroken, S., Kasuga, T., Geiser, D. M., Hibbett, D. S. and Fisher, M. C. 2000. Phylogenetic species recognition and species concepts in fungi. Fungal Genetics and Biology, 31: 21-32.

Taylor, T. N., Taylor, E. L. and Krings, M. 2009. Palaeobotany: The biology and evolution of fossil plants. 2nd edn. Elsevier Academic Press, Amsterdam.

Van Zuydam, N. R., Paciura, S., Jacobs, K., Wingfiled, M. J., Coetzee, M. P. A. and Wingfield, B. D. 2010. Barcoding and microcoding using "identiprimers" with *Leptographium* species. Mycologia, 102: 1274-1287.

Wang, H-Y., Guo, S-Y., Huang, M-R., Lumbsch, H. T. and Wei, J-C. 2010. *Ascomycota* has a faster evolutionary rate and higher species diversity than *Basidiomycota*. Science China (Life Sciences), 53: 1163-1169.

Whalley, A. J. S. 2003. Molecules vs. morphology. Mycological Research, 107: 642-643.

Wilkins, J. S. 2009. Species: A history of the idea. University of California Press, Berkeley.

*At the XVIIIth International Botanical Congress in Melbourne (Australia) in July 2011, it was decided that the dual nomenclature of pleomorphic fungi would be discontinued from 1st January 2013. From that date, a fungus will only have a single correct name, regardless of its sexual state or states; full details will be included in The International Code of Nomenclature for Algae, Fungi, and Plants scheduled for publication in mid-2012.

Chapter 2

The Systematics and Evolutionary Perspectives of Fossil Fungi

S. K. M. Tripathi

Birbal Sahni Institute of Palaeobotany, 53 University Road
Lucknow 226 007, India
E-mail: suryatripathi.2009@rediffmail.com

ABSTRACT

Being fragmentary in nature, fossil fungi lack characteristic features that are diagnostic of extant taxa. Except for some distinctive Tertiary forms, most of the fossil fungal remains can seldom be compared with modern taxa hampering their classification under the Natural System. Problems concerning the nomenclature and classification of fossil fungi have been discussed by various workers from time to time. They argue that assigning most of the fossil forms a modern name will lead to a taxonomic confusion and it will therefore be more convenient as well as logical to describe the fragmentary fossils as form genera which are described on the basis of morphological characters only. Under this scheme, fossils are given artificial generic and specific names and wherever possible their affinities with extant taxa are suggested. This chapter provides a brief introduction to the conventional systems of classification proposed for fossil fungal spores and fruiting bodies. Diagnostic characters of these form genera are also given.

Widely diversified fossil fungal remains are reported in the form of palynomorphs, mycelia or fructifications. Evidences from these fossils indicate that key events of fungal evolution took place during late Precambrian to early Palaeozoic times and all four major groups of modern fungi have now been found in Devonian to Carboniferous strata. Many coenocytic hyphae with fungal affinity are recorded from middle to late Precambrian rocks. Other stray records of fossil fungi are reported from different parts of Palaeozoic and Mesozoic eras whereas, ample varieties of fossil fungal spores, filaments, fruiting bodies and mycorrhiza have been reported globally from Cenozoic rocks. Fossils reported from Early Devonian Rhynie chert provide evidence that land plants and fungi formed symbiotic relationships at very early stage in terrestrial evolution. The evolutionary ancestry and relationships of the different fungal groups are still not well understood. The relatively new technique of molecular systematics is particularly useful in the study of the evolution of fungi. Molecular estimates of divergence times are, however, controversial and are generally much more ancient than suggested by the fossil record. The limited number of genes and species explored and the pervasive variations in evolutionary rates are the most likely sources of such discrepancies. Future work in molecular systematics is expected to tell us more about the evolution and relationships of the fungi.

INTRODUCTION

Nearly 950 widely diversified fossil fungal species representing 300 genera have so far been described (Kalgutkar and Jansonius, 2000). These fossils are reported in the form of palynomorphs, mycelia or fructifications. Fungal remains, commonly observed in the palynological preparations, have been sporadically recorded since long (Williamson, 1878, 1880; Kidston and Lang, 1921; Edwards, 1922) but their study received more attention with the amplification of palynological studies since the 1950s. During the last couple of decades, these have been investigated with greater interest involving phylogenetic, biostratigraphic and palaeoenvironmental implications (Ramanujam, 1982; Taylor and White, 1989; Berbee and Taylor, 2001). Fossil fungal remains are less common in Palaeozoic strata but become more frequent in lower Mesozoic. A great spurt in the fungal diversity is witnessed in the Tertiary Period indicating that proliferation of fungi is linked with diversification of angiosperms. Being fragmentary in nature, fossil fungi lack characteristic features that are diagnostic of extant taxa. Except for some distinctive Tertiary forms, most of the fossil fungal remains can seldom be compared with modern taxa hampering their classification under the Natural System. Problems concerning the nomenclature and classification of fossil fungi have been discussed by various workers from time to time (Elsik, 1968, 1976; Pirozynski and Weresub, 1979). These workers argued that assigning most of the fossil forms a modern name will lead to a taxonomic confusion and it will therefore be more convenient as well as logical to describe the fragmentary fossils as form genera which are described on the basis of morphological characters only. Under this scheme fossils are given artificial generic and specific names and wherever possible their affinities with extant taxa are suggested.

FUNGAL REMAINS THROUGH AGES

Recent works provide evidence that key events of fungal evolution took place during the Late Precambrian to Early Palaeozoic times and all four major groups of modern fungi have now been found in the Devonian to Carboniferous strata. Studies have shown that fungi and land plants formed symbiotic relationships at a very early stage in terrestrial evolution. Many coenocytic hyphae with fungal affinity are recorded from middle to late Precambrian rocks (Tyler and Barghoorn, 1954; Barghoorn and Tyler, 1965; Schopf, 1968; Schopf and Barghoorn, 1969). Other stray records of fossil fungi are reported from different parts of the Palaeozoic and Mesozoic eras (Kidston and Lang, 1921; Harvey et al., 1969; Krassilov, 1981; Wagner and Taylor, 1981; Stubblefield et al., 1985; White and Taylor, 1988, 1989; Taylor and White, 1989; Remy et al., 1994; Taylor et al., 1994) whereas, ample varieties of fossil fungal spores, filaments, fruiting bodies and mycorrhiza have been reported globally from the Cenozoic rocks (Kalgutkar and Jansonius, 2000; Manoharachari, 2005). Tiffney and Barghoorn (1974) and Pirozynski (1976a, 1976b) elucidated the antiquity of different fungal groups during the geological past.

Well-preserved coenocytic filaments named *Eomycetopsis* are described from the middle and late Precambrian Bitter Springs Formation of central Australia (Schopf, 1968). Fungal spores and vesicles in some early land plants from the Early Devonian Rhynie Chert, Scotland were reported by Kidston and Lang (1921). Endozoic forms resembling modern water moulds have been recorded from shells of arthropods, fish scales and corals of the Early Palaeozoic age (Elias, 1966). Mycorrhizal fungi in association with plants of the Early Devonian age have been reported by Remy et al. (1994) and Taylor et al. (1995). Spores having affinity with Ascomycetes are reported from the Late Carboniferous (Playford, 1962) and the Late Permian (Wilson, 1962) but this group became well-established and diversified during the Cretaceous and Tertiary ages. They thrived in a variety of habitats, as saprophytes, parasites, epiphytes and mycorrhizal associations (Ramanujam, 1982; Pyrozynski, 1978). Varieties of epiphyllous fungal fruiting bodies are reported from the Tertiary sediments around the world (Edwards, 1922; Cookson, 1947; Dilcher, 1965; Jain and Gupta, 1970; Rao and Ramanujam, 1976; Saxena and Misra, 1990; Rao, 2003). A clamp-bearing mycelium, in coal ball flora from the Middle Pennsylvanian, is probably the oldest fossil indicating the presence of Basidiomycetes (Dennis, 1970).

CLASSIFICATION OF FOSSIL FUNGI

As evidenced by the fossil records, Ascomycetes, the largest and most diversified group of modern fungi, got well-established during the Cretaceous time and became conspicuously abundant by the Tertiary Period (Jain, 1974; Jain and

Kar, 1979; Jansonius, 1976; Ramanujam, 1982; Kalgutkar and Jansonius, 2000; Tripathi, 2009). Fungi of this group produced ascospores and conidia which helped them to thrive in a variety of habitats as saprophytes, parasites, epiphytes and mycorrhizae. The Early Tertiary palynological assemblages are markedly characterized by increased number of epiphyllous fungi which were hosted by broad angiospermous leaves. Following the Artificial System of classification, fossil fungal remains are described under two categories–the Fungal spores and the Ascocarps (fruiting bodies of Epiphyllous fungi). Salient morphologic features of each of these groups are discussed below.

I. Fungal spores

Majority of fungal spores found in palynological preparations belong to Ascomycetes. Only few questionable spores of Basidiomycetes have been described in some fossil assemblages. Palynological assemblages are often rich in varieties of conidia. These are produced by Fungi Imperfecti and the holomorphic Ascomycetes. They may be one-celled to multi-celled and are of varied shapes.

Spores of some fungi, especially conidia and ascospores possess distinctive features leading to their identification and categorization with the extant forms. Fossil spores can be generally assigned to a natural (old) class system of Phycomycetes, Ascomycetes or Basidiomycetes if the diagnostic morphographic features are observable. Some fossil materials are assigned to the class Fungi Imperfecti where spores or isolated structures (conidia, pycnidia or isolated mycelia) are of exclusive morphology. Numerous types of fossil fungal spores are reported from the Late Cretaceous to Cenozoic sediments. These are described under "dispersed spores" which include detached spores, microscopic sporangia, hyphae or fragmented mycelia. In a classification system proposed by van der Hammen (1956), fossil fungal spores were grouped under various morphologic categories having the suffix "Sporites". Clarke (1965) proposed the suffix "Sporonites" for naming the fossil fungal spores. Considering the characters such as shape, size and symmetry of spores, absence/presence and number of apertures, septa characters and the wall features, Elsik (1976) attempted to prepare a comprehensive applicable taxonomy for the fossil spores. He proposed artificial supra-generic categories for classification of fossil fungal spores. These categories were primarily based on the cell number and presence or absence of apertures. Under these categories artificial genera and species could be conveniently described. Pirozynski and Weresub (1979) suggested a system named "Saccardoan System" for classifying the fungal spore types. This scheme is based on the shape and number of cells and fungal spores are recognized as Amerospores (monocellate), Didymospores (dicellate), Phragmospores (tri- to multicellate), Dictyospores (muriform), Scolecospores (filiform), Helicospores (spirally coiled) or Staurospores (star-like). Following this system of classification, fossil spore genera are given in Table 2.1.

Table 2.1 Diagnostic characters of fossil fungal spores

Taxa	*Diagnostic features*
	Amerospores
Basidiosporites Elsik	Spores with single offset pore, unicellate, elongate, wall psilate, shape variable
Diporisporites van der Hammen	Shape generally elongate, diporate, pores on opposite ends
Exesisporites Elsik	Unicellate, lenticular, monoporate, pore small, pore surrounded by thickening
Hypoxylonites Elsik	Oval to elongate, bilateral, psilate, provided with elongate scar, slit or furrow
Inapertisporites van der Hammen ex van der Hammen	Inaperturate, shape and size variable, wall psilate to variously ornamented
Lacrimasporonites Clarke	Spatulate to elliptical in shape, wall psilate, monoporate, pore apical
Monoporisporites van der Hammen	Spherical to sub-spherical, monoporate, wall psilate to finely punctuate
Palaeoamphisphaerella Ramanujam and Srisailam	Shape elliptical, oblong or rhomboidal with rounded ends, provided with equatorial pore
	Didymospores
Dicellaesporites Elsik	Two-celled, uniseptate, shape variable, inaperturate, wall psilate
Didymoporisporonites Sheffy and Dilcher	Dicellate, uniseptate, apex of one cell provided with pore, psilate to punctuate
Diploneurospora Jain and Gupta	Two-celled, unicellate, elliptical, upper cell prominent, thick-walled, sculptured with longitudinal ribs, lower cell smaller, hyaline with faint rib sculpture
Dyadosporites van der Hammen ex Clarke	Diporate, with a single pore at each end, wall psilate to variously sculptured
Fusiformisporites Rouse	Fusiform, inaperturate, the unit is split into equal halves by equatorial wall, bearing characteristic elongate striae, ribs, ridges or costae oriented parallel to longer axis

Table 2.1 Contd. ...

Taxa	*Diagnostic features*
	Phragmospores
Brachysporisporites Lang and Smith	Obovate, turbinate or pyriform, multi-celled, cells broader than long, gradually diminishing in size towards the attachment cell which is the smallest, with very dark, thick bands of septa similarly reducing in size.
Cannanorosporonites Ramanujam and Rao	Tetracellate, barrel-shaped, basal and terminal cells smaller than central cells
Diporicellaesporites Elsik	Elongate, diporate, one pore at each end of the spore, wall psilate to finely structured
Foveoletisporonites Ramanujam and Rao	Four or more celled, elongate, foveolate, foveolae irregularly aligned
Multicellaesporites Elsik	Three or more celled, shape variable, inaperturate, spore wall psilate
Ornasporonites Ramanujam and Rao	Fusiform, four-celled, diporate, basal and apical cells much small, one pore at each end
Pluricellaesporites van der Hammen	Three or more celled, long, monoporate, psilate to scabrate
Polycellaesporonites Chandra et al.	Elongate, multicellate, inaperturate, psilate, one end rounded, other end giving rise to a tube-like projection, cells arranged in clusters
	Dictyospores
Spinosporonites Saxena and Khare	Circular to sub-circular, inaperturate, multicellate, each cell giving rise to a robustly built spine
Staphlosporonites Sheffy and Dilcher	Shape variable, four or more irregular cells arranged in clusters along more than one axis, inaperturate, psilate to punctate
Tricellaesporonites Sheffy and Dilcher	Shape variable, tri-cellate, inaperturate, cells along more than one axis, spore wall psilate to punctuate
	Helicospores
Elsikisporonites Kumar	Tubular and coiled in shape, monopore, pore at outer end, non-septate, spore wall smooth and hyaline
Involutisporonites Clarke	Coiled, transversely septate, monoporate, psilate to variously ornamented

Table 2.1 Contd. ...

Taxa	*Diagnostic features*
	Staurospores
Frasnacritetrus Taugourdeau	Main body rectangular, spherical or oval, psilate to variously ornamented, body provided with four unicellular processes
	Miscellaneous
Alleppeysporonites Ramanujam and Rao	Spores branched, multicellate, septate, individual cell rectangular, basal and terminal cells provided with a conspicuous appendage
Appendicisporonites Saxena and Khare	Subcircular, inaperturate, psilate, multicelate, each cell provided with a long process
Rhizophagites Rosendhal	Nonseptate, thick-walled, hyphae with terminal subspherical vesicles of varying size

II. Ascocarps

Enormous diversity in fossil epiphyllous fruiting bodies (80 to 160 μm in size) are recorded from the mid Tertiary strata (Cookson, 1947; Elsik, 1978a). These were amongst the first fungal groups that were unquestionably identified in the microfossil assemblages. Because of distinctive morphological features, fossil fungal fructifications can be compared with the extant counterpart with greater accuracy than the dispersed spores. Commonly occurring as parasites on epidermis of leaves, stem and flowers of higher plants, these belong to Ascomycetes (now Ascomycotina) and are placed with the family Microthyriaceae. These have been extensively recorded from the Neocomian to Quaternary sediments around the world. Fruiting bodies of Microthyriaceous fungi are known as Ascocarps. These are provided with radiating rows of mycelial cells giving an appearance of tissues arranged in a radial fashion. Ascocarps contain asci that are surrounded by or enclosed within protective tissues and may be globose, flask-shaped or saucer-shaped open bodies. These may or may not possess an opening known as ostiole.

Fossil fungal fruiting bodies are also classified under the artificial system grouping them with Fungi Imperfecti. Several workers attempted to classify and formally describe the fossil ascocarps (Edwards, 1922; Rosendahl, 1943; Cookson, 1947; Rao, 1959; Dilcher, 1965; Venkatachala and Kar, 1969; Jain and Gupta, 1970; Elsik, 1978b; Pirozynski, 1978). Fruiting bodies of Microthyriaceous fungi are classified on the basis of mode of their dehiscence (through irregular or regular cracking pattern or by a central pore or ostiole).

Other characteristic features considered for their classification are: shape and margin of the fruiting body, presence or absence of pores in individual cells and nature of the central part of the fruiting body. Widely accepted classification scheme to describe the dispersed ascocarps is summarized in Table 2.2. This system is primarily based on porate or aporate individual cells of multicellular fruiting body. Forms with porate individual cells are kept under the genus *Callimothallus* whereas those without pores are divided into non-radiate and radiate forms. The non-radiate forms may be ostiolate or non-ostiolate. The radiate forms are further divided into genera having smooth, fimbriate or spinose margins. The radiate forms with smooth to fimbriate margins are further divided on the basis of presence or absence and the nature of ostiole. Salient characters of fossil genera described as fruiting bodies are given in Table 2.3.

Table 2.2 Classification of fossil ascocarps (after Elsik, 1978b)

<table>
<tr><td colspan="2" rowspan="2">Ascocarp cells aporate</td><td colspan="2">Ascocarp cells porate</td></tr>
<tr><td>Body radiate
(Callimothallus)</td><td>Body non-radiate
(Ratnagiriathyrites)</td></tr>
<tr><td colspan="3">Body radiate, margin smooth or irregular</td><td rowspan="2">Body radiate, margin with projecting spines
(Parmathyrites)</td></tr>
<tr><td>Ostiole distinct
1. Body made up of intertwined thin hyphae; ostiole margin thickened
(Plochmopeltinites)
2. Ostiole bordered with single/double walled cells
(Trichothyrites)</td><td>Ostiole indistinct
1. Central cells simple
(Paramicrothallites)
2. Central cells thick-walled
(Microthyriacites)</td><td>Non-ostiolate
1. Central cells modified, provided with star-shaped opening
(Asterothyrites)
2. Central cells unmodified
(Phragmothyrites)
3. Central knob present
(Microthallites)</td></tr>
<tr><td colspan="4">Ascocarp cells aporate, body non-radiate</td></tr>
<tr><td colspan="3">Ostiolate</td><td rowspan="2">Non-ostiolate
(Trichopeltinites)</td></tr>
<tr><td colspan="2">Irregular ostiole, body fan-shaped
(Brefeldiellites)</td><td>Elongated dehiscence, body multi fan-shaped
(Euthythyrites)</td></tr>
</table>

Table 2.3 Diagnostic features of fossil ascocarps (fruiting bodies)

Taxa	*Diagnostic features*
Asterothyrites Cookson	Ascomata round, flat, made up of radially arranged hyphae, cells isodiametric. Ascomata ostiolate, ostiole stellate in shape, probably formed by dissolution of central cells.
Brefeldiellites Dilcher	Hyphae produce a large rounded membranous structure with marginal fertile areas or ascomata. Central ascoma cells break away as a dehiscence mechanism.
Callimothallus Dilcher	Stroma round, radiate, no central dehiscence, individual cells may possess single pore.
Euthythyrites Cookson	Ascomata linear, elliptical to oblong, ends rounded or flattened, lateral margins uneven, dehiscence by a longitudinal slit, cells radiating from mid-vertical line, hyphopodiate, hyphopodia small.
Microthallites Dilcher	Stroma radiate, more or less round, ostiolate or non-ostiolate.
Microthyriacites Cookson	Ascomata very large (1000-1200 μm), slightly convex. Central part constituted by thick isodiametric cells, peripheral cells elongated, radial.
Paramicrothallites Jain and Gupta	Stroma radiate, more or less rounded, ostiolate, ostiole not surrounded by specialized cells.
Parmathyrites Jain and Gupta	Ascomata flattened, non-ostiolate, more or less circular, hyphae radially arranged. Peripheral cells prominent with thickened radial walls, spines peripheral. Ostiole distinct.
Phragmothyrites Edwards	Ascomata sub-circular to circular with radially arranged hyphae, hyphal cells may be differentiated forming separate regions in the fruiting body. Central cells isodiametric.
Plochmopeltinites Cookson	Ascomata of dimidiate form with ascomal membranes of sinuous plectenchyma.
Ratnagiriathyrites Saxena and Misra	Ascomata sub-circular or irregular in shape, margin thick, wavy, dark brown in colour, margin thick, wavy, non-ostiolate. Cells not arranged radially, porate. Pores generally distributed throughout stromata. Peripheral cells hexagonal, bigger, central cells small.
Trichopeltinites Cookson	Ascomata developed as thickened areas of the thallus and dehiscing by an irregular ostiole as in *Trichopeltis* Theiss (Stevens, 1925).

Table 2.3 Contd. ...

Taxa	*Diagnostic features*
Trichothyrites Rosendahl	Thyriothecia disc- or saucer-shaped, made up of almost square radiating cells. Ostiolate, ostiole placed on an erect collar, made up of 2-6 tier of thick-walled quadrilateral cells. Uppermost tier of cells may have short prolongations in some cases. Outline usually smooth but may appear lobate.

FOSSIL EVIDENCES AND EVOLUTION OF FUNGI

The evolutionary ancestry and relationships of the different fungi are not well understood. There is some fossil evidence that they existed in the Precambrian (over four billion years ago), although identification of these very early fossils is uncertain. The oldest fossil fungi known so far are probably chytrid-like forms from the Vendian Period (Late Precambrian), found in north Russia. Varieties of fossil fungi including various endotrophic mycorrhizae, the earliest ascomycetes and several chytridiomycetes have been reported from Early Devonian Rhynie chert (Kidston and Lang, 1921; Harvey et al., 1969; Hass et al., 1994; Remy et al., 1994; Taylor et al., 1994; Manoharachari, 2005). Fossil records indicate that majority of fungal groups came into existence by the Carboniferous Period. Fossil fungal remains are less common in the Palaeozoic strata but become more frequent in the Lower Mesozoic. A great deal of fungal diversity is witnessed in the Tertiary Period indicating that proliferation of fungi is linked with diversification of angiosperms.

The relatively new technique of molecular systematics is particularly useful in the study of the evolution of fungi. This technique compares the sequence of DNA segments of different species to determine evolutionary relationships. One important finding from this technique is that the plasmodial slime moulds, cellular slime moulds, and water moulds are only distantly related to the true fungi. Evolutionary relationships among organisms also are being studied by comparing the sequences of ribosomal RNAs and transfer RNAs from different organisms. Therefore, the sequences of ribosomal and transfer RNAs tend to be more conserved through evolution. Comparison of these RNAs among fungi, plants, and animals also suggests that fungi are more closely related to animals than to plants. The use of nucleotide and amino acid sequences allows improved understanding of the timing of evolutionary events of life on Earth. Molecular estimates of divergence times are, however, controversial and are generally much more ancient than suggested by the fossil record. The limited number of genes and species explored and pervasive variations in evolutionary rates are the most likely sources of such discrepancies. Future work in molecular systematics is expected to tell us more about the evolution and relationships of the fungi.

GENERAL REMARKS AND FUTURISTIC APPROACH

During the last four decades or so serious efforts were made towards the study of fossil fungi laying emphasis on phylogenetic, stratigraphic and environmental considerations. Data generated on fossil fungi during this period is significant but is only a good beginning. It will emerge as one of the exciting fields of research in the years to come. A number of fossil fungal taxa reported from Indian sediments need taxonomic revision. Molecular systematics, the newly emerging branch, will be of great significance in the study of the evolution of fungi. These studies will throw more light on searching questions like time and rate of emergence of different fungal groups and their relationships with each other.

REFERENCES

Barghoorn, E. S. and Tyler, S. A. 1965. Microorganisms from the Gunflint Chert. Science, 147: 563-577.

Berbee, M. L. and Taylor, J. W. 2001. Fungal molecular evolution: Gene trees and Geologic time, pp. 229-245. In: The Mycota VII Part B Systematics and Evolution. D. G. McLaughlin, E. G. McLaughlin and P. A. Lemke (eds.). Springer-Verlag, Berlin, Heidelberg.

Clarke, R. T. 1965. Fungal spores from Vermejo Formation coal beds (Upper Cretaceous) of central Colorado. Mountain Geologist, 2: 85-93.

Cookson, I. C. 1947. Fossil fungi from Tertiary deposits in the southern hemisphere. Part I. Proceedings of the Linnean Society of New South Wales, 72: 207-214.

Dennis, R. L. 1970. A Middle Pennsylvanian basidiomycete mycelium with clamp connections. Mycologia, 62: 578-584.

Dilcher, D. L. 1965. Epiphyllous fungi from Eocene deposits in western Tennessee, U.S.A. Palaeontographica, Abt. B, 116: 1-54.

Edwards, W. N. 1922. An Eocene Microthyriaceous fungus from Mull, Scotland. Transactions of the British Mycological Society, 8: 66-72.

Elias, M. K. 1966. Living and fossil algae and fungi, formerly known as structural parts of marine bryozoans. Palaeobotanist, 14: 5-18.

Elsik, W. C. 1968. Palynology of Palaeocene Rockdale lignite, Milam County, Texas. I. Morphology and Taxonomy. Pollen et Spores, 10: 263-314.

Elsik, W. C. 1976. Microscopic fungal remains and Cenozoic palynostratigraphy. Geoscience and Man, 15: 115-120.

Elsik, W. C. 1978a. Palynology of Paleocene Rockdale lignite, Milam County, Texas. I. Morphology and taxonomy. Pollen et Spores, 10: 263-314.

Elsik, W. C. 1978b. Classification and geologic history of the Microthyriaceous fungi. Proceedings of the IV International Palynological Conference, Lucknow (1976-77), 1: 331-342.

Hammen, T. van der. 1956. A palynological systematic nomenclature. Boletin Geologico, 4: 63-101.

Harvey, R., Lyson, A. G. and Lewis, P. N. 1969. A fossil fungus from Rhynie chert. Transactions of the British Mycological Society, 53: 155-157.

Hass, H., Taylor, T. N. and Remy, W. 1994. Fungi from the Lower Devonian Rhynie chert: Mycoparasitism. American Journal of Botany, 44: 29-39.

Jain, K. P. 1974. Fossil Fungi, pp. 38-46. In: Aspects and Appraisal of Indian Palaeobotany. K.R. Surange, R.N. Lakhanpal and D.C. Bhardwaj (eds.). Birbal Sahni Institute of Palaeobotany, Lucknow, India.

Jain, K. P. and Gupta, R. C. 1970. Some fungal remains from the Tertiaries of Kerala Coast. Palaeobotanist, 18: 177-182.

Jain, K. P. and Kar, R. K. 1979. Palynology of Neogene sediments around Quilon and Varkala, Kerala Coast, South India 1. Fungal remains. Palaeobotanist, 26: 105-118.

Jansonius, J. 1976. Palaeogene fungal spores and fruiting bodies of the Canadian Arctic. Geoscience and Man, 15: 129-132.

Kalgutkar, R. M. and Jansonius, J. 2000. Synopsis of Fungal spores, Mycelia and Fructifications. AASP Contribution Series 39: 1-423.

Kidston, R. and Lang, W. H. 1921. On Old Sandstone plants showing structure, from the Rhynie Chert Bed, Aberdeenshire. Part V. The thallophyta occurring in the peat-bed; the succession of the plants through a vertical section of the bed, and the conditions of accumulation and preservation of the deposit. Transactions of the Royal Society of Edinburgh, 52: 855-902.

Krassilov, V. 1981. *Orestovia* and the origin of vascular plants. Lethaia, 14: 235-250.

Manoharachari, C. 2005. Interaction of fungi with higher plants – some palaeobotanical glimpses. Diamond Jubilee National Conference, Birbal Sahni Institute of Palaeobotany, Lucknow, 66-67 (Abstract).

Pirozynski, K. A. 1976a. Fossil fungi. Annual Review of Phytopathology, 14: 237-246.

Pirozynski, K. A. 1976b. Fungal spores in the fossil record. Biological Memoirs (In collaboration with International Society of Applied Biology), 1: 104-120.

Pirozynski, K. A. 1978. Fungal spores through the ages – a mycologist's view. Proceedings of the IV International Palynological Conference, Lucknow, (1976-77) 1: 327-330.

Pirozynski, K. A. and Weresub, L. K. 1979. The classification and nomenclature of fossil fungi. In: The Whole Fungus, the Sexual-Asexual Synthesis. B. Kendrick (ed.). Proceedings of the 2nd International Mycological Conference, University of Calgary, Kananaskis, Alberta. 2: 653-688.

Playford, G. 1962. Lower Carboniferous microfloras of Spitsbergen. Palaeontology, 5: 550-618.

Ramanujam, C. G. K. 1982. Recent advances in the study of fossil fungi, pp. 287-301. In: Recent Advances in Cryptogamic Botany, Part II: Fossil cryptogams. D. C. Bharadwaj (ed.). The Palaeobotanical Society, Lucknow, India.

Rao, A. R. 1959. Fungal remains from some Tertiary deposits of India. Palaeobotanist, 7: 43-46.

Rao, K. P. and Ramanujam, C.G.K. 1976. A further record of Microthyriaceous fungi from the Nogene deposits of Kerala in South India. Geophytology, 6: 98-104.

Rao, M. R. 2003. *Kalviwadithyrites*, A new fungal fruiting body from Sindhudurg Formation (Miocene) of Maharashtra, India. Palaeobotanist, 52: 117-119.

Remy, W., Taylor, T. N. and Hass, H. 1994. Early Devonian fungi: A blastocladalean fungus with sexual reproduction. American Journal of Botany, 81: 690-702.

Rosendahl, C. O. 1943. Some fossil fungi from Minnesota. Bulletin of Torry Botanical Club, 70: 126-138.

Saxena, R. K. and Misra, N. K. 1990. Palynological investigation of the Ratnagiri beds of Sindhu Drug District, Maharashtra. Palaeobotanist, 38: 263-276.

Schopf, J. W. 1968. Microflora of Bitter Springs Formation, Late Precambrian, Central Australia. Journal of Palaeontology, 42: 651-688.

Schopf, J. W. and Barghoorn, E. S. 1969. Microorganisms from the late Precambrian of South Australia. Journal of Palaeontology, 43: 111-118.

Stubblefield, S. P., Taylor, T. N. and Beck, C. B. 1985. Studies of Palaeozoic fungi. V. Wood decaying fungi in *Callixylon newberryi*, the Upper Devonian. American Journal of Botany, 72: 1765-1774.

Taylor, T. N. and White, J. F., Jr. 1989. Fossil fungi (Endogonaceae) from the Triassic of Antarctica. American Journal of Botany, 76: 389-396.

Taylor, T. N., Galtier, J. and Axsmith, B. J. 1994. Fungi from Lower Carboniferous of central France. Review of Palaeobotany and Palynoloogy, 83: 253-260.

Taylor, T. N., Remy, W., Hass, H. and Kerp, H. 1995. Fossil arbuscular mycorrhizae from the Early Devonian. Mycologia, 87: 560-573.

Tiffney, B. H. and Barghoorn, E. S. 1974. The fossil record of the fungi. Occasional papers of the Farlow Herbarium of Cryptogamic Botany. 7: 1-42.

Tripathi, S. K. M. 2009. Fungi from Palaeoenvironments: Their Role in Environmental Interpretations, pp. 1-27. In: Fungi from Different Environments. J. K. Misra and S. K. Deshmukh (eds.). Science Publishers, Enfield, NH, USA.

Tyler, S. A. and Barghoorn, E. S. 1954. Occurrence of structurally preserved plants in Pre-Cambrian rocks of the Canadian Shield. Science, 119: 606-608.

Venkatachala, B. S. and Kar, R. K. 1969. Palynology of the Tertiary sediments in Kutch-2. Epiphyllous fungal remains from the borehole no. 14; The Palaeobotanist, 17: 179-183.

Wagner, C. A. and Taylor, T. N. 1981. Evidence for endomycorrhizae in Pennsylvanian age plant fossils. Science, 216: 562-563.

White, J. F., Jr. and Taylor, T. N. 1988. Triassic fungus from Antarctica with possible ascomycetous affinities. American Journal of Botany, 75: 1495-1500.

White, J. F., Jr. and Taylor, T. N. 1989. A trichomycete-like fossil from Triassic of Antarctica. Mycologia, 81: 643-646.

Williamson, W. C. 1878. On the organization of fossil plants of the Coal Measures. Part IX. Philosophical Transactions of the Royal Society of London, 169: 319-364.

Williamson, W. C. 1880. On the organization of the fossil plants of the Coal-Measures. Part X. Including an examination of the supposed radiolarians of the Carboniferous rocks. Philosophical Transactions of the Royal Society of London, 171: 493-539.

Wilson, L. R. 1962. A Permian fungus spore type from the Flowerpot Formation of Oklahoma. Oklahoma Geology Notes, 22: 91-96.

Chapter 3

From Zoospores to Molecules: The Evolution and Systematics of Chytridiomycota

Martha J. Powell* and Peter M. Letcher

Department of Biological Sciences, Box 870344, The University of Alabama
Tuscaloosa, AL 35487, United States of America
E-mail: *mpowell@bama.ua.edu
letch006@bama.ua.edu

ABSTRACT

Chytridiomycota characteristically reproduce with posteriorly uniflagellate zoospores and are known as zoosporic fungi. Electron microscopic studies and molecular analyses of gene sequences have transformed our understanding of relationships within the Chytridiomycota. From syntheses of these data, systematics of zoosporic fungi is being rewritten, including the separation of Blastocladiomycota, Neocallimastigomycota and Monoblepharidomycota from the Chytridiomycota as new phyla. This chapter traces the recent history of zoosporic fungal taxonomy and summarizes our current classification of Chytridiomycota. Since the 1970s, the concept of distinct "zoospore types" based on a constellation of characters allowed zoospore ultrastructure to define orders in the Chytridiomycota. During the 1990s, phylogenetic analyses of gene sequences provided greater resolution needed to refine chytrid relationships, and by the 2000s classification decisions, even at the species level, applied molecular-based monophyly. Unexpectedly, molecular analyses of zoosporic fungi revealed the order Chytridiales as polyphyletic and composed of multiple distinct lineages. Revisionary systematic studies have now characterized several of these lineages as new orders for taxa previously classified in the Chytridiales. How lineages of zoosporic fungi evolved is controversial, but molecular based

*Corresponding author

hypotheses of phylogeny suggest that holocarpic parasitic zoosporic fungi are most basal in the radiation of fungal evolution and hyphal forms are more primitive than simple eucarpic thallus forms.

INTRODUCTION

First recognized as parasites of algae by a European botanist (Braun, 1851, 1855), Chytridiomycota have never generated more fascination among scientists and the public than now. As basal in the evolution of fungi, they hold the key to understanding ancestral forms and factors that may have led to the radiation of fungi (Stajich et al., 2009). Increasingly, ecologists are exploring the role of chytrids in microbial loops and food webs, as well as in soil fertility (Midgley et al., 2006; Lefèvre et al., 2007; Sime-Ngando et al., 2011). The emergence of *Batrachochytrium dendrobatidis* as the first recognized chytrid pathogen of vertebrates and its annihilation of numerous species of frogs and other amphibians raise concern among conservation biologists (Longcore et al., 1999; Fisher et al., 2009).

Chytridiomycota are readily isolated from aquatic habitats and soils and function as parasites of algae, fungi, invertebrates, plants and animals and as biodegraders of pollen, cellulose, chitin and keratin (Powell, 1993; Letcher and Powell, 2001; Bills et al., 2004). Traditionally classified on the basis of thallus morphology (Whiffen, 1944; Sparrow, 1960; Karling, 1977), chytrids often are difficult to identify with light microscopy because of problems with morphological plasticity and overlap of distinguishing diagnostic characters (Powell and Koch, 1977). Thalli of Chytridiomycota are varied and can be monocentric-eucarpic (Fig. 3.3C; thallus consists of one sporangium plus rhizoids), polycentric-rhizomycelial (Fig. 3.3E; more than one sporangium per thallus with rhizoids arising from several sites), or hyphal-like (Fig. 3.2F; filamentous bearing terminal sporangia and often with a basal holdfast). Some Chytridiomycota that grow inside of host cells form a monocentric-holocarpic thallus (composed only of a sporangium). However, diagnostic structures such as apophyses (sub-sporangial swellings of the rhizoids), thallus complexity, and operculate (Fig. 3.2E) versus inoperculate sporangial discharge openings are not always consistent, especially when organisms are grown in culture (Powell and Koch, 1977).

Chytridiomycota reproduction is primarily asexual with the discharge of zoospores bearing a single, posteriorly directed whiplash flagellum (Fig. 3.2A), and it is zoospore structure that distinguishes Chytridiomycota from other organisms producing motile cells. Chytridiomycota zoospores are distinctive, even at the light microscopic level, because of their erratic darting swimming trajectory (Sparrow, 1960; Karling, 1977). At the electron microscopic level, zoospores possess a centriole along with the kinetosome, nine flagellar props joining the kinetosome to the plasma membrane, and lipid globules organized into an organellar assemblage, the microbody-lipid globule complex (Powell, 1976, 1979). Sexual reproduction occurs in some but not all genera of chytrids (Sparrow, 1960), and thick-walled resting spores may arise sexually or asexually.

Recent applications of zoospore ultrastructure and gene sequence characters to the phylogenetics of Chytridiomycota have transformed our understanding of their relationships and have revealed that they are more diverse than earlier understood (Letcher et al., 2005, 2006, 2008a,c; James et al., 2006; Simmons et al., 2009). Molecular investigations of environmental samples detect Chytridiomycota in wide-ranging habitats, temperate (Lefèvre et al., 2007; Chen et al., 2008; Lepère et al., 2008) as well as extreme environments such as anoxic deep-sea ecosystems, high elevation exposed soils, and arctic habitats (Dawson and Pace, 2002; Stoeck and Epstein, 2003; Stoeck et al., 2007; Freeman et al., 2009; Le Calvez et al., 2009). Molecular environmental studies often reveal undescribed lineages (Lefèvre et al., 2007; Lepère et al., 2008; Freeman et al., 2009; Le Calvez et al., 2009) harkening the untapped diversity of Chytridiomycota waiting to be described. As a consequence, systematics of Chytridiomycota is in a dynamic state of flux as we explore the range of their habitats and patterns of geographic distributions.

SYSTEMATICS

Recent history of Chytridiomycota classification

In a traditional classification (Sparrow, 1960), the single class Chytridiomycetes comprised three orders: the Chytridiales, considered basal because of their simple thalli, and the Blastocladiales and Monoblepharidales, considered advanced because of the production of hyphae and prominence of sexual reproduction. Emerson and Whisler (1968) proposed a fourth order, the Harpochytriales, for a unique group of organisms producing an unbranched filament with a disc-shaped holdfast. With a view that zoospore flagellar structure and number should be the primary taxonomic character for zoosporic 'fungi', this system of classification embraced Scherffel's (1925) emphasis on zoospore structure.

Early electron microscopic studies (Koch, 1961) gave the first glimpse of the amazing intricacies of zoospore fine structural architecture and illuminated the potential to use differences in zoospore structure to understand relationships. With electron microscopy growing as a research tool in the 1970s, the importance of zoospore fine structure became widely recognized as a way to explore phylogenetic relationships of Chytridiomycota. The concept of distinct "zoospore types" based on a constellation of characters (Powell, 1978b; Lange and Olson, 1979; Barr, 2001) allowed zoospore ultrastructure to define orders in the Chytridiomycota. Emphasis was placed on structures associated with the flagellar apparatus (Barr, 1980, 2001) and the organization of organelles in the microbody-lipid globule complex (MLC) (Fig. 3.1), an organellar assemblage involved in processing stored lipids for energy (Powell 1976, 1978b). In classification of Chytridiomycota, zoospore ultrastructural character suites were used to delineate orders. The first direct application of zoospore ultrastructure to taxonomic revision of Chytridiomycota was Barr's (1980) establishment of Spizellomycetales as a new order delineated from the Chytridiales, the fifth in the Chytridiomycetes. Heath et al. (1983) at first considered the anaerobic zoosporic rumen fungi to be members of this new order, but later recognized their distinctiveness as a group and erected the Neocallimastigales (Li et al., 1993) as a sixth order in the Chytridiomycetes. With new understandings of zoospore ultrastructure (Gauriloff et al., 1980), Barr (1990) proposed disbanding the order Harpochytriales and redistributing its members. Consequently, he transferred the Harpochytriaceae to the Chytridiales and erected a new family, the Oedogoniomycetaceae, in the Monoblepharidales. Later studies (Chambers, 2003; James et al., 2006), however, revealed that Emerson and Whisler's (1968) grouping of members in the Harpochytriales was phylogenetically correct. Despite the tremendous insights zoospore ultrastructural characters provided for decisions at higher taxonomic levels, they were less informative at the lower taxonomic levels of genus and species.

Molecular sequence analyses beginning in the 1990s furnished the increased sensitivity needed to resolve species-level relationships and revolutionized our classification among zoosporic fungi. For example, Paquin et al. (1997) studied mitochondrial genes and revealed that the Harpochytriaceae was actually allied with the Monoblepharidales rather than the Chytridiales (Barr, 1980). Consequently, they transferred the Harpochytriaceae from the Chytridiales to the Monoblepharidales. The first broadly-scoped analysis of zoosporic fungi (James et al., 2000) revealed unexpected diversity and distinct lineages in the Chytridiomycetes. Additional investigations (James et al., 2006) confirmed the divergence of the Blastocladiales from the Chytridiales, the polyphyly of the Chytridiales, and the phylogenetic reliability of zoospore ultastructure.

Based on molecular analyses and confirming ultrastructural distinction in characters, the Blastocladiales was elevated to the rank of Phylum Blastocladiomycota (=blastoclads) (James et al., 2006), the Phylum Chytridiomycota (=chytrids) was circumscribed (Hibbett et al., 2007), and the distinctive anaerobic zoosporic fungi (Ho and Barr, 1995) were raised to Phylum Neocallimastigomycota (=neocallimastigos) (Hibbett et al., 2007). In the Phylum Chytridiomycota, the Class Monoblepharidomycetes was emended (Doweld, 2001; Hibbett et al., 2007); thus, in the Hibbett et al. (2007) compilation of fungal classification, Chytridiomycetes and Monoblepharidomycetes made up the Chytridiomycota. However, Doweld (2001) established a phylum for Monoblepharidomycetes as well as other ranks within this new phylum with *Monoblepharis* recognized as the type. Because of the close molecular sister relationship of the Monoblepharidomycota (=monoblephs) to Chytridiomycota and similarities in presence of stacked cisternae in the Golgi apparatus and open poles during mitosis, we also include the phylum Monoblepharidomycota in the following discussion.

Ongoing revisionary systematic studies that combine molecular sequence characters and ultrastructual features are delineating monophyletic orders from the polyphyletic Chytridiales in an attempt to develop a natural system of classification for Chytridiomycota/Chytridiomycetes. There are now six orders in the Chytridiomycota, four delineated since 2006. Three other clades, *Synchytrium, Blyttiomyces,* and *Polychytrium* clades (James et al., 2006), have not yet been characterized, but they will likely be circumscribed as additional orders. In the following section we summarize the classification of Chytridiomycota for which zoospore ultrastructural and/or molecular sequence analyses can be used to assign phylogenetic affinities.

CHYTRIDIOMYCOTA

Chytridiomycetes/Chytridiales (emend Mozley-Standridge et al., 2009; Vélez et al., 2011)

Chytridium olla (Fig. 3.2B) (Braun, 1851, 1855), the first chytrid to be described, and consequently the type for the genus *Chytridium*, order Chytridiales, class Chytridiomycetes and phylum Chytridiomycota, was vital to characterize and anchor the monophyly of these groups. In a recent molecular and ultrastructural study, Vélez et al. (2011) discovered that *C. olla* phylogenetically placed in the *Chytriomyces* clade of the James et al. (2006) study and defined the limits of a monophyletic Chytridiales. As now circumscribed (Vélez et al., 2011), Chytridiales includes species with morphologically diverse thalli that are operculate or inoperculate and monocentric or polycentric (Letcher and Powell, 2005b; Letcher et al., 2005). Sexual reproduction is common in members such as *Chytriomyces* and *Siphonaria,* but rarely reported in other genera (Sparrow, 1960).

Chytridiales contains both soil inhabiting and aquatic members, but aquatic members populate the clade more than soil members do at this time. Taxa such as *Asterophlyctis* and *Podochytrium* are exclusively found in water, but others such as *Chytriomyces hyalinus* are reported from both habitats. Freeman et al. (2009) revealed a significant presence of Chytridiales in high-elevation soils, which predicts that Chytridiales are more common in soils than presently understood. Most saprophytic representatives of this clade are isolated on chitin or pollen, and some are found specifically on exuviae of insects. Many of the representatives are parasites of algae, including *Chytridium olla* (Fig. 3.2B) and *Phlyctochytrium planicorne* (Fig. 3.2D), and *Rhizidium phycophilum* appears to have a symbiotic relationship with an alga (Picard et al., 2009).

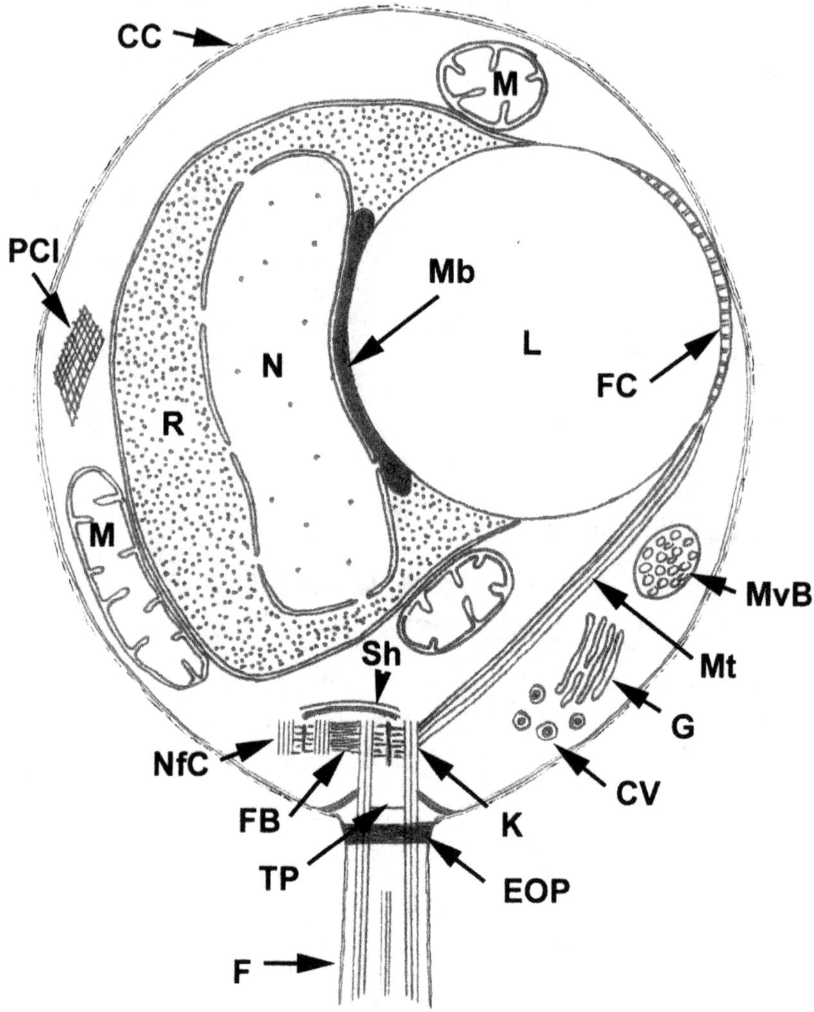

Fig. 3.1 Summary diagram illustrating ultrastructural characters of the *Chytridium olla* zoospore typical of the Chytridiales. CC – cell coat; CV – cored vesicles; EOP – electron opaque flagellar plug; F – flagellum; FB – fibrous bridge connecting the kinetosome and non-flagellated centriole; FC – fenestrated MLC cisterna (= rumposome); G – Golgi apparatus; K – kinetosome; L– lipid globule; M – mitochondrion; Mb – microbody; Mt – microtubular root; MvB – multivesicular body; NfC – non-flagellated centriole; PCI – paracrystalline inclusion; Sh – shield as a kinetosome-associated structure; TP – terminal plate.

Both zoospore ultrastructure and molecular phylogeny confirmed that the classical concept of the Chytridiales was a diverse grouping of organisms (Barr, 2001; James et al., 2006). Barr (1980) characterized a chytridialian zoospore as one with a tightly organized core consisting of the ribosomal aggregation

Table 3.1 Major characteristics of zoospore types for orders of Chytridiomycota and Monoblepharidomycota

	AP	*CC*	*MLC*	*MbC*	*PCI*	*Root*	*RA*	*N-K*
Chytridiomycetes								
Chytridiales	+	+	T	–	+	LB or –	+	–
Lobulomycetales	+	–	T	–	–	–	+	–
Rhizophydiales	–	–	T	–	–	LS or –	+	–
Cladochytriales	+	–	T	–	–	LL	+	–
Spizellomycetales	–	–	L	–	–	A	–	+
Rhizophlyctidales	–	–	L	–	–	–	+ or –	+ or –
Monoblepharidomycetes								
Harpochytriales	+	–	T	+	–	A	+	+ or –
Monoblepharidales	+ or –	–	T	+	–	A	+	–
Hyaloraphidiomycetes								
Hyaloraphidiales	–	–	L	–	–	–	–	–

Axonemal basal plug (AP): + present; – absent.
Cell coat (CC): + prominent; – not prominent or absent.
Microbody-lipid globule organization (MLC): T organelles tightly (compactly) associated; L organelles loosely associated; – absent.
Microbody backs MLC cisterna (MbC): + yes, – no.
Paracrystalline inclusion (PCI): + present; – absent.
Microtubular root (Root): – absent; A radiating anteriorly; LS lateral with 4-5 microtubules stacked with spaces between; LB lateral and bundled as in a cord; LL lateral and clustered with linkers cross bridging between evenly spaced 9-25 microtubules;
Ribosomal aggregation (RA): + present; – absent.
Nucleus spatially or structurally associated with the kinetosome (N-K): + yes; – no.

around the nucleus and closely associated microbodies, lipid globules, cisterna and mitochondria in the MLC (Microbody-Lipid Globule Complex) (Fig. 3.1). Beyond this characterization, he emphasized differences among zoospores and defined types. We now know from molecular sequence analyses that those types of chytridialian zoospores actually represent distinct lineages that we define as orders, and the Type I and Type II zoospores (Barr and Hartmann, 1976; Barr, 1980) are diagnostic of the Chytridiales: the Type I in the family Chytriomycetaceae and the Type II for the Chytridiaceae (Vélez et al., 2011). Figure 3.1 illustrates the organization of the *Chytridium olla* zoospore and is characteristic of the order Chytridiales and family Chytridiaceae (Letcher et al., 2005; Vélez et al., 2011). A paracrystalline inclusion (Fig. 3.1) is found only in zoospores of Chytridiales among Chytridiomycota (Table 3.1). Probably all zoospores have some type of extracellular matrix or cell coat, but a distinctive cell coat covering the zoospore body (Fig. 3.1) and not the flagellar axonome is readily observed in members of the Chytridiales (Dorward and Powell, 1983) and is one of the diagnostic features of this clade (Table 3.1).

Although molecular and ultrastructural analyses have now defined a monophyletic Chytridiales, some genera within the order appear to be polyphyletic, including *Chytriomyces* and *Rhizidium* (Letcher et al., 2005; Picard et al., 2009; Vélez et al., 2011). Chytridiales contains some of the oldest genera of chytrids, and many of the type species have not been collected and cultured. Thus, there remains much systematic work to be done to fully characterize this important order.

Spizellomycetales (Barr, 1980)

Barr (1980) delineated this order based on an assemblage of zoospore ultrastructural characters (Barr, 2001). Unlike most zoospores in the Chytridiales (*sensu* Barr, 1980), Spizellomycetales zoospores lacked aggregated ribosomes, organelles of the MLC were loosely associated, and microtubule roots were oriented anteriorly and not associated with the MLC (Table 3.1). Zoospores were often polymorphic, reverting from spherical to elongate to amoeboid, and sometimes swimming with the flagellum insertion anterior and the flagellum trailing posteriorly (Powell, 1978a). Pulsating (contractile) vacuoles were readily visible (Fig. 3.2A). All of its described members produce inoperculate, monocentric thalli, and the tips of rhizoids tend to be blunt or rounded (Fig. 3.3D) (Barr, 1984). Sporangia develop from the zoospore cyst in all representatives except for *Powellomyces,* where the nucleus migrates into the germ tube which develops into the sporangium (Powell and Koch, 1977). The order as now defined is monophyletic and composed of five genera (Table 3.2) (James et al., 2006; Wakefield et al., 2010).

Table 3.2 Current classification of Chytridiomycota/Chytridiomycetes

Chytridiales	family *incertae sedis*
(for illustrations see Letcher et al., 2005)	*Polyphlyctis unispina*
Chytridiaceae	*Rhizidium endosporangiatum*
Chytridium olla	**Spizellomycetales**
Chytridium lagenaria	(for descriptions see Barr, 1984)
Phlyctochytrium planicorne	Spizellomycetaceae
Chytriomycetaceae	*Spizellomyces punctatus*
Asterophlyctis sarcoptoides	*Spizellomyces plurigibbosus*
Chytriomyces appendiculatus	*Spizellomyces palustris*
Chytriomyces hyalinus	*Spizellomyces acuminatus*
Chytriomyces spinosus	*Spizellomyces kniepii*
Entophlyctis luteolus	*Spizellomyces lactosolyticus*
Obelidium mucronatum	*Spizellomyces dolichospermus*
Phlyctorhiza endogena	*Spizellomyces pseudodichotomus*
Physocladia obscura	*Gaertneriomyces semiglobifer*
Podochytrium dentatum	*Gaertneriomyces spectabile*
Rhizidium phycophilum	*Gaertneriomyces tenuis*
Rhizoclosmatium globosum	*Kochiomyces dichotomus*
Rhizoclosmatium aurantiacum	*Triparticalcar arcticum*
Siphonaria petersenii	family *incertae sedis*
	(removed from Olpidiaceae)

Table 3.2 Contd. ...

Powellomyces hirtus	Arizonaphlyctidaceae
Powellomyces variabilis	*Arizonaphlyctis lemmonensis*
Rhizophydiales	Borealophlyctidaceae
(for key see pg. 781 in Letcher et al., 2008c)	*Borealophlyctis paxensis*
Alphamycetaceae	family *incertae sedis*
Alphamyces chaetifer	*Catenomyces persicinus*
Angulomycetaceae	**Lobulomycetales**
Angulomyces argentinensis	(for comparison of taxa see pg. 455 in Simmons et al., 2009)
Aquamycetaceae	Lobulomycetaceae
Aquamyces chlorogonii	*Lobulomyces angularis*
Globomycetaceae	*Lobulomyces poculatus*
Globomyces pollinis-pini	*Clydaea vesicula*
Urceomyces sphaerocarpum	*Maunachytrium keaense*
Gorgonomycetaceae	family *incertae sedis*
Gorgonomyces haynaldii	*Chytridium polysiphoniae*
Kappamycetaceae	**Cladochytriales**
Kappamyces laurelensis	(for descriptions see Mozley-Standridge et al., 2009)
Pateramycetaceae	Cladochytriaceae
Pateramyces corrientinensis	*Cladochytrium replicatum*
Protrudomycetaceae	Nowakowskiellaceae
Protrudomyces lateralis	*Nowakowskiella elegans*
Rhizophydiaceae	*Nowakowskiella hemisphaerospora*
Rhizophydium globosum	Septochytriaceae
Rhizophydium brooksianum	*Septochytrium variabile*
Terramycetaceae	Endochytriaceae
Boothiomyces macroporosum	*Endochytrium ramosum*
Terramyces subangulosum	family *incertae sedis*
family *incertae sedis*	*Nephrochytrium aurantium*
Batrachochytrium dendrobatidis	*Allochytridium luteum*
Coralloidiomyces digitatus	*Allochytridium expandens*
Entophlyctis helioformis	*Catenochytridium* sp.
Rhizophlyctis harderi	**Orders** *incertae sedis*
Rhizophydium aestuarii, R. littoreum	Olpidiaceae
Rhizophlyctidales	*Olpidium brassicae*
(for key see pg. 1046 in Letcher et al., 2008a)	*Rozella allomycis*
Rhizophlyctidaceae	Caulochytriaceae
Rhizophlyctis rosea	*Caulochytrium protostelioides*
Sonoraphlyctidaceae	**Order and Family** *incertae sedis*
Sonoraphlyctis ranzonii	*Mesochytrium penetrans*

As a terminal clade in chytridiomycota (James et al., 2006), these chytrids appear to be exclusively soil and near-shore inhabiting fungi (Wakefield et al., 2010). Most members of this group are pollen degraders with a few growing on cellulose. Several are parasites found on nematodes, arbuscular mycorrhizae, and oospores of downy mildews (reviewed in Wakefield et al., 2010). This clade may be important in soil fertility because they are commonly isolated from agricultural soils (Wakefield et al., 2010). Chytrids are typically not considered dung fungi, but we now know that *Gaertneriomyces* and *Triparticalcar* can be cultured from herbivore dung (Wakefield et al., 2010).

In molecular phylogenetic analyses of all existing cultures of holotypes used to establish taxa in this order (Wakefield et al., 2010), *Kochiomyces, Powellomyces* and *Triparticalcar* were found to be monophyletic. However, *Gaertneriomyces* and *Spizellomyces* were polyphyletic, and some undescribed isolates formed unique clades. This study (Wakefield et al., 2010) demonstrates that the Spizellomycetales is richer in genera and species than presently described. Consistent with this perception, Freeman et al.'s (2009) study of soil fungi from high elevations using molecular and culture analyses found members of the Spizellomycetales clade were common and many were uncharacterized.

Rhizophydiales (Letcher et al., 2006)

The Rhizophydiales was the first clade to be elevated to an order from the polyphyletic Chytridiales identified in molecular sequence analyses of a wide range of chytrids (James et al., 2000, 2006; Letcher et al., 2006). The order is based on the *Rhizophydium* clade (James et al., 2006), which primarily contained isolates identified as *Rhizophydium*. *Rhizophydium* is among the most species rich genera in the Chytridiomycota with over 220 described taxa, 84 of which have been described since Sparrow's (1960) monograph (Letcher and Powell, 2005a; Letcher et al., 2006; Letcher et al., 2008b, c). Isolated from soil and marine and fresh water, the majority of the described saprotrophic species of *Rhizophydium* are biodegraders of pollen, with only four reported on keratin, one on chitin, and one on cellulose. Many are parasites of algae, and one is a root parasite of wheat (Sparrow, 1960). *Rhizophydium* is characterized by a thallus composed of a single sporangium bearing one to many inoperculate discharge pores, papillae, or tubes and a single rhizoidal axis (Fig. 3.2C) (Sparrow, 1960; Karling, 1977). The Rhizophydiales is now known to include additional thallus types: *Entophlyctis helioformis* where the nucleus migrates into the germ tube that differentiates into the sporangium with multiple rhizoidal axes, and *Batrachochytrium dendrobatidis* where the thallus is sometimes colonial (James et al., 2006). All chytrids in this clade are eucarpic and monocentric, but both operculate and inoperculate species are found.

This is an important group to characterize more extensively. *Batrachochytrium dendrobatidis*, the highly destructive pathogen of frogs (Longcore et al., 1999;

Fisher et al., 2009), radiates within this clade on a long branch, but with no close relatives (Letcher et al., 2008c). Discovering a close relative to *B. dendrobatidis* will contribute to better understanding of its phylogenetic affinities and its pathogenicity. The potential transformational role of members of the Rhizophydiales in aquatic food webs is little studied and newly recognized (reviewed in Sime-Ngando et al., 2011). Rhizophydiales as parasites of a range of organisms, including invertebrates, other chytrids and algae (notably planktonic desmids and diatoms), may have a role in natural control of aquatic populations (Powell, 1993).

The Rhizophydiales clade is the most extensively sampled clade of Chytridiomycota for molecular phylogenetic studies because it has been the focus of an effort to collect representatives in geographically and environmentally diverse habitats so as to determine species range and genetic variability (Letcher and Powell, 2005a; Letcher et al., 2006, 2008b,c). Initial efforts to revise the *Rhizophydium* clade established the historic type of the genus, *R. globosum* (Letcher et al., 2006), necessary to characterize the new order Rhizophydiales. Letcher and Powell (2005a) demonstrated the utility of the phylogenetic species concept in Rhizophydiales. *Kappamyces* was the first phylogenetic genus of a chytrid to be circumscribed primarily on molecular sequence analysis and confirmed with a unique zoospore structure (distinguished by the presence of an electron-dense cylinder in the core of the kinetosome and secondary centriole). To understand the biodiversity of Rhizophydiales better and to reveal new suites of zoospore ultrastructural characters, Letcher et al. (2008b) explored isolates that have no near relatives in phylograms and do not occur in families as circumscribed. From submersed mud at the edge of an oligotrophic lake in southern Argentina near the Andes in Patagonia, Letcher et al. (2008b) isolated a chytrid that defied the original concept that thallus structure in this clade was entirely highly conserved. This new chytrid had a thallus with a sporangium shaped like a coral, and *Coralloidiomyces digitatus* was erected for this remarkable chytrid.

Barr (1980) recognized one type of zoospore (Group III-type) for *Rhizophydium,* and it was characterized by a microtubular root composed of two to six parallel microtubules aligned one above the other and the absence of an electron-opaque plug in the base of the flagellum (Table 3.1). The simple thallus of the classical concept of *Rhizophydium,* however, belies the variety of zoospore subtypes now recognized. Recent ultrastructural studies have revealed 18 variations of this basic zoospore type (Letcher et al., 2008c), and these differences form the basis of families within the order (Table 3.2).

The Rhizophydiales is clearly a genetically diverse order (Letcher et al., 2008c). Although it now comprises 10 families and 14 genera (Table 3.2), expectations are that it is even larger. The untapped biodiversity of organisms in this clade is underscored in molecular environmental studies of arctic soils, high elevation soils and fresh water (Lefèvre et al., 2007; Stoeck et al., 2007; Lepère et al., 2008; Freeman et al., 2009). Discovery of clades in the Rhizophydiales with no known described species (Lefèvre et al., 2007; Lepère et al., 2008) highlights the need to culture and characterize more of these undescribed organisms (Letcher et al., 2008b, c).

Rhizophlyctidales (Letcher et al., 2008a)

Rhizophlyctidales was circumscribed from Spizellomycetales on the bases of molecular monophyly and zoospore ultrastructure of the morphospecies *Rhizophlyctis rosea* (Letcher et al., 2008a), the type species of the genus (Blackwell and Powell, 1999). Chytrids identified as *Rhizophlyctis rosea* are commonly observed decomposers of cellulose and are frequently collected from grass compost heaps and soils, especially disturbed soils such as lawns and crop fields (Whiffen, 1941; Willoughby, 2001). *Rhizophlyctis rosea* thalli are easily observed on cellulosic baits of soil (Bills et al., 2004) because of their bright orange to pink sporangia, extensive rhizoids radiating from multiple sites on the sporangium, and numerous scattered inoperculate discharge tubes projecting from the surface of the sporangia. Sporangial pigmentation is due to γ-carotene (Davies, 1961; Vincent and Powell, 1988), which may protect thalli from UV irradiation and contribute to their resistance to desiccation in soils (Gleason et al., 2004).

Barr (1980) classified the *Rhizophlyctis rosea* species complex in the Spizellomycetales, primarily because electron microscopy of the zoospore revealed that the nucleus was structurally associated with the kinetosome. Surprisingly, Barr and Désaulniers (1986) discovered four sub-types of zoospores among isolates of this morphospecies, envisaging the diversity of the species complex later revealed by molecular phylogenetic analyses. *Rhizophlyctis rosea* morphospecies complex was distinct from the core members of the Spizellomycetales because of details of zoospore architecture (Barr and Désaulniers, 1986). Thus, in placing *Rhizophlyctis rosea* in the Spizellomycetales, Barr (1980) was cautionary about its inclusion.

In molecular phylogenetic analyses, the *Rhizophlyctis rosea* clade is sister to the Spizellomycetales, but distantly (James et al., 2006; Letcher et al., 2008a). Analyses of molecular, morphological, and ultrastructural data of 49 isolates of

Fig. 3.2 Morphology of Chytridiomycota and Monoblepharidomycota. (A) Zoospore of *Spizellomyces kniepii* (Spizellomycetales) with posteriorly directed whiplash flagellum (F). In the cytoplasm a lipid globule (L) and contractile vacuole (CV) system are visible. (B) *Chytridium olla* (Chytridiales) with characteristic apical spine (arrow) and apophysate (Ap) rhizoidal system penetrating the oogonium of the alga, *Oedogonium*. (C) Thallus of *Terramyces subangulosum* (Rhizophydiales) growing on sweet gum pollen (P). Notice that the sporangium (S) has fully cleaved zoospores and the rhizoid (R) is a single axis that branches, typical of numerous members of the Rhizophydiales. (D) *Phlyctochytrium planicorne* (Chytridiales) with

Fig. 3.2 Contd. ...

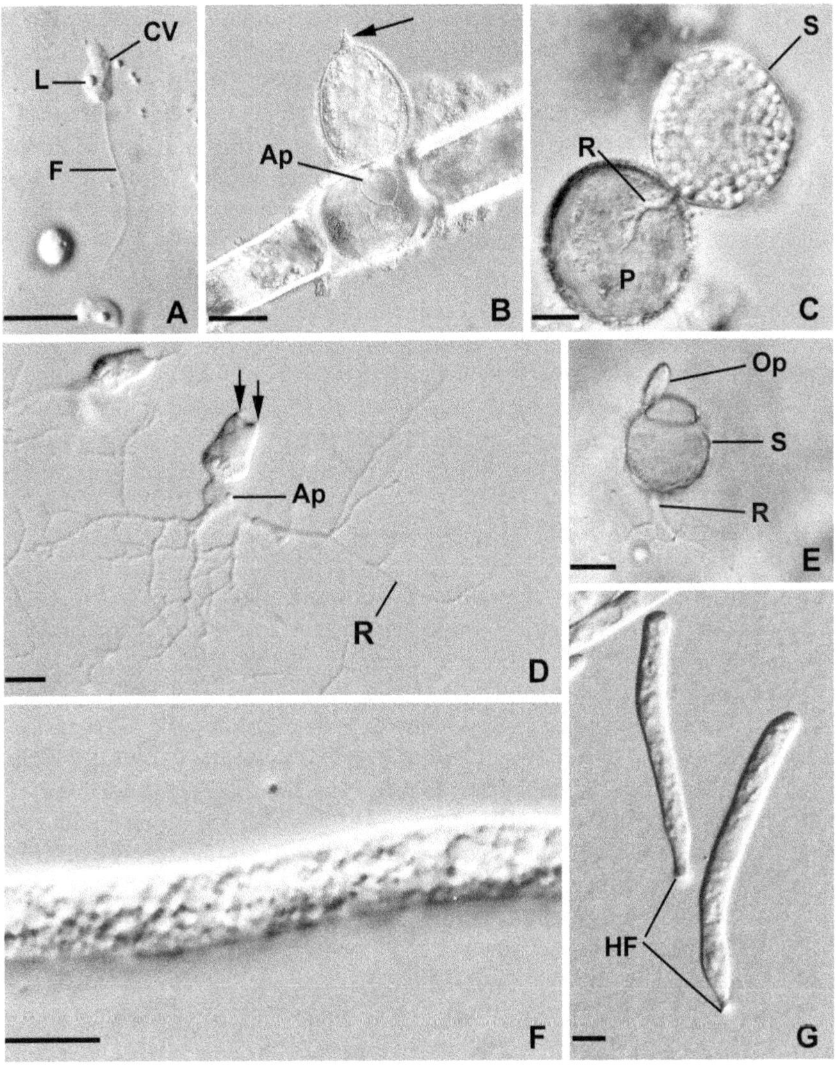

Fig. 3.2 Contd. ...

easily recognized thallus bearing a subsporangial swelling on the rhizoid, the apophysis (Ap), and spines at the apex of the sporangium (arrows). The extensive, open branched rhizoidal system which tapers toward the tips of rhizoids is characteristic of members of the Chytridiales. (E) Undescribed chytrid with lid-like operculum (Op) and empty sporangium (S) bearing rhizoids (R). (F) *Monoblepharella* sp. hypha with typical foamy-appearance of cytoplasm due to abundance of vacuoles and vesicles. (G) Short-tubular thallus of *Harpochytrium* sp. with basal hold fast (HF) attaching the thallus to the substrate. Scale bars = 10 μm.

the *Rhizophlyctis rosea* morphospecies collected from around the world revealed that this clade was genetically diverse and morphologically distinct, leading to the delineation of the new order Rhizophlyctidales (Letcher et al., 2008a). Cultures of this species complex exhibiting all four types of zoospores (Barr and Désaulniers, 1986) were recovered, and additional molecular and ultrastructural analyses of these cultures led to the description of three new families and genera (Letcher et al. 2008a), each family corresponding to a zoospore subtype Barr and Désaulniers (1986) had characterized (Table 3.2).

The vast majority of isolates in this clade are monocentric and soil-inhabiting, but *Rhizophlyctis rosea* is known to sometimes produce polycentric thalli, and the polycentric species isolated from an aquatic habitat, *Catenomyces persicinus,* appears from molecular phylogenetic analyses to belong to this order (James et al., 2006). *Catenomyces* is currently still classified in the Blastocladiomycota, and additional studies of this organism and the ultrastructure of its zoospores are needed before it is formally transferred into the Rhizophlyctidales.

Lobulomycetales (Simmons et al., 2009)

The order Lobulomycetales is based on species once classified as members of the genus *Chytriomyces* but which molecular phylogenetic studies demonstrated were not monophyletic with *C. hyalinus,* the type of the genus (James et al., 2000, 2006; Letcher and Powell, 2002; Letcher et al., 2005). In molecular phylogenetic analyses, *Chytriomyces angularis* was found in a clade distant to the Chytridiales clade (James et al., 2006). Examination of the *C. angularis* zoospore (Table 3.1) (Longcore, 1992) demonstrated that it was neither a Type I nor Type II zoospore (Barr, 1980; Barr and Hartmann, 1976), which are diagnostic of the Chytridiales (Vélez et al., 2011). *C. angularis* was also morphologically distinct from *C. hyalinus.* Rather than producing a trunk-like rhizoid (Fig. 3.3C), *C. angularis'* rhizoidal axis was thread-like (Fig. 3.3A) (Longcore, 1992). It closely resembled another species classified as *Chytriomyces, C. poculatus,* a distinctive chytrid with cupules wrapped around an obpyriform sporangium (Fig. 3.3B).

Chytriomyces angularis, C. poculatus and *Chytridium polysiphoniae,* along with undescribed taxa, formed a clade in molecular phylogenetic analyses (James et al., 2006; Simmons et al, 2009). Consequently, Lobulomycetales was established as a new order (Simmons et al., 2009), *C. angularis* and *C. poculatus* were moved to a new genus, (*Lobulomyces*) and two new genera (*Clydaea* and *Maunachytrium*) were erected for additional members of this clade (Table 3.2). No taxonomic revision was made of *C. polysiphoniae* (Simmons et al., 2009), but it was included in the new order.

Although at this time Lobulomycetales is a small order (Table 3.2), indications are that it is species rich. Molecular environmental sampling of diversity in soil and aquatic habitats often reveal undescribed members of this clade (Freeman et al., 2009; Luo et al., 2009), and it is already apparent that taxa are found in a diverse range of habitats. Members are found in under-sampled harsh environments such as alpine barren soils in Peru and Hawaii (USA), oxygen-depleted salt marshes, deep-sea hydrothermal ecosystems, and polar glacial melts (Stoeck and Epstein, 2003; Stoeck et al., 2007; Chen et al., 2008; Le Calvez et al., 2009; Luo, et al., 2009; Simmons et al., 2009). *Lobulomyces poculatus* is found on pollen baits of forest soils as well as detritus in forest canopy layer, and *L. angularis* is collected with *Sphagnum* from acidic lakes. *Chytridium polysiphoniae* is a marine parasite of brown and red algae. Because *Chytridium olla* is the type for the genus in the Chytridiales clade (Vélez et al., 2011), *C. polysiphoniae* is not a valid species of *Chytridium* and probably represents a new genus in the Lobulomycetales.

Cladochytriales (Mozley-Standridge et al., 2009)

The Cladochytriales was established for an assemblage of polycentric and monocentric chytrids commonly found decaying vegetable matter in soils and in aquatic habitats and forming a well-supported clade (James et al., 2006; Mozley-Standridge et al., 2009). Thalli in this order typically have swellings in rhizoids, as in *Septochytrium* and *Catenochytridium,* or spindle-shaped swellings along the rhizomycelium, as in *Nowakowskiella* and *Cladochytrium* (Karling, 1977). This monophyletic order illustrates well the fact that thallus type and operculate versus inoperculate zoospore discharge are convergent characters and not valid as primary taxonomic characters as used in classical morphology based taxonomy (Whiffen, 1944; Sparrow, 1960; Karling, 1977). In fact *Septochytrium* can exhibit monocentric as well as polycentric thallus forms. Zoospores of this clade superficially resemble zoospores of Chytridiales because of the organization of the ribosomal aggregation, MLC (microbody lipid complex), and axonomenal plug (Fig. 3.1, Table 3.1). However, zoospores lack a prominent cell coat and paracrystalline inclusion. The primary distinguishing feature is the lattice-like organization of the microtubule root extending between the side of the kinetosome and the fenestrated MLC cisterna (Table 3.1). In cross section of the root, lateral linkers can be seen connecting rows of microtubules (Lucarotti, 1981; see Fig. 2 in Mozley-Standridge et al., 2009).

From molecular phylogenetic analyses, seven genera are classified in this order within four families or family *incertae sedis* (Mozley-Standridge et al., 2009) (Table 3.2). Molecular phylogenetic studies revealed that several taxa were polyphyletic. For example the two described species of *Allochytridium* (Barr, 1986) placed in different subclades, raising questions about the distinction of this genus. Isolates identified as *Endochytrium* were located in two distinct clades and *Nephrochytrium* in three (Mozley-Standridge et al., 2009). Molecular analyses of lake water have revealed great diversity within this clade (called *Nowakowskiella* clade in Lepère et al., 2008). Elucidation of the systematics

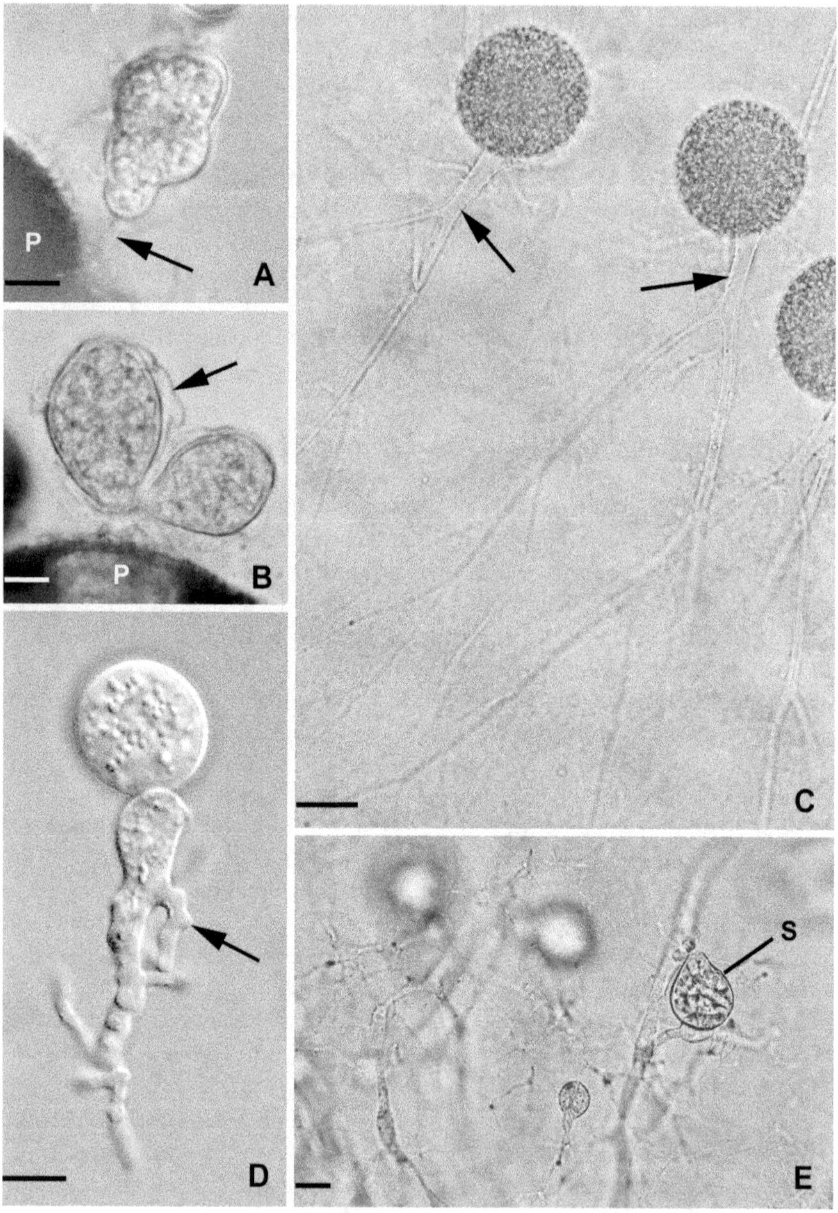

Fig. 3.3 Morphology of Chytridiomycota. (A) The sporangium of *Lobulomyces angularis* (Lobulomycetaceae) is lobose, and the rhizoid entering the pollen grain (arrow) is finely tubular. (B) Cupules (arrow) surround the obpyriform sporangium of *Lobulomyces poculatus* (Lobulomycetaceae). (C) *Chytriomyces hyalinus* (Chytridiales) produces trunk-like rhizoidal axes (arrows). Notice that branching is widely open and branches narrow toward the tips into fine processes. (D) The thallus of *Spizellomyces punctatus* (Spizellomycetales), as in other

Fig. 3.3 Contd. ...

of this clade and determination of its diversity require more extensive sampling and examination of thallus morphology, zoospore ultrastructure and gene sequences.

Incertae sedis

Progress has been made in characterizing well-resolved lineages in the former Chytridiales (Table 3.2). However, there are still lineages that are under-sampled and not circumscribed taxonomically. From the James et al. (2006) study, the *Polychytrium, Synchytrium,* and *Blyttiomyces* clades remain uncharacterized. *Mesochytrium penetrans,* an obligate parasite on a coccoid green alga, is the sole representative of a lineage placed between the Lobulomycetales and *Polychytrium* clades and is especially interesting because of its unique zoospore (Karpov et al., 2010). *Caulochytrium protostelioides,* a parasite of septate hyphal fungi earlier classified in the Spizellomycetales (Barr, 2001), placed in the Chytridiomycota but on a long branch with no near relatives (Karpov et al., 2010).

For some taxa that have historically been considered chytrids, we are currently uncertain as to their phylogenetic position and their taxonomic classification. *Olpidium brassicae, Rozella allomycis* and *Caulochytrium protostelioides* were classified in the Spizellomycetales (Barr, 1980), because their zoospores, like those of *Rhizophlyctis rosea,* have a striated rhizoplast. The rhizoplast is clearly a convergent character because in molecular phylogenetic studies *Rozella* was exterior to the Chytridiomycota + Monoblepharidomycota+ Neocallimastigomycota clade (James et al., 2006; Lara et al., 2009). Even more unexpectedly *Olpidium brassicae* diverged from all other zoosporic fungi in a lineage that radiated among zygomycetous fungi (James et al., 2006). Undoubtedly greater sampling of chytrid diversity is needed to resolve relationships, to find near relatives of members that reside on single distinct branches, and to characterize described taxa that have never been studied using ultrastructural and molecular techniques.

Monoblepharidomycota

The Monoblepharidomycota (Doweld, 2001) is monophyletic and sister to the Chytridiomycota in molecular phylogenetic analyses (James et al., 2006). Monoblephs are morphologically easily distinguished from chytrids because their thalli are either hyphal-like with a foamy appearance (Fig. 3.2F) and bear

Fig. 3.3 Contd. ...
members of this order, produces rhizoids with rounded or blunt tips (arrow). (E) Rhizomycelium of *Polychytrium* sp. (*Polychytrium* clade). Finely branched bifurcations radiating from the broadly tubular portion of the thallus increase area of nutrient absorption. Sporangia (S) occur along the rhizomycelium. Scale bars = 10 μm.

terminal sporangia, as for *Monoblepharella,* or are a reduced lunate tube-like filament, as in *Harpochytrium* (Fig. 3.2G). Thalli may produce holdfasts (Fig. 3.2G) that serve to attach the organism to substrates. Monoblephs are unique among true fungi in their oogamous mode of sexual reproduction. When present, sexual reproduction occurs between uniflagellate sperms produced in antheridia and non-motile oospheres produced in oogonia, resulting in a thick-walled oospore (Sparrow, 1960). Basal members of this clade, such as *Harpochytrium* and *Oedogoniomyces,* are only known to reproduce asexually (Emerson and Whisler, 1968). The ultrastructure of monobleph zoospores shares characteristics with chytrids, but are distinctive in the structure of the flagellar apparatus and the organization of organelles in the MLC in which the MLC cisterna covers the microbody rather than the lipid globules (Table 3.1) (Reichle, 1972; Powell, 1978b; Dorward and Powell, 1980; Gauriloff et al., 1980; Barr, 2001).

A multigene, broadly-sampled phylogenetic analysis has not been completed for the Monoblepharidomycota; and as a consequence, only minimal phylogenetically-based taxonomic revisionary work has been done for this group. Molecular phylogenetic analyses thus far support much of the classical taxonomy (Paquin et al., 1997; Chambers, 2003; James et al., 2006). Species of *Monoblepharis* (the type for the phylum) and *Monoblepharella* are monophyletic, supporting their grouping in the same family; and *Gonapodya* sp. is in another lineage, also supporting its placement in a distinct family (Table 3.3). A new revelation was that *Hyaloraphidium curvatum,* previously classified as a colorless green alga, was a basal member of this class, unique for this phylum in its production of autospores (Ustinova et al., 2000; Forget et al., 2002). Since the organism was revealed to be a fungus rather than a green alga, Doweld (2001) established a class, subclass, order and family for this organism. In the same publication, Doweld (2001) established Monoblepharidomycota as a Phylum. A summary of the updated classification of Monoblepharidomycota, using validly published taxa and applying classical morphological and developmental characters in addition to molecular and ultrastructural characters, is presented in Table 3.3.

EVOLUTION

Zoosporic Fungi and Metazoans are believed to have diverged from a shared ancestor over 1 billion years ago (Berbee and Taylor, 2007) and are members of the Opisthokonta, a supergroup of eukaryotes characterized by the production of motile cells bearing a posteriorly directed flagellum and mitochondria containing discoid or flattened cristae (when present). Opisthokont organisms that still contain the translation elongation factor 1-α protein share a 12 amino acid insert in this gene as a molecular signature (Baldauf and Palmer, 1993).

Table 3.3 Current classification of Monoblepharidomycota

Class – Monoblepharidomycetes **Harpochytriales** Harpochytriaceae *Harpochytrium* sp. Oedogoniomycetaceae *Oedogoniomyces* sp. **Monoblepharidales** Monoblepharidaceae *Monoblepharis* *Monoblepharella*	Gonapodyaceae *Gonapodya* sp. (may be polyphyletic; Chambers, 2003) **Class – Hyaloraphidiomycetes** **Hyaloraphidiales** Hyaloraphidiaceae *Hyaloraphidium curvatum*

How lineages of zoosporic fungi evolved is controversial, but molecular-based hypotheses of phylogeny suggest that holocarpic parasitic zoosporic fungi (*Rozella allomycis*) are most basal in the radiation of fungal evolution and hyphal thallus forms (monoblephs) are more primitive than eucarpic thallus forms (James et al., 2006; Stajich et al., 2009). Among Chytridiomycota, adaptation to terrestrial versus aquatic habitats may have been a driving force in speciation. More basal clades, such as Monoblepharidomycota and Chytridiales (James et al., 2006), tend to be predominantly aquatic. Terminal clades, including Rhizophlyctidales and Spizellomycetales, are most typically found in soils. Apparently along with evolution to more terrestrial habitats, sexuality is lost. Whereas sexuality is common among Monoblepharidales and Chytridiales, it has never been reported in the Spizellomycetales. Within lineages of zoosporic fungi there have been parallel changes in thallus structure, indicating these changes were adaptive for organisms in similar habitats. For example, *Hyaloraphidium curvatum* in the monoblephs and *Sporophlyctis rostrata* in the chytrids have both lost flagellated spores and reproduce with autospores. Both monocentric and polycentric thalli are found in lineages of Chytridiales, Rhizophlyctidales, and Cladochytriales. Operculate and inoperculate discharge release mechanisms are found in Chytridiales, Rhizophydiales, Lobulomycetales, and Cladochytriales.

Chytridiomycota (=chytrids) and Neocallimastigomycota (=neocallimastigos) form a monophyletic group deviating from the lineage leading to the Blastocladiomycota, which radiates among the non-flagellated spored hyphal fungi (James et al., 2006). The neocallimastigos are in a highly derived lineage. Chytrids and neocallimastigos appear to have shared a common uniflagellate aerobic ancestor, but the neocallimastigos diverged in a lineage that adapted them as obligate symbiotic anaerobes living in the digestive system of ruminate herbivores. Although neocallimastigos' thalli superficially resemble the monocentric and polycentric thalli of chytrids, thallus form is convergent among flagellated osmotrophic eukaryotes as demonstrated in comparing chytrid thalli to those of the stramenopile, Hyphochytriomycetes. In diverging from the chytrids, neocallimastigos appeared to have acquired enzymes, such as glycosyl hydrolases used to degrade fibrous cellulosic plant material, by horizontal gene transfer from rumen bacteria (Garcia-Vallvé et al., 2000) Unlike chytrids, neocallimastigos zoospores lack a nonflagellated centriole and the nine flagellar props bridging the side of the kinetosome to the plasma membrane. Their zoospores may be polyflagellate as well as uniflagellate, and their kinetosome-associated structures are unique, consisting of a complex system of circumciliary rings, a spur, struts and scoop (Li et al., 1993). Neocallimastigos contain hydrogenosomes derived from mitochondria rather than cristate mitochondria. Unlike chytrids, neocallimastigos shed their flagella at encystment and during zoospore formation flagella are produced in vesicles prior to cytoplasmic cleavage. While the mitotic poles open during mitosis in Chytridiomycota, the nucleus remains closed at the poles during mitosis in neocallimastigos (Li et al., 1993). These numerous ultrastructural differences and their molecular phylogenetic placement make a compelling case for ranking neocallimastigos as a Phylum (Hibbett et al., 2007).

Rozella is an obligate endoparasite of blastoclads, other chytrids, monoblephs and oomycetes (Sparrow, 1960). In molecular phylogenetic studies, *Rozella allomycis* placed basal to the Chytridiomycota + Monoblepharidomycota + Neocallimastigomycota clade (James et al., 2006; Lara et al., 2009; Stajich et al., 2009), and in most studies it is a sister clade to fungi. Currently *Rozella allomycis* is recognized as a member of a clade of organisms detected molecularly from environmental samples of fresh water, peat bogs and anoxic marine environments and informally labeled, "Rozellida" (Lèpere et al., 2008; Lara et al., 2009). Although earlier analyses suggested that microsporidia (plasmodial animal endoparasites) were sister to *Rozella allomycis* (James et al., 2006), more recent genetic analyses alternatively postulate microsporidia evolved among the Zygomycetes (Corradi and Keeling, 2009). No matter the phylogenetic placement of the microsporidia, a controversy exists as to whether *Rozella allomycis* is considered the deepest diverging lineage of zoosporic fungi or a lineage outside the Fungal Kingdom (James et al., 2006; Lara et al., 2009). It is necessary to resolve certain features of *Rozella allomycis* to ascertain its fit in the concept of a fungus. As Brown et al. (2009) discuss, a chitin-containing cell wall covering the vegetative phase, which allows absorptive heterotrophic nutrition but blocks phagocytotic nutrition as found in amoebae, is a shared character of the Kingdom Fungi (Cavalier-Smith, 1987; Brown et al., 2009). *Rozella* lives inside its host as an unwalled thallus expanding to fill the cellular compartment of its host before functioning as a sporangium and releasing zoospores (Held, 1981). Ultrastructural evidence suggests that one species of *Rozella,* parasitic in a chytrid parasite of *Euglena,* phagocytizes its host cytoplasm (Powell, 1984). If this nutritional mode is consistent for this group, it would place the "Rozellida" outside the fungi. Thus, characterizing the nutritional mode of *Rozella*'s assimilative phase, along with the host-parasite interface, becomes seminal in determining the limits of the Kingdom Fungi.

The Nucleariidae (a group of amoebae characterized by formation of filose pseudopodia and phagocytosis) and *Fonticula alba* (a cellular slime mold) form a clade sister to "Rozellida" + Fungi (Steenkamp et al, 2006; Lara et al., 2009). To recognize the importance of the sister relationship of the nucleariidae amoebae to fungi, Brown et al. (2009) informally proposed the supergroup name Nucletmycea to circumscribe the assemblage of *Nuclearia* amoebae, *Fonticula* cellular slime molds and true fungi from a second group of Opisthokonta, the Holozoa. This association supports the idea that plasmodial, phagocytotic forms may have been ancestral to fungi.

CONCLUSIONS

Identification of chytrids with light microscopy is often difficult because morphological characters used to distinguish taxa overlap due to thallus phenotypic plasticity, especially when chytrids are in axenic culture (Powell and Koch, 1977). Beginning in the 1970s, electron microscopic studies of chytrid zoospores generated stable morphological characters for systematic considerations. Molecular phylogenetics provided additional resolution of relationships and revealed unrealized diversity. Combined ultrastructural and molecular analyses are now used for taxonomic decisions of zoosporic fungi.

The majority of molecular studies suggest that monocentric and polycentric rhizomycelial thallus forms evolved from hyphal and sexually reproducing thallus forms of zoosporic fungi. The most primitive thallus form appears to be plasmodial endoparasites such as *Rozella allomycis*. Still many genera and species of chytrids (Sparrow, 1960; Karling, 1977) have not been studied with either electron microscopy or gene sequence analyses, inviting more participation in efforts to disentangle the story of zoosporic fungal evolution. At this time total nuclear genome sequencing has been completed for only two taxa of Chytridiomycota, *Spizellomyces punctatus* and *Batrachochytrium dendrobatidis*. Significant phylogenetic analyses of chytrid genomic data are still to come, but as more genomes of more chytrids are sequenced, phylogenomics will provide even greater understanding of the evolution and systematics of this fascinating group of fungi.

ACKNOWLEDGEMENTS

This material is based upon work supported by the National Science Foundation under Grant Numbers PEET DEB-0529694 and AFTOL DEB-0732599. Any opinions, findings, and conclusion or recommendations expressed in this material are those of the authors and do not necessarily reflect the views of the National Science Foundation.

REFERENCES

Baldauf, S. L. and Palmer, J. D. 1993. Animals and fungi are each other's closest relatives: congruent evidence from multiple proteins. Proceedings of the National Academy of Sciences, U.S.A., 90: 11558-11562.

Barr, D. J. S. 1980. An outline for the reclassification of the Chytridiales, and for a new order, the Spizellomycetales. Canadian Journal of Botany, 58: 2380-2394.

Barr, D. J. S. 1984. The classification of *Spizellomyces, Gaertneriomyces, Triparticalcar,* and *Kochiomyces.* Canadian Journal of Botany, 62: 1171-1201.

Barr, D. J. S. 1986. *Allochytridium expandens* rediscovered: morphology, physiology and zoospore ultrastructure. Mycologia, 78: 439-448.

Barr, D. J. S. 1990. Phylum Chytridiomycota, pp. 454-466. In: Handbook of Protoctista. L. Margulis, J. O. Corliss, M. Melkonian and D. J. Chapman (eds). Jones & Bartlett, Boston, MA.

Barr, D. J. S. 2001. Chytridiomycota, pp. 93-112. In: The Mycota. Vol. 7A. D. J. McLaughlin, E. G. McLaughlin and P. A. Lemke (eds.). Springer-Verlag, New York.

Barr, D. J. S. and Désaulniers, N. L. 1986. Four zoospore subtypes in the *Rhizophlyctis-Karlingia* complex (Chytridiomycetes). Canadian Journal of Botany, 64: 561-572.

Barr, D. J. S. and Hartmann, V. E. 1976. Zoospore ultrastructure of three *Chytridium* species and *Rhizoclosmatium globosum.* Canadian Journal of Botany, 54: 2000-2013.

Berbee, M. L. and Taylor, J. W. 2007. Rhynie chert: a window into a lost world of complex plant-fungus interactions. New Phytologist, 174: 475-479.

Bills, G., Christensen, M., Powell, M. and Thorn, G. 2004. Saprobic soil fungi, pp. 271-302. In: Biodiversity of Fungi: Inventory and Monitoring Methods. G. M. Mueller, G. Bills and M. S. Foster (eds.). Elsevier Academic Press, Burlington, MA.

Blackwell, W. H. and Powell, M. J. 1999. The nomenclatural propriety of *Rhizophyctis rosea.* Mycotaxon, 70: 213-217.

Braun, A. 1851. Betrachtungen über die Erscheinung der Verjüngung in der Natur, insbesondere in der Lebens- und Bildungsgeschichte der Pflanze. Leipzig.

Braun, A. 1855. Über *Chytridium,* eine Gattung einzelner Schmarotzergewächse auf Algen und Infusorien. Monatsber. Berlin Akad., 1855: 378-384.

Brown, M. W., Spiegel, F. W. and Silberman, J. D. 2009. Phylogeny of the "forgotten" cellular slime mold, *Fonticula alba,* reveals a key evolutionary branch within Opisthokonta. Molecular Biology and Evolution, 26: 2699-2709.

Cavalier-Smith, T. 1987. The Origin of Fungi and Pseudofungi, pp. 339-353. In: Evolutionary Biology of Fungi. A. D. M. Rayner, C. M. Braiser and D. Moore (eds.). Cambridge University Press, Cambridge.

Chambers, J. G. 2003. Ribosomal DNA, secondary structure, and phylogenetic relationships among the Chytridiomycota. Ph.D. dissertation. University of Alabama, Tuscaloosa, AL.

Chen, M., Chen, F., Yu, Y., Ji, J. and Kong, F. 2008. Genetic diversity of eukaryotic microorganisms in Lake Taihu, a large shallow subtropical lake in China. Microbial Ecology, 56: 572-583.

Corradi, N. and Keeling, P. J. 2009. Microsporidia: a journey through radical taxonomical revisions. Fungal Biology Reviews, 23: 1-8.

Davies, B. H. 1961. The carotenoids of *Rhizophlyctis rosea*. Phytochemistry, 1: 25-29.

Dawson, S. C. and Pace, M. R. 2002. Novel kingdom-level eukaryotic diversity in anoxic environments. Proceedings of the National Academy of Sciences, U.S.A., 99: 8324-8329.

Dorward, D. W. and Powell, M. J. 1980. Microbodies in *Monoblepharella* sp. Mycologia, 72: 549-557.

Dorward, D. W. and Powell, M. J. 1983. Cytochemical detection of polysaccharides and ultrastructure of the cell coat of zoospores of *Chytriomyces aureus* and *Chytriomyces hyalinus*. Mycologia, 75: 209-220.

Doweld, A. 2001. *Prosyllabus tracheophytorum*: Tentamen systematis plantarum vascularium (Tracheophyta). Geos, Moscow.

Emerson, R. and Whisler, H. C. 1968. Cultural studies of *Oedogoniomyces* and *Harpochytrium*, and a proposal to place them in a new order of aquatic phycomycetes. Archiv für Mikrobiologie, 61: 195-211.

Fisher, M. C., Garner, T. W. J. and Walker, S. F. 2009. Global emergence of *Batrachochytrium dendrobatidis* and amphibian chytridiomycosis in space, time, and host. Annual Review of Microbiology, 63: 291-310.

Forget, L., Ustinova, J., Wang, Z., Huss, V. A. R. and Lang, B. F. 2002. *Hyaloraphidium curvatum*: a linear mitochondrial genome, tRNA editing, and an evolutionary link to lower fungi. Molecular Biology and Evolution, 19: 310-319.

Freeman, K. R., Martin, A. P., Karki, D., Lynch, R. C., Mitter, M. S., Meyer, A. F., Longcore, J. E., Simmons, D. R. and Schmidt, S. K. 2009. Evidence that chytrids dominate fungal communities in high-elevation soils. Proceedings of the National Academy of Sciences, U.S.A., 106: 18315-18320.

Garcia-Vallvé, S., Romeu, A. and Palau, J. 2000. Horizontal gene transfer of glycosyl hydrolases of the rumen fungi. Molecular Biology and Evolution, 17: 352-361.

Gauriloff, L. P., Delay, R. J. and Fuller, M. S. 1980. Comparative ultrastructure and biochemistry of chytridiomycetous fungi and the future of the Harpochytriales. Canadian Journal of Botany, 58: 2098-2109.

Gleason, F. H., Letcher, P. M. and McGee, P. A. 2004. Some Chytridiomycota in soil recover from drying and high temperatures. Mycological Research, 108: 583-589.

Heath, I. B., Bauchop, T. and Skipp, R. A. 1983. Assignment of the rumen anerobe *Neocallimastix frontalis* to the Spizellomycetales (Chytridiomycetes) on the basis of its polyflagellate zoospore ultrastructure. Canadian Journal of Botany, 61: 295-307.

Held, A. A. 1981. *Rozella* and *Rozellopsis*: Naked endoparasitic fungi which dress-up as their hosts. The Botanical Review, 47: 451-515

Hibbett, D. S., Binder, M., Bischoff, J. F., Blackwell, M., Cannon, P. F., Eriksson, O. E., Huhndorf, S., James, T., Kirk, P. M., Lücking, R., Lumbsch, H. T., Lutzoni, F., Matheny, P. B., McLaughlin, D. J., Powell, M. J., Redhead, S., Schoch, C. L., Spatafora, J. W., Stalpers, J. A., Vilgalys, R., Aime, M. C., Aptroot, A., Bauer, R., Begerow, D., Benny, G. L., Castelbury, L. A., Crous, P. W., Dai, Y. C., Gams, W., Geiser, D. M., Griffith, G. W., Gueidan, C., Hawksworth, D. L., Hestmark, G., Kosaka, K., Humber, R. A., Hyde, K. D., Ironside, J. E., Kõljalg, U., Kurtzman, C. P., Larsson, K.-H., Lichtwardt, R., Longcore, J., Miądlikowska, J., Miller, A., Moncalvo, J.-M., Mozley-Standridge, S., Oberwinkler, F., Parmasto, E., Reeb, V., Rogers, J. D., Roux, C., Ryvarden, L., Sampaio, J. P., Schüßler, A., Sugiyama, J., Thorn, R. G., Tibell, L., Untereiner, W. A., Walker, C.,

Wang, Z., Weir, A., Weiss, M., White, M. M., Winka, K., Yao, Y.-J. and Zhang, N. 2007. A higher-level classification of the fungi. Mycological Research 111: 509-547.

Ho, Y. W. and Barr, D. J. S. 1995. Classification of anaerobic gut fungi from herbivores with emphasis on rumen fungi from Malaysia. Mycologia, 87: 655-677.

James, T. Y., Porter, D., Leander, C. A., Vilgalys, R. and Longcore, J. E. 2000. Molecular phylogenetics of the Chytriomycota supports the utility of ultrastructural data in chytrid systematics. Canadian Journal of Botany, 78: 336-350.

James, T. Y., Letcher, P. M., Longcore, J. E., Mozley-Standridge, S. E., Porter, D., Powell, M. J., Griffith, G. W. and Vilgalys, R. 2006. A molecular phylogeny of the flagellated Fungi (Chytridiomycota) and description of a new phylum (Blastocladiomycota). Mycologia, 98: 860-871.

Karling, J. S. 1977. *Chytridiomycetarum Iconographia*. Lubrecht and Cramer, Monticello, N.Y.

Karpov, S. A., Letcher, P. M., Mamkaeva, M. A. and Mamkaeva, K. A. 2010. Phylogenetic position of the genus *Mesochytrium* (Chytridiomycota) based on zoospore ultrastructure and sequences from the 18S and 28S rRNA gene. Nova Hedwigia, 90: 81-94.

Koch, W. J. 1961. Studies of the motile cells of chytrids. III. Major types. American Journal of Botany, 48: 786-788.

Lange, L. and Olson, L. W. 1979. The uniflagellate Phycomycete zoospore. Dansk Botanisk Arkiv, 33: 7-95.

Lara, E., Moreira, D. and López-Garcia, P. 2009. The environmental clade LKM11 and *Rozella* form the deepest branching clade of fungi. Protist, 161: 116-121.

Le Calvez, T., Burgaud, G., Mahé, S., Barbier, G. and Vandenkoornhuyse, P. 2009. Fungal diversity in deep-sea hydrothermal ecosystems. Applied and Environmental Microbiology, 75: 6415-6421.

Lefèvre, E., Bardot, C., Noël, C., Carrias, J., Viscogliosi, E., Amblard, C. and Sime-Ngando, T. 2007. Unveiling fungal zooflagellates as members of freshwater picoeukaryotes: evidence from a molecular diversity study in a deep meromictic lake. Environmental Microbiology, 9: 61-71.

Lepère, C., Domaizon, I. and Debroas, D. 2008. Unexpected importance of potential parasites in the composition of the freshwater small-eukaryote community. Applied and Environmental Microbiology, 74: 2940-2949.

Letcher, P. M. and Powell, M. J. 2001. Distribution of zoosporic fungi in forest soils of the Blue Ridge and Appalachian Mountains of Virginia. Mycologia, 93: 1029-1041.

Letcher, P. M. and Powell, M. J. 2002. A taxonomic summary of *Chytriomyces* (Chytridiomycota). Mycotaxon, 84: 447-487.

Letcher, P. M. and Powell, M. J. 2005a. *Kappamyces*, a new genus in the Chytridiales (Chytridiomycota). Nova Hedwigia, 80: 115-133.

Letcher, P. M. and Powell, M. J. 2005b. Phylogenetic position of *Phlyctochytrium planicorne* (Chytridiales, Chytridiomycota) based on zoospore ultrastructure and partial nuclear LSU rRNA gene sequence analysis. Nova Hedwigia, 80: 134-146.

Letcher, P. M., Powell, M. J., Chambers, J. G., Longcore, J. E., Churchill, P. F. and Harris, P. M. 2005. Ultrastructural and molecular delineation of the Chytridiaceae (Chytridiales). Mycologia, 83: 1561-1573.

Letcher, P. M., Powell, M. J., Churchill, P. F. and Chambers, J. G. 2006. Ultrastructural and molecular phylogenetic delineation of a new order, the Rhizophydiales (Chytridiomycota). Mycological Research, 110: 898-915.

Letcher, P. M., Powell, M. J., Barr, D. J. S., Churchill, P. F., Wakefield, W. S. and Picard, K. T. 2008a. Rhizophlyctidales - a new order in Chytridiomycota. Mycological Research, 112, 1031-1048.

Letcher, P. M., Powell, M. J. and Viusent, M. C. 2008b. Rediscovery of an unusual chytridiaceous fungus new to the order Rhizophydiales. Mycologia, 100: 325-334.

Letcher, P. M., Vélez, C. G., Barrantes, M. E., Powell, M. J., Churchill, P. F. and Wakefield, W. S. 2008c. Ultrastructural and molecular analyses of Rhizophydiales (Chytridiomycota) isolates from North America and Argentina. Mycological Research, 112: 759-782.

Li, J., Heath, I. B. and Packer, L. 1993. The phylogenetic relationships of the anaerobic chytridiomycetous gut fungi (Neocallimasticaceae) and the Chytridiomycota. II. Cladistic analysis of structural data and description of Neocallimasticales *ord. nov.* Canadian Journal of Botany, 71: 393-407.

Longcore, J. E. 1992. Morphology and zoospore ultrastructure of *Chytriomyces angularis* sp. nov. (Chytridiales). Mycologia, 84: 442-451.

Longcore, J. E., Pessier, A. P. and Nichols, D. K. 1999. *Batrachochytrium dendrobatidis* gen. et sp. nov., a chytrid pathogenic to amphibians. Mycologia, 91: 219-227.

Lucarotti, C. J. 1981. Zoospore ultrastructure of *Nowakowskiella elegans* and *Cladochytrium replicatum*. Canadian Journal of Botany, 59: 137-148.

Luo, W., Li, H., Cai, M. and He, J. 2009. Diversity of microbial eukaryotes in Kongsfjorden, Svalbard. Hydrobiologia, 636: 233-248.

Midgley, D. J., Letcher, P. M. and McGee, P. A. 2006. Access to organic and insoluble sources of phosphorous varies among soil Chytridiomycota. Archives for Microbiology 186, 211-217.

Mozley-Standridge, S. E., Letcher, P. M., Longcore, J. E., Porter, D. and Simmons, D. R. 2009. Cladochytriales - a new order in Chytridiomycota. Mycological Research, 113: 498-507.

Paquin, B., Laforest, M. J., Forget, L., Roewer, I., Wang, Z., Longcore, J. and Lang, B. F. 1997. The fungal mitochondrial genome project: evolution of fungal mitochondrial genomes and their gene-expression. Current Genetics, 31: 380-395.

Picard, K. T., Letcher, P. M. and Powell, M. J. 2009. *Rhizidium phycophilum*, a new species in the Chytridiales. Mycologia, 101: 696-706.

Powell, M. J. 1976. Ultrastructure and isolation of glyoxysomes (microbodies) in zoospores of the fungus *Entophlyctis* sp. Protoplasma, 89: 1-27.

Powell, M. J. 1978a. *Phlyctochytrium kniepii*, pp. 13-14. In: Lower Fungi in the Laboratory. M. S. Fuller (ed.). Athens, GA: Palfrey Contributions in Botany. No. 1, Department of Botany, University of Georgia.

Powell, M. J. 1978b. Phylogenetic implications of the microbody-lipid globule complex. BioSystems, 10: 167-180.

Powell, M. J. 1979. The structure of microbodies and their associations with other organelles in zoosporangia of *Entophlyctis variabilis*. Protoplasma, 98: 177-198.

Powell, M. J. 1984. Ultrastructure of the interface between *Polyphagus euglenae* and the unwalled thallus of its endoparasite *Rozella polyphagi*. Mycologia, 76: 1039-1048.

Powell, M. J. 1993. Looking at mycology with a Janus face. A glimpse at Chytridiomycetes active in the environment. Mycologia, 85: 1-20.

Powell, M. J. and Koch, W. J. 1977. Morphological variations in a new species of *Entophlyctis*. I. The species concept. Canadian Journal of Botany, 55: 1668-1685.

Reichle, R. E. 1972. Fine structure of *Oedogoniomyces* zoospores, with comparative observations on *Monoblepharella* zoospores. Canadian Journal of Botany, 50: 819-824.

Scherffel, A. 1925. Endophytische Phycomyceten-Parasiten der Bacillariaceen und einige neue Monadinen. Archive für Protistenkunde, 52: 1-141.

Sime-Ngando, T., Lefèvre, E. and Gleason, F. H. 2011. Hidden diversity among aquatic heterotrophic flagellates: ecological potential of zoosporic fungi. Hydrobiologia, 659: 5-22.

Simmons, D. R., James, T. Y., Meyer, A. F. and Longcore, J. E. 2009. Lobulomycetales, a new order in the Chytridiomycota. Mycological Research, 113: 450-460.

Sparrow, F. K. 1960. Aquatic Phycomycetes. 2nd edn. University of Michigan Press, Ann Arbor.

Stajich, J. E., Berbee, M. L., Blackwell, M., Hibbett, D. S., James, T. Y., Spatafora, J. W. and Taylor, J. W. 2009. The Fungi. Current Biology, 19: 840-845.

Steenkamp, E. T., Wright, J. and Baldauf, S. L. 2006. The protistan origins of animals and fungi. Molecular Biology and Evolution, 23: 93-106.

Stoeck, T. and Epstein, S. 2003. Novel eukaryotic lineages inferred from small-subunit rRNA analyses of oxygen-depleted marine environments. Applied and Environmental Microbiology, 69: 2657-2663.

Stoeck, T., Kasper, J., Bunge, J., Leslin, C., Ilyin, V. and Epstein, S. 2007. Protistan diversity in the arctic: a case of paleoclimate shaping modern biodiversity. PLoS ONE, 2: e728.

Ustinova, I., Krienitz, L. and Huss, V. A. R. 2000. *Hyaloraphidium curvatum* is not a green alga, but a lower fungus; *Amoebidium parasiticum* is not a fungus, but a member of the DRIPs. Protist, 151: 253-262.

Vélez, C. G., Letcher, P. M., Schultz, S., Powell, M. J. and Churchill, P. F. 2011. Molecular phylogenetic and zoospore ultrastructural analyses of *Chytridum olla* establish the limits of a monophyletic Chytridiales. Mycologia, 103: 118-130.

Vincent, M. A. and Powell, M. J. 1988. Effects of the pyridazinone herbicide Sandoz 6706 (metflurazon) on pigmentation of fungi. Mycologia, 80: 673-678.

Wakefield, W. S., Powell, M. J., Letcher, P. M., Barr, D. J. S., Churchill, P. F., Longcore, J. E. and Chen, S-F. 2010. A molecular phylogenetic evaluation of the Spizellomycetales. Mycologia, 102: 596-604.

Whiffen, A. J. 1941. Cellulose decomposition by saprophytic chytrids. Journal of the Elisha Mitchell Scientific Society, 57: 321-330.

Whiffen, A. J. 1944. A discussion of taxonomic criteria in the Chytridiales. Farlowia, 1: 583-597.

Willoughby, L. G. 2001. The activity of *Rhizophlyctis rosea* in soil: some deductions from laboratory observations. Mycologist, 15: 113-117.

Chapter 4

Current Systematics of Zygomycota with a Brief Review of their Biology

Gerald L. Benny

Department of Plant Pathology, 1453 Fifleld Hall, University of Florida
Gainesville, Florida 32611-0680 U.S.A.
E-mail: gbenny@ufl.edu

ABSTRACT

Recent molecular studies have resulted in the publication of a phylogenetically based classification of the fungi. The organisms formerly included in the Trichomycetes and Zygomycetes are dispersed into four subphyla containing one to four orders: 1) Entomophthoromycotina (Entomophthorales), 2) Kickxellomycotina (Asellariales, Dimargaritales, Harpellales, Kickxellales), 3) Mucoromycotina (Endogonales, Mortierellales, Mucorales) and 4) Zoopagomycotina (Zoopagales). Two families are included in Harpellales (Harpellaceae, Legeriomycetaceae) and Mucorales (Mucoraceae, Umbelopsidaceae), five in the Zoopagales (Cochlonemataceae, Helicocephalidaceae, Piptocephalidaceae, Sigmodeomycetaceae, Zoopagaceae) and six in the Entomophthorales (Ancylistaceae, Basidiobolaceae, Completoriaceae, Entomophthoraceae, Meristracraceae, Neozygitaceae). Each genus is listed under its respective family. Each taxon includes a citation and one or more critical references. Information is given on the ecology, classification and phylogeny of zygomycetous fungi. The Zygomycota are not included in a phylum because too little is known about their phylogeny to make this determination.

INTRODUCTION

The Zygomycota are those fungi that reproduce sexually by the formation of zygospores (Benjamin, 1979). Many taxa are not known to produce zygospores and, therefore, their placement in this phylum is based on other morphological or phylogenetic criteria.

The reader is referred to the references cited in this chapter or other treatments on the Zygomycota and reliable websites on the Internet. The latest version of the *Dictionary of the Fungi* (Kirk et al., 2008), the *Bibliography of Systematic Mycology* (Anon, 2008), the books—*The Mycota,* vols. VIIA and VIIB (McLaughlin et al., 2001a, b), *Biodiversity of Fungi* (Mueller et al., 2004) and *Fungal Families of the World* (Cannon and Kirk, 2007)—should be consulted for information on the Zygomycota and other fungi. The most important references for information on the Zygomycota are: 1) reviews (Benjamin, 1959, 1979; Lichtwardt, 1986; Wöstemeyer and Schimek, 2007), 2) Entomophthoromycotina (Batko, 1964a, b, c, d, 1966; Tucker, 1981; Humber, 1989; Keller and Petrini, 2005), 3) Kickxellomycotina (Young, 1999; Benny, 2001; Kurihara et al., 2001, 2004, 2008; Kurihara and Degawa, 2006), 4) Mucoromycotina (Zycha et al., 1969; Young, 1985; Benny et al., 2001), and 5) Zoopagomycotina (Drechsler, 1935-1962; in Lumsden, 1987; Benny et al., 1992; Benjamin, 1985a, b). Other important information can be obtained on the use of Zygomycota in Asian foods (Hesseltine, 1965, 1983), ultrastructure (Cole and Samson, 1979; O'Donnell, 1979), industrial processes (Osiewacz, 2002) and phylogeny (James et al., 2006; Liu et al., 2006; White et al., 2006a).

Hints for the collection, isolation and culture of members of the Zygomycota can be obtained from Mueller et al. (2004) and Benny (2008). The methods used to grow, and for the preservation and distribution of zygomycete cultures, have been published by Jong and Birmingham (2001) and Nakasone et al. (2004). Procedures for the preparation and use of herbarium specimens are discussed by Wu et al. (2004). Also, strict adherence to the latest version of the *International Code of Botanical Nomenclature* (ICBN—McNeill et al., 2006) is required when describing a new taxon in order to prevent it from being invalidated (see Benny, 1995b, c; Kwaśna and Nirenberg, 2008a, b).

The Internet can be used to find information on zygomycotan fungi. Literature can be found using PubMed (www.ncbi.nlm.nih.gov/sites/entrez). Several databases can be accessed using the CABI Bioscience Databases (www.indexfungorum.org/Index.htm). Additional information on Zygomycota can be obtained from the Trichomycetes (www.nhm.ku.edu/~fungi/) and Zygomycetes (www.zygomycetes.org) websites, and also specific parts of the Species2000 (www.sp2000.org), the Tree of Life Web Project (www.tolweb.org), the Encyclopedia of Life (www.eol.org/index) and probably many other internet sites, especially in the future.

Life cycle

The life cycle of a Zygomycete (Alexopoulos et al., 1996) begins when a spore germinates, produces vegetative hyphae, and then sporangiophores are formed that bear spore-forming sporangia, sporangiola or merosporangia. This asexual life cycle is the most common method of reproduction in the Mucorales. Sexual reproduction in heterothallic species requires two mating types, (+) and (−), to be crossed in order for zygospores to be produced whereas in homothallic fungi only a single culture is required. A germinating zygospore can produce a germ sporangium, germ sporangiolum or germ merosporangium in which spores are produced (Alexopoulos et al., 1996) whereas in water a germ tube is formed (Guo and Michailides, 1998). Michailides and Spotts (1990) presented the life cycle of *Mucor piriformis* A. Fischer in a fruit tree orchard in soil and on diseased stone and pome fruits. This species can survive 12 months in soil on endocarps and on infected fruit for 19 to 20 months. Birds, insects and rodents wound fruit and, thereby, can spread spores of *M. piriformis* from one fruit to another. When stored, an infected fruit can initiate spread of the fungus when they are in direct contact with one another. In nature many more variables are present in the life cycle of Mucorales and other members of the Zygomycota than can be observed *in vitro.*

Ecology

Members of the Zygomycota, especially taxa of the Mortierellales and Mucorales, are common components of the mycoflora of dung, soil, and other substrates that support their growth and sporulation. These fungi do not utilize cellulose or lignin and, therefore, they use less complex, soluble sources of carbon and nitrogen. Zygomycetes exhibit rapid spore germination and a similar rate of growth and sporulation. These organisms, the "sugar fungi," appear early on dung and complete their life cycle by the time the nutrient source is exhausted. Harper and Webster (1964) reported that *Mucor hiemalis* Wehmer, *M. mucedo* Fresen., *Pilaira anomala* (Ces.) A. Schröt. (Mucorales) and *Kickxella alabastrina* Coem. (Kickxellales) sporulated abundantly on rabbit dung after 2 or 3 days incubation. Webster (1970) reported that sporulation of species *of Mucor, Pilaira* and *Pilobolus* lasted 10 to 14 days. This is about the time that the facultative parasitic species *Chaetocladium brefeldii* Tiegh. & G. Le Monn. (Mucorales) and members of the mycoparasitic genus *Piptocephalis,* for example *P. freseniana* de Bary (Zoopagales), began to appear (Harper and Webster, 1964).

An extensive variety of mucoralean fungi are present on the dung of omnivorous mice and rats collected in southern California (U.S.A.) and elsewhere. These fungi include *Backusella, Benjaminiella, Chaetocladium, Cokeromyces, Dichotomocladium, Ellisomyces, Fennellomyces, Helicostylum, Kirkomyces, Mycotypha, Phascolomyces, Pirella, Thamnidium, Thamnostylum* and *Zychaea* (Benny and Benjamin, 1975, 1976, 1993; Benny et al., 1985; Benny and Schipper, 1992; Benny, 1992, 1995b, c). Species of *Coemansia* (Kickxellales), *Dimargaris, Dispira, Tieghemiomyces* (Dimargaritales), *Piptocephalis* and *Syncephalis* (Zoopagales) were also isolated from collections of this substrate (Benjamin, 1958, 1959, 1961, 1963, 1965). *Spiromyces minutus* R.K. Benj., *S. aspiralis* Benny & R.K. Benj. and *Spirodactylon aureum* (Kickxellales) also are coprophilus (Benjamin, 1959, 1963; O'Donnell et al., 1998).

Soil is also good source for members of the Zygomycota. Benny (unpubl. data 2008) has isolated numerous zygomycetous fungi from various Florida soils, including species of *Mortierella* (Mortierellales), *Absidia, Mucor, Zygorhynchus* (Mucorales), and occasionally *Actinomucor elegans* (Eidam) C.R. Benj. & Hesselt., *Blakeslea trispora* Thaxt., *Choanephora cucurbitarum* (Berk. & Ravenel) Thaxt., *Gongronella butleri* (Lendn.) Peyronel & Dal Vesco, *Poitrasia circinans* (H. Nagan. & N. Kawak.) P.M. Kirk, *Syncephalastrum racemosum* Cohn ex Schröt., and unidentified species of *Cunninghamella* and *Rhizopus* (Mucorales). Many species of *Syncephalis* (Zoopagales) are present. *Ramicandelaber brevisporus* Kurihara, Degawa & Tokum. and species of *Coemansia* (Kickxellales) and *Piptocephalis* (Zoopagales) can be isolated from some Florida soil collections.

Recently, Schmidt et al. (2008) reported that some unidentified coenocytic "snow molds" were found, under snow, growing on litter and soil in a subalpine forest in Colorado, USA. These fungi did not sporulate in culture but, based on comparison of SSU rDNA, the isolates were found to belong in the clades containing members of the Mortierellales and Mucorales. There are psychrophilic and psychrotolerent species in the following genera including: *Dissophora* (Mortierellales) and *Chaetocladium, Dicranophora, Helicostylum, Phycomyces, Spinellus, Thamnidium* and *Spinellus*. Members of two of these genera (*Dicranophora, Spinellus*) are only known in nature as mushroom parasites and, therefore, they would probably not be found in the same habitat as the "snow molds."

Spore dispersal

The Zygomycota have evolved several methods for the dispersal of spores, both active and passive. The active mode of spore dispersal is the forcible release of conidia (papilla reversion, propulsive jet of conidiophore contents, fluid discharge from a specialized cell between the conidium and conidiophore, backward discharge of contents of the conidiophore after a circumscissile break in the conidiophore vesicle) in most genera of Entomophthorales (Humber, 1981a, 1989), except *Massospora, Meristacrum* and *Zygnemomyces* (Tucker, 1981). Multispored sporangia are forcibly discharged by *Pilobolus* (Mucorales) from phototrophic, turgid, vesiculate sporangiophores (Ingold, 1971). Discharge of a mature sporangium is preceded by circumscissile rupture of the sporangiophore wall immediately below the columella (Buller, 1934). Propulsion of the sporangium is produced by the cytoplasm from the subsporangial vesicle and sporangiophore that is stretched a relatively long distance before it breaks into droplets (Page, 1964; Yafetto et al., 2008). When the *Pilobolus* sporangium hits a piece of wet vegetation the calcium oxalate crystals on the outer wall make it unwettable, it rotates, and then it becomes attached to the substrate by the cytoplasm on the sporangial base (Ingold, 1971).

The majority of the fungi in the Zygomycota have the spores dispersed by air movement. These are dry-spored species that produce an evanescent sporangial wall (*Mucor*—some spp., *Rhizopus*) and dry uni- or multispored sporangiola (for instance *Backusella, Blakeslea, Choanephora, Chaetocladium, Dichotomocladium, Mycotypha, Radiomyces, Thamnidium, Thamnostylum*) or merosporangia (*Syncephalastrum, Piptocephalis*—some spp., *Rhopalomyces*) (Ingold and Zoberi, 1964; Ingold, 1971; Zoberi, 1985).

Many taxa form a spore drop (Ingold, 1961). These organisms either produce sporangia with a persistent wall that splits longitudinally releasing the spore drop (*Blakeslea, Choanephora, Gilbertella*) or a deliquencent wall (for instance *Mucor*—some spp., the terminal sporangia of *Backusella, Thamnidium* and *Thamnostylum*) or arthrospores (*Helicocephalum*) and the propagules of these species may be dispersed by water droplets, mites, insects, small mammals or birds (Ingold and Zoberi, 1964; Michailides and Spotts, 1990).

Two genera, *Pilaira* and *Utharomyces*, facilitate spore dispersal via extensive elongation of the sporangiophore and the attachment of the sporangium to a surface by the subsporangial cytoplasm. In *Pilaira anomala* (Ces.) J. Schröt., the persistent sporangial wall separates from the sporangiophore along a circumscissile zone, releasing the sporangium and spores; the columella remains on the sporangiophore (Ingold, 1971; Benny and O'Donnell, 1978). In *Utharomyces epallocaulus* Boedijn ex P.M. Kirk & Benny, sporangial release occurs when the thin wall of the mature subsporangial vesicle breaks. The stalked, intact sporangium is cemented to an aerial hypha or other surface by the cytoplasm remaining on the inside of the upper portion of the subsporangial vesicle. Spores are released when the sporangial wall is broken (Kirk and Benny, 1980).

Other specialized modes of spore dehiscence were discussed by Ingold (1971). The wet-spored species of *Piptocephalis*, for example *P. freseniana* de Bary, form head-cells bearing spore-drops that contain merosporangia which are detached readily from the sporangiophore by air currents; this fungus is a "spore-drop shedder" (Dobbs, 1942; Ingold and Zoberi, 1964). In *Tieghemiomyces californicus* R.K. Benjamin (1959), the apex of the main sporangiophore is sticky and it readily attaches to any dry surface (insect, etc). The sporangiophore breaks below the sporulating head and, therefore, the sporangiophore and the spore-forming region can be removed intact (Ingold, 1971). *Spirodactylon aureum*

R.K. Benjamin (1959) forms coiled, spore-producing regions that may spring back if attached to a passing insect, dispersing the dry spores. Other zygomycotan fungi with elaborate sporulating heads, sterile spines and dry spores, are species of *Chaetocladium* and *Dichotomocladium* (Benny and Benjamin, 1975, 1976, 1993). They also may be adapted for animal dispersal.

ECONOMIC IMPORTANCE

The zygomycetous fungi have become more important because of their economic impact on food processing and storage, industry and as the causative agents of mycoses in animals, including humans.

Zygomycetes used in food fermentation

Hesseltine (1965, 1983) reported that the following Mucorales are used in the preparation of various Asian foods: rice—Ragi [in Indonesia used as a starter to breakdown starch in other foods] (*Amylomyces* [*Chlamydomucor*] *oryzae* Calmette, *Rhizopus oryzae* Went. & Prins. Geerl.); soybeans—Sufu [Chinese cheese] (*Actinomucor elegans* (Eidam) C.R. Benj. & Hesselt., *Mucor* sp.), Tempeh (*Rhizopus oligosporus* Saito or less commonly *R. oryzae* and *R. stolonifer* (Ehrenb.) Vuill.). The steps involved in making the aforementioned Asian foods are discussed by Nout and Aidoo (2002).

Thamnidium elegans Link and *Helicostylum pulchrum* (Preuss) Pidopl. & Milko can be isolated from cut meat aging and being tenderized in refrigerators. The colonies of these fungi growing on the cut surfaces of the meat have the name "whiskery beef" according to Hesseltine and Anderson (1957).

Recently, a patent was published to use Mucorales as a meat substitute (De Hann et al., 2000), a potential product compared with Quorn™ (Wiebe, 2004), which is made from the mycelium of *Fusarium venenatum* Nirenberg. Unfortunately, several of the suggested taxa of the Mucorales listed by De Hann et al. (2000) are potential human pathogens in the genera *Rhizomucor* and *Rhizopus*. Refinement of the potential species to be tested should be done before any mucoralean fungus can be used for food.

Zygomycoses

In humans, zygomycoses usually appear after an injury, burns, or in patients with diabetes or with an immune system compromised by drugs or HIV infection. The known species causing zygomycosis are classified in three orders of the Zygomycota. Entomophthoromycosis is caused by fungi in the Entomophthorales (including Basidiobolales here) (*Basidiobolus ranarum* Eidam,

Conidiobolus coronatus (Costantin) A. Batko and other *Conidiobolus* spp.) and Mucormycosis is caused by species in the Mortierellales (*Mortierella* spp.) and Mucorales (several taxa). The mucoralean species causing mucormycosis are *Apophysomyces elegans* P.C. Mirsa, K.J. Srivast. & Lata, *Cokeromyces recurvatus* Poitras, *Cunninghamella bertholletiae* Stadel, *Saksenaea vasiformis* S.B. Saksena, *Lichtheimia corymbifera* K. Hoffman, G. Walther & K. Voigt [in the literature as *Absidia corymbifera* (Cohn) Sacc. & Trotter or *Mycocladus corymbifer* (Cohn) Váňová], *Rhizomucor pusillus* (Lindt) Schipper, *Syncephalastrum racemosum* Cohn ex J. Schröt., and species of *Mucor* and *Rhizopus* (Ribes et al., 2000).

Plant Pathology

A few species (*Choanephora cucurbitarum* (Berk. & Ravenel) Thaxt., *Gilbertella persicaria* (E.D. Eddy) Hesselt., *Mucor piriformis* A. Fisch. (Michailides and Spotts, 1990), *Rhizopus stolonifer* (Ehrenb.) Vuill.), all Mucorales, are plant pathogens. The latter three species cause storage rots (Csernetics et al., 2005) and *C. cucurbitarum* causes fruit rot of cucurbits, beans, beets, papaya, pepper, potato, sweet potato and other crops (*Choanephora* blight or wet rot). In *cucurbitarum*, the fungus grows from the fading flower to the young squash fruit, sporulating in 24 to 48 hours by producing sporangiola. The spores of *C. cucurbitarum* are disseminated by insects (bees, beetles) and wind (Wolf, 1917).

Industrial uses of zygomycotan fungi

Zygomycetes, especially members of the Mortierellales and Mucorales are used in many industrial processes (Osiewacz, 2002). Industrial uses include the production of fumaric and lactic acid by *Rhizopus oryzae* and *Rhizopus* spp., biotin by *Rhizopus delemar* (Boidin) Wehmer & Hanazawa (Mucorales) and polyunsaturated fatty acids [linoleic acid by *Mortierella ramanniana* (Möller) Linnem. and *M. vinacea* Dixon-Stew, both of the aforementioned taxa of *Mortierella* are now species of *Umbelopsis* (Mucorales) and arachidonic acid by *Mortierella alpina* Peyronel (Mortierellales)], carotenoid metabolism has been studied in *Phycomyces blakesleeanus* Burgeff and the industrial production of β-carotene by *Blakeslea trispora* Thaxt. Several species of *Rhizopus* are used in biosorption of heavy metals.

Evolution of the classification of zygomycotan fungi

Many taxa of the Zygomycota were described in the 19th century or before but the first survey of the Mucorales s.l. was published by van Tieghem (van Tieghem and Le Monnier, 1873; van Tieghem, 1875, 1878). The fungi now included in the Entomophthorales and Mucorales s.l. were separated even in the first classification of the Mucorales (van Tieghem, 1875, 1878). The first monograph of the Entomophthorales was published 10 years later (Thaxter, 1888). Other 19th century treatments of the Zygomycetes also recognized these two orders (Fischer, 1892; Schröter, 1893a, b, c). Underwood (1899) was the first to use the correct ordinal endings when he recognized the Entomophthorales and Mucorales. In the 20th century, several treatments of the Mucorales were published that often were based on regional floras of the fungi but the Entomophthoralean fungi were not included (Lendner, 1908; Naumov, 1935, 1939; Zycha, 1935; Zycha et al., 1969; Mil'ko, 1974).

Fitzpatrick (1930) and some early editions of the *Dictionary of the Fungi* (Ainsworth and Bisby, 1943; Ainsworth, 1971) recognized the two orders. Later, Bessey (1950) included the Entomophthorales, Mucorales and Zoopagales in the Zygomycetes, although the latter order was not validated until later. Hesseltine (1955) presented his synopsis of the families and genera of the Mucorales with descriptions of the order and families, keys to taxa, type species for genera and synonyms, if applicable. Ainsworth et al. (1973) included the latter three orders in the Zygomycetes: Entomophthorales (Waterhouse, 1973), Mucorales (Hesseltine and Ellis, 1973) and Zoopagales (Duddington, 1973). The Trichomycetes (Lichtwardt, 1973b) contained four orders: 1) Amoebidiales L. Léger & Duboscq with one family, Amoebidiaceae J.L. Licht. 2) Asellariales with one family, Asellariaceae, 3) Eccrinales L. Léger & Duboscq with three familes—a) Eccrinaceae L. Léger & Duboscq, b) Palavasciaceae Duboscq, L. Léger & Manier ex Manier & Lichtw and c) Parataeniellaceae Manier & Lichtw. and Harpellales with two families—a) Genistellaceae and b) Harpellaceae. This classification of the Trichomycetes was proposed by Manier and Lichtwardt (1968) and was discussed in more detail by Lichtwardt (1973a) and it was elaborated upon later by Lichtwardt (1986). Several taxa of the Trichomycetes were validated by Manier (1968), Lichtwardt and Manier (1978) and Cannon and Kirk (2007). A revised version of Lichtwardt's (1986) monograph of the Trichomycetes is now available on the internet (Lichtwardt et al., 2001a).

Several years later, Benjamin (1979) published his concept of the Zygomycetes which contained seven orders (Dimargaritales, Endogonales, Entomophthorales, Harpellales, Kickxellales, Mucorales, Zoopagales). The Dimargaritales, an order described by Benjamin (1979), with one family, Dimargaritaceae. The Endogonales, validated by Benjamin (1979), has one family, Endogonaceae. The Entomophthorales possesses two families – a) Basidiobolaceae and b) Entomophthoraceae. The Harpellales contains two families – a) Harpellaceae and b) Legeriomycetaceae [= Genistellaceae]. The Kickxellales were validated by Benjamin (1979) with one family, Kickxellaceae. The Mucorales contained nine families – a) Choanephoraceae, b) Cunninghamellaceae, c) Mortierellaceae, d) Mucoraceae, e) Pilobolaceae, f) Radiomycetaceae, g) Saksenaeaceae, h) Syncephalastraceae and i) Thamnidiaceae. The Zoopagales, validated by Benjamin

(1979) included four families – a) Cochlonemataceae, b) Helicocephalidaceae, transferred from Mucorales, c) Piptocephalidaceae, also transferred from Mucorales and d) Zoopagaceae. Benjamin's (1979) inclusion of the Harpellales in the Zygomycetes was not followed by other mycologists. Molecular studies later demonstrated that the Harpellales and, based on morphological evidence, the Asellariales also are Zygomycetes.

The Mycota Volume VIIA (McLaughlin et al., 2001a) contains two chapters on the Zygomycota. One chapter is on the Trichomycetes (Benny, 2001) that includes the classical four orders: 1) Amoebidiales, 2) Asellariales, 3) Eccrinales and 4). Harpellales. The Amoebidiales and Eccrinales remain the same as listed by Lichtwardt (1973b – above). The Asellariales have only a single family and the Harpellales (Benny, 2001) still have two.

The Zygomycetes, as discussed by Benny et al. (2001), includes the following orders: 1) Dimargartiales, 2) Endogonales, 3) Entomophthorales, 4) Glomerales [as Glomales], 5) Kickxellales, 6) Mucorales and 7) Zoopagales. The Dimargaritales has one family, Dimargaritaceae. The Endogonales also have a single family, Endogonaceae. The Entomophthorales contain the six families discussed and described by Humber (1989) – a) Ancylistaceae, b) Basidiobolaceae, c) Completoriaceae, d) Entomophthoraceae, e) Meristacraceae and f) Neozygitaceae. The Glomerales Morton & Benny (as Glomales), following Morton and Benny (1990), included the following three families – a) Acaulosporaceae Morton & Benny, b) Gigasporaceae Morton & Benny and Glomeraceae Pirozynski & Dalpé. The Kickxellales, validated by Benjamin (1979), have a single family, Kickxellaceae. The Mucorales, the largest order of the Zygomycetes, has 13 families – a) Chaetocladiaceae, b) Choanephoraceae, c) Cunninghamellaceae, d) Gilbertellaceae, e) Mortierellaceae, f) Mucoraceae, g) Mycotyphaceae, h) Phycomycetaceae, i) Pilobolaceae, j) Radiomycetaceae, k) Saksenaeaceae, l) Syncephalastraceae and m) Thamnidiaceae. The Zoopagales have the following five families – a) Cochlonemataceae, b) Helicocephalidaceae, c) Piptocephalidaceae, d) Sigmoideomycetaceae and e) Zoopagaceae.

Recent molecular phylogenies, however, have demonstrated that members of the Amoebidiales and Eccrinales are not fungi (Benny and O'Donnell, 2000; Ustinova et al., 2000; Cafaro, 2005) but protozoans belonging to the class Mesomycetozoa (Mendoza et al., 2002); these latter two orders, however, will continue to be studied by students of the trichomyceteous Fungi.

Members of the class Zygomycetes are distributed here in seven orders, Dimargaritales, Endogonales, Entomophthorales (including Basidiobolales), Kickxellales, Mortierellales, Mucorales, and Zoopagales (Benjamin, 1979; Cavalier-Smith, 1998), and the Asellariales and Harpellales, Trichomycetes of many authors, constitute the nine orders currently accepted by me in the Zygomycota.

Two other orders, Glomerales (Morton and Benny, 1990), a segregate from the Endogonales, and the Geosiphonales (Cavalier-Smith, 1998), were included in the Zygomycota until recently. Schüssler et al. (2001) proposed the Glomeromycota for the Glomerales, Geosiphonales, and three orders (Archaeosporales, Paraglomerales, Diversisporales) formerly included in the Glomerales. The Diversisporales and two families were validly described later (Walker and Schüssler, 2004) and new taxa were added to the Archaeosporales (Walker et al., 2007). Recently, the Gigasporaceae were emended and the genus *Scutellospora* divided into several genera and three families were proposed (Oehl et al., 2008). Since the glomeralean taxa are no longer considered Zygomycota they will not be discussed further.

PHYLOGENY OF THE ZYGOMYCOTA

Several phylogenies were published showing the evolution of zygomycetous fungi before the use of molecular analyses. These analyses were based on similarities between taxa. Benjamin (1959) presented his version of the relationships of the Mucorales s.l. based on his discussion of the families (Dimargaritaceae, Kickxellaceae, Piptocephalidaceae, Syncephalastraceae) producing merosporangia (the merosporangiferous Mucorales). Indoh (1962) proposed a possible evolution of *Syncephalis*. Moss and Young (1978) presented a study of the relationships of the Asellariales, Harpellales and Kickxellales.

In the last decade and half, several cladistic analyses have been conducted on selected members of various zygomycete orders. Nagahama et al. (1995) presented a phylogeny of the Entomophthorales based on the comparison of 18S rDNA sequences. This was the first analysis to demonstrate that *Basidiobolus* was more closely related to the clade containing several members of the Chytridiomycota, including *Chytridium*, Jensen et al. (1998) added several taxa, including additional Entomophthorales. *Basidiobolus* continued to be phylogenetically distinct from the latter order.

O'Donnell et al. (1998) published a phylogeny of the Harpellales and Kickxellales using a data set composed of 18S rDNA sequences and morphology. Three clades were revealed as, Harpellales, Kickxellales and *Spiromyces*, with *Spiromyces* being the sister to the Harpellales clade.

Tanabe et al. (2000) added 18S rDNA sequences from the Dimargaritiales and Zoopagales, including *Dimargaris bacillispora* R.K. Benj., *Dispira cornuta* Tiegh., *Tieghemiomyces parasiticus* R.K. Benj. (Dimargaritiales), *Kuzuhaea moniliformis* R.K. Benj., *Piptocephalis corymbifera* Vuill., *Rhopalomyces elegans* Corda, *Syncephalis depressa* Tiegh. & G. Le Monn., *Thamnocephalis sphaerospora* R.K. Benj. & Benny and *Zoophagus insidians* Sommerst. (Zoopagales). Other sequences added to the data set were from Monoblepharidales and

Mucorales. The Dimargaritales were in one clade that is the sister order to the Kickxellales without *Spiromyces*. The Zoopagales were divided into two subclades: 1) *Kuzuhaea* and *Piptocephalis* and 2) *Rhopalomyces, Syncephaslis* and *Thamnocephalis. Zoophagus* was basal to the Harpellales-*Spiromyces* clade.

Two analyses based on more than 80 taxa from the Mortierellales and Mucorales were published the next year. O'Donnell et al. (2001) analyzed a data set consisting of the morphology and sequences from three genes (18S rRNA, 28S rRNA, ef-1α) from 81 taxa. In this analysis the taxa from the Mucorales were distributed in three clades none of which could be defined based on morphological characters. Until more information is available, the author has abandoned the traditional family system of the Mucorales, except Umbelopsidaceae, as reflected below in the Synopsis and the Zygomycetes website (/www.zygomycetes.org/). Voigt and Wöstemeyer (2001) analyzed a data set of actin and ef-1α genes derived from 82 members of the Mortierellales and Mucorales. Several clades were in the Mucorales analysis that might be interpreted as representing currently described families, including Choanephoraceae, Cunninghamellaceae, Pilobolaceae, Saksenaeaceae, Umbelopsidaceae, all with one to three genera. Many clades having multiple genera, however, could not be placed in a mucoralean family that is currently valid.

One large data set that included many sequences of the Zygomycota (two genes - SSU rDNA, LSU rDNA) was that of Lutzoni et al. (2004). Many taxa from the Mucorales were included but only a few taxa (one or two) from many of the other orders were present in the data set. Sequences were not included for Asellariales, Endogonales and Harpellales. This indicated possible phylogenetic relationships for the Chytridiomycota, Glomeromycota and Zygomycota taxa included in the analysis but more genes and taxa were added in papers published later (see below).

White et al. (2006a) presented a phylogeny of the Zygomycota based on a combined data set containing 5.8S rRSA + 18S rRNA + 28S rRNA sequences. In this analysis the Endogonales-Mortierellales-Mucorales clade was sister to the Ascomycota-Basidiomycota-Glomeromycota clade. The remaining Zygomycota were in the Basidiobolales-Dimargaritales-Entomophthorales-Harpellales-Kickxellales-Zoopagales clade. The Basidiobolales clade, which includes *Olpidium brassicae* (Woronin) P.A. Dang., was basal to the other orders in the latter clade. The Dimargaritales + *Neozygites*-Zoopagales clade is sister clade to Harpellales-Kickxellales [including *Orphella* (Harpellales) and *Spiromyces* (Kickxellales)]. The Dimargaritales is the sister to *Neozygites* (Entomophthorales), an artifact probably due to Long Branch Attraction (LBA), and *Nephridiophaga blattellae* (Crawley) Woolever (Zygomycota *incertae sedis*) and *Ramicandelaber brevisporus* Kurihara (Kickxellales). This Dimargaritales + *Neozygites* clade is sister to the Zoopagales clade. The Harpellales is the sister clade to Kickxellales + *Orphella* + *Spiromyces. Zoophagus* is consistently the basal species to the Harpellales-Kickxellales-*Orphella-Spiromyces* clade in this analysis (White et al., 2006a).

Several genera have been defined using molecular analysis, including *Absidia* and *Rhizopus* (Abe et al., 2006, 2007; Hoffmann et al., 2007; Liou et al., 2007; Liu et al., 2007; Jennessen et al., 2008). New taxa were also described or recognized in the Mucorales based on sequence analysis—*Lentamyces, Lichtheimia* and Lichtheimiaceae, and *Mycocladus* and Mycocladaceae (Hoffmann et al., 2007, 2009; Hoffmann and Voigt, 2008). Research has indicated that two traditional families of the Mucorales, Choanephoraceae and Phycomycetaceae, may need to be recognized with one additional genus added (Voigt and Olsson, 2008; Voigt et al., 2008). White (2006) added several taxa to the data set for analysis of the Harpellales. Kurihara et al. (2005) proposed a possible future division of the Kickxellales into three orders: (Kickxellales s.l., and two additional orders based on the *Ramicandelaber* and the *Spiromyces* clades, as revealed by the analysis of White et al., 2006a).

Two other phylogenetic analyses, one (Liu et al., 2006) based on three genes (18S rDNA, RPB1, RPB2, or RPB1 + RPB2) showed that *Basidiobolus ranarum* was in the clade containing the Chytridiomycota and that this species is the sister taxon to *Chytridiomyces hyalinus* (18S rDNA), *B. ranarum* is basal to the Zygomycota-Glomeromycota clade (RPB1) or it is basal to the Harpellales clade (RPB2, RPB1 + RPB2). In these analyses, Glomeromycota is the sister to the Ascomycota-Basidiomycota clade (18S rDNA) or the Zygomycota clade (RPB1, RPB2, RPB1 + RPB2). James et al. (2006) included more Zygomycota in their data set that consisted of six genes (18s rRNA, 28S rRNA, 5.8S rRNA, EF1-α, RPB1, RPB2). This analysis revealed two major clades, one composed of taxa from the Endogonales, Mortierellales and Mucorales and the second with taxa from the Dimargaritales, Harpellales, Kickxellales, Zoopagales, and Entomophthorales. The latter clade included Entomophthorales, *Basidiobolus ranarum* and its sister species, *Olpidium brassicae*. The Entomophthorales-*Basidiobolus ranarum-Olpidium brassicae* clade is the sister to the Dimargaritales-Harpellales-Zoopagales clade. In the Dimargaritales-Harpellales-Kickxellales-Zoopagales clade the Zoopagales are the sister to the clade containing the remaining orders, and the Dimargaritales is the sister to the Harpellales-Kickxellales clade.

The Microsporidia and the Zygomycota

Microsporidia are highly reduced organisms without mitochondria that are obligate parasites of Eukaryotes. These organisms produce a meront in the host and an infective spore with a chitinous wall and a polar tube (Keeling and Fast, 2002). The life cycle is discussed and illustrated by Canning (1990).

Cavalier-Smith (1993) questioned the final placement (Archezoa or Protozoa) of the Microsporidia because it had not been determined whether or not the lack of mitochondria was due to one of two possibilities: 1) they had been lost, or 2) they were never present. He (Cavalier-Smith, 1993) favored their classification in the Archezoa at this time. Microsporidia were most closely related to *Entamoeba histolytica* Schaudinn (Protozoa) or but it was unresolved, at that time, whether Microsporidia were a sister clade to the Fungi (Tanabe et al., 2002) based on an analysis of EF-1α and RPB1 sequences.

Later, Cavalier-Smith (1998) believed that Microsporidia were Fungi and not Protozoa and that they had an origin in the Trichomycetes (Zygomycota). A data set consisting of 28S rRNA (van de Peer et al., 2000) demonstrated that the Microsporidia were the sister clade to *Entomophaga aulicae* (E. Reichardt) Humber (Entomophthorales). Keeling (2003), using sequences from α-tubulin and β-tubulin genes, postulated that the origins of the Microsporidia were in the Zygomycota. In these analyses, Microsporidia were the sister-group to *Syncephalis depressa* van Tieghem & Le Monnier (Zoopagales), a haustorial mycoparasite (α-tubulin only or α-tubulin + β-tubulin data sets) or *Conidiobolus coronatus* (Costantin) A. Batko (Entomophthorales), a human pathogen (β-tubulin data set).

Several other scientists, using sequences from various genes, determined that the Microsporidia arose from within the Basidiomycota based on an 18S rRNA data set (Fischer and Palmer, 2005) or it is a sister to the Ascomycota-Basidiomycota clade using sequences from eight genes (Gill and Fast, 2006).

Other studies have indicated that Microsporidia evolved from a *Rozella*-like ancestor, an endoparasitic chytrid (James et al., 2006). Liu et al. (2006), however, believe that Microsporidia are a sister group to the Fungi. Microsporidia Balbiani was treated as a phylum of the Fungi by Hibbett et al. (2007) but it was not subdivided into orders or other intermediate taxa. Unless additional molecular data indicates otherwise, the Microsporidia are not being included in the zygomycotan Fungi in this discussion. The Microsporidia are one of the seven fungal phyla (Hibbett et al., 2007). A molecular based phylogeny of the Microsporidia was published by Vossbrink and Debrunner-Vossbrink (2005).

Phylogenetic classification of the zygomycotan Fungi

The subphyla, orders, families, and genera (with synonyms) of the Zygomycota as currently recognized by the author are presented in the following annotated synopsis [lists of doubtful or excluded genera of the Mucorales *sensu lato* can be found in Hesseltine (1955) and O'Donnell (1979) although many of these taxa are presently recognized here and in other treatments of various taxa].

The following "Synopsis" includes most, if not all, of the references that the author considers pertinent for each taxon. Articles containing usuable illustrations (as illus.) and/or keys are indicated.

SYNOPSIS OF THE ZYGOMYCOTAN FUNGI

ZYGOMYCOTA Moreau, Les Champignons 2, p 2035, 1954; no Latin diagnosis and type not designated) (Zycha et al., 1969 – illus. and keys to taxa; Hanlin, 1973 – keys to taxa; Hesseltine and Ellis, 1973 – illus. and keys to taxa; Mil'ko, 1974 – keys to taxa, Benjamin, 1979 – illus; Cole and Samson, 1979 – illus; O'Donnell, 1979 – illus. and key to taxa; von Arx, 1981 – illus. and keys to taxa; Kirk et al., 2001 – synopsis of orders and families) [NOTE – See the CABI Bioscience Databases (http://www.indexfungorum.org) Index Fungorum for a list of species, with literature citations, for each genus accepted in the zygomycotan fungi].

The Zygomycota are not accepted as a valid phylum because of lack of resolution of the basal fungal clades (James et al., 2006). When these clades are resolved then the Zygomycota will probably include at least the Mucoromycotina (Hibbett et al., 2007).

Fungal classification to subphylum and order

Recently, a consensus resulted in the publication of a phylogenetic based classification of the Fungi at the ordinal level and above (Hibbett et al., 2007). Recent papers published in the issue [volume 98(6)] devoted to fungal phylogeny in *Mycologia* (Spatafora et al., 2006), James et al. (2006), Liu et al. (2006) and others cited by Hibbett et al. (2007) were the basis for this classification. It was decided that the zygomycotan Fungi should be distributed in four subphyla: Entomophthoromycotina, Kickxellomycotina, Mucoromycotina and Zoopagomycotina. (Hibbett et al., 2007).

ENTOMOPHTHOROMYCOTINA Humber, in Hibbett et al., Mycol. Res., 111: 517, 2007 (Hibbett et al., 2007 – subphylum description).

ENTOMOPHTHORALES G. Winter, in Rabenhorst's Kryptogamen-Flora von Deutschland, Oesterreich und der Schweiz. 1(1): 74, 1880 (Batko, 1964a, b, c, d, 1966, 1974; Remaudière and Keller, 1980; Ben-Ze'ev and Kenneth, 1982; Humber, 1989).

= Ancylistales Schröter [as Ancylistineae], in A. Engler and K. Prantl's, Die natürlichen Pflanzenfamilien 1(1): 88, 1893.

= Basidiobolales Cavalier-Smith, Biol. Rev. 73: 246, 1998 (Cavalier-Smith, 1998).

Ancylistaceae J. Schröter, in A. Engler and K. Prantl's, Die natürlichen Pflanzenfamilien 1(1): 92, 1893 (Humber, 1989 – family description [cited Fisher as the author of the family]).

Ancylistes Pfitzer, Monatsber. Königl. Preuss. Akad. Wiss. Berlin, p. 396, 1872 (Tucker 1981 – key to spp.)

Conidiobolus Brefeld, Untersuchungen aus dem Gesammtgebiete der Mykologie 6: 390, 1884 emend. Humber, Mycotaxon 34: 455, 1989 (King, 1976a, b, 1977; Keller, 1987; Humber, 1989).

Subgenera

Capillidium Ben-Ze'ev & Kenneth, Mycotaxon, 14: 431, 1982.

Conidiobolus (Bref.) Ben-Ze'ev & Kenneth, Mycotaxon, 14: 429, 1982.

Delacroixia (Sacc. & Syd.) Tyrrell & MacLeod, J. Invertebr. Pathol., 7: 12, 1972.

Macrobiophthora Reukauf, Centralbl Bakteriol, Abt 1, Originale 63: 390, 1912 emend. Tucker, Mycotaxon, 13: 498, 1981 (Tucker 1981 - key to spp.).

Basidiobolaceae Engler & E. Gilg, Syllabus der Pflanzenfamilien, Ed 9 & 10 (Berlin), p. 45, 1924 (Humber, 1989 - family description).

Basidiobolus Eidam, in Cohn's Beiträge zur Biologie der Pflanzen, 4: 194, 1885 (Benjamin, 1962).

Completoriaceae Humber, Mycotaxon, 34: 453, 1989 (Humber, 1989 - family description).

Completoria Lodhe, Tagebl Versamml Deutsch Naturf Aertze, 47: 206, 1874 (Tucker, 1981 - in key with other nonentomogenous taxa).

Entomophthoraceae Nowakowski, Bot Zeit 35: 35, 1877 [Humber, 1989 - family description (listed Winter as the author of the family), Keller and Petrini, 2005 - subfamily key].

Entomophthoroideae S. Keller, in Keller and Petrini, Sydowia, 57: 28, 2005 (Keller and Petrini, 2005—subfamily description and key to genera).

Batkoa Humber, Mycotaxon, 34: 446, 1989 (Humber 1989; Keller and Petrini, 2005 - genus description and species list).

Entomophaga Batko, B Acad. Pol. Sci. Biol., 12: 325, 1964 emend Humber, Mycotaxon, 34: 447, 1989 (Keller, 1987; Humber, 1989; Keller and Petrini, 2005 - genus description and species list).

Entomophthora Fresenius, Bot. Zeit., 14: 882, 1856 (Keller, 1987, 2002, 2004, 2007; Keller et al., 1999; Keller and Petrini, 2005 - genus description and species list).

=*Empusa* subgen. *Triplosporium* Thaxter, Mem. Boston. Soc. Nat. Hist., 4: 169, 1888 (Thaxter, 1888).

= *Triplosporium* (Thaxter) Batko, B Acad. Pol. Sci. Biol., 12: 325, 1964 (Batko, 1964a).

Eryniopsis Humber, Mycotaxon, 21: 258, 1984 (Humber, 1984; Keller, 1991; Keller and Eilenberg, 1991).

Erynioideae S. Keller, in Keller and Petrini, Sydowia, 57: 33, 2005 (Keller and Petrini, 2005 – subfamily description and key to genera).

Erynia (Nowakowski ex Batko) Remaudière & Hennebert, Mycotaxon, 11: 301, 1980. emend. Humber, Mycotaxon, 34: 448, 1989 (Humber, 1981b, 1982, 1989; Keller, 1991, 1993; Keller and Petrini, 2005 – genus description and species list).

≡ *Zoophthora* Batko subgenus *Erynia* Nowakowski ex Batko, Acta Mycol., 2: 18, 1966.

≡ *Erynia* (Nowakowski ex Batko) Remaudière & Hennebert, Mycotaxon, 11: 301, 1980.

Furia (Batko) Humber, Mycotaxon, 34: 450, 1989 (Humber, 1989; Keller and Petrini, 2005 – genus description and species list).

≡ *Zoophthora* subgenus *Furia* Batko, Acta Zool., 2: 20, 1966.

≡ *Erynia* subgenus *Furia* (Batko) Li & Humber, Can. J. Bot., 62: 662, 1984.

Orthomyces Steinkraus, Humber & Oliver, in Steinkraus, Oliver, Humber and Gaylor, J. Invert. Pathol., 72: 7, 1988 (Steinkraus et al., 1998; Keller and Petrini, 2005 – genus description and species list).

Pandora Humber, Mycotaxon, 34: 451, 1989 (Humber, 1989; Keller and Petrini, 2005 – genus description and species list).

= *Erynia* subgenus *Neopandora* Ben-Ze'ev & Kenneth, Mycotaxon, 14: 460, 1982.

Strongwellsea A. Batko & J. Weiser, J. Invertebr. Pathol., 7: 463, 1965 emend. Humber, Mycologia, 68: 1054, 1976 (Humber, 1976, 1982; Keller and Petrini, 2005 – genus description and species list).

Tarichium Cohn, Beiträge zur Biologie der Pflanzen, 1: 69, 1875 (MacLeod and Müller-Kögler, 1970; Keller, 1991).

Zoophthora Batko, B Acad. Pol. Sci. Biol., 12: 323, 1964 emend. Ben-Ze'ev and Kenneth, Mycotaxon, 11: 301, 1982 (Humber, 1989; Keller, 1991; Keller and Petrini, 2005 – genus description and species list).

Massosporoideae S. Keller, in Keller and Petrini, Sydowia, 57: 47, 2005 (Keller and Petrini, 2005 – subfamily description).

Massospora Peck, 31st Rep. New York State Mus. Nat. Hist., p. 144, 1879 emend, Soper, Mycotaxon, 1: 15, 1974 (Soper, 1974; Keller and Petrini, 2005 – genus description and species list).

Meristacraceae Humber, Mycotaxon, 34: 456, 1989 (Humber, 1989 – family description).

Ballocephala Drechsler, B. Torrey Bot. Club, 78: 199 1951 (Tucker, 1981 – key to spp.).

Meristacrum Drechsler, J. Washington Acad. Sci., 30: 250, 1940 emend. B. Tucker and Humber, in Tucker, Mycotaxon, 13: 494, 1981 (Tucker, 1981).

Zygnemomyces Miura, Rep. Tottori Mycol. Inst. (Japan) 10: 520, 1973 (Miura, 1973; Tucker, 1981 - key to spp.).

Neozygitaceae I. Ben-Ze'ev, R.G Kenneth & A. Uziel, Mycotaxon, 28: 321, 1987 (Humber, 1989 - family description, Keller and Petrini, 2005 - key to genera).

Apterivorax S. Keller, in Keller and Petrini, Sydowia, 57: 47, 2005 (Keller and Petrini, 2005 - genus description and species list, Keller, 2006b - validation).

Neozygites Witlaczil, Arch. F. Mikr. Anat., 24: 599, 1885 (Keller, 1991, 1997, 2006a; Keller and Petrini, 2005 - genus description and species list).

Thaxterosporium Ben-Ze'ev & Kenneth, in Ben-Ze'ev, Kenneth and Uziel, Mycotaxon, 28: 323, 1987 (Ben-Ze'ev et al., 1987).

KICKXELLOMYCOTINA Benny, in Hibbett et al. Mycol. Res., 111: 518, 2007 (Hibbett et al., 2007 - subphylum description).

ASELLARIALES Manier, Ann. Sci. Nat. Bot., Sér XI, 11: 157, 1950 ex Manier & Lichtwardt, in Lichtwardt and Manier, Mycotaxon, 7: 442, 1978 emend. Valle and Cafaro, Mycologia, 100: 123, 2008 (Lichtwardt, 1986 - discussion, Valle and Cafaro, 2008 - order description).

Asellariaceae Manier, Ann. Sci. Nat. Bot., Sér. XI, 11: 157, 1950 ex Manier & Lichtwardt, Ann. Sci. Nat. Bot., Sér XII, 9: 526, 1968 [Manier and Lichtwardt, 1968 (1969) - family description].

Asellaria Poisson, B. Biol. Fr. Belg., 66: 233, 1932 (Valle and Cafaro, 2008).

Orchesellaria Manier, Ann. Sci. Nat. Zool., Sér. XX, 11: 139, 1958 ex Manier & Lichtwardt, Ann. Sci. Nat. Bot., Sér. XII, 9: 526, 1968 [1969] (Lichtwardt and Moss, 1984b; Valle, 2006).

Genus of unknown affinity

Baltomyces Cafaro, Mycologia, 91: 517, 1999 (Cafaro, 1999).

DIMARGARITALES R.K. Benjamin in Kendrick, The Whole Fungus 2: 607, 1979 (Benjamin, 1979 - order description).

Dimargaritaceae R.K. Benjamin, Aliso 4: 364, 1959 (Benjamin, 1959 - family description).

Dimargaris van Tieghem, Ann. Sci. Nat. Bot., Sér VI, 1: 154, 1875 (Benjamin, 1959 - genus description, 1965 - key to species, Mikawa, 1976),

Dispira van Tieghem, Ann. Sci. Nat. Bot., Sér VI, 1: 160, 1875 (Benjamin, 1959 - genus description, 1963 - key to species).

Spinalia Vuillemin, B. Soc. Mycol. Fr., 20: 32, 1904 (Vuillemin, 1904; Benjamin, 1959).

Tieghemiomyces R.K. Benjamin, Aliso, 4: 390, 1959 (Benjamin, 1959 - genus description, 1961 - key to species).

HARPELLALES Lichtwardt & Manier, Mycotaxon, 7: 442, 1978 (Lichtwardt and Manier, 1978 - order description, Lichtwardt, 1986 - discussion, Benny et al., 2001).

Harpellaceae L. Léger & Duboscq ex P.M. Kirk & P.F. Cannon, in Cannon and Kirk, Fungal Families of the World, p. 152, 2007 (Cannon and Kirk, 2007).

= Harpellaceae L. Léger & Duboscq [as "Harpellacées"], CR. Hebd. Séanc. Acad. Sci., 188: 954, 1929; nomen nudum - vernacular ending, Art. 18.4 of the ICBN, McNeill et al., 2006) (Lichtwardt, 1986 - family description).

Carouxella Manier, J.-A. Rioux & Whisler, Nat. Monspeliensia, Sér. Bot., 13: 30, 1961 ex Manier, J.-A. Roux & Lichtwardt, Nat. Monspeliensia, Sér. Bot., 16: 87, 1964 (1965) (Lichtwardt, Ferrington and López Lastra, 1999).

Harpella L. Léger & O. Duboscq, CR. Hebd. Acad. Sci., 188: 951, 1929 (Williams and Lichtwardt, 1984; Lichtwardt and Arenas, 1996).

Harpellomyces Lichtwardt & S.T. Moss, Mycotaxon, 20: 512, 1984 emend, Lichtwardt, M.M. White and Colbo, Mycologia, 93: 765, 2001 (Lichtwardt and Moss, 1984a; White and Lichtwardt, 2004; White et al., 2006b, c).

Stachylina L. Léger & M. Gauthier, CR. Hebd. Acad. Sci., 194: 2262, 1932 (Lichtwardt and Williams, 1988; Valle, 2007).

Stachylinoides Ferrington, Lichwardt & López-Lastra, in Lichtwardt, Ferrington and López-Lastra, Mycologia, 91: 1069, 1999 (Lichtwardt, Ferrington and López Lastra, 1999).

Legeriomycetaceae Pouzar, Folia Geobot. Phytotax., 7: 319, 1972 (Pouzar, 1972; Lichtwardt, 1986 - family description).

= Genistellaceae L. Léger & M. Gauthier, CR. Hebd. Acad. Sci., 194: 226, 1932 - as Genistellacées) (Pouzar, 1972).

Allantomyces M.C. Williams & Lichwardt, Can. J. Bot., 71: 1109, 1993 (Williams and Lichtwardt, 1993; Valle, White and Cafaro, 2008).

Austrosmittium Lichtwardt & M.C. Williams, in Williams and Lichtwardt, Can. J. Bot., 68: 1047, 1990 (Williams and Lichtwardt, 1990, Lichtwardt and Williams, 1992b).

Baetimyces L.G. Valle & Santamaria, Mycologia, 94: 321, 2002 (Valle and Santamaria, 2002a).

Barbatospora M.M. White, A. Siri & Lichtwardt, Mycologia, 98: 335, 2006 (White et al., 2006b).

Bojamyces Longcore, Mycologia, 81: 482, 1989 emend. L.G. Valle and Santamaria, Mycologia, 96: 1387, 2004 (Longcore, 1989; Valle and Santamaria, 2004; Valle, White and Cafaro, 2008).

Capniomyces S.W. Peterson & Lichtwardt, Mycologia, 75: 242, 1983 (Peterson and Lichtwardt, 1983; Valle, 2007).

Caudomyces Lichtwardt, Kobayasi & Inhoh, T. Mycol. Soc. Jpn., 28: 376, 1987 (Lichtwardt et al., 1987; Strongman and Xu, 2006).

Coleopteromyces L.C. Ferrington, Lichtwardt & López-Lastra, in Lichtwardt, Ferrington and López-Lastra, Mycologia, 91: 1064, 1999 (Lichtwardt et al., 1999).

Ejectosporus S.W. Peterson, Lichtwardt & M.C. Williams, Mycologia, 83: 389, 1991 emend. Strongman, Mycologia, 97: 553, 2005 (Lichtwardt, Peterson and Williams, 1991b; Strongman, 2005).

Ephemerellomyces M.M. White & Lichtwardt, Mycologia, 96: 893, 2004 (White and Lichtwardt, 2004).

Furculomyces Lichtwardt & M.C. Williams, Can. J. Bot., 70: 1196, 1992 (Lichtwardt and Williams, 1992a; Misra, White and Lichtwardt, 1999).

Gauthieromyces Lichtwardt, Mycotaxon, 17: 213, 1983 (Lichtwardt, 1983; Strongman and Xu, 2006; Misra and Tiwari, 2008; Valle, White and Cafaro, 2008).

Genistelloides S.W. Peterson, Lichtwardt & B.W. Horn, Mycologia, 73: 477, 1981 (Peterson et al., 1981; Williams and Lichtwardt, 1987; White and Lichtwardt, 2004).

Genistellospora Lichtwardt, Mycologia, 64: 168, 1972 (Lichtwardt, 1972, 1997 [1998], Lichtwardt and Arenas, 1996; Alencar et al., 2003).

Glotzia M. Gauthier, CR. Hebd. Acad. Sci., 202: 1098, 1936 ex Manier & Lichtwardt, Ann. Sci. Nat. Bot., Sér XII, 9: 528, 1969 (Lichtwardt, 1972; Lichtwardt and Williams, 1990; Williams and Lichtwardt, 1999; White and Lichtwardt, 2004).

Graminella L. Léger & M. Gauthier, CR. Hebd. Acad. Sci., 204: 29, 1937 ex Manier, Bull. Soc. Hist. Nat. Toulouse, 97: 251, 1962 (Lichtwardt and Moss, 1981; Lichtwardt and Arenas, 1996; Valle, White and Cafaro, 2008).

Graminelloides Lichtwardt, Rev. Biol. Trop., 45: 1354, 1998 (Lichtwardt, 1997 [1998]).

Lancisporomyces Santamaria, Mycologia, 89: 639, 1997 (Santamaria, 1997, Strongman and White, 2006; Valle, White and Cafaro, 2008).

Legerioides M.M. White, Mycologia, 91: 1022, 1999 (White, 1999).

Legeriomyces Pouzar, Folia. Geobot. Phytotax., 7: 319, 1972 (Lichtwardt and Williams, 1983; Williams and Lichtwardt, 1993; Valle and Santamaria, 2004; Valle, 2007; Strongman and White, 2008).

= *Genistella* L. Léger & M. Gauthier, CR. Hebd. Acad. Sci., 194: 2264, 1932 (Manier, 1962a).

Legeriosimilis M.C. Williams, Lichtwardt, M.M. White & Misra, in Williams and Lichtwardt, Mycologia, 91: 400, 1999 (Williams and Lichtwardt, 1999; White and Lichtwardt, 2004; Valle, 2007; Strongman and White, 2008).

Orphella L. Léger & M. Gauthier, Trav. Lab. Hydrobiol. Piscicult. Grenoble, 23: 70, 1931 emend. Santamaria and Girbal, Mycol. Res., 102: 174, 1998

(Lichtwardt, Peterson and Huss 1991a; Santamaria and Girbal, 1998; Valle and Santamaria, 2005; Strongman and White, 2006).

Pennella Manier, Ann. Sci. Nat. Bot., Sér XII, 4: 744, 1963 ex Manier, Ann. Sci. Nat. Bot., Sér XII, 9: 106, 1968 (Manier, 1968; Williams and Lichtwardt, 1971; Lichtwardt, 1972, 1984, 1997 [1998], Lichtwardt, White and Colbo, 2001b; Valle, 2007; Strongman and White, 2008).

Plecopteromyces Lichtwardt, Ferrington & López-Lastra, Mycologia, 91: 1066, 1999 (Lichtwardt et al., 1999; Ferrington et al., 2005).

Pseudoharpella Ferrington, M.M. White & Lichtwardt, Aquat. Insects, 25: 86, 2003 (Ferrington et al., 2003).

Pteromaktron H.C. Whisler, Can. J. Bot., 41: 897, 1963 (Whisler, 1963).

Simuliomyces Lichwardt, Mycologia, 64: 179, 1972 (Lichtwardt, 1972; Lichtwardt and Williams, 1990; Lichtwardt, Peterson and Williams, 1991b).

Smittium Poisson, Mélanges Dédiés au Professeur Lucien Daniel, p. 76, 1936 (Williams and Lichtwardt, 1984; Lichtwardt and Williams, 1990; White and Lichtwardt, 2004).

Spartiella Tuzet & Manier, Ann. Sci. Nat. Zool., Sér XI, 12: 18, 1950 ex Manier, Ann. Sci. Nat. Zool., Sér XII, 3: 523, 1962 (Manier, 1962b; Valle, 2007).

Stipella L. Léger & M. Gauthier, CR. Hebd. Acad. Sci., 194: 2263, 1932 (Valle, 2007).

Tectimyces L.G. Valle & Santamaria, Mycol. Res., 106: 842, 2002 (Valle and Santamaria, 2002b).

Trichozygospora Lichtwardt, Mycologia, 64: 173, 1972 (Lichtwardt, 1972).

Zygopolaris S.T. Moss, Lichtwardt & Manier, Mycologia, 67: 120, 1975 (Moss et al., 1975, Lichtwardt and Williams 1984, Valle, White and Cafaro, 2008).

KICKXELLALES Kreisel ex R.K. Benjamin, in Kendrick, The Whole Fungus, 2: 610, 1979 (Kriesel 1969, Benjamin 1979 – order description).

Kickxellaceae Linder, Farlowia, 1: 56, 1943 (Linder, 1943; Benjamin 1959—family description, Kurihara and Degawa, 2006 – key to genera).

Coemansia van Tieghem & Le Monnier, Ann. Sci. Nat. Bot., Sér V, 17: 392, 1873 (Bainier, 1906; Linder, 1943; Benjamin, 1958, 1959; Kurihara et al., 2000, 2008; Kwásna et al., 2002).

Dipsacomyces R.K. Benjamin, Aliso, 5: 15, 1961 (Benjamin, 1961).

Kickxella Coemans, B. Soc. Roy. Bot. Belg., 1: 155, 1862 (Linder, 1943; Benjamin, 1958).

= *Coronella nivea* Crouan, Florule de Finistère, p. 12, 1867.

= *Coemansiella alabastrina* Saccardo, Sylloge Fungorum, 2: 815, 1883.

Linderina Raper & Fennell, Am. J. Bot., 39: 81, 1952 (Raper and Fennell, 1952; Benjamin 1959; Kurihara et al., 2008).

Martensella Coemans, B. Acad. Roy. Sci. Lett. Arts Belg. Cl. Sci., Sér 2, 15: 540, 1863. [Linder, 1943; Jackson and Dearden, 1948; Benjamin, 1959. The host for *M. corticii* Thaxter is *Vesiculomyces citrinus* (Pers.) Hagström (Hagströml 1977)].

Martensiomyces Meyer, B. Soc. Mycol. Fr., 73: 189, 1957 (Meyer, 1957; Benjamin, 1959).

Mycoëmilia Kurihara, Degawa & Tokumasu, Mycol. Res., 108: 1143, 2004 (Kurihara et al., 2004).

Myconymphaea Kurihara, Degawa & Tokumasu, Mycol. Res., 105: 1398, 2001 (Kurihara et al., 2001).

Pinnaticoemansia Kurihara & Degawa, Mycoscience, 47: 205, 2006 (Kurihara and Degawa, 2006).

Ramicandelaber Y. Ogawa, S. Hayashi, Degawa & Y. Yaguchi, Mycoscience, 42: 193, 2001 (Ogawa et al., 2001; Kurihara et al., 2004).

Spirodactylon R.K. Benjamin, Aliso, 4: 408, 1959 (Benjamin, 1959; O'Donnell, 1979).

Spiromyces R.K. Benjamin, Aliso, 5: 273, 1963 (Benjamin, 1963; Mikawa, 1975; O'Donnell et al., 1998).

MUCOROMYCOTINA Benny, Mycol. Res., 111: 517, 2007 (Hibbett et al., 2007 - subphylum description).

ENDOGONALES Moreau, Encyclopédie Mycologique, 23: 1231, 1953. ex R.K. Benjamin, in Kendrick, The Whole Fungus, 2: 599, 1979 emend Morton & Benny, Mycotaxon, 37: 473, 1990 (Benjamin, 1979--order validation, Morton and Benny, 1990 - order description and emendation).

Endogonaceae Paoletti, in P.A. Saccardo, Sylloge Fungorum, 8: 905, 1889 emend Morton & Benny, Mycotaxon, 37: 473, 1990 (Morton and Benny, 1990 - family emendation, Wu and Lin, 1997 - key to genera).

Endogone Link, Ges. Naturf. Freunde Berlin Mag., 3: 33, 1809 (Thaxter ,1922; Gerdemann and Trappe, 1974, Yao et al., 1995a, 1996 - key to species).

Pteridiospora C.-g. Wu & Lin, Mycotaxon, 64: 180, 1997 (Wu and Lin, 1997 - key to species).

Sclerogone Warcup, Mycol. Res., 94: 176, 1990 (Warcup, 1990; Yao et al., 1996).

Youngiomyces Y.J. Yao, in Yao, Pegler and Young, Kew B, 50: 355, 1995 (Yao et al., 1995b, 1996 - key to species).

MORTIERELLALES Cavalier-Smith, Biol. Rev., 73: 246, 1998 (Cavalier-Smith, 1998).

Mortierellaceae A. Fischer, in L. Rabenhorst, Kryptogamen-Flora von Deutschland, Oesterreich und der Schweiz, 1(4): 268, 1892 (Fischer 1892; Benjamin, 1978, 1979; Cavalier-Smith, 1998).

Aquamortierella Embree & Indoh, B. Torrey Bot. Club, 94: 464, 1967 (Embree and Indoh, 1967; Indoh, 1967)

Dissophora Thaxter, Bot. Gaz., 58: 361, 1914 (Thaxter, 1914; Gams and Carreiro, 1989 – key to species, Benny, 1995a).

Gamsiella (R.K. Benjamin) Benny & M. Blackwell, Mycologia, 96: 147, 2004 (Benjamin, 1978, Benny and Blackwell, 2004).

≡ *Mortierella* Coemans subgen. *Gamsiella* R.K. Benjamin, Aliso, 9: 157, 1978 (Benjamin, 1978).

Lobosporangium M. Blackwell & Benny, in Benny and Blackwell, Mycologia, 96: 144, 2004 (Benny and Blackwell, 2004).

= *Echinosporangium* Malloch, Mycologia, 59: 327, 1967 non *Echinosporangium* Kylin (Malloch, 1967; Benny and Blackwell, 2004).

Modicella Kanouse, Mycologia, 28: 60, 1936 (Thaxter, 1922 – as *Endogone malleola* Harkn. and *E. reniformis* Bres., Kanouse 1936 – genus description, Gerdemann and Trappe 1974 – key to species, Benny, Gibson and Kimbrough, 1987).

Mortierella Coemans, B. Acad. Roy. Sci. Lett. Arts Belg., Sér 2, 15: 536, 1863 (Linnemann, 1969; Gams 1976, 1977 – key to sections and species, Degawa and Tokumasu, 1997, 1998a, b; Degawa and Gams, 2004).

= *Haplosporangium* Thaxter, Bot. Gaz., 58: 362, 1914 (Thaxter, 1914).

= *Azygozygum* Chesters, T. Brit. Mycol. Soc., 18: 213, 1933 (Chesters, 1933).

= *Actinomortierella* Chalabuda, Nov. Syst. Plant non Vascul., 1968: 129, 1968 (Chalabuda, 1968).

Genus of unknown affinity

Nothadelphia Degawa & W. Gams, Stud. Mycol., 50: 569, 2004 (Degawa and Gams, 2004).

MUCORALES Fries, Systema Mycologicum, 3: 296, 1832 (Benjamin, 1979). Note that there is little family structure listed in this treatment. As more taxa are sequenced and analyzed in multigene data sets then many of these families may be recognized again or others proposed. Recent publications have argued for the recognition of the revised families Choanephoraceae (Voigt and Olsson, 2008) and Phycomycetaceae (Voigt et al., 2008). The families Mycocladaceae and Lichtheimiaceae were recently described (Hoffmann et al., 2007, 2009).

Mucoraceae Dumortier (as 'Mucoreae"), Commentationes Botanicae, p. 69, 1822 (Hesseltine and Ellis, 1973; Benjamin, 1979).

= Pilobolaceae Corda (as "Pilobolideae"), Anleitung zum Studium der Mykologie, p. 17, 1842 (Grove, 1934; Kirk and Benny, 1980).

= Chaetocladiaceae A. Fischer, in Rabenhorst's Kryptogamen-Flora von Deutschland, Oesterreich und der Schweiz 1(4), p. 283, 1892 emend. Benny and R.K. Benjamin, Mycologia, 85: 670, 1993 (Benny and Benjamin, 1993).

= Choanephoraceae J. Schröter, in Engler & Prantl, Die natürlichen Pflanzenfamilien 1(1), p. 131, 1894 (Kirk, 1984; Voigt and Olsson, 2008).

= Gilbertellaceae Benny, Mycologia, 83: 151, 1991 (Benny, 1991; Voigt and Olsson, 2008 – Gilbertellaceae a synonym of Choanephoraceae).

= Thamnidiaceae Fitzpatrick, The Lower fungi, p. 242, 1930 (Benny and Benjamin, 1975, 1976, 1993; Benny, Kirk and Samson, 1985; Benny and Schipper, 1992; Benny, 1992, 1995b, c).

= Syncephalastraceae R.K. Benjamin, Aliso, 4: 327, 1959 (Benjamin, 1959).

≡ Syncephalastraceae Naumov, Opredetitel' Mucoronÿkh, Ed. 2, p. 16, 1935., nom. inval., without a Latin diagnosis, Art. 32.1(c), 36.1 of the ICBN (McNeill et al., 2006).

= Cunninghamellaceae R.K. Benjamin, Aliso, 4: 415, 1959 emend. Benny, R.K. Benjamin and P.M. Kirk, Mycologia, 84: 639, 1992 (Benjamin, 1959; Benny, Benjamin and Kirk, 1992).

≡ Cunninghamellaceae Naumov, Opredetitel' Mucoronÿkh, Ed. 2, p. 20, 1935; nom. inval., without a Latin diagnosis, Art. 32.1(c), 36.1 of the ICBN (McNeill et al., 2006).

= Saksenaeaceae Hesseltine & J.J Ellis, in Ellis and Hesseltine, Mycologia, 66: 87, 1974 (Ellis and Hesseltine, 1974).

= Radiomycetaceae Hesseltine & J.J. Ellis, in Ellis and Hesseltine, Mycologia, 66: 91, 1974 (Ellis and Hesseltine, 1974).

= Dicranophoraceae J.H. Mirza, in Mirza, Khan, Begum and Shagufta, Mucorales of Pakistan, p. 19, 1979 (Mirza et al., 1979).

= Absidiaceae von Arx, Sydowia, 35: 21, 1982 [1983] (von Arx, 1982 [1983]).

= Phycomycetaceae von Arx, Sydowia, 35: 22, 1982 (1983) [von Arx, 1982 (1983); Voigt et al., 2008].

= Mycotyphaceae Benny & R.K. Benjamin, in Benny, Kirk & Samson, Mycotaxon, 22: 122, 1985 (Benny, Kirk and Samson, 1985).

= Mycocladaceae K. Hoffmann, S. Discher & K. Voigt, Mycol. Res., 111: 1177, 2007 (Hoffmann et al., 2007).

= Lichtheimiaceae K. Hoffmann, G. Walter & K. Voigt, Mycol. Res., 113: 277, 2009 (Hoffmann et al., 2009).

Absidia van Tieghem, Ann. Sci. Nat. Bot., Sér. VI, 4: 350, 1878 (Hesseltine and Ellis. 1961; 1964, 1966; Ellis and Hesseltine, 1965, 1966; Schipper, 1990).

= *Tieghemiella* Berl. & de Toni, in P.A. Saccardo, Sylloge Fungorum, 7: 215, 1888.

= *Proabsidia* Vuillemin, B. Soc. Mycol. Fr., 19: 116, 1903.

Actinomucor Schostakowitsch, Ber. Deutsch. Bot. Ges., 16: 155, 1898 (Benjamin and Hesseltine, 1957; Zheng and Liu, 2005; Khan et al., 2008).

= *Glomerula* Bainier, B. Soc. Mycol. Fr., 19: 154, 1903.

Amylomyces Calmette, Ann. I. Pasteur., 6: 611, 1892 (Ellis, Rhodes and Hesseltine, 1976).

Apophysomyces Misra, in Misra, Srivastava and Lata, Mycotaxon, 8: 377, 1979 (Misra, Srivastava and Lata, 1979).

Backusella Hesseltine & J.J. Ellis, in Ellis and Hesseltine, Mycologia, 61: 863, 1969 (Benny and Benjamin, 1975).

Benjaminiella von Arx, The Genera of Fungi Sporulating in Pure Culture, 3rd. ed., p. 60, 1981 (Benny and Benjamin, 1976; Benny, Kirk and Samson, 1985; Kirk, 1989).

≡ *Benjaminia* Pidoplichko & Mil'ko, Atlas mukoral'vykh gribov [Atlas of Mucoralean Fungi], p. 96, 1971 non *Benjaminia* Ahmad – a later homonym, Art. 53.1 of the ICBN (McNeill et al., 2006) (Pidoplichko and Mil'ko, 1971).

Blakeslea Thaxter, Bot Gaz 58: 353, 1914 (Thaxter, 1914; Kirk, 1984; Ho and Chang, 2003).

Chaetocladium Fresenius, Beiträge zur Mykologie, p. 97, 1863 (Benny and Benjamin, 1976).

Chlamydoabsidia Hesseltine & J.J. Ellis, Mycologia, 58: 761, 1966 (Hesseltine and Ellis, 1966).

Choanephora Currey, J. Linn. Soc. Bot., 13: 578, 1873 (Thaxter, 1903, 1914; Kirk, 1984; Sakai et al., 2000 – pollenation mutualism).

Circinella van Tieghem & Le Monnier, Ann. Sci. Nat. Bot., Sér V, 17: 298, 1873 (Hesseltine and Fennell, 1955; Hesseltine and Ellis, 1961).

≡ *Circinumbella* van Tieghem & Le Monn., CR. Hebd. Acad. Sci., 74: 999, 1872.

Cokeromyces Shanor, in Shanor, Poitras and Benjamin, Mycologia, 42: 272, 1950 (Shanor et al., 1950; Benny and Benjamin, 1976).

Cunninghamella Matruchot, Ann. Mycol., 1: 46, 1903 (Zheng and Chen, 2001).

= *Actinocephalum* Saito, Bot. Mag. (Tokyo), 19: 36, 1905.

= *Saitomyces* Ricker, J. Mycol., 12: 61, 1906.

= *Muratella* Bainier & Sartory, B. Soc. Mycol. Fr., 29: 129, 1913.

Dichotomocladium Benny & R.K. Benjamin, Aliso, 8: 338, 1975 (Benny and Benjamin, 1975, 1993 – key to spp.).

Dicranophora Schröter, Jahresber. Schles. Ges. Vaterl. Cultur., 64: 84, 1886 (Dobbs, 1938; Voglmayr and Krisai-Greilhuber, 1996).

Ellisomyces Benny & R.K. Benjamin, Aliso, 8: 330, 1975 (Benny and Benjamin, 1975).

Fennellomyces Benny & R.K. Benjamin, Aliso, 8: 328, 1975 (Hesseltine and Fennell, 1955 – ill. as *Circinella linderi*, Benny and Benjamin, 1975; Misra, Gupta and Lata, 1979).

Gilbertella Hesseltine, B. Torrey Bot. Club, 87: 24, 1960 (Benny, 1991).

Gongronella Ribaldi, Rev. Biol. Gen., n.s. 44: 164, 1952 (Hesseltine and Ellis, 1961, 1964).

Halteromyces Shipton & Schipper, Anton van Lee. J. M. S., 41: 337, 1975 (Shipton and Schipper, 1975).

Helicostylum Corda, Icones Fungorum hucusque Cognitorum, 5: 18, 1842 emend Benny, Mycologia, 87: 257, 1995 (Benny, 1995c).

= *Haynaldia* Schulzer von Müggenberg, Verh. Zool.-Bot. Ges. Wien., 17: 37, 1866 fide Upadhyay (1973).

= *Bulbothamnidium* Klein, Verh. Zool.-Bot. Ges. Wien., 20: 557, 1870 fide Upadhyay (1973).

= *Chaetostylum* van Tiegh. & Le Monn., Ann. Sci. Nat. Bot., Sér V, 17: 328, 1873 fide Upadhyay (1973).

Hesseltinella Upadhyay, Persoonia, 6: 111, 1970 (Benny and Benjamin, 1991 [1992]).

Hyphomucor Schipper & Lunn, in Schipper, Mycotaxon, 27: 83, 1986 (Schipper, 1986b).

Kirkomyces Benny, Mycologia, 87: 922, 1995 (Benny, 1995b, c)

= *Kirkia* Benny, Mycologia, 87: 261, 1995 non *Kirkia* Oliv, in Hooker's Icones Plantarum: ad. T. 1036, 1868. – a later homonym [Benny, 1995c; Art. 53.1 of the ICBN (McNeill et al. , 2006)].

Lentamyces K. Hoffm. & K. Voigt, Plant Biology, 11: 550, 2009. Plant Biology, 10, 2008 [early view](DOI: 10.1111/j.1438-8677.2008.00145.x) (Hoffmann and Voigt, 2008).

Lichtheimia Vuillemin, B. Soc. Mycol. Fr., 19: 126, 1903 (Hoffmann et al., 2009).

= *Pseudoabsidia* Bainier, B. Soc. Mycol. Fr., 19: 155, 1903.

= *Pseudoabsidia* Saccardo, Sylloge Fungorum, 17: 50, 1905 fide Hesseltine (1955).

= *Protoabsidia* Naumov, *Opredelitel' Mukorovykh (Mucorales)*, Ed. 2, pp. 23 and 78, 1935 *nomen nudum*, without a Latin diagnosis, Art. 36.1 of the ICBN, McNeill et al. (2006).

Mucor Fresen., Beiträge zur Mykologie, p. 7, 1850 (nom. conserv., Art. 14.1 of the ICBN, McNeill et al., 2006 – see Kirk, 1986) (Schipper, 1973, 1975, 1976, 1978a).

= *Hygrophora* Tode, Fungi Mecklenburgenses Selecti, Fasc II, p. 5, 1791: E.M. Fries, Systema Mycologicum, 3(2): 313, 1832, fide Kirk (1986).

= *Mucor* P. Micheli: Fr., Systema Mycologicum, 3: 317, 1832 (nom. rej., Art. 56.1 of the ICBN, McNeill et al., 2006, see Kirk, 1986 – a synonym of *Rhizopus*.)

– *Circinomucor* von Arx, Sydowia, 35: 17, 1983 [von Arx, 1982 (1983)].

Mycocladus Beauverie, Ann. Univ. Lyon. Sér 2, Sci. Méd. 3: 162, 1900 [Beauverie, 1900, Hoffmann et al., 2007, 2009; a subgenus of *Absidia* according to Hesseltine and Ellis (1964)].

≡ *Absidia* van Tiegh. subgen. *Mycocladus* (Beauv.) Hesseltine & Ellis, Mycologia, 56: 569, 1964.

Mycotypha Fenner, Mycologia, 24: 196, 1932 (Benny and Benjamin, 1976; Benny, Kirk and Samson, 1985).

Parasitella Bainier, B. Soc. Mycol. Fr., 19: 153, 1903 (Schipper, 1978b).

Phascolomyces Boedijn, Sydowia, 12: 349, 1958 (1959). [Boedijn 1958 (1959), Benny and Benjamin, 1976; Jeffries and Young, 1978].

Phycomyces Kunze, in Kunze and Schmidt, Mykologische Hefte 2, p. 113, 1823 (Benjamin and Hesseltine, 1959).

Pilaira van Tieghem, Ann. Sci. Nat. Bot., Sér VI, 1: 51, 1875 (Grove. 1934; Zheng and Liu, 2009).

Pilobolus Tode, Schriften Berlin Ges. Naturf. Freunde, 5: 47, 1784 (Grove, 1934; Hu et al., 1989).

Pirella Bainier, Étude sur les Mucorinées, p. 83, 1882 (Benny and Schipper, 1992).

Poitrasia P.M. Kirk, Mycol. Pap., 152: 51, 1984 (Kirk, 1984).

Protomycocladus Schipper & Samson, Mycotaxon, 50: 487, 1994 (Schipper and Samson, 1994).

Radiomyces Embree, Am. J. Bot., 46: 25, 1959 (Benny and Benjamin, 1991).

= *Radiomycopsis* Pidoplichko & Milko, Atlas mukoral'vykh gribov [Atlas of the Mucorales], Kiev, p. 9, 1971.

Rhizomucor (Lucet & Cost.) Wehmer ex Vuillemin, Les Champignons parasites et les mycoses de l'hommes, p. 19, 1921 (Schipper, 1978b; Zheng and Chen, 1991, 1993; Zheng and Jiang, 1995).

≡ *Mucor* Micheli ex L.: Fr. sect. *Rhizomucor* Lucet & Cost., CR. Hebd. Acad. Sci., 129: 1033, 1899 fide Schipper, 1978b).

≡ *Mucor* Micheli ex L.: Fr. subgen. *Rhizomucor* (Lucet & Cost.) Sacc. & Syd. in P.A. Saccardo, Sylloge Fungorum, 16: 385, 1902 fide Schipper (1978b).

≡ *Rhizomucor* (Lucet & Cost.) Wehmer, in F. Lafar, Handbuch der technischen Mykologie, 4: 459, 1907; fide Schipper (1978b).

Rhizopodopsis Boedijn, Sydowia, 12: 330, 1959 [Boedijn, 1958 (1959)].

Rhizopus Ehrenberg, Nova Acta Phys.-Med. Acad. Caes. Leop.-Carol. Nat. Cur., 10: 198, 1820 (Schipper, 1984; Schipper and Stalpers, 1984; Zheng et al., 2007).

= *Ascophora* Tode, Fungi Mecklenburgenses Selecti, Fasc I, p. 13, 1791: E.M. Fries, Systema Mycologicum, 3(2): 313, 1832 fide Kirk (1986).

Saksenaea Saksena, Mycologia, 45: 434, 1953 (Saksena, 1953).

Siepmannia Nirenberg & Kwaśna, in Kwaśna and Nirenberg, Mycologia, 100: 260, 2008 (nom, inval., Art. 37.1 of the ICBN, McNeill et al., 2006) ex Nirenberg & Kwaśna, in Kwaśna and Nirenberg, Pol. Bot. J., 53: 187 2008. (Kwaśna and Nirenberg, 2008a, b).

Spinellus van Tieghem, Ann. Sci. Nat. Bot., Sér VI, 1: 66, 1875 (van Tieghem. 1875; Zycha et al., 1969).

Sporodiniella Boedijn, Sydowia, 12: 336, 1959 (Evans and Samson, 1977; Chien and Hwang, 1997).

Syncephalastrum Schröter, in Cohn's Kryptogamen-Flora von Schlesiens 3(2): 217, 1886 (Benjamin, 1959; Zheng et al., 1988; Schipper and Stalpers, 1983).

Syzygites Ehrenberg, Silvae Mycologicae Berolinensis, p. 25, 1818 (Hesseltine, 1957; Kovacs and Sundberg, 1999).

= *Sporodinia* Link, Species Plantarum, 6(1): 94, 1824.

≡ *Sporodinia* Link ex Wallroth, Flora Cryptogamica Germanica, 2: 317, 1833.

= *Azygites* Moug. & Fr.: Fr., Systema Mycologicum, 3: 330, 1832, *nomen conserv.*, Art. 13.1(d) of the ICBN (McNeill et al., 2006).

≡ *Azygites* Moug. & Fr., Systema Orbis Vegetabilis, 1: 364, 1825.

Thamnidium Link, Ges. Naturf. Freunde Berlin Mag. Neuesten Entdeck. Gesammten Naturk., 3: 31, 1809 (non *Thamnidium* Tuck. ex Schwend - see Kirk et al., 2001, 2008; Benny (1992).

≡ *Thamnidium* Link ex Gray, *A Natural Arrangement of British Plants,* 1: 560; 1821, fide Benny (1992).

≡ *Thamnidium* Link ex Wallroth, Flora Cryptogamica Germanica, 4(2): 324, 1833; fide Benny (1992).

= *Melidium* Eshweiler: Fr., Systema Mycologicum, 3: 330, 1832; fide Benny (1992), *nomen conserv.*, Art. 13.1(d) of the ICBN (McNeill et al., 2006).

≡ *Melidium* Eshweiler, De Fructificatione Generis Rhizomorphae Commentatio, p. 33, 1822; fide Benny, 1992).

Thamnostylum von Arx & Upadhyay, in von Arx, The genera of fungi sporulating in pure culture, p. 247, 1970 (Benny and Benjamin, 1975).

Thermomucor A. Subrahamanyam, B.S. Mehrotra & M.J. Thirumalachar, Georgia J. Sci., 35: 1, 1977 (Schipper, 1979).

Utharomyces Boedijn, Sydowia, 12: 340, 1959 (Kirk and Benny, 1980).

Zychaea Benny & R.K. Benjamin, Aliso, 8: 334, 1975 (Benny and Benjamin, 1975).

Zygorhynchus Vuillemin, B. Soc. Mycol. Fr., 19: 116, 1903 (Hesseltine et al., 1959; Schipper 1986a; Zheng, 2002).

Umbelopsidaceae W Gams & W. Meyer, in Meyer & Gams, Mycol. Res., 107: 348, 2003 (Meyer and Gams, 2003 – family description).

Umbelopsis R.E. Amos & H.L. Barnett, Mycologia, 58: 807, 1966 (Meyer and Gams, 2003; Sugiyama et al., 2003, Mahoney et al., 2004).

= *Micromucor* (W. Gams) von Arx, Sydowia, 35: 19, 1984.

ZOOPAGOMYCOTINA Benny, Mycol. Res., 111: 518, 2007 (Hibbett et al., 2007 - subphylum description).

ZOOPAGALES Bessey ex R.K. Benjamin, 1979 (in Kendrick, The Whole Fungus, 2: 590) (Bessey 1950: Dayal, 1973/1974 - key to most species, Benjamin, 1979 - order description).

Cochlonemataceae Duddington (as 'Cochlonemaceae'), in Ainsworth, Sparrow & Susmann, The Fungi 4B: 233, 1974 (Duddington, 1973 - family description and key to genera).

Amoebophilus Dangeard, Le Botaniste, 11: 57, 1910 (Drechsler, 1959; Barron, 1983).

Aplectosoma Drechsler, Mycologia, 43: 173, 1951 (Drechsler, 1951).

Bdellospora Drechsler, Mycologia, 27: 25, 1935 (Drechsler, 1935a).

Cochlonema Drechsler, Mycologia, 27: 18, 1935 (Drechsler, 1935a, 1946).

Endocochlus Drechsler, Mycologia, 27: 14, 1935 (Drechsler, 1935a, 1936).

Euryancale Drechsler, Mycologia, 31: 410, 1939 (Drechsler, 1939; Saikawa and Aoki, 1995).

Possible member of the family

Aenigmatomyces R.F. Castañeda & W.B. Kendrick, Mycologia, 85: 1023, 1993 (Castañeda Ruiz and Kendrick, 1993; Degawa, 2002a, b).

Helicocephalidaceae Boedijn, Sydowia, 12: 355, 1959 [Boedijn, 1958 (1959) - family description, Ellis, 1963].

Brachymyces Barron, Can. J. Bot., 58: 2450, 1980 (Barron, 1980).

Helicocephalum Thaxter, Bot. Gaz., 16: 201, 1891 (Thaxter, 1891b; Drechsler, 1934, 1943; Barron, 1975).

Rhopalomyces Corda, Pract-Flora europaeischer Schimmelbildungen, p. 3, 1839 (Thaxter, 1891a; Drechsler, 1955).

Piptocephalidaceae J. Schröter (as Piptocephalidei), in Cohn, Krypt-Fl. Schlesien (Breslau) 3(1): 215, 1886 (Benjamin, 1959 - family description).

Kuzuhaea R.K. Benjamin, Bot. J. Linn. Soc., 91: 122, 1985 (Benjamin, 1985a).

Piptocephalis de Bary, Abh. Senckenberg Naturf. Ges., 5: 356, 1865 (Benjamin, 1959, 1966; Gräfenhan 1998).

Syncephalis van Tieghem & Le Monnier, Ann. Sci. Nat. Bot., Sér V, 17: 372, 1873 (Thaxter, 1897; Benjamin, 1959, 1966, 1985b; Ho and Benny, 2007, 2008).

Sigmoideomycetaceae Benny, R.K. Benjamin & P.M. Kirk, Mycologia, 84: 620, 1992 (Benny, Benjamin and Kirk, 1992 - family description and key to genera and species).

Reticulocephalis Benny, R.K. Benjamin & P.M. Kirk, Mycologia, 84: 633, 1992 (Benny et al., 1992).

Sigmoideomyces Thaxter, Bot. Gaz., 16: 22, 1891 (Thaxter, 1891a; Benny et al., 1992).

Thamnocephalis Blakeslee, Bot. Gaz., 40: 165, 1905 (Blakeslee, 1905; Benny et al., 1992).

Zoopagaceae Drechsler, Mycologia, 30: 154, 1938 emend. Duddington, The Fungi, p. 233, 1973 (Drechsler, 1938; Duddington, 1973 – family description and key to some genera, Benjamin, 1979).

Acaulopage Drechsler, Mycologia, 27: 185, 1935 (Drechsler, 1935b, 1942).

Cystopage Drechsler, Mycologia, 33: 251, 1941 (Drechsler, 1941, 1959).

Stylopage Drechsler, Mycologia, 27: 197, 1935 (Drechsler, 1935b, 1947).

Zoopage Drechsler, Mycologia, 27: 30, 1935 (Drechsler, 1936, 1938).

Zoophagus Sommerstorff, Österr. Bot. Zeit., 61: 372, 1911 emend. Dick, Mycol. Res., 94: 352, 1990 (Karling, 1936; Drechsler, 1962 – illus. as *Acaulopage pectospora* Drechsler, Dick, 1990).

DISCUSSION

The Zygomycota have assumed increased importance in the last several decades because of their use in industry and in medicine as the causative agents of zygomycosis in animals, including humans.

Historically, knowledge of the zygomycotan fungi has evolved because of the extensive local floras that were published in the 19th century by the European mycologists van Tieghem (van Tieghem and Le Monnier, 1873; van Tieghem, 1875, 1878), Bainier (1882), Fischer (1892) and Schröter (1889). Even at that time, the entomophthoralean and mucoralean taxa were included in separate orders (Fischer, 1892; Schröter, 1893a, b, c; Underwood, 1899). European mycologists continued publishing major works on the Zygomycota in the early 20th century, either as local floras (Hagem, 1908; Lendner, 1908,) or as taxonomic treatises (Naumov, 1935, 1939; Zycha, 1935). Dr. Charles Drechsler, U.S.D.A., Beltsville, Mayland, U.S.A., published the majority of the papers (Lumsden, 1987) that constitute the current state of knowledge of the species diversity for the Cochlonemataceae and Zoopagaceae (Zoopagales).

After World War II, extensive studies on the Trichomycetes were published by French mycologists (Duboscq et al., 1948; Manier 1950 [1951], 1969 [1970]). Manier (1968) validated trichomycete taxa. Manier and Lichtwardt (1968 [1969]) published a paper describing the classification of the Trichomycetes that was to be followed for the next 45 years. Dr. R.W. Lichtwardt (1973a, b, 1986), an American mycologist, and his students, published many papers (see References) on these fungi.

During this time, Dr. C.W. Hesseltine, U.S.D.A., Peoria, Illinois, U.S.A., alone or alongwith Drs. C.R. Benjamin or J.J. Ellis, published monographs on many genera of the Mucorales (see References). Concurrently, Dr. R.K. Benjamin, is presenting his work on the merosporangiferous Mucorales, fungi that reproduce asexually by producing sporangiospores in merosporangia, that are now included in four orders (Dimargaritales, Kickxellales, Mucorales, Zoopagales). Other papers have been published on the Zygomycetes from India, Japan, Napal, Pakistan, Peoples Republic of China and Taiwan. A relatively recent series of publications, which is still ongoing, have been published on the Zygomycetes isolated in Brazil.

Illustrations for these papers, especially those published during the early 20th century, and earlier, were line drawings, some of the best being produced by Thaxter (1888, 1897, 1922), Drechsler (1935a), Benjamin (1958, 1959) and Benny and Benjamin (1975, 1976). Later papers often were illustrated by the inclusion of photographs, either singly or in plates, although drawings may also have been included. During the last three decades of the 20th century many publications were illustrated with scanning electron micrographs (SEM). SEM is especially useful for demonstrating the surface structure of the aerial hyphae and reproductive structures of the Zygomycetes. Transmission electron microscopy (TEM) has been used to study the ultrastructure of internal cell ontogeny, for instance the merosporangial ontogeny of *Linderina pennispora* Raper & Fennell (Benny and Aldrich, 1975). The use of drawings and/or SEM photographs is often essential for illustrating new taxa of Zygomycota because the characteristic fruiting structures are often too large to be observed in a single optical section using a light microscope. The author has found that a plate that contains several drawings, including a habit sketch, is very useful when trying to identify many members of the Zygomycota.

There is research, both classical and molecular, that still should be conducted on the members of the Zygomycota that readily grow and sporulate in culture (Dimargaritales, Kickxellales, Mortierellales, Mucorales, Zoopagales [Piptocephalidaceae]). In the last 15 years phylogenetic studies have resulted in the realignment of several genera, including *Absidia* s.l. (*Absidia* s.s., *Lentamyces, Lichtheimia, Mycocladus* – Hoffmann et al., 2007, 2009; Hoffmann and Voigt, 2008) and *Rhizopus* s.s. (*Rhizopus* s.l., including *Amylomyces,* see the Index Fungorum [http://www.IndexFungorum.org] – Abe et al., 2006; Liu et al., 2006, 2007; Jennessen et al., 2008). Two major monographs were published this decade on *Cunninghamella* (Zheng and Chen, 2001) and *Rhizopus* (Zheng et al., 2007).

A few genera of the Mucorales are polyphyletic based on the results of two phylogenetic studies that included sequences of at least one species of each genus in culture that was recognized at that time (O'Donnell et al., 2001; Voigt and Wöstemeyer, 2001). Four species of *Mucor* were sequenced by O'Donnell et al. (2001) and they were distributed in three subclades of clade 3c in this analysis.

There should be as many *Mucor* species sequenced, as possible, in order to determine the phylogeny of this genus. Additional genera in the Mucorales that are polyphyletic are *Backusella* and *Rhizomucor* (O'Donnell et al., 2001; Voigt and Wöstemeyer, 2001). Other genera that should be studied include *Mortierella* (Mortierellales), *Circinella* and *Zygorhynchus.*

The author has treated the majority of the families of the Mucorales, except Umbelopsidaceae, as synonyms of the Mucoraceae (see also the Zygomycetes website, http://www.zygomycetes.org) based of the results of the two published phylogenies of the Mucorales (O'Donnell et al., 2001; Voigt and Wöstemeyer, 2001). There are recent published proposals to recognize the Choanephoraceae and Phycomycetaceae but with the addition of one genus to each family (Voigt and Olsson, 2008; Voigt et al., 2008). There, however, are several traditional families that could be recognized, again in a nontraditional format (O'Donnell et al., 2001; Voigt and Wöstemeyer 2001) but several unnamed clades need to be studied to determine their status. Finally, additional phylogenetic studies should be done to clarify the relationships of the recognized four subphyla of the zygomycotan Fungi (Hibbett et al., 2007).

REFERENCES

Abe, A., Oda, Y., Asano, K. and Sone, T. 2006. The molecular phylogeny of the genus *Rhizopus* based on rDNA sequences. Biosci. Biotech. Bioch., 70: 2387-2393.

Abe, A., Oda, Y., Asano, K. and Sone, T. 2007. *Rhizopus delemar* is the proper name for *Rhizopus oryzae* fumaric-malic acid producers. Mycologia, 99: 714-722.

Ainsworth, G. C. 1971. Ainsworth & Bisby's Dictionary of the Fungi. 6th edn. Commonwealth Mycological Institute, Kew, Surrey, United Kingdom.

Ainsworth, G. C. and Bisby, G. R. 1943. A Dictionary of the Fungi. The Imperial Mycological Institute, Kew, Surrey, United Kingdom.

Ainsworth, G. C., Sparrow, F. K. and Sussman, A. S. (eds.). 1973. The Fungi: An Advanced Treatise. Vol. IVB. A Taxonomic Review with Keys: Basidiomycetes and Lower Fungi. Academic Press, New York, U.S.A.

Alencar, Y. B., Ríos-Velásquez, C. M., Lichtwardt, R. W. and Hamada, N. 2003. Trichomycetes (Zygomycota) in the digestive tract of arthropods in Amazonas, Brazil. Mem. I. Oswaldo Cruz, 98: 799-810.

Alexopoulos, C. J., Mims, C. W. and Blackwell, M. 1996. Introductory Mycology. 4th edn. John Wiley and Sons, Inc., New York, U.S.A.

Anonymous. 2008. Bibliography of Systematic Mycology. CABI, Wallingford, U.K. [biannual publication].

Bainier, G. 1882. Étude sur les Mucorinées. Thèse. F. Pichon et A. Cotillon, Imprimeurs, Paris. France.

Bainier, G. 1906. Mycothèque de l'Ecole de Pharmacie.—VIII. Recherches sur les *Coemansia* et sur *l'Acrostalagmus nigripes* n. sp. B. Soc. Mycol. Fr., 22: 216-223.

Barron, G. L. 1975. Nematophagus fungi: *Helicocephalum*. T. Brit. Mycol. Soc., 65: 309-310.

Barron, G. L. 1980. A new genus of the Zygomycetes. Can. J. Bot., 58: 2450-2453.

Barron, G. L. 1983. A new *Amoebophilus* (Zygomycetes) ectoparasitic on amoebae. Can. J. Bot., 61: 3091-3094.

Batko, A. 1964a. On the new genera: *Zoophthora* gen. nov., *Triplosporium* (Thaxter) gen. nov. and *Entomophaga* gen. nov. (Phycomycetes: Entomophthorales). B. Acad. Pol. Sci. Biol., 12: 323-326.

Batko, A. 1964b. Remarks on the genus *Entomophthora* Fresenius 1856 non Nowakowski 1883 (Phycomycetes: Entomophthgorales). B. Acad. Pol. Sci. Biol., 12: 319-321.

Batko, A. 1964c. Remarks on the genus *Lamia* Nowakowski 1883 vs. *Culicola* Nieuwland 1916 (Phycomycetes: Entomophthoraceae). B. Acad. Pol. Sci. Biol., 12: 399-402.

Batko, A. 1964d. Some new combinations in the fungus family Entomophthoraceae (Phycomycetes). B. Acad. Pol. Sci. Biol., 12: 403-406.

Batko, A. 1966. On the subgenera of the fungus *Zoophthora* Batko 1964 (Entomophthoraceae). Acta. Mykol., 2: 15-21.

Batko, A. 1974. Filogeneza a struktury taksonomiczne Entomophthoraceae [Phylogeny and taxonomic structure of Entomophthoraceae [English Trans. TT 77-54076]. Polska Akademia Nauk Ossolineum 1974: 209-305.

Beauverie, J. 1900. *Mycocladus verticillatus* (gen. nov. sp. nov.). Ann. Univ. Lyon., Sér 2, Sci., Méd., 3: 162-180.

Benjamin, C. R. and Hesseltine, C. W. 1957. The genus *Actinomucor*. Mycologia, 49: 240-249.

Benjamin, C. R. and Hesseltine, C. W. 1959. Studies on the genus *Phycomyces*. Mycologia, 51: 751-771.

Benjamin, R. K. 1958. Sexuality in the Kickxellaceae. Aliso, 4: 149-169.

Benjamin, R. K. 1959. The merosporangiferous Mucorales. Aliso, 4: 321-433.

Benjamin, R. K. 1961. Addenda to "The merosporangiferous Mucorales." Aliso, 5: 11-19.

Benjamin, R. K. 1962. A new *Basidiobolus* that forms microspores. Aliso, 5: 223-233.

Benjamin, R. K. 1963. Addenda to "The merosporangiferous Mucorales" II. Aliso, 5: 273-288.

Benjamin, R. K. 1965. Addenda to "The merosporangiferous Mucorales" III. *Dimargaris*. Aliso, 6: 1-10.

Benjamin, R. K. 1966. The merosporangium. Mycologia, 58: 1-42.

Benjamin, R. K. 1978. *Gamsiella*, a new subgenus *of Mortierella* (Mucorales: Mortierellaceae). Aliso, 9: 157-170.

Benjamin, R. K. 1979. Zygomycetes and their spores, pp. 573-616. In: The Whole Fungus: the Sexual-Asexual Synthesis. Vol. 2. B. Kendrick (ed.). National Museums of Canada, Ottawa, Canada.

Benjamin, R. K. 1985a. A new genus of the Piptocephalidaceae (Zoopagales) from Japan. Bot. J. Linn. Soc., 91: 117-133.

Benjamin, R. K. 1985b. A novel new species of *Syncephalis* (Zoopagales: Piptocephalidaceae) from California that forms hypogenous merosporangia. Aliso, 11: 1-15.

Benny, G. L. 1991. Gilbertellaceae, a new family of the Mucorales (Zygomycetes). Mycologia, 83: 150-157.

Benny, G. L. 1992. Observations on Thamnidiaceae (Mucorales). V. *Thamnidium*. Mycologia, 84: 834-842.

Benny, G. L. 1995a. Classical morphology in zygomycete taxonomy. Can. J. Bot. (Suppl 1), 73: S725-S730.

Benny, G. L. 1995b. *Kirkomyces*, a new name for *Kirkia* Benny. Mycologia, 87: 922.

Benny, G. L. 1995c. Observations on Thamnidiaceae. VII. *Helicostylum* and a new genus *Kirkia*. Mycologia, 87: 253-264.

Benny, G. L. 2001. Zygomycota: Trichomycetes, pp. 147-160. In: The Mycota. Vol VII. Systematics and Evolution. Part A. D. G. McLaughlin, E. G. McLaughlin and P. A. Lemke (eds.). Springer-Verlag, Berlin, Germany.

Benny, G. L. 2008. The methods used by Dr. R. K. Benjamin, and other mycologists, to isolate Zygomycetes. Aliso, 26: 37-61.

Benny, G. L. and Aldrich, H. C. 1975. Ultrastructural observations on septal and merosporangial ontogeny in *Linderina pennispora* (Kickxellales; Zygomycetes). Can. J. Bot., 53: 2325-2335.

Benny, G. L. and Benjamin, R. K. 1975. Observations on Thamnidiaceae (Mucorales). New taxa, new combinations, and notes on selected species. Aliso, 8: 301-351.

Benny, G. L. and Benjamin, R. K. 1976. Observations on Thamnidiaceae (Mucorales). II. *Chaetocladium, Cokeromyces, Mycotypha*, and *Phascolomyces*. Aliso, 8: 391-424.

Benny, G. L. and Benjamin, R. K. 1991 [1992]. The Radiomycetaceae (Mycorales; Zygomycetes). III. A new species of *Radiomyces*, and cladistic analysis and taxonomy of the family; with a discussion of the evolutionary ordinal relationships in Zygomycotina. Mycologia, 83: 713-735.

Benny, G. L. and Benjamin, R. K. 1993. Observations on Thamnidiaceae (Mucorales). VI. Two new species of *Dichotomocladium* and the zygospores of *D. hesseltinei* (Chaetocladiaceae). Mycologia, 85: 660-671.

Benny, G. L., Benjamin, R. K. and Kirk, P. M. 1992. A reevaluation of Cunninghamellaceae (Mucorales). Sigmoideomycetaceae fam. nov. and *Reticulocephalis* gen. nov.; cladistic analysis and description of two new species. Mycologia, 84: 615-641.

Benny, G. L. and Blackwell, M. 2004. *Lobosporangium*, a new name for *Echinosporangium* Malloch, and *Gamsiella*, a new genus for *Mortierella multidivaricata*. Mycologia, 96: 143-149.

Benny, G. L., Gibson, J. L. and Kimbrough, J. W. 1987. The taxonomic position of *Modicella*, p. 311. In: Mycorrhizae in the Next Decade. Practical applications and research priorities. D. M. Sylvia, L. L. Hung and J. H. Graham (eds.). Institute of Food and Agricultural Sciences, University of Florida, Gainesville, Florida, U.S.A.

Benny, G. L., Humber, R. A. and Morton, J. B. 2001. Zygomycota: Zygomycetes, pp. 113-146. In: The Mycota. Vol VII. Systematics and Evolution. Part A. D. G. McLaughlin, E. G. McLaughlin and P. A. Lemke. (eds.). Springer-Verlag, Berlin, Germany.

Benny, G. L., Kirk, P. M. and Samson, R. A. 1985. Observations on Thamnidiaceae (Mucorales). III. Mycotyphaceae fam. nov. and a re-evaluation of *Mycotypha* sensu Benny & Benjamin illustrated by two new species. Mycotaxon, 22: 119-148.

Benny, G. L. and O'Donnell, K. 1978. *Pilaira anomala*, pp. 129-130. In: Lower Fungi in the Laboratory. M. S. Fuller (ed.). Palfrey Contributions in Botany No. 1. Department of Botany, University of Georgia, Athens, Georgia, U.S.A.

Benny, G. L. and O'Donnell, K. 2000. *Amoebidium parasiticum* is a protozoan, not a trichomycete. Mycologia, 92: 1133-1137.

Benny, G. L. and Schipper, M. A. A. 1992. Observations on Thamnidiaceae (Mucorales). IV. *Pirella*. Mycologia, 84: 52-63.

Ben-Ze'ev, I. and Kenneth, R. G. 1982. Features-criteria of taxonomic value in the Entomophthorales. I. A revision of the batkoan classification. Mycotaxon, 11: 393-455.

Ben-Ze'ev, I. S., Kenneth, R. G. and Uziel, A. 1987. A reclassification of *Entomophthora turbinata* in *Thaxterosporium* gen. nov., Neozygitaceae fam. nov. (Zygomycetes: Entomophthorales). Mycotaxon, 28: 313-326.

Bessey, E. A. 1950. Morphology and Taxonomy of Fungi. The Blakiston Company, Philadelphia. U.S.A.

Blakeslee, A. F. 1905. Two conidia-bearing fungi. *Cunninghamella* and *Thamnocephalis* n. gen. Bot. Gaz., 40: 161-170.

Boedijn, K. B. 1958 [1959]. Notes on the Mucorales of Indonesia. Sydowia, 12: 321-362.

Buller, A. H. R. 1934. *Pilobolus* and the ocellus function of its subsporangial swelling, pp. 47-168. In: Researches on Fungi, Vol. 6. A. H. R. Buller (ed.). Longmans Green and Company London, U.K.

Cafaro, M. J. 1999. *Baltomyces*, a new genus of gut-inhabiting fungus in an isopod. Mycologia, 91: 517-519.

Cafaro, M. J. 2005. Eccrinales (Trichomycetes) are not fungi, but a clade of protists at the early divergence of animals and fungi. Mol. Phylogenet. Evol., 35: 21-34.

Canning, E. U. 1990. Phylum Microspora, pp. 53-72. In: Handbook of Protoctista. L. Margulis, J. O. Corliss, M. Melkonian and D. J. Chapman (eds.). Jones and Barlett Publishers, Boston, Massachussetts, U.S.A.

Cannon, P. F. and Kirk, P. M. 2007. Fungal Families of the World. CAB International, Egham, Surrey, U.K.

Castañeda Ruiz, R. F. and Kendrick, B. 1993. *Aenigmatomyces*, an enigmatic new genus of fungi from Algonquin Park, Canada. Mycologia, 85: 1023-1027.

Cavalier-Smith, T. 1993. Kingdom Protozoa and its 18 Phyla. Microbiol. Rev., 57: 953-994.

Cavalier-Smith, T. 1998. A revised six-kingdom system of life. Biol. Rev., 73: 203-266.

Chalabuda, T. V. 1968. Systematica familiae Mortierellaceae. Nov. Syst. Plant. non Vascul., 1968: 120-131.

Chesters, C. G. C. 1933. *Azygozygum chlamydosporum* nov. gen. et sp. A phycomycete associated with a diseased condition of *Antirrhinum majus*. T. Brit. Mycol. Soc., 18: 199-214.

Chien, C.-Y. and Hwang, B.-C. 1997. First record of the occurrence of *Sporodiniella umbellata* (Mucorales) in Taiwan. Mycoscience, 38: 343-346.

Cole, G. T. and Samson, R. A. 1979. Patterns of Development in Conidial Fungi. Pitman Publishing Limited, London, U.K.

Csernetics, Á., Péteri, Z., Linka, B. and Takó, M. 2005. Physiological and genetic variability of Zygomycetes causing post-harvest decay. Acta. Phytopathol. Hun., 40: 267-277.

Dayal, R. 1973/1974. Key to Phycomycetes predaceous or parasitic in nematodes or amoebae. I. Zoopagales. Sydowia, 27: 293-301.

Degawa, Y. 2002a. 655. Elucidating the 'enigma' of *Aenigmatomyces ampullisporus* which is now found as a zygomycetous fungus parasitizes a spermatophore of collembolan insect, p. 198. In: IMC7 Book of Abstracts. Organizing Committee (eds.). International Mycological Congress 7, Oslo, Norway.

Degawa, Y. 2002b. On *Aenigmatomyces ampullisporus*. Newslet. Kanagawa Prefect. Mus. Nat. Hist., 8(4): 1-4. [nh.kanagawa-museum.jp/tobira/8-4/degawa.html (20 August 2008)]

Degawa, Y. and Gams, W. 2004. A new species of *Mortierella*, and an associated sporangiiferous mycoparasite in a new genus, *Nothadelphia*. Stud. Mycol., 50: 567-572.

Degawa, Y. and Tokumasu, S. 1997. Zygospore formation in *Mortierella capitata*. Mycoscience, 38: 387-394.

Degawa, Y. and Tokumasu, S. 1998a. Two new homothallic species of *Mortierella, M. cogitans,* and *M. microzygospora*, and their zygospore formation. Mycologia, 90: 1040-1046.

Degawa, Y. and Tokumasu, S. 1998b. Zygospore formation in *Mortierella umbellata*. Mycol. Res., 102: 593-598.

De Hann, A., Geraats, S. G. M., Kruyssen, F. J. and Bijl, H. L. 2000. Mucorales fungi for use in preparation of foodstuffs. (http://www.wipo.int/pctdb/en/wo.jsp?wo=2000015045&IA =EP1999007722&DISPLAY=STATUS).

Dick, M. W. 1990. The systematic position of *Zoophagus insidians*. Mycol. Res., 94: 347-354.

Dobbs, C. G. 1938. The life history and morphology of *Dicranophora fulva* Schröt. T. Brit. Mycol. Soc., 21: 167-192.

Dobbs, C. G. 1942. On the primary dispersal and isolation of fungal spores. New Phytol., 41: 63-69.

Drechsler, C. 1934. A new species of *Helicocephalum*. Mycologia, 26: 33-37.

Drechsler, C. 1935a. Some conidial Phycomycetes destructive to terricolous amoebae. Mycologia, 27: 6-40.

Drechsler, C. 1935b. Some non-catenulate conidial Phycomycetes preying on terricolous amoebae. Mycologia, 27: 176-205.

Drechsler, C. 1936. New conidial Phycomycetes destructive to terricolous amoebae. Mycologia, 28: 363-389.

Drechsler, C. 1938. New Zoopagaceae capturing and consuming soil amoebae. Mycologia, 30: 137-157.

Drechsler, C. 1939. Five new Zoopagaceae destructive to rhizopods and nematodes. Mycologia, 31: 388-415.

Drechsler, C. 1941. Four Phycomycetes destructive to nematodes and rhizopods. Mycologia, 33: 248-269.

Drechsler, C. 1942. New species of *Acaulopage* and *Cochlonema* destructive to soil amoebae. Mycologia, 34: 274-297.

Drechsler, C. 1943. A new non-helicoid bisporus *Helicocephalum* parasitizing nematode eggs. Mycologia, 35: 134-141.

Drechsler, C. 1946. Three new Zoöpagaceae subsisting on soil amoebae. Mycologia, 38: 120-143.

Drechsler, C. 1947. Three zoöpagaceous fungi that capture and consume soil-inhabiting rhizopods. Mycologia, 39: 253-281.

Drechsler, C. 1951. Various zoöpagaceous fungi subsisting on protozoans and eelworms. Mycologia, 43: 161-185.

Drechsler, C. 1955. A new species of *Rhopalomyces* occurring in Florida. B. Torrey Bot. Club, 82: 473-479.

Drechsler, C. 1959. Several Zoopagaceae subsisting on a nematode and some terricolous amoebae. Mycologia, 51: 787-823.

Drechsler, C. 1962. A nematode-capturing phycomycete with distally adhesive branches and proximally imbedded fusiform conidia. Am. J. Bot., 49: 1089-1095.

Duboscq, O., Léger, L. and Tuzet, O. 1948. Contribution á la connaissance des Eccrinides: les Trichomycètes. Arch. Zool. Exp. Gen., 86: 29-144.

Duddington, C. L. 1973. Zoopagales, pp. 231-234. In: The Fungi. Vol. 4B. A Taxonomic Review with Keys: Basidiomycetes and Lower Fungi. G. C. Ainsworth, F. K. Sparrow and A. S. Sussman (eds.). Academic Press, New York, U.S.A.

Ellis, J. J. 1963. A study of *Rhopalomyces elegans* in pure culture. Mycologia, 55: 183-198.

Ellis, J. J. and Hesseltine, C. W. 1965. The genus *Absidia*: globose-spored species. Mycologia, 57: 222-235.

Ellis, J. J. and Hesseltine, C. W. 1966. Species of *Absidia* with ovoid sporangiospores. II. Sabouraud., 5: 59-77.

Ellis, J. J. and Hesseltine, C. W. 1974. Two new families of Mucorales. Mycologia, 66: 87-95.

Ellis, J. J., Rhodes, L. J. and Hesseltine, C. W. 1976. The genus *Amylomyces*. Mycologia, 68: 131-143.

Embree, R. W. and Indoh, H. 1967. *Aquamortierella*, a new genus in the Mucorales. B. Torrey Bot. Club, 94: 464-467.

Evans, H. C. and Samson, R. A. 1977. *Sporodiniella umbellata*, an entomogenous fungus of the Mucorales from cocoa farms in Ecuador. Can. J. Bot., 55: 2981-2984.

Ferrington, L. C. Jr., Lichtwardt, R. W., Hayford, B. and Williams, M. C. 2005. Symbiotic Harpellales (Trichomycetes) in Tasmanian aquatic insects. Mycologia, 97: 254-262.

Ferrington, L. C. Jr., White, M. M. and Lichtwardt, R. W. 2003. A new genus of Trichomycetes (*Pseudoharpella arcolamylica*) from *Dixia fluvica* Peters (Diptera: Dixidae). Aquat. Insects, 25: 85-94.

Fischer, A. 1892. Phycomycetes. In: Kryptogamen-Flora von Deutschland, Oesterreich und der Schweiz. Zweite Auflage. L. Rabenhorst (ed.) Band I, Abth. IV. Eduard Kummer, Leipzig, Germany.

Fischer, W. M. and Palmer, J. D. 2005. Evidence from small-subunit ribosomal RNA sequences for a fungal origin of Microsporidia. Mol. Phylogenet. Evol., 36: 606-622.

Fitzpatrick, H. M. 1930. The Lower Fungi. Phycomycetes. McGraw-Hill Book Co., Inc., New York, U.S.A.

Gams, W. 1976. Some new or noteworthy species of *Mortierella*. Persoonia, 9: 111-140.

Gams, W. 1977. A key to the species of *Mortierella*. Persoonia, 9: 381-391.

Gams, W. and Carreiro, M. M. 1989. Two new species of *Mortierella* and rediscovery of Thaxter's *Dissophora decumbens*. Stud. Mycol., 31: 85-91.

Gerdemann, J. W. and Trappe, J. M. 1974. The Endogonaceae in the Pacific Northwest. Mycologia Mem., 5: 1-76.

Gill, E. E. and Fast, N. M. 2006. Assessing the Microsporidia-Fungi relationship: combined phylogenetic analysis of eight genes. Gene, 375: 103-109.

Gräfenhan, T. 1998. Taxonomic revision of the genus *Piptocephalis* (Fungi). M.S. Thesis. Ernst-Moritz-Arndt-Universität, Greifswald, Germany.

Grove, W. B. 1934. A systematic account and arrangement of the Pilobolidae, pp. 190-224. In: Researches on Fungi. Vol. 6. A. H. R. Buller (ed.). Longmans Green and Company London, U.K.

Guo, L.-Y. and Michailides, T. J. 1998. Factors affecting the rate and mode of germination of *Mucor piriformis* zygospores. *Mycol. Res.*, 102: 815-819.

Hagem, O. 1908. Untersuchungen über Norwegische Mucorineen. I. Videnskabs-Selskabets Skrifter I. Math.-Naturv. Klasse, 1908: 5-50.

Hagström, E. 1977. *Vesiculomyces* Hagström gen. nov. segregated from *Gloeocystidiellum* (Corticiaceae). Bot. Notiser, 130: 53-54.

Hanlin, R. T. 1973. Keys to the Families, Genera and Species of the Mucorales. J. Cramer, Lehre, Lichtenstein.

Harper, J. E. and Webster, J. 1964. An experimental analysis of the coprophilous fungal succession. T. Brit. Mycol. Soc., 47: 511-530.

Hesseltine, C. W. 1955. Genera of Mucorales with notes on their synonymy. Mycologia, 47: 344-363.

Hesseltine, C. W. 1957. The genus *Syzygites* (Mucoraceae). Lloydia, 20: 228-237.

Hesseltine, C. W. 1965. A millennium of fungi, food, and fermentation. Mycologia, 57: 149-197.

Hesseltine, C. W. 1983. Microbiology of oriental fermented foods. Annu. Rev. Microbiol., 37: 575-601.

Hesseltine, C. W. and Anderson, P. 1957. Two genera of molds with low temperature growth requirements. B. Torrey Bot. Club, 84: 31-45.

Hesseltine, C. W., Benjamin, C. R. and Mehrotra, B. S. 1959. The genus *Zygorhynchus*. Mycologia, 51: 173-194.

Hesseltine, C. W. and Ellis, J. J. 1961. Notes on Mucorales, especially *Absidia*. Mycologia, 53: 406-426.

Hesseltine, C. W. and Ellis, J. J. 1964. The genus *Absidia: Gongronella* and cylindrical-spored species of *Absidia*. Mycologia, 56: 568-601.

Hesseltine, C. W. and Ellis, J. J. 1966. Species of *Absidia* with ovoid sporangiospores. I. Mycologia, 58: 761-785.

Hesseltine, C. W. and Ellis, J. J. 1973. Mucorales, pp. 187-217. In: The Fungi. Vol. 4b. A Taxonomic Review with Keys: Basidiomycetes and Lower Fungi. G. C. Ainsworth, F. K. Sparrow and A. S. Sussman (eds.). Academic Press, New York, U.S.A.

Hesseltine, C. W. and Fennell, D. I. 1955. The genus *Circinella*. Mycologia, 47: 193-212.

Hibbett, D. S., Binder, M., Bischoff, J. F., Blackwell, M., Cannon, P. F., Eriksson, O. E., Huhndorf, S., James, T., Kirk, P. M., Lücking, R., Lumbsch, H. T., Lutzoni, F., Matheny, P. B., McLaughlin, D. J., Powell, M. J., Redhead, S., Schoch, C. L., Spatafora, J. W., Stalpers, J. A., Vilgalys, R., Aime, M. C., Aptroot, A., Bauer, R., Begerow, D., Benny, G. L., Castelbury, L. A., Crous, P. W., Dai, Y. C., Gams, W., Geiser, D. M., Griffith, G. W., Gueidan, C., Hawksworth, D. L., Hestmark, G., Kosaka, K., Humber, R. A., Hyde, K. D., Ironside, J. E., Kõljalg, U., Kurtzman, C. P., Larsson, K.-H., Lichtwardt, R., Longcore, J., Miądlikowska, J., Miller, A., Moncalvo, J.-M., Mozley-Standridge, S., Oberwinkler, F., Parmasto, E., Reeb, V., Rogers, J. D., Roux, C., Ryvarden, L., Sampaio, J. P., Schüßler, A., Sugiyama, J., Thorn, R. G., Tibell, L., Untereiner, W. A., Walker, C., Wang, Z., Weir, A., Weiss, M., White, M. M., Winka, K., Yao, Y.-J. and Zhang, N. 2007. A higher-level classification of the fungi. Mycol. Res., 111: 509-547.

Ho, C.-Y. and Chang, L.-L. 2003. Notes on Zygomycetes of Taiwan. (III): Two *Blakesea* species (Choanephoraceae) new to Taiwan. Taiwania, 48: 232-238.

Ho, H.-M. and Benny, G. L. 2007. Two new species of *Syncephalis* from Taiwan, with a key to the *Syncephalis* species found in Taiwan. Bot. Stud., 48: 319-324.

Ho, H.-M. and Benny, G. L. 2008. A new species of *Syncephalis* from Taiwan. Bot. Stud., 49: 45-48.

Hoffmann, K., Discher, S. and Voigt, K. 2007. Revision of the genus *Absidia* (Mucorales, Zygomycetes) based on physiological, phylogenetic, and morphological characters; thermotolerant *Absidia* spp. form a coherent group, Mycocladiaceae fam. nov. Mycol. Res., 111: 1169-1183.

Hoffmann, K. and Voigt, K. 2009. *Absidia parricida* plays a dominant role in biotrophic fusion parasitism among mucoralean fungi (Zygomycetes): *Lentamyces*, a new genus for *A. parricida* and *A. zychae*. Plant Biology, 11: 537-554. (early view) (DOI: 10.1111/j.1438-8677.2008.00145.x).

Hoffmann, K., Walther, G. and Voigt, K. 2009. *Mycocladus* vs. *Lichtheimia*: a correction (Lichtheimiaceae fam. nov., Mucorales, Mucoromycotina). Mycol. Res., 113: 277-278.

Hu, F.-m., Zheng, R.-y. and Chen, G.-q. 1989. A redelimitation of the species of *Pilobolus*. Mycosystema, 2: 111-133.

Humber, R. A. 1976. The systematics of the genus *Strongwellsea* (Zygomycetes: Entomophthorales). Mycologia, 68: 1042-1060.

Humber, R. A. 1981a. An alternative view of certain taxonomic criteria used in the Entomophthorales. Mycotaxon, 13: 191-240.

Humber, R. A. 1981b. *Erynia* (Zygomycetes: Entomophthorales): validations and new species. Mycotaxon, 13: 471-480.

Humber, R. A. 1982. *Strongwellsea* vs. *Erynia*: The case for a phylogenetic classification of the Entomophthorales (Zygomycetes). Mycotaxon, 15: 167-184.

Humber, R.A. 1984. *Eryniopsis*, a new genus of the Entomophthoraceae (Entomophthorales). Mycotaxon, 21: 257-264.

Humber, R. A. 1989. Synopsis of a revised classification for the Entomophthorales (Zygomycotina). Mycotaxon, 34: 441-460.

Indoh, H. 1962. Studies on Japanese Mucorales. I. On the genus *Syncephalis*. Sci. Rep. Tokyo Bunrika Daigaku, Sect B, 11: 1-26.

Indoh, H. 1967. On an aquatic "mucor" found in Japan. T. Mycol. Soc. Jpn., 8: 28.

Ingold, C. T. 1961. The stalked spore-drop. New Phytol., 60: 181-183.

Ingold, C. T. 1971. Spore liberation in Mucorales, pp. 61-91, In: Fungal Spores: Their Liberation and Dispersal. C. T. Ingold (ed.). Clarendon Press, Oxford, U.K.

Ingold, C. T. and Zoberi, M. H. 1964. The asexual apparatus of Mucorales in relation to spore liberation. T. Brit. Mycol. Soc., 46: 115-134.

Jackson, H. S. and Dearden, E. R. 1948. *Martensella corticii* Thaxter and its distribution. Mycologia, 40: 168-176.

James, T. Y., Kauff, F., Schoch, C. L., Matheny, P. B., Hofstetter, V., Cox, C. J., Celio, G., Gueidan, C., Fraker, E., Miadlikowska, J., Lumbsch, H. T., Rauhut, A., Reeb, V., Arnold, A. E., Amtoft, A., Stajich, J. E., Hosaka, K., Sung, G.-H., Johnson, D., O'Rourke, B., Crockett, M., Binder, M., Curtis, J. M., Slot, J. C., Wang, Z., Wilson, A. W., Schüßler, A., Longcore, J. E., O'Donnell, K., Mozley-Standridge, S., Porter, D., Letcher, P., Powell, M. J., Taylor, J. W., White, M. M., Griffith, G. W., Davies, D. R., Humber, R. A., Morton, J. B., Sugiyama, J., Rossman, A. Y., Rogers, J. D., Pfister, D. H., Hewitt, D., Hansen, K., Hambleton, S., Shoemaker, R. A., Kohlmeyer, J., Volkmann-Kohlmeyer, B., Spotts, R. A., Serdani, M., Crous, P. W., Hughes, K. W., Matsuura, K., Langer, E., Langer, G., Untereiner, W. A., Lücking, R., Büdel, B., Geiser, D. M., Aptroot, A., Diederich, P., Schmitt, I., Schultz, M., Yahr, R., Hibbett, D. S., Lutzoni, F., McLaughlin, D. J., Spatafora, J. W. and Vilgalys, R. 2006. Reconstructing the early evolution of Fungi using a six-gene phylogeny. Nature, 443: 818-822.

Jeffries, P. and Young, T. W. K. 1978. Ultrastructure of the dormant and germinating sporangiospore of *Phascolomyces articulosus* (Mucorales). Can. J. Bot., 56: 747-753.

Jennessen, J., Schnürer, J., Olsson, J., Samson, R. A. and Dijksterhuis, J. 2008. Morphological characteristics of sporangiospores of the tempe fungus *Rhizopus oligosporus* differentiate it from other taxa of the *R. microsporus* group. Mycol. Res., 112: 547-563.

Jensen, A. B., Gargas, A., Eilenberg, J. and Rosendahl, S. 1998. Relationships of the insect-pathogenic order Entomophthorales (Zygomycota, Fungi) based on phylogenetic analysis of nuclear small subunit ribosomal DNA sequences (SSU rDNA). Fungal Genet. Biol., 24: 325-334.

Jong, S.-C. and Birmingham, J.M. 2001. Cultivation and preservation of fungi in culture, pp. 193-202. In: The Mycota. Vol. VIIB. Systematics and Evolution. D. J. McLaughlin, E. G. McLaughlin and P. A. Lemke (eds.). Springer-Verlag, Berlin, Germany.

Kanouse, B. B. 1936. Studies of two species of *Endogone* in culture. Mycologia, 28: 47-62.

Karling, J. S. 1936. A new predaceous fungus. Mycologia, 28: 307-320.

Keeling, P. J. 2003. Congruent evidence from α-tubulin and β-tubulin gene phylogenies for a zygomycete origin of Microsporidia. Fungal Genet. Biol., 38: 298-309.

Keeling, P. J. and Fast, N. M. 2002. Microsporidia: Biology and evolution of highly reduced intracellular parasites. Annu. Rev. Microbiol., 56: 93-116.

Keller, S. 1987. Arthropod-pathogenic Entomophthorales of Switzerland. I. *Conidiobolus, Entomophaga* and *Entomophthora*. Sydowia, 40: 122-167.

Keller, S. 1991. Arthropod-pathogenic Entomophthorales of Switzerland. II. *Erynia, Eryniopsis, Neozygites, Zoophthora*, and *Tarichium*. Sydowia, 43: 39-122.

Keller, S. 1993. Taxonomic considerations on some species of *Erynia* (Zygomycetes, Entomophthorales) attacking flies (Diptera). Sydowia, 45: 252-263.

Keller, S. 1997. The genus *Neozygites* (Zygomycetes, Entomophthorales) with special reference to species found in tropical regions. Sydowia, 49: 118-146.

Keller, S. 2002. The genus *Entomophthora* (Zygomycetes, Entomophthorales) with a description of five new species. Sydowia, 54: 157-197.

Keller, S. 2004. Validation of the description of some species of *Entomophthora* (*Zygomycetes, Entomophthorales*). Sydowia, 56: 241-242.

Keller, S. 2006a. Entomophthorales attacking aphids with a description of two new species. Sydowia, 58: 38-74.

Keller, S. 2006b. Validation of the combination *Apterivorax acaricida* (Petch) S. Keller. Sydowia, 58: 75.

Keller, S. 2007. Arthropod-pathogenic Entomophthorales from Switzerland. III. First additions. Sydowia, 59: 75-113.

Keller, S. and Eilenberg, J. 1993. Two new species of Entomophthoraceae (Zygomycetes, Entomophthorales) linking the genera *Entomophaga* and *Eryniopsis*. Sydowia, 45: 264-274.

Keller, S., Kalsberg, V. and Eilenberg, J. 1999. Redescription of *Entomophthora muscae* (Cohn) Fresenius. Sydowia, 51: 197-209.

Keller, S. and Petrini, O. 2005. Keys to the identification of the arthropod pathogenic genera of the families Entomophthoraceae and Neozygitaceae (Zygomycetes), with descriptions of three new subfamilies and a new genus. Sydowia, 57: 23-53.

Khan, Z. U., Ahmad, S., Mokaddas, E., Chandy, R., Cano, J. and Guarro, J. 2008. *Actinomucor elegans* var. *kuwaitiensis* isolated from the wound of a diabetic patient. Anton. van Lee. J. M. S., 94: 343-352.

King, D. S. 1976a. Systematics of *Conidiobolus* (Entomophthorales) using numerical taxonomy. I. Biology and cluster analysis. Can. J. Bot., 54: 45-65.

King, D. S. 1976b. Systematics of *Conidiobolus* (Entomophthorales) using numerical taxonomy. II. Taxonomic considerations. Can. J. Bot., 54: 1285-1296.

King, D. S. 1977. Systematics of *Conidiobolus* (Entomophthorales) using numerical taxonomy. III. Descriptions of recognized species. Can. J. Bot., 55: 718-729.

Kirk, P. M. 1984. A monograph of the Choanephoraceae. Mycol. Pap., 152: 1-61.

Kirk, P. M. 1986. (815) Proposal to conserve *Mucor* Fresenius over *Mucor* Micheli ex L. and (816) Proposal to conserve *Rhizopus* Ehrenberg over *Ascophora* Tode (Fungi) with notes on the nomenclature and taxonomy of *Mucor, Ascophora, Hydrophora* and *Rhizopus*. Taxon, 35: 371-377.

Kirk, P. M. 1989. A new species of *Benjaminiella* (Mucorales: Mycotyphaceae). Mycotaxon, 35: 121-125.

Kirk, P. M. and Benny, G. L. 1980. The genus *Utharomyces* Boedijn (Pilobolaceae: Zygomycetes). T. Brit. Mycol. Soc., 75: 123-131.

Kirk, P. M., Cannon, P. F., Davis, J. C. and Stalpers, J. A. [eds.]. 2001. Ainsworth & Bisby's Dictionary of the Fungi. 9th edn. CAB International, Wallingford, U.K.

Kirk, P. M., Cannon, P. F., Minter, D. W. and Stalpers, J. A. [eds.]. 2008. Ainsworth & Bisby's Dictionary of the Fungi. 10th edn. CAB International, Wallingford, U.K.

Kovacs, R. L. and Sundberg, W. L. 1999. *Syzygites megalocarpus* (Mucorales, Zygomycetes) in Illinois. T. Illinois Acad. Sci., 93: 181-190.

Kreisel, H. 1969. Grunzüge eines natürlichen Systems der Pilze. J. Cramer, Lehre, Lichtenstein.

Kurihara, Y. and Degawa, Y. 2006. *Pinnaticoemansia*, a new genus of Kickxellales, with a revised key to the genera of Kickxellales. Mycoscience, 47: 205-211.

Kurihara, Y., Degawa, Y. and Tokumasu, S. 2001. A new genus *Myconymphaea* (Kickxellales), with peculiar septal plugs. Mycol. Res., 105: 1397-1402.

Kurihara, Y., Degawa, Y. and Tokumasu, S. 2004. Two novel kickxellalean fungi, *Mycoëmilia scoparia* gen. sp. nov. and *Ramicandelaber brevisporus* sp. nov. Mycol. Res., 108: 1143-1152.

Kurihara, Y., Ogawa, Y., Degawa,Y. and Tokumasu, S. 2005. A proposal for the division of the order Kickxellales based on the comparison with its related orders. MSA Abstracts. Inoculum, 56(4): 34.

Kurihara, Y., Sukarno, N., Ilyas, M., Yuniarti, E., Mangunwardoyo, W., Park, J.-Y., Saraswati, R., Widyastuti, Y. and Ando, K. 2008. Indonesian Kickxellales: Two species of *Coemansia* and *Linderina*. Mycoscience, 49: 250-257.

Kurihara, Y., Tokumasu, S. and Chien, C.-Y. 2000. *Coemansia furcata* sp. nov. and its distribution in Japan and Taiwan. Mycoscience, 41: 579-583.

Kwaśna, H. and Nirenberg, H. I. 2008a. *Siepmannia*, a new genus in the Mucoraceae. Mycologia, 100: 259-274.

Kwaśna, H. and Nirenberg, H. I. 2008b. Validation of the genus *Siepmannia* (Mucoraceae) and its four species. Pol. Bot. J., 53: 187-188.

Kwaśna, H., Richardson, M. J. and Bateman, G. L. 2002. Morphological variation in the *Coemansia spiralis* complex. Mycol. Res., 106: 252-256.

Lendner, A. 1908. Les Mucorinées de la Suisse. Beiträge zur Kryptogamenflora der Schweiz. III(I). K.-J. Wyss, Bern, Switzerland.

Lichtwardt, R. W. 1972. Undescribed genera and species of Harpellales (Trichomycetes) from the guts of aquatic insects. Mycologia, 64: 167-197.

Lichtwardt. R. W. 1973a. The Trichomycetes: what are their relationships? Mycologia, 65: 1-20.

Lichtwardt, R. W. 1973b. Trichomycetes, pp. 237-243. In: The Fungi: An Advanced Treatise. Vol. IVB. G. C. Ainsworth, F. K. Sparrow and A. S. Sussman. (eds.). Academic Press, Inc., New York, U.S.A.

Lichtwardt, R. W. 1983. *Gauthieromyces*, a new genus of Harpellales based on *Genistella microspora*. Mycotaxon, 17: 213-215.

Lichtwardt, R. W. 1984. Species of Harpellales living within the guts of aquatic Diptera larvae. Mycotaxon, 19: 529-550.

Lichtwardt, R. W. 1986. The Trichomycetes. Fungal Associates of Arthropods. Springer-Verlag, New York.

Lichtwardt, R. W. 1997 [1998]. Costa Rican gut fungi (Trichomycetes) infecting lotic insect larvae. Rev. Biol. Trop., 45: 1349-1383.

Lichtwardt, R. W. and Arenas, J. 1996. Trichomycetes in aquatic insects from southern Chile. Mycologia, 88: 844-857.

Lichtwardt, R. W., Cafaro, M. J. and White, M. M. 2001a. The Trichomycetes. Fungal Associates of Arthropods. Revised Edition. (http://www.nhm.ku.edu/fungi/monograph/mono.htm/) (2 April 2009)

Lichtwardt, R. W., Ferrington, L. C., Jr. and López Lastra, C. 1999. Trichomycetes in Argentinean aquatic insect larvae. Mycologia, 91: 1060-1082.

Lichtwardt, R. W., Kobayasi, Y. and Indoh, H. 1987. Trichomycetes of Japan. T. Mycol. Soc., Jpn., 28: 359-412.

Lichtwardt, R. W. and Manier, J.-F. 1978. Validation of the Harpellales and Asellariales. Mycotaxon, 7: 441-442.

Lichtwardt, R. W. and Moss, S. T. 1981. Vegetative propagation in a new species of Harpellales, *Graminella microspora*. T. Brit. Mycol. Soc., 76: 311-316.

Lichtwardt, R. W. and Moss, S. T. 1984a. *Harpellomyces eccentricus*, an unusual Harpellales from Sweden and Wales. Mycotaxon, 20: 511-517.

Lichtwardt, R. W. and Moss, S. T. 1984b. New Asellariales (Trichomycetes) from the hindguts of aquatic isopods and springtails. Mycotaxon, 20: 259-274.

Lichtwardt, R. W., Peterson, S. W. and Huss, M. J. 1991a. *Orphella hiemalis*: a new and rare trichomycete occurring in winter-emerging stoneflies (Plectoptera, Capniidae). Mycologia, 83: 214-219.

Lichtwardt, R. W., Peterson, S. W. and Williams, M. C. 1991b. *Ejectosporus*, an unusual new genus of Harpellales in winter-emerging stonefly nymphs (Capniidae) and a new species of *Paramoebidium* (Amoebidiales). Mycologia, 83: 389-396.

Lichtwardt, R. W., White, M. M. and Colbo, M. H. 2001b. Harpellales in Newfoundland aquatic insect larvae. Mycologia, 93: 764-773.

Lichtwardt, R. W. and Williams, M. C. 1983. A new *Legeriomyces* (Harpellales) with variable trichospore size. Mycologia, 75: 757-761.

Lichtwardt, R. W. and Williams, M. C. 1984. *Zygopolaris borealis*, a new gut fungus (Trichomycetes) living in aquatic mayfly larvae. Can. J. Bot., 62: 1283-1286.

Lichtwardt, R. W. and Williams, M. C. 1988. Discovery of sexual reproduction in an unusual species of *Stachylina* (Trichomycetes). Mycologia, 80: 400-405.

Lichtwardt, R. W. and Williams, M. C. 1990. Trichomycete gut fungi in Australian aquatic insect larvae. Can. J. Bot., 68: 1057-1074.

Lichtwardt, R.W. and Williams, M.C. 1992a. *Furculomyces*, a new homothallic genus of Harpellales (Trichomycetes) from Australian midge larvae. Can. J. Bot., 70: 1196-1198.

Lichtwardt, R. W. and Williams, M. C. 1992b. Tasmanian trichomycete gut fungi in aquatic insect larvae. Mycologia, 84: 384-391.

Linder, D. H. 1943. The genera *Kickxella*, *Martensella*, and *Coemansia*. Farlowia, 1: 49-77.

Linnemann, G. 1969. I. *Mortierella* Coemans 1863, pp. 155-241. In: Mucorales eine Bescreibung aller Gattungen und Arten dieser Pilzgruppe. H. Zycha, R. Siepmann and G. Linnemann (eds.). J. Cramer, Lehre, Lichtenstein.

Liou, G.-Y., Chen, S.-R., Wei, Y.-H., Lee, F.-L., Fu, H.-M., Yuan, G.-F. and Stalpers, J. A. 2007. Polyphasic approach to the taxonomy of the *Rhizopus stolonifer* group. Mycol. Res., 111: 196-203.

Liu, X.-Y., Huang, H. and Zheng, R.-Y. 2007. Molecular phylogenetic relationships within *Rhizopus* based on combined analysis of ITS rDNA and pyrG gene sequences. Sydowia, 59: 235-253.

Liu, Y. J., Hodson, M. C. and Hall, B. D. 2006. Loss of the flagellum happened only once in the fungal lineage: phylogenetic structure of kingdom Fungi inferred from RNA polymerase II subunit genes. BMC. Evol. Biol., 6: 74, 13 p. doi: 10.1186/1471-2148-6-74.

Longcore, J. E. 1989. *Bojamyces repens*: a new genus and species of Harpellales (Trichomycetes) from a lentic mayfly. Mycologia, 81: 482-486.

Lumsden, R. D. 1987. Charles Drechsler, 1892-1896. Mycologia, 79: 345-352.

Lutzoni, F., Kauff, F., Cox, C. J., McLaughlin, D. J., Celio, G., Dentinger, B., Padamsee, M., Hibbett, D., James, T. Y., Baloch, E., Grube, M., Reeb, V., Hofstetter, V., Schoch, C., Arnold, A. E., Miadlikowska, J., Spatafora, J., Johnson, D., Hambleton, S., Crockett, M., Shoemaker, R., Sung, G.-H., Lücking, R., Lumbsch, T., O'Donnell, K., Binder, M., Diederich, P., Ertz, D., Gueidan, C., Hansen, K., Harris, R. C., Hosaka, K., Lim, Y.-W., Matheny, B., Nishida, H., Pfister, D., Rogers, J., Rossman, A., Schmitt, I., Sipman, H., Stone, J., Sugiyama, J., Yahr, R. and Vilgalys, R. 2004. Assembling the fungal tree of life: progress, classification, and evolution of subcellular traits. Am. J. Bot., 91: 1446-1480.

MacLeod, D. M. and Müller-Kögler, E. 1970. Insect pathogens: species originally described from their resting spores mostly as *Tarichium* species (Entomophthorales: Entomophthoraceae). Mycologia, 62: 33-66.

Mahoney, D., Gams, W., Meyer, W. and Starink-Willemse, M. 2004. *Umbelopsis dimorpha* sp. nov., a link between *U. vinacea* and *U. versiformis*. Mycol. Res., 108: 107-111.

Malloch, D. 1967. A new genus of Mucorales. Mycologia, 59: 326-329.

Manier, J.-F. 1950 (1951). Recherches sur les Trichomycètes. Ann. Sci. Nat. Bot., Sér XI, 11: 53-162.

Manier, J.-F. 1962a. Présence de Trichomycètes dans le rectum des larves d'ephémères des Torrents du Massif de Néovieille (Hautes-Pyrénées). B. Soci. Hist. Nat. Toulouse, 97: 241-254.

Manier, J.-F. 1962b. Revision du genre *Spartiella* Tuzet et Manier 1950 (sa place dans la classe des Trichomycètes). Ann. Sci. Nat. Zool., Sér. XII, 4: 517-525.

Manier, J.-F. 1968. Validation de Trichomycètes par leur diagnose Latine. Ann. Sci. Nat. Bot., Sér. XII, 9: 93-108.

Manier, J.-F. 1969 (1970). Trichomycètes de France. Ann Sci Nat Bot, Sér XII, 10: 565-672.

Manier, J.-F. and Lichtwardt, R. W. 1968 [1969]. Révision de la systématique de la Trichomycètes. Ann. Sci. Nat. Bot., Sér. XII, 9: 519-532.

McLaughlin, D. J., McLaughlin, E. G. and Lemke, P. A. (eds.). 2001a. The Mycota. Vol. VIIA. Systematics and Evolution. Part A. Spinger-Verlag, Berlin, Germany.

McLaughlin, D. J., McLaughlin, E. G. and Lemke, P. A. (eds.). 2001b. The Mycota. Vol. VIIB. Systematics and Evolution. Part B. Spinger-Verlag, Berlin, Germany.

McNeill, J., Barrie, F. R., Burdet, H. M., Demoulin, V., Hawksworth, D. L., Marhold, K., Nicolson, D. H., Prado, J., Silva, P. C., Skog, J. E., Wiersema, J. H. and Turland, N. J. (eds.). 2006. International Code of Botanical Nomenclature (Vienna Code). [Regnum Vegetabile Vol. 126]. A.R.G. Gantner Verlag KG, Ruggell, Liechtenstein.

Mendoza, L., Taylor, J. W. and Ajello, L. 2002. The class Mesomycetozoa: a heterogeneous group of microorganisms at the animal-fungal boundary. Annu. Rev. Microbiol., 56: 315-344.

Meyer, J. 1957. *Martensiomyces pterosporus* nov. gen. nov. sp. nouvelle Kickxellacée isolee du sol. B. Soc. Mycol. Fr., 73: 189-201.

Meyer, W. and Gams, W. 2003. Delimitation of *Umbelopsis* (Mucorales, Umbelopsidaceae fam. nov.) based on ITS sequence and RFLP data. Mycol. Res., 107: 339-350.

Michailides, T. J. and Spotts, R. A. 1990. Post-harvest diseases of pome and stone fruits caused by *Mucor piriformis* in the Pacific Northwest and California. Plant Dis., 74: 537-543.

Mikawa, T. 1975. Materials for the fungus flora of Japan (18). T. Mycol. Soc. Jpn., 16: 146-148.

Mikawa, T. 1976. Materials for the fungus flora of Japan (21). T. Mycol. Soc. Jpn., 17: 4-8.

Mil'ko, A. A. 1974. Identification of Mucoralean Fungi. Acad. Sci. Ukrainian S. S. R., Kiev. (In Russian.)

Mirza, J. H., Khan, S. M., Begum, S. and Shagufta, S. 1979. Mucorales of Pakistan. University of Agriculture: Faisalabad, Pakistan.

Misra, J. K. and Tiwari, V. K. 2008. A new species *Gauthieromyces* and range extension for other Harpellales to India. Mycologia, 100: 94-98.

Misra, J. K., White, M. M. and Lichtwardt, R. W. 1999. *Furculomyces septentrionalis* reveals an unexpected distribution for this genus of Harpellales. Mycologia, 91: 703-706.

Misra, P. C., Gupta, N. N. and Lata, K. 1979. A new species of *Fennellomyces* (Mucorales). Mycotaxon, 10: 251-254.

Misra, P. C., Srivastava, K. J. and Lata, K. 1979. *Apophysomyces*, a new genus of the Mucorales. Mycotaxon, 8: 377-382.

Miura, K. 1973. Three entomophthoralean parasites of nematodes collected from Japan. Rep. Tottori Mycol., I 10: 517-522.

Morton, J. B. and Benny, G. L. 1990. Revised classification of arbuscular mycorrhizal fungi (Zygomycetes): a new order, Glomales, two new suborders, Glomineae and Gigasporineae and two new families, Acaulosporaceae and Gigasporaceae, with an emendation of Glomaceae. Mycotaxon, 37: 471-491.

Moss, S. T., Lichtwardt, R. W. and Manier, J.-F. 1975. *Zygopolaris*, a new genus of Trichomycetes producing zygospores with polar attachment. Mycologia, 67: 120-127.

Moss, S. T. and Young, T. W. K. 1978. Phyletic considerations of the Harpellales and Asellariales (Trichomycetes, Zygomycotina) and the Kickxellales (Zygomycetes, Zygomycotina). Mycologia, 70: 944-963.

Mueller, G. M., Bills, G. F. and Foster, M. S. (eds.). 2004. Biodiversity of Fungi. Inventory and Monitoring Methods. Elsevier Academic Press, Amsterdam, The Netherlands.

Nagahama, T., Sato, H., Shimazu, M. and Sugiyama, J. 1995. Phylogenetic divergence of the entomophthoralean fungi: evidence from nuclear 18S RNA gene sequences. Mycologia, 87: 203-209.

Nakasone, K. K., Peterson, S. W. and Jong, S.-C. 2004. Preservation and Distribution of Cultures, pp. 37-47. In: Biodiversity of Fungi: Inventory and Monitoring Methods. G.

M. Mueller, G. F. Bills and M. S. Foster (eds.). Elsevier Academic Press, Amsterdam, The Netherlands.

Naumov, N. A. 1935. Opredelitel Mukorovykh (Mucorales). 2nd edn. Bot. Inst. Acad. Sci. U.S.S.R., Moscow and Leningrad. Russia.

Naumov, N. A. 1939. Clés des Mucorinées. Encylopédie Mycologique, 9: 1-137.

Nout, M. J. R. and Aidoo, K. E. 2002. Asian fungal fermented food, pp. 23-47. In: The Mycota. Vol. X. Industrial Applications. H. D. Osiewacz (ed.). Springer-Verlag, Berlin, Germany.

O'Donnell, K. L. 1979. Zygomycetes in Culture. Palfrey Contributions in Botany. No. 2. Department of Botany, University of Georgia, Athens, Georgia.

O'Donnell, K., Cigelnik, E. and Benny, G. L. 1998. Phylogenetic relationships among the Harpellales and Kickxellales. Mycologia, 90: 624-639.

O'Donnell, K., Lutzoni, F. M., Ward, T. J. and Benny, G. L. 2001. Evolutionary relationships among mucoralean fungi (Zygomycota): Evidence for family polyphyly on a large scale. Mycologia, 93: 286-296.

Oehl, F., de Souza, F. A. and Sieverding, E. 2008. Revision of *Scutellospora* and description of five new genera and three new families in the arbuscular mycorrhiza-forming Glomeromycetes. Mycotaxon, 106: 311-360.

Ogawa, Y., Hayashi, S., Degawa, Y. and Yaguchi, Y. 2001. *Ramicandelaber*, a new genus of the Kickxellales, Zygomycetes. Mycoscience, 42: 193-199.

Osiewacz, H. D. 2002. (ed.). The Mycota. Vol. X. Industrial Applications. Springer-Verlag, Berlin, Germany.

Page, R. M. 1964. Sporangium discharge in *Pilobolus*: a photographic study. Science, 146: 925-927.

Peterson, S. W. and Lichtwardt, R. W. 1983. *Capniomyces stellatus* and *Simuliomyces spica*: new taxa of Harpellales (Trichomycetes) from winter-emerging stoneflies. Mycologia, 75: 242-250.

Peterson, S. W., Lichtwardt, R. W. and Horn, B. W. 1981. *Genistelloides hibernus*: a new trichomycete from a winter-emerging stonefly. Mycologia, 73: 477-485.

Pidoplichko, N. M. and Mil'ko, A. A. 1971. Atlas mukoral'vykh gribov [Atlas of the Mucorales]. Izdat. 'Naukova Dumka,' Kiev, Ukraine.

Pouzar, Z. 1972. *Genestella* Léger et Gauthier vs. *Genestella* Ortega; a nomenclatural note. Folia Geobota. Phytotax., 7: 319-320.

Raper, K. B. and Fennell, D. I. 1952. Two noteworthy fungi from Liberian soil. Am. J. Bot., 39: 79-86.

Remaudière, G. and Keller, S. (as Remaudiere). 1980. Revision systematique des genres d'Entomophthoraceae a potentiale entomopàthogene. Mycotaxon, 11: 323-338.

Ribes, J. A., Vanover-Sams, C. L. and Baker, D. J. 2000. Zygomycetes in human disease. Clin. Microbiol. Rev., 13: 236-301.

Saikawa, M. and Aoki, Y. 1995. Taxonomic studies on *Euryancale marsipospora* and *E. saccispora*. Nova Hedwigia, 60: 571-581.

Sakai. S., Kato, M. and Nagamasu, H. 2000. *Artocarpus* (Moraceae)—gall midge pollination mutualism mediated by male-flower parasitic fungus. Am. J. Bot., 87: 440-445.

Saksena, S. B. 1953. A new genus of the Mucorales. Mycologia, 45: 426-436.

Santamaria, S. 1997. *Lancisporomyces*, a new genus of Trichomycetes with lance-shaped zygospores. Mycologia, 89: 639-642.

Santamaria, S. and Girbal, J. 1998. Two new species of *Orphella* from Spain. Mycol Res., 102: 174-178.

Schipper, M. A. A. 1973. A study on variability in *Mucor hiemalis* and related species. Stud. Mycol., 4: 1-40.

Schipper, M. A. A. 1975. *Mucor mucedo, Mucor flavus* and related species. Stud. Mycol., 10: 1-33.

Schipper, M. A. A. 1976. *On Mucor circinelloides, Mucor racemosus* and related species. Stud. Mycol., 12: 1-40.

Schipper, M. A. A. 1978a. On certain species of *Mucor* with a key to all accepted species. Stud. Mycol., 17: 1-52.

Schipper, M. A. A. 1978b. On the genera *Rhizomucor* and *Parasitella*. Stud. Mycol., 17: 53-71.

Schipper, M. A. A. 1979. *Thermomucor* (Mucorales). Anton. van Lee. J. M. S., 45: 275-280.

Schipper, M. A. A. 1984. A revision of the genus *Rhizopus*. I. The *Rhizopus stolonifer*-group and *Rhizopus oryzae*. Stud. Mycol., 25: 1-19.

Schipper, M. A. A. 1986a. Notes on *Zygorhynchus* species. Persoonia, 13: 97-105.

Schipper, M. A. A. 1986b. *Hyphomucor*, a new genus in the Mucorales for *Mucor assamensis*. Mycotaxon, 27: 83-86.

Schipper, M. A. A. 1990. Notes on Mucorales – I Observations on *Absidia*. Persoonia, 14: 133-149.

Schipper, M. A. A. and Samson, R. A. 1994. Miscellaneous notes on Mucoraceae, Mycotaxon, 50: 475-491.

Schipper, M. A. A., Samson, R. A. and Stalpers, J. A. 1975. Zygospore ornamentation in the genera *Mucor* and *Zygorhynchus*. Persoonia, 8: 321-328.

Schipper, M. A. A. and Stalpers, J. A. 1983. Spore ornamentation and species concept in *Syncephalastrum*. Persoonia, 12: 81-85.

Schipper, M. A. A. and Stalpers, J. A. 1984. A revision of the genus *Rhizopus*. II. The *Rhizopus microsporus*-group. Stud. Mycol., 25: 20-34.

Schmidt, S. K., Wilson, K. L., Meyer, A. F., Gebauer, M. M. and King, A. J. 2008. Phylogeny and ecophysiology of opportunistic "snow molds" from a subalpine forest ecosystem. Microbiol. Ecol., 56: 681-687.

Schröter, J. 1889. VIII. Ordn. Zygomycetes. Brefeld, 111. Gatt. *Endogone*. Link 1809, pp. 197-225, 260. In: Kryptogamen-Flora von Schleisen III Band, 1 Hälfte. Dr. F. Cohn (ed.). Pilze. J. U. Kern's Verlag (Max Müller), Breslau, Germany.

Schröter, J. 1893a. Ancylistineae, pp. 88-92. In: Die natürlichen Pflanzenfamilien. I. Teil 1. A. Engler and K. Prantl (eds.). Abth. Wilhelm Engelmann, Leipzig, Germany.

Schröter, J. 1893b. Entomophthorineae, pp. 134-141. In: Die natürlichen Pflanzenfamilien. I. Teil 1. A. Engler and K. Prantl (eds.). Abth. Wilhelm Engelmann, Leipzig, Germany.

Schröter, J. 1893c. Mucorineae, pp. 119-134. In: Die natürlichen Pflanzenfamilien. I. Teil 1. A. Engler and K. Prantl (eds.). Abth. Wilhelm Engelmann, Leipzig, Germany.

Schüßler, A., Schwarzott, D. and Walker, C. 2001. A new fungal phylum, the Glomeromycota: phylogeny and evolution. Mycol. Res., 105: 1413-1421.

Shanor, L., Poitras, A. W. and Benjamin, R. K. 1950. A new genus of the Choanephoraceae. Mycologia, 42: 271-278.

Shipton, W. A. and Schipper, M. A. A. 1975. *Halteromyces*, a new genus in the Mucorales. Anton. van Lee. J. M. S., 41: 337-342.

Soper, R. S. 1974. The genus *Massospora*, entomopathogenic for cicadas, Part I, Taxonomy of the genus. Mycotaxon, 1: 13-40.

Spatafora, J. W., Hughes, K. W. and Blackwell, M. (eds.). 2006. A Phylogeny for Kingdom Fungi. Deep Hypha issue. Mycologia, 98: 829-1103.

Steinkraus, D. C., Oliver, J. B., Humber, R. A. and Gaylor, M. J. 1998. Mycosis of banded whitefly (*Trialeurodes abutilonea*) (Homoptera: Aleyrodidae) caused by *Orthomyces aleyrodis* gen & sp. nov. (Entomophthorales: Entomophthorales). J. Invertebr. Pathol., 72: 1-8.

Strongman, D.B. 2005. Synonymy of *Ejectosporus magnus* and *Simuliomyces spica*, and a new species *Ejectosporus trisporus*, from winter-emerging stoneflies. Mycologia, 97: 552-561.

Strongman, D. B. and White, M. M. 2006. New species of *Lancisporomyces, Orphella*, and *Paramoebidium*, endosymbionts of stonefly nymphs from streams in Nova Scotia, Canada. Can. J. Bot., 84: 1478-1495.

Strongman, D. B. and White, M. M. 2008. Trichomycetes from lentic and lotic aquatic habitats on Ontario, Canada. Botany, 86: 1449-1466.

Strongman, D. B. and Xu, S. 2006. Trichomycetes from China and the description of three new *Smittium* species. Mycologia, 98: 479-487.

Sugiyama, M., Tokumasu, S. and Gams, W. 2003. *Umbelopsis gibberispora* sp. nov. from Japanese leaf litter and a clarification of *Micromucor ramannianus* var. *angulisporus*. Mycoscience, 44: 217-226.

Tanabe, Y., O'Donnell, K. Saikawa, M. and Sugiyama, J. 2000. Molecular phylogeny of parasitic Zygomycota (Dimargaritales, Zoopagales) based on nuclear small subunit ribosomal DNA sequences. Mol. Phylogenet. Evol., 16: 253-262.

Tanabe, Y., Watanabe, M. M. and Sugiyama, J. 2002. Are Microsporidia really related to fungi? A reappraisal based on additional gene sequences from basal fungi. Mycol. Res., 106: 1380-1391.

Thaxter, R. 1888. The Entomophthoreae of the United States. Mem. Boston Soc. Nat. Hist., 4: 133-201.

Thaxter, R. 1891a. On certain new or peculiar North American Hyphomycetes. I. *Oedocephalum, Rhopalomyces* and *Sigmoideomyces* n. g. Bot. Gaz., 16: 14-26.

Thaxter, R. 1891b. On certain new or peculiar North American Hyphomycetes. II. *Helicocephalum, Gonatorrhodiella, Desmidiospora* nov. genera and *Everhartia lignatilis* n. sp. Bot. Gaz., 16: 201-205.

Thaxter, R. 1897. New or peculiar Zygomycetes. 2. *Syncephalastrum* and *Syncephalis*. Bot. Gaz., 24: 1-14.

Thaxter, R. 1903. Mycological notes, 1-2. Rhodora, 5: 97-108.

Thaxter, R. 1914. New or peculiar Zygomycetes. 3: *Blakeslea, Dissophora* and *Haplosporangium*, nova genera. Bot. Gaz., 58: 353-366.

Thaxter, R. 1922. A revision of the Endogoneae. P. Am. Acad. Arts. Sci., 57: 291-351.

Tucker, B. E. 1981. A review of the nonentomogenous Entomophthorales. Mycotaxon, 13: 481-505.

Underwood, L. M. 1899. Moulds, Mildews, and Mushrooms; A Guide to the Systematic Study of the Fungi and Mycetozoa and Their Literature. H. Holt and Company, New York, New York, U.S.A.

Upadhyay, H. P. 1973. *Helicostylum* and *Thamnostylum* (Mucorales). Mycologia, 65: 733-751.

Ustinova, I, Krienitz, L. and Huss, V. A. R. 2000. *Hyalorаphidium curvatum* is not a green alga but a lower fungus; *Amoebidium parasiticum* is not a fungus, but a member of the DRIPs. Protist, 151: 253-262.

Valle, L. G. 2006. Asellariales (Trichomycetes) from the Iberian peninsula. Fungal Divers., 21: 167-179.

Valle, L. G. 2007. New species and summary of Iberian Harpellales. Mycologia, 99: 442-455.

Valle, L. G. and Cafaro, M. J. 2008. First report of zygospores in Asellariales and a new species from the Caribbean. Mycologia, 100: 122-131.

Valle, L. G. and Santamaria, S. 2002a. *Baetimyces*, a new genus of Harpellales and first report of *Legeriomyces ramosus* from the northeastern Iberian Peninsula. Mycologia, 94: 321-326.

Valle, L. G. and Santamaria, S. 2002b. *Tectimyces*, a new genus of Harpellales on mayfly nymphs (Leptophlebiidae) in Spain. Mycol. Res., 106: 841-847.

Valle, L. G. and Santamaria, S. 2004. *Bojamyces transfuga* sp. nov. and new records of Trichomycetes from mayfly larvae in Spain. Mycologia, 96: 1386-1392.

Valle, L. G. and Santamaria, S. 2005. Zygospores as evidence of sexual reproduction in the genus *Orphella*. Mycologia, 97: 1335-1347.

Valle, L. G., White, M M. and Cafaro, M. J. 2008. Harpellales in the digestive tracts of Ephemeroptera and Plecoptera nymphs from Veracruz, Mexico. Mycologia, 100: 149-162.

van de Peer, Y., Ben Ali, A. and Meyer, A. 2000. Microsporidia: accumulating molecular evidence that a group of amitochondriate and suspectedly primitive eukaryotes are just curious fungi. Gene, 246: 1-8.

van Tieghem, P. 1875. Nouvelles recherches sur les Mucorinées. Ann. Sci. Nat. Bot., Sér VI, 1: 5-175.

van Tieghem, P. 1878. Troisième mémoire sur les Mucorinées. Ann. Sci. Nat. Bot., Sér VI, 312-399.

van Tieghem, P. and Le Monnier, G. 1873. Recherches sur les Mucorinées. Ann. Sci. Nat. Bot., Sér V, 17: 261-399.

Voigt, K., Hoffmann, K., Einax, E., Eckart, M., Papp, T., Vágvölgyi, C. and Olsson, L. 2008. Revision of the family structure of the Mucorales (Mucoromycotina, Zygomycetes) based on multigene-genealogies: Phylogenetic analyses suggest a bigeneric Phycomycetaceae with *Spinellus* as sister group to *Phycomyces*, pp. 313-332. In: Current Advances in Molecular Mycology. Y. Gherbawy, R. L. Mach and M. K. Rai (eds.). Nova Science Publishers, Hauppauge, New York, U.S.A.

Voigt, K. and Olsson, I. 2008. Molecular phylogenetic and scanning electron microscopical analysis places the Choanephoraceae and the Gilbertellaceae in a monophyletic group within the Mucorales (Zygomycetes, Fungi). Acta Biol. Hun., 59: 365-383.

Voigt, K. and Wöstemeyer, J. 2001. Phylogeny and origin of 82 Zygomycetes from all 54 genera of the Mucorales and Mortierellales based on combined analysis of actin and translation elongation factor EF-1α genes. Gene, 270: 113-120.

Volgmayr, H. and Krisai-Greilhuber, I. 1996. *Dicranophora fulva*, a rare mucoraceous fungus growing on boletes. Mycol. Res., 100: 583-590.

von Arx, J. A. 1981. The Genera of Fungi Sporulating in Pure Culture. 3rd edn. J. Cramer (ed.). Vaduz.

von Arx, J. A. 1983 (1982). On Mucoraceae *s. str.* and other families of the Mucorales. Sydowia, 35: 10-26.

Vossbrinck, C. R. and Debrunner-Vossbrinck, B. A. 2005. Molecular phylogeny of the Microsporidia: ecological, ultrastructural and taxonomic considerations. Folia Parasit., 52: 131-142.

Vuillemin, P. 1904. Le *Spinalia radians* g. et sp. nov. et la série des Dispirées. B. Soc. Mycol. Fr., 20: 26-33.

Walker, C. and Schüßler, A. 2004. Nomenclatural clarifications and new taxa in the Glomeromycota. Mycol. Res., 108: 981-982.

Walker, C., Vestberg, M., Demercik, F., Stockinger, H., Saito, M., Sawaki, H., Nishmura, I. and Schüßler, A. 2007. Molecular phylogeny and new taxa in the Archaeosporales (Glomeromycota): *Ambispora fennica* gen. sp. nov., Ambisporaceae fam. nov., and emendation of *Archaeospora* and Archaeosporaceae. Mycol. Res., 111: 137-153.

Warcup, J. H. 1990. Taxonomy, culture and mycorrhizal associations of some zygosporic Endogonaceae. Mycol. Res., 94: 173-178.

Waterhouse, G. M. 1973. Entomophthorales, pp. 219-229. In: The Fungi. Vol. IVB. A Taxonomic Review with Keys: Basidiomycetes and Lower Fungi. G. C. Ainsworth, F. K. Sparrow and A. S. Sussman (eds.). Academic Press, Inc., New York, New York, U.S.A.

Webster, J. 1970. Coprophilous fungi. T. Brit. Mycol. Soc., 54: 161-180.

Whisler, H. C. 1963. Observations on some new and unusual enterophilous Phycomycetes. Can. J. Bot., 41: 887-900.

White, M. M. 1999. *Legerioides*, a new genus of Harpellales in isopods and other Trichomycetes from New England, U.S.A. Mycologia, 91: 1021-1030.

White, M. M. 2006. Evolutionary implications of a rRNA-based phylogeny of Harpellales. Mycol Res., 110: 1011-1024.

White, M. M., James, T. Y., O'Donnell, K., Cafaro, M. J., Tanabe, Y. and Sugiyama, J. 2006a. Phylogeny of the Zygomycota based on nuclear ribosomal sequence data. Mycologia, 98: 872-884.

White, M. M. and Lichtwardt, R. W. 2004. Fungal symbionts (Harpellales) in Norwegian aquatic insect larvae. Mycologia, 96: 891-910.

White, M. M., Siri, A. and Lichtwardt, R. W. 2006b. Trichomycete insect symbionts in the Great Smoky Mountains National Park and vicinity. Mycologia, 98: 333-352.

Wiebe, M.G. 2004. Quorn™ myco-protein — overview of a successful fungal product. Mycologist, 18: 17-20.

Williams, M. C. and Lichtwardt, R. W. 1971. A new *Pennella* (Trichomycetes) from *Simulium* larvae. Mycologia, 63: 910-914.

Williams, M. C. and Lichtwardt, R. W. 1984. Two *Stachylina* and two *Smittium* species (Trichomycetes) from Montana. Mycologia, 76: 204-210.

Williams, M. C. and Lichtwardt, R. W. 1987. Two new trichomycete species from *Zapada* spp. (stonefly) nymphs with an unusual distribution. Mycologia, 79: 473-478.

Williams, M. C. and Lichtwardt, R. W. 1990. Trichomycete gut fungi in New Zealand aquatic insect larvae. Can. J. Bot., 68: 1045-1056.

Williams, M. C. and Lichtwardt, R. W. 1993. A new monotypic fungal genus, *Allantomyces*, and a new species of *Legeriomyces* (Trichomycetes, Harpellales) in the hindgut of a western Australian mayfly nymph (*Tasmanocoenis*). Can. J. Bot., 71: 1109-1113.

Williams, M. C. and Lichtwardt, R. W. 1999. Two new Harpellales living in Ephemeroptera nymphs in Colorado Rocky Mountain streams. Mycologia, 91: 400-404.

Wolf, F. A. 1917. A squash disease caused by *Choanephora cucurbitarum*. J. Agr. Res., 8: 319-327.

Wöstemeyer, J. and Schimek, C. 2007. Trisporic acid and mating in Zygomycetes, pp. 431-443. In: Sex in Fungi: Molecular Determination and Evolutionary Implications. J. Heitman, J. W. Kronstad, J. W. Taylor and L. A. Casselton (eds.). ASM Press, Washington, D.C., U.S.A.

Wu, C.-g. and Lin, S.-J. 1997. Endogonales in Taiwan: A new genus with unizygosporic sporocarps and a hyphal mantle. Mycotaxon, 64: 179-188.

Wu, Q., Thiers, B. M. and Pfister, D. H. 2004. Preparation, preservation, and use of fungal specimens in herbaria, pp. 23-36. In: Biodiversity of Fungi: Inventory and Monitoring Methods. G. M. Mueller, G. F. Bills and M. S. Foster (eds.). Elsevier Academic Press, Amsterdam, The Netherlands.

Yafetto, L., Carroll, L., Cui, Y., Davis, D. J., Fischer, M. W. F., Henterly, A. C., Kessler, J. D., Kilroy, H. A., Shidler, J. B., Stolze-Rybczynski, J. L., Sugawara, Z. and Money, N. P. 2008. The fastest flights in nature: high-speed spore discharge mechanisms among fungi. *PLoS One* 3(9): e3237. Doi: 10.1371/journal.pone.0003237. (also with Video S4 of *Pilobolus kleinii* sporangial discharge, 50,000 fps. Doi:10.1371/journal.pone.0003237.s003).

Yao, Y.-J., Pegler, D. N. and Young, T. W. K. 1995a. New species of *Endogone* (Endogonales). Kew B., 50: 359-365.

Yao, Y.-J., Pegler, D. N. and Young, T. W. K. 1995b. *Youngiomyces*, a new genus in Endogonales (Zygomycotina). Kew B., 50: 349-357.

Yao, Y.-J., Pegler, D. N. and Young, T. W. K. 1996. Genera of Endogonales. Royal Botanic Garden, Kew, Richmond, Surrey, United Kingdom.

Young, T. W. K. 1985. Ultrastructure of mucoralean sporangiospores. Bot. J. Linn. Soc., 91: 151-165.

Young, T. W. K. 1999. Taxonomy of Kickxellaceae and radiation of the asexual apparatus. Kew B., 54: 651-661.

Zheng, R.-Y. 2002. *Zygorhynchus multiplex*, a new species from paddy soil. Mycotaxon, 84: 367-378.

Zheng, R.-Y. and Chen, G.-Q. 1991. A non-thermophilic *Rhizomucor* causing human primary cutaneous mucormycosis. Mycosystema, 4: 45-57.

Zheng, R.-Y. and Chen, G.-Q. 1993. Another non-thermophilic *Rhizomucor* causing human primary cutaneous mucormycosis. Mycosystema, 6: 1-11.

Zheng, R.-Y. and Jiang, H. 1995. *Rhizomucor endophyticus* sp. nov., an endophytic Zygomycetes from higher plants. Mycotaxon, 56: 455-466.

Zheng, R.-Y. and Chen, G.-Q. 2001. A monograph of *Cunninghamella*. Mycotaxon, 80: 1-75.

Zheng, R.-Y., Chen, G.-Q. and Hu, F.-M. 1988. Monosporous varieties of *Syncephalastrum*. Mycosystema, 1: 35-52.

Zheng, R.-Y., Chen, G.-Q. Huang, H. and Liu, X.-Y. 2007. A monograph of *Rhizopus*. Sydowia, 59: 273-372.

Zheng, R.-Y. and Liu, X.-Y. 2005. *Actinomucor elegans* var. *meitauzae*, the correct name for *A. taiwansensis* and *Mucor meitauzae* (Mucorales, Zygomycota). Nova Hedwigia, 80: 419-431.

Zheng, R.-Y. and Liu, X.-Y. 2009. Taxa of *Pilaira* (Mucorales, Zygomycota) from China. Nova Hedwigia, 88: 255-267.

Zoberi, M. H. 1985. Liberation of asexual propagules in the Mucorales. Bot. J. Linn. Soc., 91: 167-173.

Zycha, H. 1935. Mucorineae, pp. 1-264. In: Kryptogamenflora der Mark Brandenburg. Band VIa. Gebrüder Borntraeger, Leipzig, Germany.

Zycha, H., Siepmann, R. and Linnemann, G. 1969. Mucorales eine Beschreibung aller Gattungen und Arten dieser Pilzgruppe. J. Cramer, Lehre, Lichtenstein.

Chapter 5

Evolution of Trichomycetes

Robert W. Lichtwardt

Department of Ecology & Evolutionary Biology
University of Kansas, 1200 Sunnyside Avenue
Lawrence, Kansas 66045-7534, USA
E-mail: licht@ku.edu

ABSTRACT

Trichomycetes as a fungal class live worldwide in the guts of a variety of arthropods (Nematocera, Plecoptera, Ephemeroptera, Trichoptera, Collembola, Isopoda) with ancient evolutionary histories. These fungi have adapted themselves successfully to the guts of individual types of hosts, both morphologically and physiologically. It is hypothesized that the history of the fungal symbionts is also ancient, and that they may have begun their development as trichomycetes in the earlier stages of host evolution. As the arthropod hosts eventually became distributed to modern continents and islands, they carried with them their fungal partners. Some disjunct distributions apparently led to speciation of the host, and at times speciation of the fungi. Evidence suggests that the current wide distribution patterns of particular hosts and their fungi may sometimes be attributed to continental drift. It is very likely, based on collecting experiences in various parts of the world (Lichtwardt et al., 2001), that the vast majority of Trichomycetes are yet to be discovered. Some of these will undoubtedly be important missing taxa that will help molecular phylogenetic efforts to better infer the evolutionary history of these interesting symbiotic fungi.

INTRODUCTION

Trichomycetes (Zygomycota) consist of fungi that live within the guts of insects and other arthropods, currently with two orders, Harpellales and Asellariales. These will be the subject of this chapter. Until recently there were four orders within the class, but DNA sequence data indicate that the Amoebidiales and Eccrinales are, in fact, sister clades of protists that fall within the Mesomycetozoea (Benny and O'Donnell, 2000; Cafaro, 2005; White et al., 2006a), a group of mostly unculturable parasites that belong to a region in the tree of life where animals and fungi diverged.

Harpellales are symbionts of freshwater insects where they live most of the time as commensals within the hindgut or midgut of larvae, but under special circumstances may have a parasitic or mutualistic relationship with their larval or adult hosts (Horn and Lichtwardt, 1981; Lichtwardt, 2001; White et al., 2006b, Lichtwardt, 2008). The Asellariales live within the hindgut of springtails (Collembola) or in marine, freshwater or terrestrial isopods (Crustacea).

EVOLUTIONARY CONSIDERATIONS

Lacking fossil evidence, hypotheses relating to the evolution of Trichomycetes are somewhat circumstantial and to a large extent speculative. Their worldwide distribution and the ancestral primitiveness of their hosts strongly suggest that the symbiotic relationship of the partners is ancient and very successful. All evidence indicates that, unlike many other fungi with airborne propagules, trichomycetes do not disperse outside of their hosts except locally in order to transmit asexual trichospores (Fig. 5.1 A, B) or sexually produced zygospores (Fig. 5.1 C, D) that pass the infection from one individual host to another. Dispersal over greater distances may occur when hosts are carrying the fungi internally, or in cases where the Harpellales fungus has invaded the ovaries of adult females and produced fungal cysts (Fig. 5.2) (White et al., 2006b) that are "oviposited" by the flying female in a new habitat.

Antiquity of host groups

It is probably significant that the hosts of Harpellales have ancient origins. With two exceptions (an isopod and a beetle larva), all of the described 37 genera and 212 species of Harpellales inhabit freshwater larval stages of stoneflies (Plecoptera), mayflies (Ephemeroptera), caddis worms (Trichoptera) or Nematocera. The latter is the most primitive suborder of Diptera, consisting of families such as mosquitoes, black flies, nonbiting and biting midges, solitary midges, craneflies, meniscus midges and moth flies. Ancestors of these groups

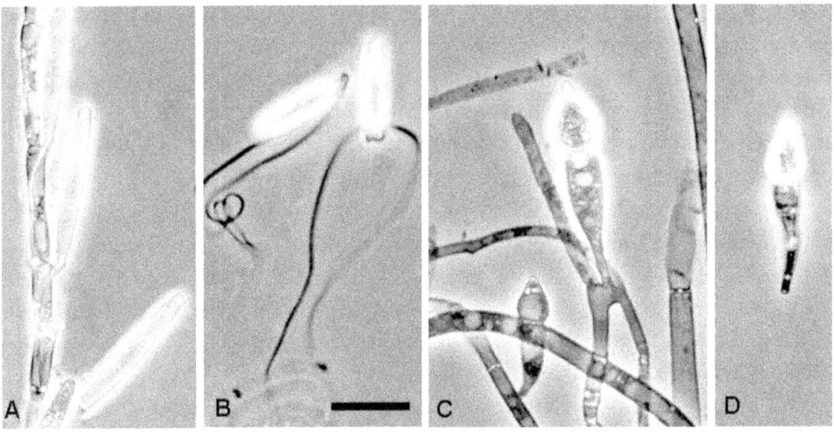

Fig. 5.1 *Plecopteromyces patagoniensis* from the hindgut of a nymph in the Southern Hemisphere stonefly family Gripopterygidae. A, B. Trichospores developing on a fertile branch, and two detached trichospores each with two appendages, respectively. The number of trichospore appendages is a generic character. Appendages aid entanglement of trichospores in the stream substrate and help to keep them in the vicinity of feeding larvae to increase the chance of being ingested. C, D. Zygospores, attached and detached, produced after hyphal conjugation. This is one of four basic zygospore types in the Harpellales. Bar = 20 μm for all figures. Photos from Lichtwardt et al., 1999.

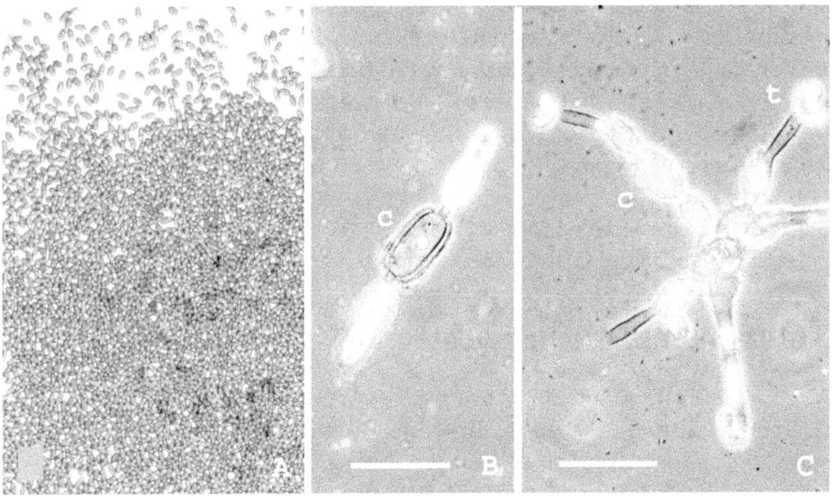

Fig. 5.2 Ovarian cysts from black flies (Simuliidae). A. Dense mass of cysts of an unidentified Harpellales removed from the ovary of an adult female *Prosimulium* sp. from Newfoundland, Canada (photomicrograph by M.M. White). B. Cyst of *Harpella melusinae* from an adult *Simulium* sp. from New York, USA, germinating to produce two hyphae. C. Cysts from another New York *Simulium* sp. adult forming coiled trichospores at the tips of the hyphae. c = cyst, t = trichospore. Bars = 40 μm.

are believed to have begun evolving some 190-250 million years ago. Given the number of genera and species of Harpellales currently known, the fungi have required time to develop special adaptive morphological and physiological features to ensure survival (Lichtwardt, 2008). It is hypothesized that harpellid fungi began to evolve during the earlier stages of development of these kinds of freshwater arthropods.

Likewise, arthropod hosts of the lesser studied Asellariales (3 genera, 14 known species) have a long history of evolution. The primitive springtails (Hexapoda, Collembola) have origins in the Early Devonian, some 400 million years ago, based on fossil evidence (Grimaldi and Engel, 2005). The fossil records of isopods, the other hosts of Asellariales, begin in the Carboniferous period some 300 million years ago. As in the Harpellales, one might assume, given the wide geographical distribution and the variety of hosts to which the fungi have successfully adapted, that the symbiotic relationships of the relatively few currently known Asellariales is also ancient.

There is little doubt that many new taxa of Trichomycetes are yet to be discovered, and some may shed some important insight into the evolution of this group. Evidence to date indicates a wide range of host specificity among Harpellales, depending on the species, and different individual distribution patterns. Some Harpellales appear to be locally endemic, whereas others have very wide distributions. Three species that are unusual in this respect are *Smittium culisetae* Lichtw., *S. culicis* Manier, and *S. simulii* Lichtw. Each of these common and widespread Smittiums has been found in larvae belonging to three to six different families or Nematocera. Although some harpellids are known in only one genus or one species of host, generally specificity is at the family level. A good example is the genus *Harpella* Léger & Duboscq with five known species that has a worldwide distribution wherever black flies (Simuliidae) are found. Some of these species appear to be restricted to particular geographic regions, but *H. melusinae* Léger & Duboscq has been found on most continents and many islands (Lichtwardt et al., 2001).

Distribution of Trichomycetes

Widespread distribution patterns of trichomycete hosts, carrying within them their fungal partners, may be attributed to passive or phoretic dispersal (wind, water, or transported by other animals) or by human activity in more recent times. An example of the latter is the Hawaiian Islands where mosquitoes were unknown prior to 1826 (Hardy, 1960). In that year and subsequent years ships brought mosquitoes to the islands, possibly in water tanks or barrels. The author and later a colleague, Marvin Williams, did a limited collection of mosquito larvae on two of the islands and found all of the populations infected with *S. culisetae* (unpubl.), one of the two species of *Smittium* Poisson common in mosquitoes.

The ability of many species of trichomycete hosts to disperse over distances and on their own is often limited. Marine environments, being interconnected, may differ from disjunct terrestrial or freshwater habitats, and allow greater passive dispersal of the arthropods and their fungi. Figure 5.3 is an example of the widespread distribution of some marine isopod populations comprised of several different species of *Ligia* that live in specific types of habitats along shorelines of continents and islands. All of these sampled populations were found to be infected with *Asellaria ligiae* ex Manier (Asellariales). One might well visualize that over considerable time by various means, perhaps following storms and rafting on flotsam, isopods carrying the fungus internally separated from their populations, accidentally landed on a suitable shoreline habitat and established a new population; and in some instances speciated but the fungus did not. Alternatively, cryptic species might be masked by convergent morphology, which lends particular interest to ongoing molecular systematic studies.

Fig. 5.3 Distribution of populations of several different species of the seashore marine isopod genus *Ligia*, each infected with the same Asellariales species, *Asellaria ligiae*.

If, as hypothesized, the fungus-insect relationships are ancient, tectonic events might account for the widespread occurrence today of certain genera and species. Some hosts of Trichomycetes may have been evolving on the supercontinent of Pangea that existed some 250 million years ago and began to disassemble, followed later by the breakup of the southern landmass, Gondwana. As continental drift progressed over time, arthropods and fungi may have been carried on these drifting land masses, resulting in some of the arthropod-fungus distributions we see today. As shown in Fig. 5.4, species of the harpellid genus *Plecopteromyces* Lichtw., Ferrington & López Lastra have been found only in the Southern Hemisphere stonefly family Gripopterygidae.

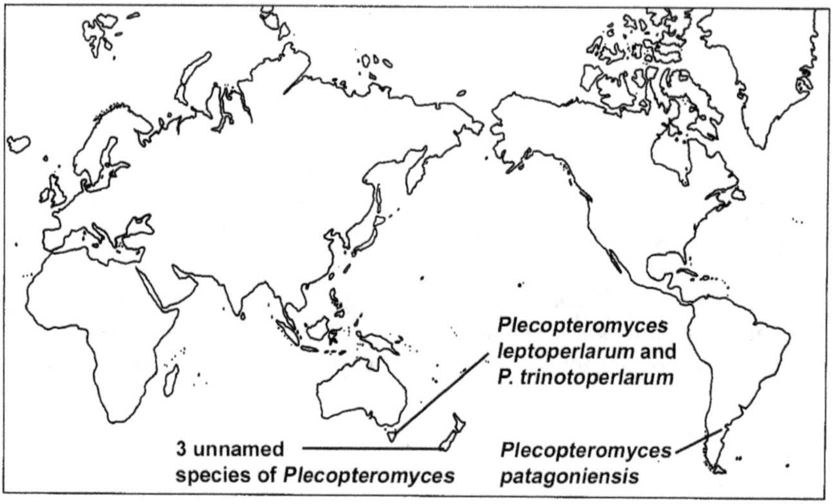

Fig. 5.4 Distribution of six different species of *Plecopteromyces* (Harpellales), all inhabitants of nymphs of several genera and different species of Gripopterygidae from Tasmania, Australia, South Island, New Zealand, and the Patagonia region of Argentina. Such a distribution is suggestive of a Gondwanan origin.

One species of *Plecopteromyces* was found in the Patagonia region of Argentina (Fig. 5.1) (Lichtwardt et al., 1999), two in Tasmania, Australia (Ferrington et al., 2005), and several unpublished species on South Island, New Zealand. This is the classical distribution one expects from organisms with a Gondwanan origin.

Another example that possibly indicates a distribution related to continental drift is the seldom found species of harpellids in the genus *Carouxella* (Fig. 5.5) (Manier et al., 1961, Lichtwardt and Williams, 1990, Lichtwardt et al., 1999). They inhabit larvae of the biting midge (Ceratopogonidae) genus *Dasyhelea*. Populations of different species of *Dasyhelea* were found living in very small bodies of water consisting of rock holes and (in Argentina) cemetery vases. The adult midges have low vagility, and were unlikely to have become distributed by long distance dispersal on their own to locations as shown in Fig. 5.5. These populations contained the same genera but different species of both insect and fungus. The most logical explanation is that over considerable time, as populations drifted apart on different land masses, both the host and its fungal symbiont became dispersed and speciated.

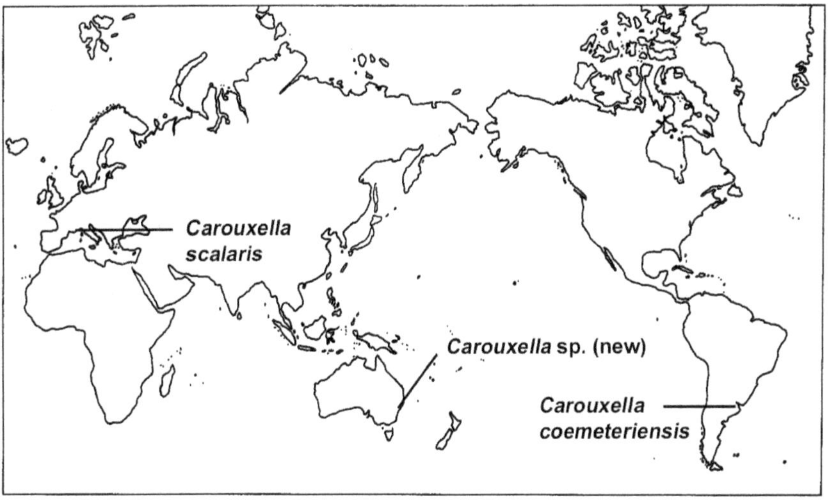

Fig. 5.5 Populations of different species of *Carouxella* (Harpellales) in larvae of different species of *Dasyhelea* (Ceratopogonidae): *Carouxella scalaris* in *D. lithotelmatica* from southern France; *Carouxella* sp. (new) in a new, unnamed species of *Dasyhelea* from New South Wales, Australia; and *C. coemeteriensis* in *D. necrophila* from La Plata, Argentina.

ACKNOWLEDGMENTS

This chapter has been supported by a grant from the National Science Foundation, DEB-0918169. The author thanks Dr. Merlin M. White for suggestions that improved the article.

REFERENCES

Benny, G. L. and O'Donnell, K. 2000. *Amoebidium parasiticum* is a protozoan, not a Trichomycete. Mycologia, 92: 1133-1137.

Cafaro, M. J. 2005. Eccrinales (Trichomycetes) are not fungi, but a clade of protists at the early divergence of animals and fungi. Mol. Phylogen. Evol., 35: 21-34.

Ferrington, L. C., Jr., Lichtwardt, R. W., Hayford, B. and Williams, M. C. 2005. Symbiotic Harpellales (Trichomycetes) in Tasmanian aquatic insects. Mycologia, 97: 254-262.

Grimaldi, D. and Engel, M. S. 2005. Evolution of the Insects. Cambridge University Press, New York.

Hardy, D. E. 1960. Insects of Hawaii. Vol. 10 Diptera: Nematocera - Brachyura. Univ. Hawaii Press, Honolulu.

Horn, B. W. and Lichtwardt, R. W. 1981. Studies on the nutritional relationship of larval *Aedes aegypti* (Diptera: Culicidae) with *Smittium culisetae* (Trichomycetes). Mycologia, 73: 724-740.

Lichtwardt, R. W. 2001.Trichomycetes: Fungi in relationship with insects and other arthropods, pp. 575-588. In: Symbiosis. J. Seckbach (ed.). Kluwer Academic Publishers, The Netherlands.

Lichtwardt, R. W. 2008. Trichomycetes and the arthropod gut, pp. 3-19. In: The Mycota, Human and Animal Relations. A. A. Brakhage and P. F. Zipfel (eds.). Springer-Verlag, Berlin.

Lichtwardt, R. W. and Williams, M. C. 1990. Trichomycete gut fungi in Australian aquatic larvae. Can. J. Bot., 68: 1057-1074.

Lichtwardt, R. W., Ferrington Jr., L. C. and López Lastra, C. 1999. Trichomycetes in Argentinean aquatic insect larvae. Mycologia, 91: 1060-1182.

Lichtwardt, R. W., Cafaro, M. J. and White, M. M. 2001. The Trichomycetes, fungal associates of arthropods. Published on the Internet www.nhm.ku.edu/~fungi

Manier, J.-F., Rioux, J. A. and Whisler, H. C. 1961. *Rubetella inopinata* n.sp. et *Carouxella scalaris* n.g., n.sp., Trichomycètes parasites de *Dasyhelea lithotelmatica* Strenzke, 1951 (Diptera Ceratopogonidae). Nat. Monspeliensia Ser. Bot., 13: 25-38.

White, M. M., James, T. Y., O'Donnell, K., Cafaro, M. J., Tanabe, Y. and Sugiyama, J. 2006a. Phylogeny of the Zygomycota based on nuclear ribosomal sequence data. Mycologia, 98: 872-884.

White, M. M., Lichtwardt, R. W. and Colbo, M. H. 2006b. Confirmation and identification of parasitic stages of obligate endobionts (Harpellales) in blackflies (Simuliidae) by means of rRNA sequence data. Mycol. Res., 110: 1070-1079.

Chapter 6

Systematics of *Stachylina* and *Smittium* – the Two Largest Genera of Harpellales, Zygomycota

J. K. Misra*

Bhartiya Sanskriti Evam Shiksha Samiti
(Indian Cultural and Educational Society)

Saroj Lalji Mehrotra Bhartiya Vidya Bhavan Girls Degree College
6 Vineet Khand, Gomti Nagar, Lucknow 226 010, India

*Correspond to: 16/1287, Indira Nagar, Lucknow 226 016, India
E-mail: jitrachravi@gmail.com

ABSTRACT

The morphological, distributional and host range data of the species of the largest Harpellales genera, *Stachylina* and *Smittium*, published up to mid 2010 are consolidated from the available literature, including species that occur in India. Keys for identification of the species of the two genera are provided, and the need for having a uniform pattern for describing a species and, where possible, integrating the molecular data of a species with that of its morphology has been emphasized for future studies.

INTRODUCTION

Trichomycetes, an ecological group of cryptic fungi, currently is comprised of only two orders, Harpellales and Asellariales, and three families (Lichtwardt, 1986; Lichtwardt et al., 2001a). During the last few years the investigations relating to this group have gained momentum. European and American researchers have discovered many new genera and species within the order Harpellales. Unfortunately, there are only a few recent reports from Asian regions (Misra and Tiwari, 2002, 2008; Strongman and Xu, 2006; Strongman et al., 2010; Wang et al., 2010).

Harpellales is the biggest order of the Trichomycetes consisting of 39 genera and more than 222 species. The two genera, *Stachylina* (Harpellaceae) and *Smittium* (Legeriomycetaceae) have the largest number of species, 30 and 79 respectively, including one variety. This chapter embodies the salient morphological details of the species of two genera, and their hosts and geographical records. Keys for their species are provided. Since there is a lack of uniformity in describing the species of these frequently encountered genera, a suggestion is made to describe all details of a species whether they are seen or not in a particular collection, so that any future investigator may know whether a specific morphological character has been seen/not seen or ignored by the past discoverer.

Stachylina Léger & Gauthier, 1932

Thalli unbranched, cylindrical to fusiform, attached to peritrophic matrix of the host either by a small secreted holdfast or by penetrating the membrane. Trichospores oval to biconical, with or without a collar, with a single appendage. Zygospores known for only one species (*St. reflexa*)

Type species: *Stachylina macrospora* Léger & Gauthier - C. R. Acad. Sci. Paris 194: 2262-2265, 1932.

(=*Stachylina intermedia* Poisson, 1936, *nom. nud.*, in Mélanges dédiés au Professeur Lucien Daniel. Université de Rennes, pp. 75-86, Figs. 1, 2)

Key to species

1a. Thallus base penetrating the peritrophic matrix of the host.......................2
1b. Thallus base not penetrating the peritrophic matrix of the host..............7
2a. Trichospores up to 20 µm long...3
2b. Trichospores more than 20 µm long...4
3a. Trichospores with a short appendage.....................................***St.. minuta***
3b. Trichospores without appendage..***St. minima***
4a. Trichospores up to 30 µm long...5
4b. Trichospores more than 30 µm long...6
5a. Penetrating base of the thallus foot-like...............................***St. pedifer***
5b. Penetrating base of the thallus bulbous........................***St. queenslandiae***
6a. Trichospores 30-50 µm long...***St. penetralis***

6b. Trichospores 55-90 μm long...***St. tianensis***
7a. Trichospores with collar..8
7b. Trichospores without collar..16
8a. Trichospores with short collar (< 1 μm long)......................................9
8b. Trichospores with long collar (> 1 μm long).......................................12
9a. Trichospores 1-4 per thallus...10
9b. Trichospores more than 4 per thallus...11
10a. Trichospores 22-50 μm long with very long and fine appendage....***St. lentica***
10b. Trichospores 42-70 μm long with unknown appendage........***St. stenospora***
11a. Trichospores with one heavy appendage.......................***St. gravicaudata***
11b. Trichospores with one appendage splitting into two..............***St. platensis***
12a. Trichspores up to 40 μm long..13
12b. Trichospores more than 40 μm long...14
13a. Trichospores 25-35 μm long with one prominent appendage....***St. euthena***
13b. Trichospores 31-40 μm long with unknown appendage.........***St. paludosa***
14a. Trichospores 78-108 μm long with one short stout appendage with knob...... ....***St. dolichospora***
14b. Trichospores with prominent appendage without knob...............................15
15a. Trichospores 35-49 μm long..............................***St. chironomidarum***
15b. Trichospores 40-72 μm long...................................***St. grandispora***
16a. Host non-Chironomidae..17
16b. Host Chironomidae..19
17a. Trichospores 17-20 μm long and host Ceratopogonidae...... ....***St. ceratopogonidarum***
17b. Trichospores more than 20 μm long..18
18a. Trichospores 24-32 μm long, with very-very long appendage and host Psychodidae...***St. lotica***
18b. Trichospores 20-38 μm long, with unknown appendage and host Thaumaleidiae..***St. thaumaleidarum***
19a. Trichospores more than 8 produced per thallus......................................20
19b. Trichospores up to 8 produced per thallus..23
20a. Thallus diameter up to 14 μm 21
20b. Thallus diameter 14-25 μm, trichospores seen produced 11-18 in number...***St. magna***
21a. Trichospores up to 50 per thallus and less than 6 μm wide......***St. prolifica***
21b. Trichospores more than 6 μm wide...22
22a. Thallus with tapered base and small secreted holdfast...........***St. jujuyensis***
22b. Thallus with discoid secreted holdfast.................................***St. robusta***

23a. Trichospores up to 4 per thallus..24

23b. Trichospores more than 4 per thallus...25

24a. Only 1-2 trichospores produced per thallus, 29-40 × 6-8 μm with single appendage...***St. paucispora***

24b. Only 2-4 trichospores produced per thallus, 25-40 × 7-10 μm, appendage ribbon like near the base..***St. nana***

25a. Biconical zygospore produced, terminal trichospores reflexed.....***St. reflexa***

25b. Zygospores not produced..26

26a. Trichospores 15-25 μm long...27

26b. Trichospores more than 39-50 μm long...29

27a. Trichospores 25 × 5-6 μm, appendage unknown.........................***St. longa***

27b. Trichospores with one appendage..28

28a. Trichospores 15-23 × 5-7 μm with holdfast and tapering base...... ....***St. acutibasilaris***

28b. Trichospores 17.5-20 × 4.5 μm, holdfast disklike..................***St. manicata***

29a. Trichospores 39-47 × 10-12 μm with unknown appendage.......***St. litoralis***

29b. Trichospores 40-50 × 7-8 μm with appendage many times longer than a trichospore..***St. macrospora***

Description

1. ***Stachylina acutibasilaris*** M. M. White & Lichtw., 2004

Thalli 78-155 × 10-18 μm, with tapering basal cell, holdfast small, producing 4-8 or more trichospores. Trichospores ellipsoidal with a slight median swelling, with a single appendage, without a collar, 15-23 × 5-7 μm. Zygospores unknown.

Host: *Brillia* cf. *longifurca* Kieffer (Diptera, Chironomidae) larvae.
Record: Norway (White and Lichtwardt, 2004).

2. ***St. ceratopogonidarum*** Lichtwardt & Arenas, 1996

Thalli long-fusiform, 90-120 × 6-8 μm, holdfast small secreted. Trichospores elongate-ovoid 17-20 × 5-8 μm, without a collar. Zygospores unknown.

Host: *Forcypomyia* (?) sp. larva (Ceratopogonidae).
Record: Southern Chile (Lichtwardt and Arenas, 1996).

3. ***St. chironomidarum*** Lichtwardt, 1972

Thalli usually 180-250 μm long × 4-5 μm diam., holdfast inconspicuous disk-like, producing 2-6 or more generative cells. Trichospores long-ellipsoidal, 35-49 × 4-5 μm, with a very slender collar 1-2 μm long and a single appendage. Zygospores unknown.

Host: Bloodworms (Diptera, Chironomidae).
Distribution: U.S.A. (Lichtwardt, 1972), India (Misra and Tiwari, 2002).

4. ***St. dolichospora*** Ferrington, Hayford & Lichtw., 2005

Thalli more than 400 long, 8-11 μm diam. Trichospores almost cylindrical, measuring 78-108 × 8-10 μm, with a collar about 2 μm long, and a short, stout appendage with a terminal knob.

Host: *Riethia* sp. larvae (Diptera, Chironomidae)
Record: Australia (Ferrington et al., 2005).

5. ***St. euthena*** Manier & Coste, 1971

Thalli 130-200 × 7-8 μm, holdfast small disk-like. Trichospores 8 per thallus, fusiform with a median swelling, 25-35 × 7-8 μm, collar ephemeral, 2.5-3 μm long, appendage single and prominent. Zygospores unknown.

Hosts: In *Chironomus plumosus* complex and *Psectrotanypus varius* Fabr. Larvae (Diptera, Chronomidae).
Record: France (Coste-Mathiez, 1970; Manier and Coste, 1971), Germany (Roland Kirschner, unpubl.), Spain (Valle, 2007).

6. ***St. grandispora*** Lichtwardt, 1972

Thalli less than 100 or more than 250 × 6-10 μm, holdfast inconspicuous. Trichospores 2-16 per thallus or more, long-ellipsoidal, 40-72 × 6-10 μm (or more), with a collar 1-3 μm long and a single, prominent and very long appendage. Zygospores unknown.

Hosts: Chironomidae (Diptera) larvae, including species of *Tanytarsus, Tendipes, Paratendipes, Polypedilum, Chironomus,* and *Cricotopus.*
Record: U.S.A., England, Sweden (Lichtwardt, 1986), U.S.A. (White et al., 2006), Japan (Lichtwardt et al., 1987), Australia (Lichtwardt and Williams, 1990), New Zealand (Williams and Lichtwardt, 1990), Argentina (Lichtwardt et al., 1999,

2000; Ferrington et al., 2005; Siri and López Lastra, 2010), India (Misra and Tiwari, 2002), Canada (Strongman, 2007; Strongman and White, 2008), Spain (Valle, 2007; Valle and Santamaria, 2009), Dominican Republic (Valle and Cafaro, 2010).

7. ***St. gravicaudata*** Siri, M. M. White & Lichtw., 2006

Thalli 160-250 × 5-10 μm, bearing 4-8 trichospores, 25-31 × 4-5 μm, with a short collar and one heavy appendage. Zygospores unknown.

Host: *Paralauterborniella* sp. (Diptera, Chironomidae).
Record: U.S.A. (White et al., 2006)

8. ***St. jujuyensis*** Mazzucchelli, López Lastra & Lichtw., 2000

Thalli about 10 μm diam, base tapered, holdfast small secreted. Trichospores many, ellipsoidal, 30-37 × 8-9 μm.

Host: Chironomidae larvae (Diptera, Chironomidae).
Record: Argentina (Lichtwardt et al., 2000).

9. ***St. lentica*** M. M. White & Lichtw., 2004

Thalli 50-150 × 4-9 μm, holdfast small rounded, producing 1-4 trichospores. Trichospores long-ellipsoidal, 22-50 × 4-7 μm, with a very short collar, and a very long and fine appendage. Zygospores unknown.

Hosts: *Chironomus* sp. larva, *Brillia* cf.. *longifurca* Kieffer, *Polypedilum* sp., *Metriocnemus eryngiotelmatus* Donato & Paggi larva (Diptera, Chironomidae)
Record: Norway (White and Lichtwardt, 2004), Argentina (Siri et al., 2008; Siri and López Lastra, 2010).

10. ***St. litoralis*** Lichtw., White & Colbo, 2001b

Thalli 135-200 × 10-12 μm, base bulbous. Trichospores 4-8 per thallus, 39-47 × 10-12 μm, without a collar.
Host: *Telmatogeton japonicus* larva (Diptera, Chironomidae).
Record: Canada, Newfoundland (Lichtwardt et al., 2001b), Crozal Islands [(Indian Ocean) (Reeves et al., 2004)].

11. ***St. longa*** Léger & Gauthier, 1932

Thalli 100 µm or more × 9-10 µm, bearing 6-8 trichospores, 25 × 5-6 µm, without a collar. Zygospores unknown.

Host: *Tanytarsus* sp. larva (Diptera, Chironomidae).
Record: France (Léger and Gauthier, 1932; Gauthier, 1961; Lichtwardt, 1986).

12. ***St. lotica*** Williams & Lichtw., 1984

Thalli 100 × 8-10 µm, producing up to 8 trichospores. Trichospores ellipsoidal to subbiconical, 24-32 × 8-10 µm, with appendage up to 10 times of spore length, without a collar. Zygospores unknown.

Host: *Maruina* sp. larva (Diptera, Psychodidae)
Record: U.S.A. (Williams and Lichtwardt, 1984), Argentina (Siri and López Lastra, 2010).

13. ***St. macrospora*** Léger & Gauthier, 1932 [= ***St. intermedia*** Poisson, 1936, ***nom. nud.***]

Thalli 100-200 × 8-10 µm, holdfast broad disk-like, bearing 2-8 long ovoid trichospores, 40-50 × 7-8 µm, without a collar, with a single appendage many times longer than a trichospore. Zygospores unknown.

Host: Larval Diamesinae (Diptera, Chironomidae), including *Diamesa* sp. and *Syndiamesa macronyx* Kieff.
Record: France (Léger and Gauthier, 1932, 1935), Alaska (Kobayasi et al., 1967, 1969), India (Misra and Tiwari, 2008).

14. ***St. magna*** Indoh, Lichtwardt & Kobayasi, 1987

Thalli 400-900 µm × 14-25 µm, producing 40-65 generative cells with 11-18 trichospores seen attached at a time. Trichospores long- ellipsoidal, 30-80 × 6.5-15 µm, without a collar. Zygospores unknown.

Host: Unidentified Chironomidae larva (Diptera, Chironomidae).
Record: Japan (Lichtwardt et al., 1987).

15. ***St. manicata*** Williams & Lichtw., 1984

Thalli up to 100 × 6.5-8 μm, bearing up to 8 trichospores. Holdfast disk-like, trichospores ellipsoidal to subbiconical, 17.5-20 × 4.5 μm, with a small sleeve-like collar, 1-1.5 × 0.5 μm surrounding the single appendage, which narrows toward the distal end and is at least 5 times the spore length. Zygospores unknown.

Hosts: Larval *Polypedilum* sp. and/or *Pseudochironomus* sp. (Diptera, Chironomidae).
Record: U.S.A. (Williams and Lichtwardt, 1984).

16. ***St. minima*** Williams & Lichtw., 1990

Thalli 32-46 × 6-8 μm, bulbous base, penetrating the peritrophic matrix. Trichospores 1-4, ovoid, 11-18 × 6-8 μm, without a collar, distal trichospore subterminal. Zygospores unknown.

Host: Orthocladiinae larvae (Diptera, Chironomidae).
Distribution: South Island, New Zealand (Williams and Lichtwardt, 1990), Argentina (López Lastra et al., 2005).

17. ***St. minuta*** Gauthier ex Lichtwardt, 1984a
(= ***Stachylina minuta***, Gauthier, 1961, ***nom. nud.***)

Thalli 30-55 × 5-6 μm, fusiform, penetrating the peritrophic matrix by their bulbous base, trichospores 1-4 per thallus, terminal cell sterile. Trichospores ellipsoidal, 15-16 × 5-6 μm, produced on a narrow cylindrical outgrowth from the generative cell, detached ones without collar, with a single short appendage. Zygospores unknown.

Host: Tanytarsini larvae (Diptera, Chironomidae).
Record: France (Gauthier, 1961; Lichtwardt, 1984).

18. ***St. nana*** Lichtwardt, 1984

Thalli cymbiform to fusiform, 60-80 μm long, holdfast small disk-like, producing 2-4 oval to biconical trichospores, 25-40 × 7-10 μm, with a single appendage, ribbon-like near the base, without collar. Zygospores unknown.

Host: Unidentified larva (Diptera, Chironomidae)

Record: France (Lichtwardt, 1984), Japan (Lichtwardt et al., 1987), South Island and New Zealand (Williams and Lichtwardt, 1990; Lichtwardt and Williams, 1992c), Spain (Valle and Santamaria, 2009), Canada (Strongman, 2007; Strongman and White, 2008).

19. *St. paludosa* Lichtw., 1994

Thalli 80-150 × 7-9 µm, with about 8 generative cells, holdfast 1 × 2 µm. Trichospores long ellipsoidal with a slight median bulge, 31-40 × 6-8 µm, with a collar about 1 µm long. Zygospores unknown.

Host: *Polypedilum* sp. and *Chironomus* sp. (Diptera, Chironomidae).
Record: Costa Rica (Lichtwardt, 1994).

20. *St. paucispora* Lichtw, 2003

Thalli fusiform, 70-85 × 6-7 µm. Trichospores 1-2 per thallus, long-ellipsoidal, 29-40 × 6-8 µm, with a single appendage and without a collar. Conjugating thalli seen, no zygospores observed.

Host: *Cricotopus* larva (Diptera, Chironomidae, Orthocladiinae)
Record: Amazonas, Brazil (Alencar et al., 2003).

21. *St. pedifer* Williams & Lichtw., 1983 (in Lichtwardt & Williams, 1983)

Thalli up to 110 × 7-12 µm, basal cell with a foot-like extension that penetrates the peritrophic matrix. Trichospores ovoid with a slight median swelling, 2-8 per thallus, 20-30 × 7-10 µm, with one long narrowing appendage, without collar. Terminal trichospore subapical.

Zygospores (Type I, 48-69 × 11-19 µm, zygosporophore 1-2 × 5 µm) have been induced to develop *in vitro* by Beard and Adler (2003) by altering the pH of the culture medium (pH 10 using KOH solution.). Since they have not seen the detached zygospores, hence they may not be the matured ones.

Host: *Boreoheptagyia lurida* Garrett (Diptera, Chironomidae).
Record: U.S.A. (Lichtwardt and Williams, 1983; White et al., 2006), Norway (White and Lichtwardt, 2004), Spain (Valle, 2007), Canada (Strongman, 2007; Strongman and White, 2008).

22. ***St. penetralis*** Lichtw., 1984

Thalli 70-180 × 8-10 μm, base bulbous, penetrating the peritrophic matrix, producing 2-12 long-ellipsoidal trichospores, 30-50 × 8-12 μm, with a single appendage and no collar. Zygospores unknown.

Host: *Diamesa* spp. (Diptera, Chironomidae).
Record: Japan (Lichtwardt et al., 1987), France and Swiss Alps (Lichtwardt, 1973, 1984), China (Strongman and Xu, 2006), Spain (Valle, 2007), Canada (Strongman and White, 2008), Argentina (Siri and López Lastra, 2010).

23. ***St. platensis*** López Lastra, Lichtw. & Ferrington, 1999

Thalli 6-10 μm diam., consisting of 4 to more than 32 generative cells that may be short when many. Trichospores ellipsoidal with a slight median bulge, 25-36 × 5-7.8 μm, with a very short collar and one appendage that sometimes appears to split into two.

Host: *Chironomus* larva (Diptera, Chironomus).
Record: Argentina (Lichtwardt et al., 1999), Argentina (Siri and López Lastra, 2010).

24. ***St. prolifica*** Lichtw. Kobayasi & Indoh, 1987

Thalli 450 × 8-12 μm, producing up to 50 trichospores, 22-36 μm × 5-5.5 μm, without a collar.

Host: *Chironomus* sp. larva (Diptera, Chironomidae).
Record: Japan (Lichtwardt et al., 1987), Spain (Valle, 2007).

25. ***St. queenslandiae*** Lichtw., 1990

Thalli 60-100 × 7-8 μm, bulbous base penetrating the peritrophic matrix and producing up to 16 trichospores per thallus. Trichospores oval, 20-23 × 8-9 μm, apical trichospore subterminal. Zygospores unknown.

Host: Orthocladiinae larva (Chironomidae).
Record: Australia (Lichtwardt and Williams, 1990).

26. ***St. reflexa*** Lichtw. & M. C. Williams, 1988

Thalli 80-160 × 8-10 μm, producing 1-7 generative cells, basal cell sterile, holdfast inconspicuous. Trichospores long-ellipsoidal, 21-41 × 5-8 μm, without an appendage, remaining attached to the generative cells which articulate singly or in units of 2-4. Terminal trichospore typically reflexed. Zygopsores biconical, 49-55 × 7-8 μm, attached laterally and perpendicularly midway between the poles to a short zygosporophore.

Host: Bloodworms, *Cryptochironomus* sp. (Diptera, Chironomidae).
Record: U.S.A. (Lichtwardt and Williams, 1988).

27. ***St. robusta*** Lichtw. & M. C. Williams, 1999

Thalli 12-14 μm diam. Holdfast small secreted and discoid, generative cells 15 or more in number. Trichospores ellipsoidal with a slight median bulge, 30-37 × 8-9 μm, without a collar. Zygospores unknown.

Host: *Psectrocladius* sp. larva (Diptera, Chironomidae).
Record: U.S.A. (Lichtwardt and Williams, 1999), Spain (Valle, 2007).

28. ***St. stenospora*** Siri, M. M. White & Lichtw., 2006

Thalli 4.5-7 μm diam., trichospores 1-4, subcylindrical with a slight median bulge, 42-70 × 4-7 μm, with a short collar. Zygospores unknown.

Host: *Phaenopsectra* sp. larvae (Diptera: Chironomidae).
Record: U.S.A. (White et al., 2006).

29. ***St. thaumaleidarum*** Lichtw. & M. C. Williams, 1990

Thalli 75-200 × 5-8 μm, generative cells 2-16 in number, the terminal one slightly tapered. Trichospores elongate-ovoid, 20-38 × 5.5-8 μm, without a collar, apical trichospore subterminal. Zygospores unknown.

Host: *Austrothaumalea* sp. (Diptera, Thaumaleidae).
Record: Australia (Lichtwardt and Williams, 1990).

30. ***St. tianensis*** J. Wang, S. Q. Xu & Strongman, 2010

Thalli 130-350 × 10-15 μm, lobulate, bulbous base penetrates the peritrophic matrix, trichospores 4-25 in number, long-ellipsoidal, 55-90 × 7-10 μm, with a short collar and a single appendage. Zygospores unknown.

Host: Chironomid larvae (Diptera, Chironomidae).
Record: Xinjiang, China (Wang et al., 2010).

Smittium Poisson, 1936

Thalli variously branched. Trichospores ellipsoidal to cylindrical, with or without a collar and a single appendage. Zygospores when present biconical to fusiform, obliquely and submedially attached to the zygosporophore, detached ones with a collar and a single appendage. Found attached to the hindgut of larval insects.

Type species: ***Smittium arvernense*** Poisson

Key to species

1a. Zygospores known...2
1b. Zygospores unknown...30
2a. Zygospores biconical...3
2b. Zygospores fusiform and of other shape..19
3a. Two types of trichospores produced...4
3b. One type of trichospores produced...9
4a. Zygospores less than 70 μm long..5
4b. Zygospores more than 70 μm long..6
5a. Longer trichospores 31-38 × 4.5-6.5 μm with a 4-6 μm long collar..... ....***S. esteparum***
5b. Longer trichospores 33.5-46 × 8.5-14 μm with a 11-18 μm long collar........ ....***S. shaanxiense***
6a. Longer trichospores up to 40 μm long...7
6b. Longer trichospores more than 40 μm long......................................8
7a. Zygospores 78-82 × 13-18 μm..***S. biforme***
7b. Zygospores 81-98 × 9-11 μm..***S. orthocladii***
8a. Smaller trichospores 10-12 × 5.5-6.5 μm with a 2.5-4 μm long collar........... ....***S. dimorphum***

8b. Smaller trichospores 12-16.5 × 5.5-6 μm with a 12.5-18 × 1.5-2 μm cylindrical collar.... ....***S. pseudodimorphum***

9a. Zygospores up to 60 μm long...10

9b. Zygospores more than 60 μm long...14

10a. Zygospores up to 43 μm long..11

10b. Zygospores more than 43 μm long..12

11a. Trichospores 13-26 × 1.6-3 μm with a short collar...............***S. brasiliense***

11b. Trichospores 9-18 × 3.5-5.5 μm with a 1.5-3 μm long collar... ....***S. bulbosporophorus***

12a. Trichospores upto 32 μm long...13

12b. Trichospores upto 33-37 × 6.5-7 μm..............................***S. mucronatum***

13a. Trichospores 15-32 × 4-8 μm with a 3-9 μm companulate collar and monopodially branched thalli...***S. culicis***

13b. Trichospores 11-30 × 3-7 μm with 1-2.5 μm collar and verticillately branched thalli...***S. culisetae***

14a. Zygospores less than 90 μm in length...15

14b. Zygospores more than 90 μm in length...18

15a. Collar of trichospores up to 10 μm long..16

15b. Collar of trichospores more than 10 μm long......................................17

16a. Trichospores 17-27 × 5-8 μm with 5-10 μm long collar.........***S. fecundum***

16b. Trichospores 21-25 × 6.5-7.5 μm with 5-9 μm long collar...........***S. ouselli***

17a. Trichospores 23-44 × 10-14 μm with 10-19 μm cylindrical collar....***S.alpinum***

17b. Trichospores 17-32 × 7-9.5 μm with 10-15 μm long collar...***S. coloradense***

18a. Trichospores 20-26 × 3-4.5 μm with 0.5-1 μm long collar..............***S. rarum***

18b. Trichospores 40-55 × 6-10 μm with 10-17 × 4 μm long collar.... ....***S. longisporum***

19a. Zygospores fusiform in shape...20

19b. Zygospores other than fusiform in shape..27

20a. Thalli verticillately branched...21

20b. Thalli not verticillately branched...23

21a. Trichospores dimorphic and zygospores 160-190 × 15-18 μm with 17-30 × 10-13 μm collar...***S. heterosporum***

21b. Trichospores only of one type..22

22a. Zygospores 51-70 × 9-12 μm......................................***S. dipterorum***

22b. Zygospores 110-150 × 10-12 μm with 9-10 × 6 μm long collar................. ....***S. megazygosporum***

23a. Zygospores less than 90 μm long..24

23b. Zygospores more than 90 μm long..26

24a. Trichospores less than 22 μm long..25
24b. Zygospores 51-52 × 10-11 μm, trichospores 21-41 × 4-6 μm..*S. cylindrosporum*
25a. Zygospores 50-60 × 6-7 μm with 5-6 × 2-3 μm collar.................*S. pusillum*
25b. Zygospores 66-75 × 10-13 μm....................................*S. precipitiorum*
26a. Zygospores 95-115 × 10-12 μm with 6-10 × 6 μm collar...........*S. bisporum*
26b. Zygospores 90-128 × 9-12 μm..*S. rupestre*
27a. Zygospores less than 60 μm in length...28
27b. Zygospores 60-76 × 8-10 μm...*S. fruticosum*
28a. Trichospores more than 10 μm in length...29
28b. Trichospores 8-10 × 2.2-3 μm, zygospores cylindrical pointed, 42-55 × 4-4.5 μm...*S. microsporum*
29a. Zygospores 30-35 × 8-10 μm...*S. arvernense*
29b. Zygospores 48-51 × 3.5-4 μm..*S. insulare*
30a. Thalli monopodially or verticillately branched..31
30b. Thalli other than monopodially and verticillately branched.......................47
31a. Thalli monopodially branched, trichospores 15-24 × 7-9 with 1.5-6.5 μm campanulate or cylindrical collar...*S. arcticum*
31b. Thalli vertcillatelly branched...32
32a. Two types of trichospores reported, larger ones 30-49 × 3.5-5.5 μm while smaller ones 14.5-24 × 2.5-3.5 μm...*S. hecatei*
32b. One type of trichospores reported...33
33a. Trichospores more than 30 μm long..34
33b. Trichospores up to 30 μm long..36
34a. Trichospores 34-36 × 2.5-3.5 μm, 3 μm short collar with thin appendage...*S. naiadis*
34b. Trichospores 20-45 μm long..35
35a. Thallus base bulbous, trichospores 2.5-4.5 μm wide and collar 1-2 μm long...*S. gravimetallum*
35b. Thallus base not bulbous, trichospores 3-6 μm wide and collar 2-4 μm long...*S. elongatum*
36a. Holdfast conspicuous...37
36b. Holdfast inconspicuous..43
37a. Secreted holdfast all along the surface of the thallus, trichospores 26-30 × 3.5-4.5 μm, campanulate collar 4.5-5.5 × 2.5-3.5 μm........................*S. prostratum*
37b. Secreted holdfast not prostrate...38
38a. Secreted small holdfast...39
38b. Secreted holdfast knob-like, horseshoe shaped, disk-like or swollen...........40

39a. Base horseshoe shaped, thalli with 2-8 generative cells, trichospores 18-29 × 2-3.5 µm.. ***S. gracilis***

39b. Trichospores 17-32 × 1.5-3.5 µm with collar 2 µm long and fine appendage .... ***S. kansense***

40a. Secreted holdfast knob-like, horseshoe shaped................................41

40b. Secreted holdfast disk-like or swollen...42

41a. Secreted holdfast knob-like, trichospores 22.5-33 × 5-6.5 µm, 3 µm long narrow collar and thin appendage... ***S. caribense***

41b. Secreted holdfast horseshoe shaped, 2-8 generative cells/fertile branch, trichospores 18-30 × 2.5 µm, with short collar.......................... ***S. delicatum***

42a. Secreted holdfast disk-like, thallus base broad, 4-6 generative cells/fertile branch, trichospores 23-27 × 2-2.5 µm, cylindrical collar 5.5-6 × 2 µm........... .... ***S. inexpectans***

42b. Secreted holdfast swollen, trichospores 12-14 × 2.5-3 µm.......... ***S. paludis***

43a. Thallus base horseshoe shaped, trichospores 12-19 × 3-5 µm, collar 1.5-3 µm.. ***S. lentaquaticum***

43b. Thallus base not horseshoe shaped...44

44a. No. of generative cells reported...45

44b. No. of generative cells not reported..46

45a. Generative cells 1-2, trichospores 19-27 × 3-4 µm with a short cylindrical collar.. ***S. urbanum***

45b. Generative cells 5-6, trichospores 25-30 × 3-3.5 µm with 3-3.5 µm narrow collar.. ***S. typhellum***

46a. Trichospores 10-30 × 2.5-6 µm with 1-3 µm long collar and a single short appendage.. ***S. commune***

46b. Trichospores 14-30 × 2-3 µm with 2-3 µm long collar..... ***S. phytotelmatum***

47a. No of generative cells reported...48

47b. No. of generative cells not reported...64

48a. So far reported from Tupilid larval host...49

48b. So far reported from other than Tupilid larval host (Chiro, Simulii, etc.)....50

49a. Trichospores 15-20 × 3.5-5 µm with 2-3.2 µm long collar and generative cells upto 14 or more.. ***S. tipulidarum***

49b. Trichospores 28.5-36 × 5-6 µm with 5-6 µm tubular collar and generative cells 2-4.. ***S. chinliense***

50a. Trichospores up to 19 µm long...51

50b. Trichospores more than 19 µm long..53

51a. Trichospores 12-14 × 7-9 µm with finely punctate wall, generative cells 1-2 per fertile branch... ***S. incrassatum***

51b. Trichospores without punctate wall...52

52a. Trichospores 9-19 × 2-2.5 µm with 0.5-1 × 1.5 µm short collar, generative cells 4-12 per fertile branch..***S. brevisporum***

52b. Trichospores 9-14 × 3.5 µm with 2-6 µm long collar, 6 or more generative cells per fertile branch..***S. parvum***

53a. Thalli fascicled or in clumps...54

53b. Thalli not in fascicle or clumps...56

54a. Trichospores more than 6 µm wide, 20-28 µm long with 1-4 generative cells per fertile branch, base tapered..***S. culicisoides***

54b. Trichospores up to 6 µm wide...55

55a. Trichospores 20-25 × 5-6.5 µm with 3-4 µm long collar and 1-3 generative cells per fertile branch...***S. compactum***

55b. Trichospores 18-29 × 2-3.5 µm with very short collar and 2-6 generative cells per fertile branch..***S. fasciculatum***

56a. No. of generative cells more than 4 per fertile branch........................57

56b. No. of generative cells less than 4 per fertile branch..........................61

57a. Whole thallus fertile, trichospores 20-36 × 7-10 µm with 5-10 × 2.5 µm long collar...***S. cellaspora***

57b. Whole thallus not fertile...58

58a. No. of generative cells 4-8 per fertile branch.................................59

58b. No. of generative cells more than 8 per fertile branch........................60

59a. Holdfast disk-like, base swollen, trichospores20-23 × 2.5-3 µm...***S. chironomi***

59b. Holdfast refractive, base swollen, trichospores 16-30 × 3-7 µm with cylindrical collar...***S. simulii***

60a. Base of the thallus formed of 6 cells arranged in a ring, trichospores 18-33 × 4-8 µm with 3-6 µm long collar..***S. annulatum***

60b. Base of the thallus consisting of cylindrical cells, trichospores 19-23 × 6-8 µm with 5 µm long campanulate collar..................................***S. simulatum***

61a. Trichospores 50-60 × 7-8.5 µm with 10-14 × 5 µm collar......***S. gigasporus***

61b. Trichospores less than 50 µm in length...62

62a. Base of the thallus penetrating the peritrophic matrix forming bulbous swellings, trichospores 33-45 × 7-8.2 µm with 6-8 × 3-4.6 µm collar.............
....***S. perforatum***

62b. Base of the thallus not bulbous...63

63a. Basal cell tapering, trichospores 31-55 × 6-8 µm with a short collar...
....***S. pennelli***

63b. Basal cell cylindrical, trichospores verrucose, 26-42 × 6-7.5, with cylindrical 10-18 µm long collar...***S. macrosporum***

64a. Two types of trichospores produced, holdfast bulbous, larger ones 32-54 × 5-11 µm with 9-13 µm long collar, smaller 13.5-18.5 × 5-7 µm.................
....***S. ditrichosporum***

64b. Two types of trichospores not produced..65
65a. Trichospores more than 30 μm long..66
65b. Trichospores less than 30 μm long...68
66a. Trichospores larger 36-47 × 9-10 μm with 9-10 μm long campanulate collar... ....*S. magnosporum*
66b. Holdfast disk-like, trichospores smaller.......................................67
67a. Base ramified, trichospores 30-35 × 5-6 μm with 1-2 μm collar...... ....*S. basiramosum*
67b. Thalli profusely branched, trichospores 31-37 × 3-3.5 μm with a short blunt 3 × 3 μm collar..*S. tynense*
68a. Trichospores less than 20 μm in length.......................................69
68b. Trichospores more than 20 μm in length......................................72
69a. Trichospores less than 4.5 μm wide..70
69b. Trichospores more than 4.5 μm wide...71
70a. Known to kill the host (mosquitoes), trichospores 10-18 × 3.5-4.5 μm with 0.5-2.5 × 1-2 μm flared collar......................................*S. morbosum*
70b. Holdfast simple, trichospores 10-15 × 2.5-3.5 μm with 1.5-2 μm collar......... ....*S. minutisporum*
71a. Trichospores 10-20 × 4-9 μm with 5-20 μm long cylindrical collar...... ....*S. caudatum*
71b. Trichospores 16-21 × 3-6 μm with about 2 μm long collar.......*S. imitatum*
72a. Trichospores without visible collar, trichospores 28-30 × 3.4 μm, base tapering..*S. aciculare*
72b. Trichospores with visible collar...73
73a. Holdfast knob-like...74
73b. Holdfast not knob-like..75
74a. Trichospores 20-24 × 5.5-7.2 μm with 4 μm long collar, base swollen........... ....*S. bullatum*
74b. Trichospores 26-29.5 × 5 6 μm with campanulate long collar.......... ....*S. nodifixum*
75a. Base tapered/pointed..76
75b. Base not tapered/pointed...77
76a. Trichospores 21-30 × 4-6 μm with 1-2 μm flared collar and very fine spiralled appendage..*S. acutum*
76b. Trichospores 23-32 × 3.5-4 μm with short collar, and a single fine appendage... ....*S. fastigatum*
77a. Trichospores 17-26 × 2.3-2.8 μm with 3-4 μm cylindrical collar... ....*S. angustum*

77b. Trichospores 15-28 × 3-5 μm with 1-3 μm long collar.............. ....***S. tronadorium***

Description

1. *Smittium aciculare* Lichtw., 1990

Thalli, 150-350 μm long, branched sparsely or unbranched, acicular, sometimes fascicled, base tapered without obvious mucilage. Trichospores long fusiform-ellipsoidal, 28-30 × 3.4 μm, without a visible collar. Zygospores unknown.

Host: Larval Simuliidae (*Simulium clathrinum* Mackerras & Mackerras or *Austrosimulium mirabile*), *Simulium trombetense.*

Record: Queensland, Australia (Lichtwardt and Williams, 1990), Amazonas, Brazil (Alencar et al., 2003).

2. *S. acutum* Lichtw. & Grigg, 1998

Thalli branched, basal cells tapered and pointed. Trichospores 21-30 × 4-6 μm, with a collar 1-2 μm long, flared outward, and a very fine spiralled to tortuous appendage. Zygospores unknown.

Host: Chironomidae larvae.

Record: U.S.A. (Lichtwardt and Grigg, 1998).

3. *S. alpinum* Lichtw., 1984

Long generative cells up to 3 per fertile branch. Trichospores oval to biconical, 23-44 × 10-14 μm, collar cylindrical 10-19 × 2-4 μm. Zygospores biconical, 63-83 × 14-18 μm with a single appendage and a collar 15-28 × 4-7 μm attached laterally at one end of the zygospore.

Host: Larval *Diamesa* spp. (Diptera, Chironomidae).

Record: Glacier National Park, Montana, U.S.A., Sweden, France and Switzerland (Lichtwardt, 1984, 1986), Spain (Valle and Santamaria, 2004).

4. *S. angustum* Williams & Lichtw., 1992

Thalli growing in discrete radiating colonies. Trichospores subcylindrical, 17-26 × 2.3-2.8 μm, collar cylindrical 3-4 μm long. Zygospores unknown.

Host: *Cricotopus* sp. (Chironomidae, Orthocladiinae).

Record: Western Australia (Lichtwardt and Williams, 1992b).

5. *S. annulatum* Lichtw., 1997

Thalli small and compact, arising from a prominent basal structure consisting of about six cells arranged in a ring. Eight or more generative cells per fertile branch. Trichospores elongate-oval, 18-33 × 4-8 μm, collar 3-6 μm long. Zygospores unknown.

Host: *Simulium* sp. larvae (Simuliidae).
Record: Costa Rica (Lichtwardt, 1997).

6. *S. arcticum* Y. Kobayasi, 1969

Thalli 300-400 μm or longer. Basal cell short cylindrical 8-10 μm diam., producing nondivergent monopodial branches, 3.5-5 μm diam. Generative cells 1-5 per fertile branch. Trichospores oval 15-24 × 7-9 μm, with a short campanulate or cylindrical collar 1.5-6.5 μm long, and a single appendage. Zygospores unknown.

Host: Chironomidae (Diptera) larvae.
Record: Peters Lake, Alaska (Kobayasi et al., 1969).

7. *S. arvernense* Poisson, 1936

Thalli up to 1.2 mm long, sparsely branched, holdfast disk-like.Trichospores ellipsoidal, 20-25 × 5 μm, with a short collar and a single short fine appendage. Zygospores 30-35 × 8-10 μm, with a small oblique submedial collar ~2 μm long.
This is a type species.

Host: *Smittia* sp. (Diptera, Chironomidae) larvae.
Record: France (Poisson, 1936; Lichtwardt, 1986).

8. *S. basiramosum* Siri, 2010

Thalli up to 350 μm long, sparsely branched at the base, holdfast disk-like. Trichospores subcylindrical, 30-35 × 5-6 μm, collar 1-2 μm, appendage single. Zygospores unknown.

Host: Larvae of *Polypedilum* sp. (Diptera: Chironomidae).
Record: Argentina (Siri and López Lastra, 2010).

9. *S. biforme* M. M. White & Lichtw., 2004

Thalli produce two types of trichospores, long-ellipsoidal ones 34-42 × 9-12 μm, with a campanulate collar; and oval trichospores, 13-15 × 6-7.5 μm, with a cylindrical collar. Zygospores biconical (Type II), 78-82 × 13-18 μm.

Host: Larval *Diamesa aberrata* Lundbeck and possibly *D. bertrami* Edwards (Chirinomidae, Diamesinae).
Record: Norway (White and Lichtwardt, 2004).

10. *S. bisporum* Manier & Coste, 1971

Thalli 500-600 μm, holdfast small. Trichospores 1-2 per fertile branch, subcylindrical with a slight median swelling, 35-40 × 6-7 μm, collar 6-7 × 3-4 μm. Zygospores fusiform, 95-115 × 10-12 μm, with a collar 6-10 × 6 μm, and a single long appendage.

Host: *Psectrotanypus varius* Fabr. (Diptera, Chironomidae) larvae.
Record: France (Coste-Mathiez, 1970; Manier and Coste, 1971).

11. *S. brasiliense* Alencar, Lichtw., Ríos-Velásquez & Hamada, 2003

Thalli in clumps, branching verticillate. Holdfast horseshoe shaped. Trichospores ellipsoidal, 13-26 × 1.6-3 μm with a very short collar and a single appendage. Zygospores biconical, 32-43 × 4-5 μm, collar at one end and a single appendage.

Host: *Simulium goeldii, S. ulyssesi,* and *S. rorotaense* larvae (Diptera: Simuliidae).
Record: Amazonas region, Brazil (Alencar et al., 2003)

12. *S. brevisporum* L. G. Valle & Santam., 2004

Thalli profusley branched. Holdfast disk-like. Generative cells 4-12 per fertile branch. Trichospores long-ellipsoidal 9-19 × 2-2.5 μm, with a collar 0.5-1 × 1.5 μm and a short thick appendage. Zygospores unknown.

Host: *Cricotopus* spp. and other larvae (Chironomidae, Orthocladiinae).
Record: Iberian Peninsula-Spain (Valle and Santamaria, 2004), Canada (Strongman, 2007; Strongman and White, 2008).

13. *S. bulbosporophorus* L. G. Valle & Santam., 2004

Thalli sparsely branched, basal cells broad. Holdfast secreted disk-like. Generative cells 2-8 per fertile branch. Trichospores obovate-ellipsoidal 9-18 × 3.5-5.5 µm, with a long collar, 1.5-3 × 2 µm, and a single long appendage. Zygospores biconical, 35-40 × 7-8 µm, with a globose collar, 3.5-4 × 3.5 µm, and an appendage. Zygosporophore attached at the last quarter of the zygospore length.

Host: Larvae of Chironomidae, Orthocladiinae and Diamesiinae.
Record: Spain (Valle and Santamaria, 2004).

14. *S. bullatum* Lichtw. & Williams, 1992

Young thalli consisting of an axis of wider cells. Basal cell swollen. Holdfast, a small knob-like projection. Trichospores biconical, 20-24 × 5.5-7.2 µm, with a collar about 4 µm long. Zygospores unknown.

Host: *Naocladius forsythi* Boothroyd larva (Chironomidae, Orthocladiinae).
Record: New Zealand (Lichtwardt and Williams, 1992c).

15. *S. caribense* L. G. Valle & Cafaro, 2010

Thalli branched, rarely verticillate, basal cell secreting a conical or knobby holdfast. Generative cells 1-4 per fertile branch. Trichospores ellipsoidal, 22.5-33 × 5-6.5, with a thin appendage and a narrow collar, 3 µm long. Zygospores unknown.

Host: Chironomid larvae (Diptera: Chironomidae).
Record: Dominican Republic (Valle and Cafaro, 2010).

16. *S. caudatum* Lichtw. & Grigg, 1998

Trichospores oval to long-oval, 10-20 × 4-9 µm, produced on long cylindrical extensions of their generative cells, released trichospore with cylindrical collar 5-20 µm long. Basal cells up to 12-16 µm diam. Zygospores unknown.

Host: Chironomidae larvae.

Record: U.S.A. (Lichtwardt and Grigg, 1998).

17. *S. cellaspora* Williams, 1982

Thalli upto 300 μm long, sparsely branched, almost all cells becoming fertile. Trichospores ellipsoidal, 20-36 × 7-10 μm, with a collar, 5-10 × 2.5 μm, single long appendage often coiled. Zygospores unknown.

Host: *Sympotthastia* sp. and other Chironomidae (Diptera) larvae.

Record: U.S.A. (Williams, 1982), Argentina (López Lastra et al., 2005).

18. *S. chinliense* Strongman & Shengquan Xu, 2006

Thalli sparsely branched, holdfast foot-like, fertile branches bearing 2-4 trichospores at their tips. Trichospores cylindrical to subcylindrical, 28.5-36 × 5-6 μm, collar tubular, 5-6 × 4 μm, with a single appendage. Zygospores unknown.

Host: Tipulid larva (*Antocha* sp.).

Record: China, Shaanxi Province (Strongman and Xu, 2006).

19. *S. chironomi* Manier, 1970b (1969b)
[= ***Typhella chironomi*** (Tuzet & Manier) Manier & Mathiez, 1965, ***nom. nud.***]
[= ***Genistella chironomi*** Tuzet & Manier, 1953 ***nom. nud.***]

Thalli with main axes up to 800 μm long. Holdfast disk-like. Main axis and larger basal branches somewhat swollen at their base, 2.5-5 μm diam, tapering abruptly to 1.5-2 μm diam. Fertile upper branches of main axis each with 4-8 generative cells producing subcylindrical trichospores with slight median swelling, 20-23 × 2.5-3 μm, collar 1-2 μm long × 1.5-2.5 μm wide. Zygospores unknown.

Host: Orthocladiinae larvae (Diptera, Chironomidae).

Record: France [(Manier, 1962, 1969 (1970); Tuzet and Manier, 1953], possibly in England (Moss, 1972), Spain (Santamaria and Girbal, 1997; Valle and Santamaria, 2004).

20. ***S. coloradense*** Lichtw. & Williams, 1987

Basal cell tapered toward the base, sometimes with small lateral projections. Trichospores elongate-ellipsoidal 17-32 × 7-9.5 µm, with collar tapering towards the distal end, 10-15 × *ca.* 3 µm, appendage single that tends to curl and is greater than 6 times the spore length.

Beard and Adler (2003) described zygospores in this species. The zygospores are Type II, 68-85 × 11-17 µm, collar narrowing, 15 × 5 µm.

Host: Larval Chironomidae [(*Eukiefferiella* sp., *Cricotopus bicinctus* or *Orthocladius* (*Eudactylocladius*), *Cricotopus* sp.)] and possibly Simuliidae (*Prosimulium* sp.).

Record: U.S.A. (Williams and Lichtwardt, 1987; Beard and Adler, 2003), Norway (White and Lichtwardt, 2004).

21. ***S. commune*** Lichtw. & Grigg, 1998

Thalli with verticillate branching, fertile branches producing few to many trichospores. Trichospores narrow-ellipsoidal with a slight median bulge, 10-30 × 2.5-6 µm, with a short collar 1-3 µm long and a single short appendage. Zygospores unknown.

Host: Larval Chironomidae.

Record: U.S.A. (Lichtwardt and Grigg, 1998).

22. ***S. compactum*** Lichtw. & Williams, 1992

Thalli in compact clusters. Trichospores short-cylindrical with swollen midregion, 1-3 per fertile branch, 20-25 × 5-6.5 µm, collar 3-4 µm long. Zygospores unknown.

Host: Chironomidae larvae.

Record: Tasmania, Australia (Lichtwardt and Williams, 1992a).

23. ***S. culicis*** Manier, 1970b (1969b)
[= ***Orphella culici*** Tuzet & Manier, 1947, ***nom. nud.***]
[= ***Rubetella culicis*** Tuzet, Rioux & Manier, 1961, ***nom. nud***.]

Thalli with divergent, often monopodial branches, holdfast inconspicuous, base sometimes pseudorhizoidal. Terminal branches are fertile and often arched.

Generative cells 4-6 per fertile branch. Trichospores oval 15-32 × 4-8 μm, collar more or less campanulate, 3-9 μm long.

Zygospores, unknown originally, have been found in Spanish collection by Valle & Santamaria (2004) in *Eukiefferiella* sp. (Orthocladiinidae). Zygospores biconical, 47-55 × 11-12 μm, with a collar, 4.5-8 × 3.5-5.5 μm. Zygospores developing thalli bear smaller and lesser number of trichospores (14.5-19.5 × 5-6 μm) per fertile branch. Zygospores seen at the top of a peduncle from one of the conjugants.

Host: Many genera and species of larval Culicidae (Diptera), rarely in Simuliidae and Chironomidae larvae, *Eukiefferiella* sp. (Orthocladiinidae).

Record: France (Tuzet and Manier, 1947; Tuzet et al., 1961; Manier et al., 1964; Manier, 1969 (1970)), U.S.A., Tunisia, Canada, New Zealand (Lichtwardt, 1986; Lichtwardt and Williams, 1992c; Strongman and White, 2008), Chile (Lichtwardt and Arenas, 1996), Argentina (Lichtwardt et al., 1999; López Lastra et al., 2005; Siri and López Lastra, 2010), Spain (Valle and Santamaria, 2004), Norway (White and Lichtwardt, 2004), China (Strongman et al., 2010).

24. ***S. culicisoides*** Lichtw., 1997

Thalli 200-500 μm long. Bases tapered, sometimes in fascicles, occasionally with minute projections from the wall. Generative cells 1-4 per fertile branch. Trichospores oval, 20-28 × 6-10 μm, collar 5-10 μm. Zygospores unknown.

Host: *Cardiocladius* sp., *Cricotopus* sp., *Chironomus* sp. and other Chironomidae larvae, and *Simulium* sp. (Simuliidae).

Record: Costa Rica (Lichtwardt, 1997).

25. ***S. culisetae*** Lichtw., 1964
[= ***Smittium inopinatum*** Manier, 1969 (1970)
= ***Rubetella inopinata*** Manier, Rioux & Whisler, 1961, ***nom. nud.***]

Thalli large, holdfast inconspicuous, branching verticillate, sporulation profuse. Trichospores long-ovoid, 11-30 × 3-7 μm, 4-10 per fertile branch, collar short 1-2.5 μm long often flared outward, fine short appendage. Zygospores rare, biconical, 46-58 × 5.5-8 μm, with a collar 3.5-4.5 μm, medially and perpendicularly attached to the body of zygospore.

Host: Species of Culicidae (Diptera) larvae, rarely in Simuliidae, Chironomidae and Ceratopogonidae (Diptera) hindguts. Once found in Ephemeroptera larva also.

Record: U.S.A. (Lichtwardt, 1964; Williams, 1983a,b; White et al., 2006), Japan, Australia, New Zealand (Lichtwardt, 1986), France (Manier et al., 1961, 1964; Manier, 1969 (1970), Costa Rica (Lichtwardt, 1994), Spain (Santamaria and Girbal, 1997; Valle and Santmaria, 2004, 2009), Brazil (Alencar et al., 2003), Argentina (López Lastra, 1997; López Lastra et al., 2005), Canada (Strongman and White, 2008), Dominican Republic (Valle and Cafaro, 2010).

26. ***S. cylindrosporum*** Lichtw. & Arenas, 1996

Thalli profusely branched, basal cells straight. Distal branches verticillate. Trichospores cylindrical, sometimes with a slight median bulge, 21-41 × 4-6 μm. Collar about 4-6 μm long. Appendage not spiralled. Zygospores are known from Argentinian collection (Lichtwardt et al., 1999) that measure 51-52 × 10-11 μm.

Host: *Paraheptagyia cinerascens* (Edwards) larva (Diamesinae).

Record: Chile (Lichtwardt and Arenas, 1996), Argentina (Lichtwardt et al., 1999).

27. ***S. delicatum*** Lichtw., 1990

Thalli delicate, sparsely branched, branching verticillate, branches 6 μm at the base while about 3 μm near the tip. Holdfast cell horseshoe-shaped. Generative cells 2-8 per fertile branch. Trichospores subcylindrical, 18-30 × 2.5 μm, with a very short collar, occasionally arising adventitiously from a cell along a main branch. Zygospores unknown.

Host: *Chironomus alternans* Walker, *Cladopelma* sp. (Diptera: Chironomidae, Chironominae).

Record: New South Wales-Australia (Lichtwardt and Williams, 1990), U.S.A. (White et al., 2006).

28. ***S. dimorphum*** Lichtw. & Williams, 1983

Thalli produce two types of trichospores. Long-ellipsoidal trichospores measure 38-50 × 5.5-6.5 μm with a long collar 2.5-5 μm. Oval trichospores measure 10-12 × 5.5-6.5 μm with a collar 2.5-4 μm long. Zygospores biconical measuring 71-96 × 13-14 μm, with an obliquely angled collar 12-17 × 5-7 μm, with a single appendage.

Host: *Boreoheptagyia lurida* (Garrett) larvae (Diptera, Chironomidae).
Record: U.S.A. (Lichtwardt and Williams, 1983).

29. *S. dipterorum* Lichtw., 1997

Thalli with verticillate branches, generative cells up to 8 per fertile branch. Holdfast simple secreted. Trichospores long-ellipsoidal, almost cylindrical, 10-26 × 2-4 μm, with a very fine appendage, collar 1-3 μm long. Zygospores 51-70 × 9-12 μm.

Host: *Cardiocladius* sp., *Cricotopus* sp., *Orthocladius* (*Euorthocladius*) (Chironomidae), *Simulium samboni*, *S. callidium* and unidentified spp. (Simuliidae).
Record: Costa Rica (Lichtwardt, 1997), Spain (Valle and Santamaria, 2004), Dominican Republic (Valle and Cafaro, 2010).

30. *S. ditrichosporum* Strongman, 2007

Thalli densely branched, terminal branchlets sporulating richly. Holdfast bulbous. Trichospores dimorphic, larger trichospores, long-ellipsoidal 32-54 × 5-11 μm, smaller ones oval, 13.5-18.5 × 5-7 μm. Collar long, 9-13 μm, with one appendage in both the types of trichospores. Zygospores unknown.

Host: Larval Chironomidae.
Record: Canada (Strongman, 2007).

31. *S. elongatum* Lichtw., 1972

Thalli profusely branched, wider hyphae verticillately branched, holdfast inconspicuous. Trichospores long-ellipsoidal, 20-44 × 3-6 μm, with a 2-4 μm long collar, a single sometimes spiraled, appendage. Zygospores unknown.

Host: Larval *Diamesa* (Diptera, Chironomidae).
Record: U.S.A. (Lichtwardt, 1972).

32. *S. esteparum* Ferrington, Lichtw. & López Lastra, 1999

Thalli with two types of trichospores, subcylindrical ones measuring 31-38 × 4.5-6.5 μm with a collar 4-6 μm long, ellipsoidal ones with a median bulge,

12-14 × 4-5 μm with a collar 2 μm long. Zygospores about 50 × 10 μm, attached about half-way between mid region and lower end of zygospore.

Host: Larvae of *Cricotopus* sp., *Eukiefferiella* sp. (Diptera: Chironomidae).
Record: Argentina (Lichtwardt et al., 1999).

33. ***S. fasciculatum*** Lichtw., 1994

Thalli in clumps up to 750 μm long. Branches tapered, terminating into fine sterile curved to circinate tips. Branches in the hindgut terminating in series of 2-6 generative cells, trichospores subcylindrical, 18-29 × 2-3.5 μm with a very short (< 1 μm) collar and short appendage. Zygospores unknown.

Host: *Chironomus* sp. (Chironomidae, Chironomini).
Record: Costa Rica (Lichtwardt, 1994).

34. ***S. fastigatum*** Lichtw. & Williams, 1992a

Thalli up to 500 μm long, basal cells pointed, branches erect. Trichospores subcylindrical with a median swelling, 23-32 × 3.5-4 μm, with a short collar and a single fine appendage. Zygospores unknown.

Host: Chironomidae larvae.
Record: Tasmania, Australia (Lichtwardt and Williams, 1992a).

35. ***S. fecundum*** Lichtw. & M. C. Williams, 1999

Thalli producing prolific short generative cells. Basal branches produce small tuberculate projections from the wall. Trichospores ellipsoidal with a slight median bulge, 17-27 × 5-8 μm, with a 5-10 μm long collar. Zygospores unknown in original form, but reported later in a Spanish collection. Zygospores biconical, 70-85 × 18-19 μm, with the zygosporophore eccentrically placed at approximately one-third the distance from the apex, collar 9-12 μm.

Host: Larvae of *Psectrocladius* sp. (Diptera, Chironomidae).
Record: U.S.A. (Lichtwardt and Williams, 1999), Spain (Valle and Santamaria, 2004).

36. *S. fruticosum* Williams & Lichtw., 1992b

Thalli forming bushy colonies up to 500 μm diam. Trichospores ellipsoidal with a median swelling, 21-28 × 4.5-5 μm, collar 3.2-4.8 μm long. Zygospores (Type II) 60-76 × 8-10 μm, attached to zygosporophore about 1/3 distance from one end.

Host: *Cricotopus* sp. larvae (Chironomidae, Orthocladiinae).
Record: Western Australia (Lichtwardt and Williams, 1992b).

37. *S. gigasporus* Williams & Lichtw., 1984

Thalli profusely branching near base. Trichospores 1-3 per fertile branch, elongate-ellipsoidal to subcylindrical measuring 50-60 × 7-8.5 μm, with a collar 10-14 × 5 μm, and a helical very long fine appendage. Zygospores unknown.

Hosts: Larval *Pagastia* sp. (Diptera: Chironomidae).
Record: U.S.A. (Williams and Lichtwardt, 1984).

38. *S. gracilis* L. G. Valle & Santam., 2004

Thalli with 2-6 branches arising from a horseshoe-shaped basal cell, secreted small holdfast. Secondary and tertiary branches verticillately arranged at the upper zone and tapering. Generative cells 2-8 per fertile branch. Trichospores subcylindrical, 18-29 × 2-3.5 μm, with inconspicuous collar, 1 × 1 μm, and a very thin appendage. Zygospores unknown.

Host: *Chironomus* sp. larvae (Chironomidae, Chironomini, Diamesinae).
Record: Spain (Valle and Santamaria, 2004), Canada (Strongman, 2007).

39. *S. gravimetallum* Lichtw., Ferrington & Hayford, 2000

Thalli verticillately branched, basal cell bulbous. Trichospores ellipsoidal, with a median bulge, 20-45 × 2.5-4.5 μm, collar 1-2 μm long, with a very fine appendage. Zygospores unknown.

Host: *Dicrotendipes fumidus* (Chironomidae).
Record: Kansas, U.S.A. (Ferrington et al., 2000).

40. *S. hecatei* L. G. Valle & Santam, 2004

Thalli branched, verticillate branches in the middle and upper regions. Basal cells give rise to bulbous lateral protuberances. Holdfast secreted. Fertile branches with 8-14 or more generative cells. Trichospores subcylindrical to long-ellipsoidal and are of two types: type α, 30-49 × 3.5-5.5 μm, with a collar, 3.5-6 × 3.5 μm; type β, 14.5-24 × 2.5-3.5 μm, with a collar, 1.5-2.5 × 2-2.5 μm. Zygospores unknown.

Host: Chironomidae Diamesini (Diamesinae) larvae, Orthocladiinae larvae.
Record: Spain (Valle and Santamaria, 2004), China (Strongman and Xu, 2006).

41. *S. heterosporum* L. G. Valle & Santam., 2004

Thalli branched, branches verticillate in the middle and upper regions of the gut. Holdfast simple, discoid or slightly campanulate. Trichospores dimorphic: type α, long-ellipsoidal, 36-74 × 7.5-13 μm, with a delicately granulated surface, collar slightly narrowing toward the end, 10-20 × 3-5.5 μm, filiform appendage; type β trichospores smaller, ovate-ellipsoidal or pyriform, 14-27 × 8.5-11 μm, collar 8-9 × 1.5 μm, appendage identical to that of type α. At the lowermost and medial areas of fertile branches, the generative cells may not arise directly from the main axis but from the lateral peduncles, variable in size (18-50 μm). Zygospores fusiform, 160-190 × 15-18 μm, with a collar, 17-30 × 10-13 μm, and a large appendage. Zygosporophore attached near one of the polar ends.

Host: Orthocladiinae (*Cricotopus bicinctus* Meig.), Diamesiinae (*Sympotthastia* spp., *Potthastia* spp.), Chironomidae larvae.
Record: Spain (Valle and Santamaria, 2004).

42. *S. imitatum* Lichtw. & Arenas, 1996

Thalli in clusters, branching compactly. Holdfast region not horseshoe-shaped. Trichospores 16-21 × 3-6 μm, cylindrical but swollen in the middle; collar about 2 μm long. Zygospores unknown.

Host: *Simulium* sp. larvae (Simuliidae).
Record: Chile (Lichtwardt and Arenas, 1996), Argentina (López Lastra et al., 2005).

43. ***S. incrassatum*** Y. Kobayasi, 1971

Thalli profusely branched near the base, 100-130 μm, branches, 4-7 μm diam., often constricted at septa. Trichospores ellipsoidal, 12-14 × 7-9 μm with thick and finely punctate wall, developing from a 7-10 × 3-4 μm outgrowth of the generative cell, 1-2 per fertile branch, released trichospores with a short collar and a single long appendage. Zygospores unknown.

Host: Chironomidae (Diptera) larvae.
Record: Angmagssalik, Greenland (Kobayasi et al., 1971).

44. ***S. inexpectans*** L. G. Valle & Santam., 2004

Thalli pinnate to verticillately branched. Holdfast simple, disk-like, and secreted. Fertile branches with 4-6 generative cells. Trichospores subcylindrical, often with a median bulge, 23-27 × 2-2.5 μm, with a cylindrical collar, 5.5-6 × 2 μm. Zygospores unknown.

Host: Orthocladiinae Chironomidae larvae.
Record: Spain (Valle and Santamaria, 2004).

45. ***S. insulare*** Strongman, 2007

Thalli sparsely branched. Holdfast inconspicuous. Trichospores sparse, produced distally, normally 2 per fertile branch, 15-20 × 3-4 μm, slightly swollen medially, with a short (2.5-4 μm) collar and a single appendage. Zygospores (Type II) elongate, thin, 48-51 × 3.5-4 μm, nearly cylindrical, narrowing abruptly at both ends, attached obliquely to the zygosporophore.

Host: Midge larvae (Chironomidae).
Record: Canada (Strongman, 2007; Strongman and White, 2008).

46. ***S. kansense*** Lichtw. & Grigg, 1998

Thalli sometimes verticillately branched, with a small secreted holdfast. Trichospores subcylindrical with a median bulge, 17-32 × 1.5-3.5 μm, with a collar (2 μm long) and a very fine appendage. Zygospores unknown.

Host: Larval Chironomidae.
Record: U.S.A. (Lichtwardt and Grigg, 1998).

47. ***S. lentaquaticum*** Siri, M. M. White & Lichtw., 2006

Thalli up to 600 μm long, branching verticillate, basal cells horseshoe shaped. Trichospores long-ellipsoidal, with a slight median bulge, 12-19 × 3-5 μm, with a 1.5-3 μm long collar. Zygospores unknown.

Hosts: Bloodworms, *Chironomus* sp. larvae (Diptera: Chironomidae).
Record: U.S.A. (White et al., 2006).

48. ***S. longisporum*** Williams, Lichtw. & Peterson, 1982

Trichospores long fusiform-ellipsoidal, 40-55 × 6-10 μm, with a long 10-17 × 4 μm collar and a long appendage. 1-2 trichospores per fertile branch. Zygospores fusiform-biconical, 102-113 × 13-17 μm, with a collar 28-32 × 5 μm at the lower tip of the zygospore and much longer appendage.

Host: *Cricotopus* sp. and other Chironomidae (Diptera) larvae.
Record: U.S.A., Sweden (Williams et al., 1982).

49. ***S. macrosporum*** Y. Kobayasi, 1969

Thalli arising from a cylindrical basal cell, branching variously, axial hyphae many, 6.5-13 μm diam. Fertile branches 3-6.5 μm diam, with 1, rarely 2, long generative cells producing outgrowths 13-20 × 2.5-4 μm bearing fusiform-ellipsoidal trichospores 26-42 × 6-7.5 μm with a finely or coarsely verrucose surface, released trichospores have a cylindrical collar 10-18 μm long and a single appendage. Zygospores unknown.

Host: Chironomidae (Diptera) larvae.
Record: Alaska, U.S.A. (Kobayasi et al., 1969).

50. ***S. magnosporum*** Ferrington, Hayford & Lichtw., 2005

Thalli with a secreted holdfast. Trichospores ellipsoidal, 36-47 × 9-10 μm, collar campanulate measuring 9-10 × 4-5 μm, less conspicuous appendage. Zygospores unknown.

Host: Larvae of Thaumaleidae (Diptera).
Record: Australia (Ferrington et al., 2005).

51. *S. megazygosporum* Manier & Coste, 1971

Thalli in tufts, verticillately branched. Trichospores nearly cylindrical with a slight median swelling, 36-49 × 3.5-5 μm, with a campanulate collar 3.5-4 × 5-6 μm and a prominent single appendage. Zygospores fusiform, 110-150 × 10-12 μm, with a collar 9-10 × 6 μm located on the lower fourth of the zygospore.

Host: Larvae of *Syncricotopus rufiventris* Meig. (Diptera: Chironomidae).
Record: France (Manier and Coste, 1971), Spain (Valle and Santamaria, 2004).

52. *S. microsporum* Williams & Lichtw., 1992b

Trichospores produced in series of up to 15 on recurved fertile branches, ellipsoidal, 8-10 × 2.2-3 μm, with a short collar and inconspicuous appendage. Zygospores (Type II) cylindrical with pointed ends, 42-55 × 4-4.5 μm, attached to zygosporophore 6-10 μm from one end.

Host: *Tanytarsus* sp. (Chironomidae, Tanytarsini).
Record: Western Australia (Lichtwardt and Williams, 1992b).

53. *S. minutisporum* Lichtw., Siri & M. M. White, 2006

Thalli without verticillate branching, holdfast simple. Trichospores elliptical, with a slight median bulge, 10-15 × 2.5-3.5 μm, with a collar 1.5-2.5 μm long. Zygospores unknown.

Host: *Ochlerotatus japonicus* Theobold larvae (Diptera: Culicidae).
Record: U.S.A. (White et al, 2006).

54. *S. morbosum* Sweeney, 1981a

Thalli in a clump of 0.5-2 mm diam, branches 2.5-4.5 μm diam. Generative cells 4-8 μm long, producing narrowly ellipsoidal trichospores 10-18 × 3.5-4.5 μm, collar slightly flared, 0.5-2.5 × 1-2 μm. Zygospores unknown.

Host: Larval *Anopheles hilli* Woodhill & Lee and *A. annulipes* Walker, also in some species of *Aedes* and *Culex* (Diptera: Culicidae).
Record: Australia (Sweeney, 1981a, b), Japan (Sato et al., 1989), Argentina (Garcia et al., 1994).

55. *S. morbosum* var. *rioplatensis*, López Lastra, 1990

Trichospores ellipsoidal, 12-23 × 3-4.7 μm. Zygospores unknown.

Host: Larval *Culex dolosus* (Diptera)
Record: Argentina (López Lastra, 1990)

56. *S. mucronatum* Manier & Mathiez ex Manier, 1969 (1970) [= *Smittium mucronatum* Manier & Mathiez, 1965, *nom. nud.*]

Trichospores elongate-ellipsoidal, 33-37 × 6.5-7 μm, with a minute apical nipple, collar cylindrical to campanulate, 7.5-9 × 3.3-4 μm, single long and fine appendage. Zygospores biconical, 44-60 × 11-13.5 μm, collar 11-13.5 × 3.4-5.5 μm attached to lower third of the zygospore, with a single fine appendage.

Host: Larval *Psectrocladius sordidellus* (Zett.) Edw. (Diptera, Chironomidae).
Record: France (Manier, 1969 (1970); Manier and Mathiez, 1965), U.S.A. (Lichtwardt and Williams, 1999), Norway (White and Lichtwardt, 2004), Canada (Strongman and White, 2008).

57. *S. naiadis* Strongman & Shengquan Xu, 2006

Thalli diffuse arising from a simple cluster of basal cells covered with mucilage, secondary branches verticillate. Trichospore cylindrical, long and narrow, 34-36 × 2.5-3.5 μm, with a short 3 μm collar, and a thin single appendage. Collars show 1 or 2 thickenings. Zygospores unknown.

Host: Bloodworms (Chironomidae).
Record: China, Shaanxi Province (Strongman and Xu, 2006).

58. *S. nodifixum* Strongman & Shengquan Xu, 2006

Thalli compact arising from a knobby holdfast, branching sparse. Trichospore subcylindrical 26-29.5 × 5-6 μm with a campanulate 5-6 μm long collar. Zygospores unknown.

Host: Bloodworms and other midge species (Chironomidae).
Record: China, Shaanxi Province ((Strongman and Xu, 2006).

59. ***S. orthocladii*** Manier, 1969 (1970), **emend.** Lichtw., 1984

Thalli compactly branched, smaller basal thalli often fascicled and sometimes enveloped in a brownish gelatinous sheath. Generative cells 2-10 per fertile branch. Trichospores are of two types: long-ellipsoidal, 25-40 × 6-8 µm with a cylindical collar 5-10 µm long, and small and oval, 8-10 × 5-6 µm with a cylindrical collar 5-10 µm long. Zygospores biconical, subcylindrical in the middle, 81-98 × 9-11 µm, with a collar 12-18 µm long attached close to one end of the zygospore, and bearing a single appendage.

Host: Larval *Orthocladius* spp., *Diamesa* sp., and other lotic Chironomidae (Diptera).

Record: France [Manier, 1969 (1970), Lichtwardt, 1984].

60. ***S. ouseli*** Williams & Lichtw., 1984

Thalli producing 1-3 trichospores per fertile branch. Trichospore long-ellipsoidal, 21-25 × 6.5-7.5 µm, collar 5-9 µm long. Zygospore biconical, 77-88 × 12-15 µm, collar obliquely angled, attached to the zygospore wall one-third to one-fourth the length from one end, appendage single.

Host: Larval *Eukiefferiella* sp. (Diptera: Chironomidae).

Record: U.S.A. (Williams and Lichtwardt, 1984).

61. ***S. paludis*** Williams & Lichtw., 1990

Thalli with mostly verticillate branchlets, trichospore 12-14 × 2.5-3.0 µm. Holdfast cell swollen, incurved at the point of attachment to the peritrophic matrix. Zygospores unknown.

Host: *Tanytarsus* nr. *inextentus* larvae (Diptera: Chironomidae).

Record: Tasmania, Australia (Lichtwardt and Williams, 1990).

62. ***S. parvum*** Lichtw., 1997

Branches compact, fertile branches bearing up to 6 or more generative cells. Trichospore oval, 9-14 × 3.5 µm, collar 2-6 µm long. Zygospores unknown.

Host: *Cricotopus* sp. (Chironomidae).

Record: Costa Rica (Lichtwardt, 1997).

63. *S. pennelli* Lichtw., 1984

Thalli up to 300 µm long, branching sparsely, mucilaginous basal cell tapered to a point. Trichospores in series of 2-4, long fusiform-ellipsoidal, 31-55 × 6-8 µm, with a short rounded collar 3.4-4 µm long. Terminal trichospores often produced subapically. Zygospores unknown.

Host: Larval *Prosimulium exigens* Dyar & Shannon, *P. onychodactylum* Dyar & Shannon, *Simulium defoliarti* Stone & Peterson, and *Simulium* sp. (Diptera, Simuliidae).

Record: U.S.A. (Lichtwardt, 1984).

64. *S. perforatum* Williams & Lichtw., 1987

Basal cell commonly penetrating the hindgut lining forming bulbous swellings. Two basal cells, 8-10 µm wide, often present before branching is initiated. Trichospores commonly 1-2, sometimes more per fertile branch, elongate-ellipsoidal to subcylindrical with a swollen mid region, 33-45 × 7-8.2 µm, collar 6-8 × 3-4.6 µm with a single curling appendage several times the spore length. Zygospores unknown.

Host: Larval Chironomidae [*Orthocladius* (*Euorthocladius*) sp., *Diamesa* spp.]

Record: U.S.A. (Williams and Lichtwardt, 1987)

65. *S. phytotelmatum* Lichtw., 1994

Thalli sparsely branched, main branches bearing sets of single to often verticillate short branchlets with series of short generative cells. Trichospore subcylindrical, 14-30 × 2-3 µm, collar usually 2-3 µm long. Zygospores unknown.

Host: Species of *Chironomus* Meigen, *Polypedilum* Kieffer and spp. of *Tanytarsus* (Chironomidae, Chironominae, Tanitarsini)

Record: Northeastern Costa Rica (Lichtwardt, 1994), Argentina (Siri et al., 2008; Siri and López Lastra, 2010).

66. *S. precipitiorum* M. M. White & Lichtw., 2004

Trichospores ellipsoidal, 13-22 × 3.5-4.5 µm. Zygospores 66-75 × 10-13 µm with a collar.

Host: Chironomidae larvae (*Orthocladius* [*Eudactylocladius*] *fuscimanus* (Kieffer) and *Diamesa* cf. *aberrata* Lundbeck or *Diamesa* sp.).

Record: Norway (White and Lichtwardt, 2004).

67. ***S. prostratum*** L. G. Valle & Santam., 2004

Thalli prostrate, secreted holdfast all along the thallial surface. Branches verticillately ramified at the upper end of the thalli. Fertile branches with 2-6 generative cells. Trichospore subcylindrical to elongate-ellipsoidal, 26-30 × 3.5-4.5 μm, collar campanulate, 4.5-5.5 × 2.5-3.5 μm. Zygospores unknown.

Host: Chironomid Orthocladiinae larvae.

Record: Spain (Valle and Santamaria, 2004)

68. ***S. pseudodimorphum*** L. G. Valle & Santam., 2004

Thalli branched, branches verticillate. Basal cell simple or bilobulated, with a secreted holdfast. Trichospores dimorphic: type α, subcylindrical, 45-66 × 4-7 μm, with a thickened apex, collar cylindrical or with convergent margins, 9-11.5 × 3.5 μm; type β, broadly ellipsoidal, 12-16.5 × 5.5-6 μm, collar cylindrical, 12.5-18 × 1.5-2 μm. Both trichospores types occur on the same thallus but on different branches, the trichospores of type β being less common. Zygospores biconical, 82-97 × 14-20 μm, released zygospores with a collar of 16.5-20 × 0.5-7 μm, and an appendage.

Host: Chironomidae Diamesinae larvae.

Record: Spain (Valle and Santamaria, 2004).

69. ***S. pusillum*** Manier & Coste, 1971

Thalli 200-300 μm long, attached to the host cuticle by a pseudorhizal base, branches 4-6 μm diam. Trichospores 4-6 per fertile branch, ellipsoidal, 14-22 × 2.5-3 μm, collar 2-3.5 × 1.2-1.5 μm. Zygospores fusiform, 50-60 × 6-7 μm, collar 5-6 × 2-3 μm.

Host: Larvae of *Procladius* sp. (Diptera, Chironomidae).

Record: France (Manier and Coste, 1971), Spain (Valle and Santamaria, 2004).

70. *S. rarum* Lichtw., 1990

Trichospore subcylindrical, swollen in the middle, 20-26 × 3-4.5 μm, with a collar 0.5-1 μm long and a single appendage. Zygospores about 100 × 10 μm, collar attached about 25 μm from one end.

Host: Unidentified Chironomidae larvae (Diptera), also *Austrothaumalea* sp. (Diptera: Thaumaleidae).
Record: New Zealand (Williams and Lichtwardt, 1990).

71. *S. rupestre* Lichtw., 1990

Thalli with a cluster of basal cells, 2-4 generative cells per fertile branch. Trichospore subcylindrical with a median swelling, 34-39 × 4-6 μm, collar 4-6 μm long, with an appendage initially very spiralled. Zygospores fusiform, 90-128 × 9-12 μm, with the collar located about ¼ the distance from one end; zygosporophore ½-⅓ the length of the zygospore.

Host: Unidentified larvae of Orthocladiinae (Diptera: Chironomidae).
Record: New South Wales – Australia (Lichtwardt and Williams, 1990), China (Wang et al., 2010).

72. *S. shaanxiense* Juan Wang, Strongman, & S. Q. Xu, 2010

Thalli dense and profusely branched, holdfast inconspicuous. Trichospores dimorphic; large trichospores, medially swollen, 33.5-46 × 8.5-14 μm, with a long collar (11-18 μm), and a single thin appendage. Small trichospores, oval, 9-12 × 4.5-7 μm, with a collar (11-14 μm), and a single appendage. Zygospores (Type II), biconical, 56.5-67 × 7.5-11.5 μm, with a collar (15.5-19.5 × 3-4 μm) and a single long, thin appendage.

Host: Midge larva (Diptera: Chironomidae).
Record: China (Strongman et al., 2010)

73. *S. simulatum* Lichtw. & Arenas, 1996

Base of thallus consisting of cylindrical cells. Generative cells more than 6 per fertile branch. Trichospore ovoid, 19-23 × 6-8 μm, with a 5 μm long campanulate collar. Zygospores unknown.

Host: *Aphrophila bidentata* Alexander larva (Tipulidae).
Record: Chile (Lichtwardt and Arenas, 1996).

74. *S. simulii* Lichtw., 1964
[= ***Rubetella simulii*** Manier, 1963, ***nom. nud.***]

Basal cell swollen, holdfast refractive. Generative cells 4-6 per fertile terminal branch. Trichospore cylindrical but swollen in the middle, 16-30 × 3-7 μm, collar cylindrical, 1.5-4.2 μm long, single short appendage. Zygospores unknown.

Host: Many species of larval Simuliidae, Chironomidae (Diptera), more rarely in Culicidae and Tipulidae larvae.
Record: U.S.A. (Lichtwardt, 1964), Japan, Australia, New Zealand, England, Sweden (Lichtwardt, 1986), France (Manier, 1963), Norway (White and Lichtwardt, 2004), China (Strongman and Xu, 2006), Spain (Girbal and Santamaria, 1998; Valle and Santamaria, 2009), Canada (Strongman, 2007; Strongman and White, 2008).

75. *S. tipulidarum* Williams & Lichtw., 1987

Thalli with sparse divergent branching in the hindgut of Tipulidae. Holdfast simple. Trichospores up to 14 or more in a series on a fertile branch. Trichospore elongate-ellipsoidal, 15-20 × 3.5-5 μm, collar 2-3.2 × *ca.* 2 μm, with a single appendage at least 3 times the spore length, tending to curl. Zygospores unknown.

Host: Larva of *Elliptera astigmatica* Alexander (Tipulidae).
Record: U.S.A. (Williams and Lichtwardt, 1987).

76. *S. tronadorium* Lichtw., Ferrington & López Lastra, 1999

Thalli densely branched. Trichospore long-ellipsoidal with a slight medial bulge, 15-28 × 3-5 μm, collar 1-3 μm long. Zygospores unknown.

Host: *Paraheptagyia* larva (Diamesinae), *Cricotopus* sp. (Orthocladiinae larvae).
Record: Argentina (Lichtwardt et al., 1999).

77. S. *tynense* Strongman, 2007

Thalli profusely branched, secreted disk-like holdfast. Trichospore long-ellipsoidal, 31-37 × 3-3.5 μm, with a short blunt collar (about 3 × 3 μm) and a single appendage. Zygospores unknown.

Host: Larval Chironomid.
Record: Canada (Strongman, 2007).

78. *S. typhellum* Manier & Coste, 1971

Thalli up to 1.2 mm long, secondary branches verticillate, about 4 μm diam. Generative cells 5-6 per fertile branch, 7-10 μm. Trichospore cylindrical with median swelling, 25-30 × 3-3.5 μm, collar narrow 3-3.5 × 1.5-2 μm. Zygospores unknown.

Host: Larval *Chironomus* sp. (*plumosus* complex) (Diptera, Chironomidae).
Record: France (Manier and Coste, 1971), Spain (Valle and Santamaria, 2004)

79. *S. urbanum* López Lastra, Mazzucchelli & Lichtw., 2000

Thalli compact, branching verticillate. Trichospore long-ellipsoidal, 19-27 × 3-4 μm, with a short cylindrical collar, produced singly on terminal branches, more commonly 1-2 on a short fertile branchlet. Zygospores unknown.

Host: Bloodworm (Chironomidae) larva.
Record: Argentina (Lichtwardt et al., 2000).

NEED FOR FUTURE RESEARCH

As one can see from the records of these two genera, described mainly from Europe and the Americas, and hence there is a need to inventory these fungi from other regions of the world. At the same time, there is also a need to follow a definite pattern when putting together the morphological data observed for a genus or species. For example, one should start the description with the holdfast, basal cell, whole thallus and branching, number of generative cells and trichospores attached to a fertile branchlet, the size and shape of the trichospores, their collar and appendage, and details of the zygospores, if present. This would help the future workers in the study of a described or undescribed species.

Furthermore, where possible, molecular data or other features should also be used to integrate with the morphological observations, both for a new or an already described species. It is now becoming imperative to use a polyphasic approach to determine a species or genus, because of the many new species that are being described, where often their spore sizes overlap and create a real problem for a new investigator.

ACKNOWLEDGEMENT

The author is grateful to Professor R. W. Lichtwardt, Department of Ecology & Evolutionary Biology, University of Kansas, Lawrence, KS 66045-7534, USA for reading the manuscript and extending suggestions. The author is thankful to Dr. S. R. Singh for extending help in the preparation of the Keys and also to those who kindly provided their published materials for his use. The references to Indian collections made here were possible because of the financial support by the Department of Science and Technology, Govt. of India, New Delhi, through a project (DST No. SP/SO/A 75/98) sanctioned to the author while he was at the Mycological Research Unit, Department of Botany, Sri J. N. P. G. College, Lucknow, India.

REFERENCES

Alencar, Y. B., Ríos-Velásquez, C. M., Lichtwardt, R. W. and Hamada, N. 2003. Trichomycetes (Zygomycota) in the digestive tract of arthropods in Amazonas, Brazil. Mem. Inst. Oswaldo Cruz, 98: 799-810.

Beard, C. E. and Adler, P. H. 2003. Zygospores of selected Trichomycetes in larval Diptera of the families Chironomidae and Simuliidae. Mycologia, 95: 317-320.

Coste-Mathiez, F. 1970. Parasites de larves de Chironomides (Diptères, Nematocères) des environs de Montpellier. Docteur de Spécialité Thesis, University of Montpellier.

Ferrington, L. C., Jr., Lichtwardt, R. W. and Hayford, B. 2000. *Smittium gravimetallum* (Trichomycetes: Harpellales), a new species of gut fungus from *Dicrotendipes fumidus* (Johannsen) (Diptera: Chironomidae) in a metal-polluted stream, pp. 253-257. In: Late 20th Century Research on Chironomidae: An Anthology from the 13th International Symposium on Chironomidae. O. Hoffrichter (ed.). Shaker Verlag, Aachen.

Ferrington, L. C., Jr., Lichtwardt, R. W., Hayford, B. and Williams, M. C. 2005. Symbiotic Harpellales (Trichomycetes) in Tasmanian aquatic insects. Mycologia, 97: 254-262.

García, J. J., Campos, R. E. and Maciá, A. 1994. Prospeccion de enemigos naturales de Culicidae (Diptera) de la selva marginal de Punta Lara (Prov. de Buenos Aires, Republica Argentina). Revista de la Academia Colombiana de Ciencias Exactas, Físicas y Naturales, 19: 209-215.

Gauthier, M. 1961. Une nouvelle espèce de *Stachylina*: *St. minuta* n. sp., parasite des larves de Chironomides Tanytarsiens. Travaux du Laboratoire d' Hydrobiologie et de Pisciculture de l'Université de Grenoble, 53: 1-4.

Girbal, J. and Santamaria, S. 1998. Trichomycetes (Fungi, Zygomycotina) comensals de larves de Simuliidae (Diptera) a la Península Ibèrica. Acta Botanica de Barcelona, 45: 49-56.

Kobayasi, Y., Hiratsuka, N., Korf, R. P., Tubaki, K., Aoshima, K., Soneda, M. and Sugiyama, J. 1967. Mycological studies of the Alaskan Arctic. Annual Report of the Institute for Fermentation Osaka, 3: 1-138.

Kobayasi, Y., Hiratsuka, N., Otani, Y., Tubaki, K., Udagawa, S. and Soneda, M. 1969. The Second Report on the Mycological Flora of the Alaskan Arctic. Bulletin of the National Science Museum (Tokyo), 12: 311-430.

Kobayasi, Y., Hiratsuka, N., Otani, Y., Tubaki, K., Udagawa, S., Sugiyama, J. and Konno, K. 1971. Mycological studies of the Angmagssalik Region of Greenland. Bulletin of the National Sciences Museum (Tokyo), 14: 1-96.

Léger, L. and Gauthier, M. 1932. Endomycètes nouveaux des larves aquatiques d'Insectes. Comptes Rendus Hebdomadaires des Séances de l'Académie des Sciences Paris, 194: 2262-2265.

Léger, L. and Gauthier, M. 1935. La spore des Harpellacees (Léger et Duboscq) Champignons parasites des Insectes. Travaux du Laboratoire d' Hydrobiologie et de Pisciculture de l' Université de Grenoble, 27: 3-6.

Lichtwardt, R. W. 1964. Axenic culture of two new species of branched Trichomycetes. Amer. J. Bot., 51: 836-842.

Lichtwardt, R. W. 1972. Undescribed genera and species of Harpellales (Trichomycetes) from the guts of aquatic insects. Mycologia, 64: 167-197.

Lichtwardt, R. W. 1973. The Trichomycetes: what are their relationships? Mycologia, 65: 1-20.

Lichtwardt, R. W. 1984. Species of Harpellales living within the guts of aquatic Diptera larvae. Mycotaxon, 19: 529-550.

Lichtwardt, R. W. 1986. The Trichomycetes: Fungal associates of arthropods. Springer-Verlag, New York.

Lichtwardt, R. W. 1994. Trichomycete fungi living in the guts of Costa Rican phytotelm larvae and other lentic dipterans. Rev. Biol. Trop., 42: 31-48.

Lichtwardt, R. W. 1997. Costa Rican gut fungi (Trichomycetes) infecting lotic insect larvae. Revista De Biologia Tropical, 45: 1349-1383.

Lichtwardt, R. W. and Arenas, J. 1996. Trichomycetes in aquatic insects from southern Chile. Mycologia, 88: 844-857.

Lichtwardt, R. W., Cafaro, M. J. and White, M. M. 2001a. The Trichomycetes, fungal associates of arthropods. Published on the Internet www.nhm.ku.edu/~fungi.

Lichtwardt, R. W., Ferrington, L. C., Jr. and López Lastra, C. 1999. Trichomycetes in Argentinean aquatic insect larvae. Mycologia, 91: 1060-1082.

Lichtwardt, R. W. and Grigg, R. D. 1998. Four new *Smittium* species inhabiting the hindgut of Chironomidae larvae. Mycologia, 90: 427-433.

Lichtwardt, R. W., Kobayasi, Y. and Indoh, H. 1987. Trichomycetes of Japan. Trans. Mycol. Soc. Japan, 28: 359-412.

Lichtwardt, R. W., López Lastra, C. and Mazzucchelli, M. G. 2000. Fungi living in the guts of larval aquatic insects in northwestern Argentina. Mycologia, 92: 332-340.

Lichtwardt, R. W., White, M. M. and Colbo, M. H. 2001b. Harpellales in Newfoundland aquatic insect larvae. Mycologia, 93: 764-773.

Lichtwardt, R. W. and Williams, M. C. 1983. Two unusual Trichomycetes in an aquatic midge larva. Mycologia, 75: 728-734.

Lichtwardt, R. W. and Williams, M. C. 1988. Discovery of sexual reproduction in an unusual new species of *Stachylina* (Trichomycetes). Mycologia, 80: 400-405.

Lichtwardt, R. W. and Williams, M. C. 1990. Trichomycete gut fungi in Australian aquatic larvae. Can. J. Bot., 68: 1057-1074.

Lichtwardt, R. W. and Williams, M. C. 1992a. Tasmanian Trichomycete gut fungi in aquatic insect larvae. Mycologia, 84: 384-391.

Lichtwardt, R. W. and Williams, M. C. 1992b. Western Australian species of *Smittium* and other Trichomycetes in aquatic insect larvae. Mycologia, 84: 392-398.

Lichtwardt, R. W. and Williams, M. C. 1992c. *Smittium bullatum* from a New Zealand midge larva and new records of other trichomycete gut fungi. Can. J. Bot., 70: 1193-1195.

Lichtwardt, R. W. and Williams, M. C. 1999. Three Harpellales that live in one species of aquatic chironomid larva. Mycologia, 91: 396-399.

López Lastra, C. C. 1990. Primera cita de *Smittium morbosum* var. *rioplatensis* var. nov. (Trichomycetes: Harpellales) patogeno de 5 especies de mosquitos (Diptera: Culicidae) en la Republica Argentina. Rev. Arg. Micologia, 13: 14-18.

López Lastra, C. C. 1997. Primera cita de *Smittium culisetae* y *S. culicis* (Trichomycetes: Harpellales) en larvas de mosquitos (Diptera: Culicidae) de la República Argentina. Boletín de la Sociedad Argentina de Botánica, 33: 43-46.

López Lastra, C. C., Scorsetti, A. C., Marti, G. A. and Coscarón, S. 2005. Trichomycetes living in the guts of aquatic insects of Missiones and Tierra del Fuego, Argentina. Mycologia, 97: 320-328.

Manier, J.-F. 1962. Présence de Trichomycètes dans le rectum des larves d'Éphémères des torrents du Massif du Néouvieille (Hautes-Pyrénées). Bulletin de la Société d'Histoire Naturelle de Toulouse, 97: 241-254.

Manier, J.-F. 1963. Trichomycètes de larves de Simulies (Harpellales du proctodeum). Annales des Sciences Naturelles Botanique et Biologie Végétale, Paris Série 12, 4: 737-750.

Manier, J.-F. 1969-1970. Trichomycètes de France. Annales des Sciences Naturelles Botanique et Biologie Végétale, Paris Série 12, 10: 565-672.

Manier, J.-F. and Coste, F. 1971. Trichomycètes Harpellales de larves de Diptères Chironomidae; création de cinq nouvelles espèces. Bulletin de la Société Mycologique de France, 87: 91-99.

Manier, J.-F. and Mathiez, F. 1965. Deux Trichomycètes Harpellales Génistellacées, parasites de larves de Chironomides. Annales des Sciences Naturelles Botanique et Biologie Végétale, Paris Série 12, 6: 183-196.

Manier, J.-F., Rioux, J.-A. and Whisler, H. C. 1961. *Rubetella inopinata* n. sp. et *Carouxella scalaris* n. g., n. sp., Trichomycètes parasites de *Dasyhelea lithotelmatica* Strenzke, 1951 (Diptera, Ceratopogonidae). Naturalia Monspeliensia, Série Botanique, 13: 25-38.

Manier, J.-F., Rioux, J.-A. and Juminer, B. 1964. Présence en Tunisie de deux Trichomycètes parasites de larves de Culicides. Archives de l' Institut Pasteur de Tunis, 41: 147-152.

Misra, J. K. and Tiwari, V. K. 2002. Harpellales (Trichomycetes) do occur in India: first report of two species of *Stachylina*. Kavaka, 30: 71-75.

Misra, J. K. and Tiwari, V. K. 2008. A new species of *Gauthieromyces* and range extensions for other Harpellales in India. Mycologia, 100: 94-98.

Moss, S. T. 1972. Occurrence, Cell Structure and Taxonomy of the Trichomycetes, with special Reference to Electron Microscope Studies of Stachylina. Ph.D. Dissertation, University of Reading, Reading.

Poisson, R. 1936. Sur un Endomycète nouveau: *Smittium arvernense* n. g., n. sp., parasite intestinal de larves de *Smittia* sp. (Diptères Chironomides) et description d'une nouvelle espèce du genre *Stachylina* Lég. et Gauth. 1932. pp. 75-86. In: Mélanges dédiés au Professeur Lucien Daniel. Université de Rennes. Rennes.

Reeves, W. K., Beard, C. E. and Craig, D. A. 2004. Trichomycete symbiotes of *Crozetia seguyi*, a primitive black fly. J. Invert. Pathol., 87: 67-69.

Santamaria, S. and Girbal, J. 1997. Contribución al conocimiento de los Trichomycetes (Fungi, Zygomycotina) Ibéricos. Anales Jard. Bot. Madrid, 55: 219-223.

Sato, H., Shimada, N. and Aoki, J. 1989. Light and electron microscopy of *Smittium morbosum* (Trichomycetes), newly recorded from Japan. Trans. Mycol. Soc. Japan, 30: 51-59.

Siri, Augusto and López Lastra, C. C. 2010. Diversity of trichomycetes in larval flies from aquatic habitats in Argentina. Mycologia, 102: 347-362.

Siri, A., Marti, G. A. and López Lastra, C. C. 2008. Prevalence of Harpellales from Chironomidae larvae in phytotelmata from Punta Lara Forest, Argentina. Mycologia, 100: 381-386.

Strongman, D. B. 2007. Trichomycetes in aquatic insects from Prince Edward Island, Canada. Can. J. Bot., 85: 949-963.

Strongman, D. B., Wang, Juan and Xu, Shengquan. 2010. New trichomycetes from western China. Mycologia, 102: 174-184.

Strongman, D. B. and White, M. M. 2008. Trichomycetes from lentic and lotic aquatic habitats in Ontario, Canada. Botany, 86: 1449-1466.

Strongman, D. B. and Xu, Shengquan. 2006. Trichomycetes from China and the description of three new *Smittium* species. Mycologia, 98: 479-487.

Sweeney, A. W. 1981a. An undescribed species of *Smittium* (Trichomycetes) pathogenic to mosquito larvae in Australia. Trans. Brit. Mycol. Soc., 77: 55-60.

Sweeney, A. W. 1981b. Fungal pathogens of mosquito larvae, pp. 403-424. In: Pathogenesis of Invertebrate Microbial Diseases. E. W. Davidson (ed.). Allanheld, Osmun & Co., Totowa, New Jersey.

Tuzet, O. and Manier, J.-F. 1947. *Orphella culici* n. sp., Entophyte parasite du rectum des larves de *Culex hortensis* Fclb. Comptes Rendus Hebdomadaires des Séances de l'Académie des Sciences Paris, 225: 264-266.

Tuzet, O. and Manier, J.-F. 1953. Recherches sur quelques Trichomycètes rameux. *Asellaria armadillidii* n. sp. *Genistella choanifera* n. sp. *Genistella chironomi* n. sp. *Spartiella barbata* Tuzet et Manier. Annales des Sciences Naturelles Zoologie et Biologie Animale, Série 11, 15: 373-391.

Tuzet, O., Rioux, J.-A. and Manier, J.-F. 1961. *Rubetella culicis* (Tuzet et Manier 1947), Trichomycète rameux parasite de l'ampoule rectale des larves de Culicides (morphologie et spécificité). Vie et Milieu, 12: 167-187.

Valle, L. G. 2007. New species and summary of Iberian Harpellales. Mycologia, 99: 442-455.

Valle, L. G. and Cafaro, Matias J. 2010. First report of Harpellales from the Dominican Republic (Hispaniola) and the insular effect on gut fungi. Mycologia, 102: 363-373.

Valle, L. G. and Santamaria, Sergi. 2004. The genus *Smittium* (Trichomycetes, Harpellales) in the Iberian Peninsula. Mycologia, 96: 682-701.

Valle, L. G. and Santamaria, Sergi. 2009. Arthropod gut symbionts from the Balearic Islands: Majorca and Cabrera. Diversity and biogeography. Anales del Jardin Botánico de Madrid, 66S1: 109-120.

Wang, Juang, Xu, Shengquan and Strongman, D. B. 2010. Two new Harpellales inhabiting the digestive tracts of midge larvae and other trichomycetes from Tian Shan Mountains, China. Mycologia, 102: 135-141.

White, Merlin M. and Lichtwardt, R. W. 2004. Fungal symbionts (Harpellales) in Norwegian aquatic insect larvae. Mycologia, 96: 891-910.

White, Merlin M., Siri, Augusto and Lichtwardt, R. W. 2006. Trichomycete insect symbionts in Great Smoky Mountains National Park and vicinity. Mycologia, 98: 333-352.

Williams, M. C. 1982. *Smittium cellaspora*, a new Harpellales (Trichomycetes) from a chironomid hindgut. Mycotaxon, 16: 183-186.

Williams, M. C. 1983a. Spore longevity of *Smittium culisetae* (Harpellales, Legeriomycetaceae). Mycologia, 75: 171-174.

Williams, M. C. 1983b. Zygospores in *Smittium culisetae* (Trichomycetes) and observations on trichospore germination. Mycologia, 75: 251-256.

Williams, M. C. and Lichtwardt, R. W. 1984. Two *Stachylina* and two *Smittium* species (Trichomycetes) from Montana. Mycologia, 76: 204-210.

Williams, M. C. and Lichtwardt, R. W. 1987. Three new species of *Smittium* (Trichomycetes) with notes on range extensions. Mycologia, 79: 832-838.

Williams, M. C. and Lichtwardt, R. W. 1990. Trichomycete gut fungi in New Zealand aquatic insect larvae. Can. J. Bot., 68: 1045-1056.

Williams, M. C., Lichtwardt, R. W. and Peterson, S. W. 1982. *Smittium longisporum*, a new Harpellales (Trichomycetes) from chironomid guts. Mycotaxon, 16: 167-171.

Chapter 7

Phylogenetic Analysis within Genera *Morchella* (Ascomycota, Pezizales) and *Macrolepiota* (Basidiomycota, Agaricales) Inferred from nrDNA ITS and EF-1α Sequences

Gayane S. Barseghyan[1,*], Anush Kosakyan[1,2,], Omoanghe S. Isikhuemhen[3], Maryna Didukh[3,4], and Solomon P. Wasser[1,4]**

[1]Institute of Evolution and Department of Evolutionary & Environmental Biology, Faculty of Science & Science Education, University of Haifa Mt. Carmel, Haifa 31905, Israel

[2]Laboratory of Soil Biology, University of Neuchâtel, Rue Emile Argand 11 CH-2009 Neuchâtel, Switzerland

[3]Mushroom Biology and Fungal Biotechnology Laboratory, School of Agriculture and Environmental Sciences, North Carolina Agricultural and Technical State University, Greensboro, NC 27411, USA

[4]M.G. Kholodny Institute of Botany, National Academy of Sciences of Ukraine, 2 Tereshchenkivska St., Kiev 01601, Ukraine

Corresponding authors: E-mails: *gbarseghyan@gmail.com **akosakyan@yahoo.com

ABSTRACT

The relationships among species within *Morchella* and *Macrolepiota* were studied using nrDNA ITS and EF-1α sequences from 50 samples belonging to genera *Gyromitra, Distiotis, Morchella,* and *Verpa*; and 58 samples belonging to genera *Macrolepiota, Leucoagaricus,* and *Chlorophyllum* species from different geographical areas. Maximum parsimony, maximum likelihood, and the Bayesian approach were conducted in order to build the phylogeny of these genera. Both genera appeared to be monophyletic; however, several ingroups appeared to be polyphyletic. The relationship of species within the genus *Morchella* was investigated using both molecular (nrDNA ITS and EF-1α) and morphological characteristics in order to re-evaluate the taxonomic position of several controversial species for which the taxonomic status is still unclear. Results showed that the genus *Morchella* is a complex of several species, which was separated into two main distinct groups: yellow (*M. crassipes*, *M. esculenta* s. str. and *M. spongiola*) and black morels [(*M. elata* s. str., *M. costata*, *M. gigas*, and *M. vulgaris* (= *M. conica*)]. Results supported evidence that *Morchella elata* and *M. angusticeps* are identical taxa, which was proposed in earlier publications. Morphological data together with molecular data did not support the proposals that *M. esculenta* and *M. crassipes* are the same species. After phylogenetic studies, the list of *Morchella* species of Israel was re-evaluated, and are currently represented by *Morchella elata*, *M. esculenta*, *M. rufobrunnea*, and *M. vulgaris* (=*M. conica*). The relationship of *Macrolepiota* species to one another within the genus was investigated using both molecular (nrDNA ITS and EF-1 alpha sequences) and morphological characteristics. The results supported evidence to transfer *M. rachodes* into the genus *Chlorophyllum*, which was proposed by earlier publications. *M. fuliginosa* and *M. rhodosperma* appear to be identical. Morphological data together with molecular data did not support the proposals to treat *M. mastoidea* and its closely related species as one taxon. The list of *Macrolepiota* species of Israel was re-evaluated, which is currently represented by *M. affinis*, *M. excoriata*, *M. excoriata* var. *rubescens*, *M. fuligineosquarrosa*, *M. konradii*, *M. mastoidea*, *M. procera*, *M. procera* var. *mediterranea*, *M. prominens*, *M. psammophila*, *M. rickenii*, and *M. subsquarrosa*. *Macrolepiota oliveri* and *M. rachodes* were transferred to the genus *Chlorophyllum*. It was shown that for *Morchella*, EF-1α is informative, whereas for *Macrolepiota* – this is hardly so. With the topology presented for *Macrolepiota* based on EF-1α, it is logical to conclude that this marker is not that suitable for the resolution of species complexes that are closely related in this particular taxon, genus *Macrolepiota*. However, assuming that the complexes should be treated as the same species is probably premature.

INTRODUCTION

The number of fungal species described is about 100,000 (Kirk et al., 2008), but the total number of fungal species is estimated to be at least 1.5 million (Hawksworth, 1991, 2001). The gap between the number of species described and the estimated total number of fungi, nevertheless, indicates that a very large number of undescribed fungi and cryptic species exist all over the world.

Traditionally, four major phyla are distinguished within the kingdom Fungi: Chytridiomycota, Zygomycota, Ascomycota, and Basidiomycota. However, it has become evident that this traditional scheme does not reflect the phylogenetic relationships among fungi.

Many molecular systematics approaches to constructing natural and phylogenetic classifications of Ascomycetes and Basidiomycetes (White et al., 1990; Sugiyama, 1998; van de Peer et al., 2000; Berbee and Taylor, 2001; James et al., 2006; Hibbett et al., 2007). By understanding evolutionary relationships, several monophyletic groups were newly identified.

The monophyletic sister groups, Ascomycota and Basidiomycota, are well-characterized phyla using the molecular phylogenetic approach (Sugiyama, 1998; van de Peer et al., 2000; Berbee and Taylor, 2001), which revealed that both phyla share a hypothetical common ancestor. Therefore, they were treated as the subphyla Ascomycotina and Basidiomycotina and combined into the phylum Dikaryomycota (Tehler et al., 2000) or even into the subkingdom Dikarya (James et al., 2006; Hibbett et al., 2007).

Accumulation of molecular data of various fungal groups is being increasingly used in systematics, but the sole use of molecular data in handling classification and identification issues of fungal groups, while ignoring morphological data, poses problems. Hence, there is a pressing need for a consensus on the classification and identification of new species within ascomycetes and basidiomycetes using both molecular and morphological data. One way to achieve this goal is to combine phenotypical and molecular information, of course with emphasis on molecular data, which tend to be more stable and resilient to changes induced by external influences. Furthermore, getting informative phenotypic data is often highly problematic due to the lack of sufficient distinguishing morphological characters, large intra-specific variability of the characters and their dependence upon physiological growth parameters (Hibbett et al., 2007). On the other hand, precise morphological identification is essential to ensure the correct assignment of molecular data to a fungal organism. In combination with DNA barcoding, which applies DNA sequences as taxon-specific molecular markers, it is possible to achieve a more precise delimitation of species and subspecies, and to identify cryptic species. A necessary requirement for a reliable DNA-barcode based species identification is a broad range of species and subspecies sampled, and well-defined taxonomic species descriptions (Meyer and Paulay, 2005). Barcodes are not suitable to define species but are useful to assign unidentified specimens to a known species or to elucidate new or cryptic species.

The molecular revolution in fungal taxonomy commenced in the early 1990s, with analyses of PCR-amplified ribosomal RNA genes (White et al., 1990). Today, fungal molecular systematics is a mature discipline in which multi-locus datasets, extensive taxon sampling, and rigorous analytical approaches are standard (Hibbett et al., 2007). There are several works employing multiple genes to identify evolutionary links in major groups of fungi. Approximately 80% of more than 100,000 fungal rRNA gene sequences are now deposited in the GenBank (James et al., 2006; Hibbett et al., 2007).

Eukaryotic ribosomal RNA genes (known as ribosomal DNA or rDNA) are found as parts of repeat units that are arranged in tandem arrays, located at the chromosomal sites known as nuclear organizing regions (NORs). Each repeat unit consists of a transcribed region (having genes for 18S, 5.8S and 26S rRNAs and the external transcribed spacers, i.e. ETS1 and ETS2) and a non-transcribed spacer (NTS) region. In the transcribed region, internal transcribed spacers (ITS) are found on either side of 5.8S rRNA gene and are described as ITS1 and ITS2 (Baldwin, 1992; Baldwin et al., 1995; Hershkovitz et al., 1996; Dubouzet and Shinoda, 1999).

The ITS region is now perhaps the most widely sequenced DNA region in fungi. It has typically been most useful for molecular systematics at the species level, and even within species (e.g., to identify geographic races). Because of its higher degree of variation than other gene regions of rDNA (for small- and large-subunit rRNA), variation among individual rDNA repeats can sometimes be observed both within the ITS and IGS regions. In addition to the standard ITS1+ITS4 primers used by most laboratories, several taxon-specific primers have been described that allow selective amplification of fungal sequences (Gardes and Bruns, 1993).

The elongation factor 1 alpha (EF-1α) has become widely employed as a phylogenetic marker for studying eukaryotic evolution. It is a highly conserved ubiquitous protein involved in translation that has been suggested to have desirable properties for phylogenetic inference (Roger et al., 1999).

Elongation factor 1-alpha (EF-1α) is a protein found in all eukaryotic lineages. EF-1α is very abundant in the cytosol, where it is involved in the GTP-dependent binding of aminoacyl-tRNAs to the A-site of the ribosomes in the second step of translation from mRNAs to proteins. Its universal occurrence and highly conserved amino acid sequence makes it an ideal phylogenetic tool with which to determine very ancient relationships, such as the relative branching order of the most primitive Eukaryotes (see, e.g., Nordnes et al., 1994; Hashimoto et al., 1995; Baldauf et al., 2000). Furthermore, the nucleotide sequences of EF-1α genes were used at a much lower level to study phylogenies of families and genera, using the degenerated third position of codons (Cho et al., 1995; Mitchell et al., 1997). EF-1α has been used for resolving the phylogenetic relationships between animal phyla and classes as a good means to confirm or cast doubt on results based on other genes (Regier and Shultz, 1997). Also, it was used in phylogenetic analyses of Pezizomycotina in combination with another four nuclear loci (SSU rDNA, LSU rDNA, RPB1, RPB2 and EF-1α) (Spatafora et al., 2006).

Despite numerous phylogenetic studies (Gargas and Taylor, 1995; Spatafora, 1995; Landvik, 1996; Landvik et al., 1997; Johnson and Vilgalys, 1998; Harrington et al., 1999; Moncalvo et al., 2000, 2002; Vellinga, 2002; Vellinga et al., 2003; Bodensteiner et al., 2004; Larsson et al., 2004; Binder et al., 2005; Hansen and Pfister, 2006) the taxonomic position of some species requires further investigations.

Two genera, namely, *Morchella* (Pezizales, Morchellaceae), and *Macrolepiota* (Agaricales, Agaricaceae), containing putative species were taken from phyla Basidiomycota and Ascomycota, respectively, for phylogenetic studies.

Ascomycota is the largest phylum of the kingdom Fungi and according to the last edition of the "Dictionary of Fungi" includes approximately 64,163 described species worldwide (Kirk et al., 2008). It is characterized by the production of ascospores in a specialized sac-shaped sporangium, the ascus (Alexopoulos et al., 1996). The current classification of Ascomycota (Outline of Ascomycota, Lumbsch and Huhndorf, 2007, now hosted at http://www.fieldmuseum.org/myconet/outline.asp) includes three subphyla: Taphrinomycotina, Saccharomycotina, Pezizomycotina. Taphrinomycotina is divided into four classes, namely, Neolectomycetes, Pneumocystidomycetes, Schizosaccharomycetes, Taphrinomycetes. Saccharomycotina comprises the single class Scacharomycetes and the order Saccharomycetales and includes the majority of ascomycetous yeast species (e.g. *Candida albicans, Saccharomyces cerevisiae*). Pezizomycotina is the largest subphylum of the Ascomycota with more than 33,000 described species which includes all filamentous, ascoma-producing species.

The Operculate Discomycetes (order Pezizales) is an early diverging lineage within the Pezizomycotina. Before molecular phylogeny, the classification of these fungi was based on the layering of walls, structure of the apical apparatus and histochemistry of the ascus (Brummelen, 1994). The introduction of molecular markers, such as nrDNA sequences, followed by the development and refinement of computer based phylogenetic programs for their analysis, has provided new possibilities for reconstructing the phylogeny of the Operculate Discomycetes.

Several recent phylogenetic studies, using the 18S rRNA, SSU, RPB-2, β-tubulin, and LSU nrDNA sequence data have shown that the Pezizales are a monophyletic group within the Discomycetes (Gargas and Taylor, 1995; Spatafora, 1995; Landvik, 1996; Landvik et al., 1997; Harrington et al., 1999; Hansen and Pfister, 2006).

The SSU nrDNA gene has proved particularly useful in the estimation of familial/ordinal and high level phylogeny in fungi (e.g. Bruns et al., 1991, 1992; Bowman and Knock, 1992; Berbee and Taylor, 1992; Wilmotte et al., 1993). Landvik et al. (1997) detected three major lineages within the Pezizales using SSU nrDNA: (A) Pezizaceae and Ascobolaceae, (B) Helvellaceae, Morchellaceae,

Tuberaceae, and (C) Sarcoscyphaceae, Sarcosomataceae, Ascodesmidaceae, Glaziellaceae, and Pyronemataceae.

In the past 10 years, molecular phylogenetic studies have focused on the Pezizales and have provided significant insight into the evolution of these fungi. The circumscription of the order and the delimitation of families have been clarified or are being clarified. At the generic and species level, many problems remain to be solved.

Morchella is one of the taxonomically problematic genera within Pezizales, and has been the focus of interest of many researchers all over the world. Descriptions of the genus *Morchella* can be found in Smith et al. (1981), Jacquetant (1984), Arora (1986), Weber (1988) and Hanlin and Hahn (1990). *Morchella* taxonomy above the level of species is non-controversial, but when species are discussed, calling the field "problematic" is an understatement (Weber, 1995). The Index Fungorum online database (http://indexfungorum.org/Names/Names.asp) lists 197 species and subspecies worldwide (CABI and others n.d.). Hallen et al. (2001) estimated that more than 100 *Morchella* species have been described based on their morphological features. Features of *Morchella* that are often described include: color of ridges, pits, and stem; configuration of the ridges; spore size and shape; attachment of the head to the stem; texture of head and stem; discoloration and bruising; stem wall thickness or layering; and changes in such features as a morel ages. The problem is that species are highly polymorphic and such observable features often integrate among described or presumed species. Some mycologists suggest that the high degree of polymorphism may be induced by environmental and climatic factors (Pilz et al., 2007). While some researchers split the moiety of described nomenclatural taxa into three to six species (Overholts, 1934; Seaver, 1942; Weber, 1988), other mycologists recognize as many as 50 species (Kimbrough, 1970; Korf, 1973; Smith and Weber, 1980; Pacioni, 1981), reflecting the enormous typological variability of the fruit bodies according to habitats, and specific abiotic and biotic conditions under which they grow (Bresinsky and Stangl, 1961; Bresinsky et al., 1972).

Several taxonomists conducting phylogenetic studies have begun to utilize the variation found in sequences of the nuclear ribosomal DNA genes (nrDNA). These variations are used to determine relatedness between species as well as among closely related species (Bruns et al., 1991). Many mycologists tried to clarify the systematics within *Morchella,* based on methods such as restriction fragment length polymorphism (RFLP) (Magee et al., 1987; Kohn et al., 1988; Cubeta et al., 1991; Hibbett and Vilgalys, 1991; Bunyard et al., 1994), and direct sequencing (Woese and Olsen, 1986; Medlin et al., 1988; White et al., 1990), to investigate the gene coding for the production of 18S, 5.8S, 28S, and 5S nrDNA, which has allowed assessments and comparisons of the phylogenetic relationships of the fungi over a wide taxonomic level (Sogin, 1990). Wipf et al.

(1996, 1999) tried to clarify the systematics within *Morchella* based on sequence data for the internal transcribed spacer (ITS) region of the nuclear ribosomal DNA (nrDNA), and thereby confirmed the separation of *Adnatae* (yellow morels) and *Distantes* (black morels) made by Boudier (1897) in one of the first systematic revisions of the genus. A study by O'Donnell et al. (1997) confirmed the monophyletic status for Morchellaceae and the genus *Morchella*, respectively, using nrDNA of the small and large ribosomal subunits. Within the genus, O' Donnell et al. (2003) identified two clades and a total of 28 *Morchella* species based on multigene phylogeny of 600 morel specimens. Thirteen species were placed into the yellow-tan-gray "esculenta" clade and 15 into the black "elata" clade. Twenty-four of the 28 species were found on only one continent, North America which appeared to be the ancestral home of *Morchella*.

Numerous species described in the United Kingdom by Dennis (1981) seemed to be varieties of two species, namely, *M. esculenta* with light colored reticulating ridges and *M. elata*, with dark, more or less parallel, longitudinal ridges. In Nordic countries (Denmark, Finland, Norway, and Sweden), Jacquetant (1984) found that it was possible, according to microscopic characteristics, to delimit several problematic *Morchella* species. In the Ukraine, according to Smickaya (1980), only six species were described as distinct independent species. Moreover, *Mitrophora semilibera* (DC.: Fr.) Lév. was described as synonymous with *Morchella semilibera*.

According to Krieglsteiner (1993) only three species were described in Germany: *Morchella elata* Fr. (including *M. conica* Pers., *M. costata* (Vent.) Pers., *M. deliciosa* Fr., *M. hortensis* Boud.), *M. esculenta* (L.) Pers. (including *M. crassipes* (Vent.) Pers.: Fr., *M. rotunda* (Fr.) Boud., *M. umbrina* Boud., *M. spongiola* Boud., *M. vulgaris* (Pers.) Boud.) and *M. gigas* (Batsch: Fr.) Pers. (including *M. hybrida* Pers., *M. rimosipes* DC., and *Mitrophora semilibera* (DC.: Fr.) Lév.). In Switzerland, Breitenbach and Kränzlin (1984) kept *M. elata* and *M. conica* as distinct species and separate four varieties within *M. esculenta* (var. *rigida* Krombh., var. *rotunda* Fr., var. *umbrina* (Boud.) S. Imai, and var. *vulgaris* Pers.), without providing the morphological features, that set the two species apart. These authors separated *M. gigas* as a monotypic genus, i.e. *Mithrophora semilibera*, distinguished by only having a cap partially attached to the stipe, whereas in all other *Morchella* species, the cap is almost completely attached. A critical analysis of the known species of *Morchella* from Mexico was made by Guzmán and Tapia (1998), who presented a list of species that included *M. angusticeps*, *M. costata*, *M. elata*, *M. esculenta* (*M. conica, M. crassipes,* and *M. rotunda*), and *M. umbrina*. The blushing species also included *M. guatemalensis* (from Mexico and Guatemala), *M. rufobrunnea* (State of Veracruz), and *M. rigidoides* (New Guinea), which are confined to the tropics or subtropics.

Basidiomycota is one of the two large phyla that, together with the Ascomycota, comprise the subkingdom Dikarya within the Kingdom Fungi.

The order Agaricales is one of the most diverse and largest orders of the Basidiomycota. The taxonomic history of Agaricales began with Persoon (1801) who wrote "Synopsis Methodica Fungorum". Persoon used macroscopic observations to place Agaricales in the order Hymenothecii, which contained all the Basidiomycetes except Gasteromycetes and some Ascomycetes. Later , the work of Fries "Systema Mycologicum" (1822), which was based on the same principle character, the configuration of hymenial layer used by Persoon, allowed Fries to divide Agaricales and Aphyllophorales into five families: the Clavariaceae with erect fruit bodies and amphigenous hymenium, the Thelephoraceae with smooth hymenium, the Hydnaceae, the Polyporaceae and Agaricaceae with toothed, tubular, and lamellate hymenophores, respectively (Kim and Jung, 2000). Fries relied on spore deposit color – white, pink, brown, purple-brown and black – to divide the gilled mushrooms (agarics) into several series. The genera were delimited on the basis of macroscopic characters of basidiocarp and color of the spore print, thus making it possible to identify a lot of specimens to a genus on the simplest observation. Fries's macroscopic system, which initially recognized 12 genera of fleshy mushroom-forming fungi, was practical for the purpose of identification. It was relatively unchallenged until Fayod (1889) surveyed the anatomy and microscopic features of many agarics.

Microscopic studies of basidiocarp structure initiated by Fayod and Patouilard, showed a number of Fries's groupings to be artificial, and new genera and families have been proposed. Consequently, 108 genera of agaric fungi were recognized (Kirk et al., 2008).

Kühner (1980) sustained and developed Fayod's classification by investigation of utility of cytological characters and used these to help shape a notable departure from previous classifications. He divided the agraic fungi into five orders: Tricholomatales (including some gilled taxa of the Polyporales), Agaricales sensu stricto, Pluteales, Russulales and Boletales. Kühner's treatment of multiple orders of agarics is not widely recognized, but neither has it been evaluated explicitly based on molecular data (Matheny et al., 2006).

Singer (1986) primarily integrated anatomical characters and spore micromorphology but studied a greater diversity of agarics from the neotropics and the Southern Hemisphere. Singers's monumental work "The Agaricales in Modern Taxonomy" treated three major groups within the Agaricales s.l.: Agaricineae, Boletineae, and Russulineae including 229 genera. These groups are accepted in modern treatments based on molecular characters, as the Euagarics clade, Boletae clade and Russuloid clade (Hibbett and Thorn, 2001).

Many others have influenced higher-level classification of mushrooms and their allies or proposed various evolutionary hypotheses for the Agaricales during the past 50 years (e.g. Pegler and Young, 1969; Horak, 1968; Heim, 1971; Petersen, 1971; Jülich, 1981; Moser, 1983). For instance, Agaricales s.l. includes

87 genera belonging to 19 families and 5 orders in the Kühner's system (1980), and 46 families and 11 orders in the Jülich's system (Jülich, 1981).

Numerous phylogenetic studies address the evolutionary lineages within Agaricales s.l. using evidence from nuclear and mitochondrial ribosomal RNA genes (16S-18S, 5.8S, 26-28S, 5S) or the spacers between them, protein coding sequences, and randomly chosen coding sequences. Nowadays, nuclear ribosomal DNA genes (25-28S – large subunit (LSU), 5S, 18S genes, internal transcribed spacer (ITS), and internal gene spacer (IGS)) are the most frequently used in fungal molecular systematics. Some of the important findings of these studies were that the forms of fruit body and hymenophore types have been phylogenetically misleading (Hibbett et al., 1997). And, that many families and genera of agarics were not monophyletic (Moncalvo et al., 2000, 2002) and also that the ecological traits have been underused in identification of natural groups (Moncalvo et al., 2002) (cit. in Matheny et al., 2006).

The large-scale Homobasidiomycetes phylogeny based on analysis of nuc-SSU and mt-SSU rDNA sequences examined by Hibbett and Thorn (2001) revealed eight clades: polyporoid, euagarics, boletae, russuloid, hymenochaetoid, cantharelloid, and gomphoid-phallid. The three clades such as euagarics, boletae and russuloid are supporting Singer's (1986) concept of Agaricales s.l.

Moncalvo et al. (2002) present an example of elegant work on euagarics phylogeny using nLSU rDNA sequences. Their large-scale tree of Homobasidiomycetes supports monophyly of six clades: euagarics, boletae, hymenochaetoid, gomphoid-phalloid and canthareloid clades, the basal position of the cantarelloid clade, and the reciprocal monophyly of Polyporaceae and corticoid clades within the poliporoid clade (Zmitrovich and Wasser, 2004).

Other broad molecular phylogenetic studies (Bodensteiner et al., 2004; Larsson et al., 2004; Binder et al., 2005) have demonstrated evolutionary relationships among non-gilled Basidiomycetes, such as resupinate and cyphelloid forms, with members of the Agaricales. Molecular systematics studies have revealed the connection between non-gilled and gasteroid representatives in various clades with gilled relatives (Binder et al., 1997; Peintner et al., 2001; Hallen et al., 2003; Matheny and Bougher, 2005). In some instances the priority of popular family and generic level names has been contested (Norvell, 2001; Redhead et al., 2001). Some classification systems (Kirk et al., 2008) began to incorporate findings of early research, but adjustments are necessary because more groups have been studied in detail and more regions sequenced (Matheny et al., 2006).

Matheny et al. (2006) have investigated the phylogeny of the Agaricales based on a multilocus analysis of a six-gene region supermatrix. Analyses of RPB1, RPB1-intron 2, RPB2 and 18S, 25S, and 5.8S ribosomal RNA genes recovered six major clades, which were recognized informally and labeled as the agaricoid, tricholomatoid, marasmioid, pluteoid, hygrophoroid, and plicaturopsidoid clades. About 11 origins of the ectomycorrhizal habit appear to have evolved in the Agaricales, with possibly as many as nine origins in the agaricoid plus tricholomatoid clade alone.

In the large-scale work "A higer-level phylogenetic classification of fungi" provided by Hibbett et al. (2007), the order Agaricales [equivalent to euagarics clade (Hibbett and Thorn, 2001)], order Atheliales [equivalent to boletae clade (Larsson et al., 2004; Binder et al., 2005)], and order Boletales [equivalent to bolete clade (Hibbett and Thorn, 2001; Binder et al., 2006)] were placed to the subclass Agaricomycetidae (class Agaricomycetes).

The family Agaricaceae has been a focus of interest for many scientists all over the world because of peculiarities in its representatives, ecology, and biochemistry. It refers to the order Agaricales s. str. of the subclass of higher Basidiomycetes (Homobasidiomycetidae). In the order system of gill fungi, Agaricaceae takes its place between light-spore (Catathelasmataceae, Amanitaceae) and dark-spore (Bolbitiaceae, Strophariaceae) families. According to the modern notion, family Agaricaceae includes genera from several families of the order Agaricales s.l. having various colorations of spore prints (Singer, 1962, 1975, 1986; Moser, 1967, 1978, 1983; Wasser, 1980, 1985 a, b; Kirk et al., 2008).

Agaricaceae Fr. sensu Singer (1986) includes 25 genera, most of them distributed over several continents and organized in four tribes: *Agariceae* Pat., *Leucocoprineae* Singer, *Lepiotae* Fayod, and *Cystodermatae* Singer.

To date, according to Kirk et al. (2008), this family comprises 85 genera and 1,340 species, including several genera with gasteroid and secotioid basidiomata.

Based on molecular studies, the family Agaricaceae was shown to be monophyletic (Johnson and Vilgalys, 1998; Johnson, 1999; Moncalvo et al., 2002; Vellinga, 2004a). The molecular data supports Singer's morphological concept of the family (Singer, 1986).

Several works on synthesis of molecular based phylogeny and the morphological classification of lepiotaceous fungi were done by Johnson and Vilgalys (1998), de Kok and Vellinga (1998), Johnson (1999), Hopple and Vilgalys (1999), Vellinga (1999, 2002, 2003 a,b, c, 2004 a, b), Moncalvo et al. (2000), Vellinga and de Kok (2002), Vellinga and Yang (2003), and Vellinga et al. (2003). LSU analyses revealed that genera *Lepiota, Macrolepiota, Leucoagaricus, Leucocoprinus,* and *Melanophyllum* are polyphyletic (Johnson and Vilgalys, 1998; Hopple and Vilgalys, 1999; Moncalvo et al., 2000). Joined analysis of LSU and ITS of tribes Agariceae, Lepioteae and Leucocoprineae showed them to be polyphyletic (Vellinga, 2003a, b), indicating that morphology-based systems are artificial. For instance, genus *Melanophyllum* placed into the tribe Agariceae in morphology-based systems (Singer, 1986; Wasser, 2002) turned out to belong to Lepioteae, close to *Lepiota, Cystolepiota,* and *Pulverolepiota* genera (Vellinga, 2003b).

Despite the fact that molecular data supports Singer's morphological concept of Agaricaceae, several changes in this family occurred after Singer (1986) which are: the re-evaluation of *Rugosospora* Heinem. (Guzmán et al., 1989; Franco-Molano, 1995); the exclusion of the tribe Cystodermatae from Agaricaceae (Johnson and Vilgalys, 1998; Vellinga, 2004a); the inclusion of *Coprinus comatus* (O.F. Müll.: Fr.) Pers., *C. sterquilinus* (Fr.: Fr.) Fr. and *C. spadiceisporus* Bogart in this family (Redhead et al., 2001); the synonymization of *Macrolepiota* Singer section *Laevistipedes* (Pázmány) Bon with *Chlorophyllum* Massee (Vellinga, 2002, 2003a; Vellinga and de Kok, 2002; Vellinga et al., 2003); the synonymization of *Volvolepiota* Singer with *Macrolepiota* (Vellinga and Yang, 2003; Vellinga et al., 2003); the transfer of *Melanophyllum* Velen. to the same clade as *Lepiota* (Pers.) Gray (Vellinga, 2003b); the inclusion of the gasteroid genera of Lycoperdales and Tulostomatales (Agerer, 2002; Moncalvo et al., 2002; Vellinga, 2004a); the transfer of *Cystoagaricus* Singer to *Psathyrellaceae* (Singer) Redhead, Moncalvo and Vilgalys (Vellinga, 2004a); and the confirmation of the monophyly of *Leucoagaricus* (Locq.) ex Singer, *Leucocoprinus* Pat. and *Sericeomyces* Heinem., which can be treated as a single genus (Vellinga, 2004a) (cit. in Wartchow et al., 2008).

The genus *Macrolepiota* has always been understood differently by taxonomists (Bon, 1981; Moser, 1983; Singer, 1986; Candusso and Lanzoni, 1990; Vellinga, 2003a,b,c). This problem is obvious from a perusal of existing systems of the genus. The way in which the species is grouped depends on the specific traits considered most important at the species or genus level. The existing systems are advantageous for identification, but they can not resolve the conflict of relationships within the group. Different authors have drawn different limitations for the genus (Fig. 7.1).

There is no definitive evidence about species number for the genus *Macrolepiota*. According to Moser (1983), the genus includes 11 species, according to Singer (1986), it includes 12 species, Candusso and Lanzoni (1990) report 20 species, and Kirk et al. (2008) reveal 30 species. The Index Fungorum online database (http://www.indexfungorum.org/Names/Names.asp) lists about 47 *Macrolepiota* species worldwide (CABI and others n.d.).

The genus *Macrolepiota* was established by Singer in 1948. Bon (1981) recognized two sections within the genus *Macrolepiota*, namely *Macrosporae* and *Procerae*. The former section is characterized by an absence of clamp connections, a pileal surface finely granular or with small concolorous squamules, broad squamulate surfaces are formed in a radial direction toward the periphery, might be in a stellate or cog-wheel manner; annulus simple or thickened at the lower surface towards the exterior. For the latter section, clamps are present in various tissues of the fruit body, though inconsistently, scales on the pileus are deposited concentrically, annulus often double or complex, sometimes only thickened towards the exterior. Later on, Singer (1986) broadly

Singer 1986

- Genus *Macrolepiota* Singer
  - Sect. *Macrolepiota*
  - Sect. *Macrosporae* (Singer) Bon

Candusso and Lanzoni 1990

- Genus *Macrolepiota* Singer
  - Sect. *Macrolepiota*
    - Subsect. *Macrolepiota*
    - Subsect. *Laevistipes* Pazmany
    - Subsect. *Microsquamatae* (Pazmany) Bellu and Lanzoni
  - Sect. *Macrosporae* (Singer) Bon

Bon 1996

- Genus *Macrolepiota* Singer
  - Sect. *Macrolepiota* (=Procerae Fr.)
  - Sect. *Laevistipedes* (Pazmany) Bon
  - Sect. *Macrosporae* (Singer) Bon
    - Subsect. *Excoriatae* Bon
    - Subsect. *Microsquamatae* (Pazmany) Bellu and Lanzoni

Vellinga 2001

- Genus *Macrolepiota* Singer
  - Sect. *Macrolepiota*
  - Sect. *Laevistipedes* (Pazmany) Bon

Fig. 7.1 Intrageneric classification of the genus *Macrolepiota* proposed by different authors.

redefined the two sections using data on the presence and frequency of clamp connections: section *Macrolepiota* (=*Procerae*) – some clamp connections present in the trama of the carpophores, or at least at the lower part of the rind of the stipe and at the base of many basidia; and section *Macrosporae* – clamp connections on the hyphae of the trama consistently absent, or present only at the base of cheilocystidia and along the hymenopodium.

Based on the phylogenetic analysis of Johnson and Vilgalys (1998) and Vellinga et al. (2003), *Macrolepiota* was shown to be polyphyletic.

Vellinga (2003c) separated the genus into two clades: corresponding to the genus *Macrolepiota* proper (sensu stricto) and *Chlorophyllum*. *Macrolepiota* s.s. includes the type species *Macrolepiota procera* (Scop.) Singer, *M. mastoidea*,

M. clelandii and allies. Other species formerly included in *Macrolepiota*, e.g. *M. rachodes*, *M. olivieri*, and *M. globosa* were transferred into the genus *Chlorophyllum*. *Macrolepiota nympharum* was assigned to the genus *Leucoagaricus*.

However, the position of many individual species among these complexes is still confusing. Thus, the question remains whether these taxa are indeed independent in evolutionary lineages and how can they be delimited. The groups that require further study are *Macrolepiota mastoidea* group - *M. affinis* (Velen.) Bon, *M. fuligineosquarrosa* Malençon, *M. gracilenta* (Krombh.) Wasser, *M. konradii* (Huijsman ex P.D. Orton) M.M. Moser, *M. mastoidea* (Fr.) Singer, *M. phaeodisca* Bellù, *M. prominens* (Fr.) M.M. Moser, *M. psammophila* Guinb., *M. rickenii* (Velen.) Bellù and Lanzoni, *M. subsquarrosa* (Locq.) Bon, *M. procera* group - *M. fuliginosa* (Barla) Bon, *M. procera* (Scop.) Singer, *M. permixta* (Barla) Pacioni, *M. rhodosperma* (P.D. Orton) Migl., and *M. rachodes* group - *M. rachodes* (Vittad.) Singer, *M. olivieri* (Barla) Wasser, *M. bohemica* (Wichanský) Krieglst., *M. venenata* Bon, *M. abruptibulba* (R. Heim) Heinem.

With the aforesaid background, the authors carried out studies to investigate polymorphism within the genera *Morchella* and *Macrolepiota* based on ribosomal nrDNA (ITS and EF-1α sequences) data and to establish whether certain taxa of these genera have independent lineages. Attempts have also been made to evaluate if the phylogenetic data can be correlated with the classification based on the morphology and the identity of some of the dubious specimens can be verified. This chapter embodies the results of these investigations.

MATERIALS AND METHODS

Taxon sampling

From Ascomycota, three samples of *Gyromitra* (represented by one taxon), one sample of *Disciotis* (one taxon), 44 samples of *Morchella* (nine taxa), two samples of *Verpa* (one taxon); and from Basidiomycota one sample of *Chlorophyllum* (one taxon), four samples of *Leucoagaricus* (one taxon), and 53 samples of *Macrolepiota* (18 taxa) collected in different geographical areas were used in this study. Eleven sequences of *Morchella*, 11 sequences of *Macrolepiota*, and one sequence of *Chlorophyllum* from GenBank were included in the datasets as well. Fruit body collections and cultures are deposited at the Herbarium and Culture Collection of the Institute of Evolution, University of Haifa, Israel (HAI), Herbarium of Yerevan State University (YSU), Armenia, and Herbarium of Komarov Botanical Institute, Russia. Information about the specimen used are provided in Tables 7.1 and 7.2.

Table 7.1 Samples of *Gyromitra, Disciotis*, *Morchella* and *Verpa* species used in the study

Taxon	*Origin and herbarium, culture collection, GenBank accession numbers*	*Source*	*ITS length (bp)*	*EF-1α length (bp)*
Disciotis venosa (Pers.: Fr.) Boud.	DQ491503; GenBank	–	957	–
Gyromitra slonevskii Heluta	HAI-D-077; Ukraine	F	1155	623
G. slonevskii Heluta	HAI-D-078; Ukraine	F	1149	734
G. slovenskii Heluta	HAI-D-079; Ukraine	F	1134	871
Morchella angusticeps Peck	HAI-C- 918; India	C	732	1056
M. costata (Vent.) Pers.	LE 180066; Germany	F	779	620
M. costata (Vent.) Pers.	DQ257333; GenBank	–	739	–
M. costata (Vent.) Pers.	EF080997; GenBank	–	737	–
M. costata (Vent.) Pers.	DQ257334; GenBank	–	739	–
M. crassipes (Vent.) Pers.	HAI-C-920; Germany	C	1129	1131
M. crassipes (Vent.) Pers.	HAI-C-1059; Germany	C	1113	1100
M. crassipes (Vent.) Pers.	HAI-C-1135; France	C	1118	1136
M. crassipes (Vent.) Pers.	HAI-C-1136; Germany	C	890	926
M. crassipes (Vent.) Pers.	HAI-C-1137; Ukraine	C	863	988
M. crassipes (Vent.) Pers.	LE 180067; Ukraine	F	1265	628
M. crassipes (Vent.) Pers.	LE 180087; Estonia	F	1267	1219
M. crassipes (Vent.) Pers.	EU701001; GenBank	–	1209	–
M. elata Fr.	HAI-D-017; Israel	F	746	1193
M. elata Fr.	HAI-D-023; Israel	F	752	592
M. elata Fr.	HAI-C-1133; Ukraine	C	756	981
M. elata Fr.	HAI-C-1180; Ukraine	C	774	922
M. elata Fr.	EF017946; GenBank	–	737	–
M. esculenta (L.) Pers.	HAI-D-011; Israel	F	1159	732
M. esculenta (L.) Pers.	HAI-D-041; Israel	F	1047	1209
M. esculenta (L.) Pers.	HAI-C-787; USA	C	1095	1143
M. esculenta (L.) Pers.	HAI-C-812; USA	C	1124	1102
M. esculenta (L.) Pers.	HAI-C-917; France	C	1046	1144
M. esculenta (L.) Pers.	HAI-C-1134; France	C	1155	1308
M. esculenta (L.) Pers.	YSU-88-I; Armenia	F	1096	629
M. esculenta (L.) Pers.	EU086776; GenBank	–	1127	–

Table 7.1 Contd. ...

Taxon	*Origin and herbarium, culture collection, GenBank accession numbers*	*Source*	*ITS length (bp)*	*EF-1α length (bp)*
M. gigas Pers.	HAI-C-1181; Germany	C	839	1150
M. gigas Pers.	LE180140; Kazakhstan	F	867	618
M. gigas Pers.	AJ543743; GenBank	–	828	–
M. rufobrunnea Guzmán & F. Tapia	HAI-D-044; Israel	F	805	–
M. rufobrunnea Guzmán & F. Tapia	DQ355921; GenBank	–	859	–
M. spongiola Boud.	HAI-C-926; Ukraine	C	1138	1139
M. spongiola Boud.	HAI-C-927; Ukraine	C	1009	1004
M. spongiola Boud.	YSU-79-II; Armenia	F	555	730
M. spongiola Boud.	YSU-79-III; Armenia	F	1216	1206
M. vulgaris (Pers.) Boud. (= *M. conica* Pers.)	HAI-D-022; Israel	F	757	1191
M. vulgaris (Pers.) Boud. (= *M. conica* Pers.)	HAI-D-043; Israel	F	681	715
M. vulgaris (Pers.) Boud. (= *M. conica* Pers.)	HAI-D-076; Israel	F	969	1047
M. vulgaris (Pers.) Boud. (= *M. conica* Pers.)	HAI-C-1058; Germany	C	742	1124
M. vulgaris (Pers.) Boud. (= *M. conica* Pers.)	HAI-C-1132; Germany	C	684	1114
M. vulgaris (Pers.) Boud. (= *M. conica* Pers.)	HAI-C-1217; Israel	C	893	823
M. vulgaris (Pers.) Boud. (= *M. conica* Pers.)	LE 180036; Canada	F	744	603
M. vulgaris (Pers.) Boud. (= *M.conica* Pers.)	LE 180063; Russia	F	776	–
M. vulgaris (Pers.) Boud. (= *M. conica* Pers.)	EF080999; GenBank	–	737	–
Verpa bohemica (Krombh.) J. Schröt.	HAI C 1932; Ukraine	C	888	1162
Verpa bohemica (Krombh.) J. Schröt.	AM269502; GenBank	–	881	–

HAI-C – Culture collection of Haifa University; HAI-D – Herbarium of Haifa University, Israel; YSU – Herbarium of Yerevan State University, Armenia; LE – Herbarium of Komarov Botanical Institute, Russia; F – fruit body; C – vegetative mycelium.

Table 7.2 Samples of *Macrolepiota, Leucoagaricus* and *Chlorophyllum* species used in the study

Taxon	*Origin and herbarium, culture collection, GenBank accession numbers*	*Source*	*ITS length (bp)*	*EF-1α length (bp)*
Macrolepiota affinis (Velen.) Bon	HAI-5; Israel	F	699	1175
M. dolichaula (Berk. & Broome) Pegler & R.W. Rayner	HAI-37; Thailand	F	747	1134
M. dolichaula (Berk. & Broome) Pegler & R.W. Rayner	DQ221111; GenBank	–	653	–
M. excoriata (Schaeff.) Wasser	HAI-C-952; Israel	C	737	1127
M. excoriata (Schaeff.) Wasser	HAI-347; Israel	F	734	1128
M. excoriata (Schaeff.) Wasser	HAI-253; Israel	F	737	1128
M. fuligineosquarrosa Malençon	HAI-331; Israel	F	666	1134
M. fuliginosa (Barla) Bon	AY243598; GenBank	–	675	–
M. gracilenta (Krombh.) Wasser	LE-9866; Russia	F	719	–
M. konradii (Huijsman ex P.D. Orton) M.M. Moser	HAI-294; Israel	F	737	1134
M. konradii (Huijsman ex P.D. Orton) M.M. Moser	HAI-C-298; Israel	C	736	–
M. konradii (Huijsman ex P.D. Orton) M.M. Moser	HAI-111; Israel	F	729	1134
M. konradii (Huijsman ex P.D. Orton) M.M. Moser	HAI-201; Israel	F	737	600
M. konradii (Huijsman ex P.D. Orton) M.M. Moser	HAI-C-550; Israel	C	732	1133
M. konradii (Huijsman ex P.D. Orton) M.M. Moser	HAI-205; Israel	F	733	1133
M. konradii (Huijsman ex P.D. Orton) M.M. Moser	LE-9920; Russia	F	734	1172
M. konradii (Huijsman ex P.D. Orton) M.M. Moser	HAI-C-11; Israel	C	737	–

Table 7.2 Contd. ...

Taxon	*Origin and herbarium, culture collection, GenBank accession numbers*	*Source*	*ITS length (bp)*	*EF-1α length (bp)*
M. konradii (Huijsman ex P.D. Orton) M.M. Moser	HAI-31; Israel	F	733	1056
M. konradii (Huijsman ex P.D. Orton) M.M. Moser	HAI-C-989; Thailand	C	739	1094
M. konradii (Huijsman ex P.D. Orton) M.M. Moser	AY243603; GenBank	–	693	–
M. mastoidea (Fr.) Singer	HAI-233; Israel	F	705	1101
M. mastoidea (Fr.) Singer	HAI-233-1; Israel	F	732	1034
M. mastoidea (Fr.) Singer	HAI-123; Israel	F	646	556
M. mastoidea (Fr.) Singer	HAI-222; Israel	F	733	1133
M. mastoidea (Fr.) Singer	LE-242079; Russia	F	–	601
M. mastoidea (Fr.) Singer	AY243604; GenBank	–	693	–
M. olivieri (Barla) Wasser = *Chlorophyllum olivieri* (Barla) Vellinga	AY081248; GenBank	–	677	–
M. procera (Scop.) Singer	HAI-23; Israel	F	723	1127
M. procera (Scop.) Singer	HAI-30; Israel	F	725	–
M. procera (Scop.) Singer	YSU-2; Armenia	F	729	600
M. procera (Scop.) Singer	HAI-605; Ukraine	F	729	600
M. procera (Scop.) Singer	LE-257592; Russia	F	634	600
M. procera (Scop.) Singer	HAI-C-942, Slovenia	C	725	1130
M. procera (Scop.) Singer	HAI-72; Israel	F	725	1052
M. procera (Scop.) Singer	AY243589; GenBank	–	685	–
M. prominens (Fr.) M.M. Moser	HAI-554; Israel	F	733	–
M. psammophila Guinb.	HAI-4; Israel	F	738	1134
M. psammophila Guinb.	HAI-C-1187; Israel	C	731	1134
M. psammophila Guinb.	HAI-C- 500; Israel	C	788	605

Table 7.2 Contd. ...

Taxon	*Origin and herbarium, culture collection, GenBank accession numbers*	*Source*	*ITS length (bp)*	*EF-1α length (bp)*
M. psammophila Guinb.	HAI-14; Israel	F	738	–
M. psammophila Guinb.	LE-241979; Russia	F	738	605
M. psammophila Guinb.	AY243599; GenBank	–	693	–
M. puellaris (Fr.) M.M. Moser	YSU-1; Armenia	F	731	–
M. rachodes (Vitad.) Singer = *Chlorophyllum rachodes* (Vitad.) Vellinga	YSU-3; Armenia	F	702	–
M. rachodes (Vitad.) Singer = *Chlorophyllum rachodes* (Vitad.) Vellinga	LE-257595; Russia	F	750	593
M. rachodes (Vitad.) Singer = *Chlorophyllum rachodes* (Vitad.) Vellinga	AY081240; GenBank	–	706	–
M. rhodosperma (P.D. Orton) Migl.	AY243596; GenBank	–	642	–
M. rickenii (Velen.) Bellù & Lanzoni	HAI-SP-3; Israel	F	738	1063
M. rickenii (Velen.) Bellù & Lanzoni	HAI-SP-80; Israel	F	737	1134
M. rickenii (Velen.) Bellù & Lanzoni	HAI-SP-8; Israel	F	735	1008
M. subsquarrosa (Locq.) Bon	HAI-C-339; Israel	C	729	1134
M. subsquarrosa (Locq.) Bon	HAI-13; Israel	F	738	608
M. subsquarrosa (Locq.) Bon	AY243605; GenBank	–	693	–
M. zeyheri (Berk. & Singer) Heinem.	LE-9887; Russia	F	325	586
Chlorophyllum molybdites (G. Mey.) Massee	AY061209; GenBank	–	655	–
Leucoagaricus leucothites (Vittad.) Wasser	HAI-C-316; USA	C	752	1128
L. leucothites (Vittad.) Wasser	HAI-C-398; Ukraine	C	747	1140
L. leucothites (Vittad.) Wasser	HAI-12; Israel	F	704	188
L. leucothites (Vittad.) Wasser	HAI-41; Israel	F	751	1163

HAI-C – Culture collection of Haifa University; HAI, HAI-SP – Herbarium of Haifa University, Israel; YSU – Herbarium of Yerevan State University, Armenia; LE – Herbarium of Komarov Botanical Institute, Russia; F – fruit body; C – vegetative mycelium.

DNA extraction

Pieces of dry herbarium material, dried mycelium obtained from submerged cultivation and fresh mycelia scraped from agar plates with a sterile scalpel were placed in 1.5 ml Eppendorf tubes and extracted using the CTAB procedure (Zolan and Pukkila, 1986). Dried DNA pellet was then diluted in 40 μl 0.1M TE buffer, and its concentration was measured using NanoDrop®ND-1000 spectrophotometer (NanoDrop Technologies, Montchanin, Delaware, U.S.A.). Concentration of DNA was adjusted to 100ng/μl for downstream applications.

PCR amplification and purification

Diluted DNA (100 ng) from all samples was used as a template for PCR. Two gene regions, internal transcribed spacer (ITS) and the translation elongation factor 1 alpha (EF-1α), were amplified. The internal transcribed spacer region (ITS1-5.8S-ITS2) amplification of nrDNA was done using primer ITS1F in combination with ITS4 (White et al., 1990; Table 7.3).

Table 7.3 List of primers used in the study

Regions of nrDNA	*Primer name*	*Sequence (5′->3′)*
ITS region	ITS1F	TCCGTAGGTGAACCTGCGG
	ITS4	GCTGCGTTCTTCATCGATGC
EF-1α region	EF1-983F	GCYCCYGGHCAYCGTGAYTTYAT
	EF1-1567R	ACHGTRCCRATACCACCRATCTT
	EF1-2218R	ATGACACCRACRGCRACRGTYTC

The PCR reaction was mixed in a total volume of 25 μL containing 1.5 μL genomic DNA (6 ng), 1.2 U Taq polymerase, 2.5 μL 10×Taq buffer, 0.5 mM of each primer, 0.3 mM dNTPs (dATP, dTTP, dCTP, dGTP), 2.5 mM $MgCl_2$. PCR cycling conditions were as follows: an initial denaturation step of 94°C for 3 min, followed by 25 cycles of denaturation at 95°C for 45 sec, primer annealing at 56°C for 40 sec, primer extension at 72°C for 1 min, and a final elongation step at 72°C for 6 min.

EF-1α was amplified using the primers EF1-983F, EF1-1567R, and EF1-2218R (Rehner and Buckley, 2005; Table 7.3). The PCR reaction was mixed in a total volume of 20 μL containing 1.5 μL genomic DNA. The 1 U Taq polymerase, 2.5 μL 10×Taq buffer, 0.52 mM of each primer, 0.25 mM dNTPs (dATP, dTTP, dCTP, dGTP), and 2.6 mM $MgCl_2$. Reaction conditions were: an initial denaturation step of 95°C for 3 min, followed by 26 cycles of denaturation at 95°C for 45 sec, primer annealing at 55°C for 40 s, and an elongation step at 72°C for one minute. Final elongation was 7 minutes at 72°C.

All PCR products were visualized in 1% agarose gels (wt/v), stained with SYBR®Green I, and viewed under ultra-violet light. Sizes of PCR amplicons were estimated against Bioline DNA Easy Ladder I (BIOLINE). Prior to DNA sequencing, PCR products were cleaned using an Exo-SAP (USB) following manufacturer's instructions. Cleaned PCR products were sequenced with the Big Dye Terminator V3.1 Cycle Sequencing Kit (ABI, Foster city, CA, USA), and analyzed on an ABI PRISM® 3700 DNA Analyzer (ABI, Foster city, CA, USA).

ALIGNMENT

Obtained sequences were aligned using the program Clustal W package integrated in BioEdit sequence alignment editor (Hall, 1999) for *Macrolepiota* species and MUSCLE v3.6 (Edgar, 2004) for *Morchella* species. Manual adjustments were made for alignment optimization in BioEdit.

Phylogenetic analyses

Maximum parsimony and maximum likelihood

Both *Morchella* and *Macrolepiota* datasets were analyzed using maximum parsimony (MP), maximum likelihood (ML), and Bayesian approach. MP and ML analyses were performed using PAUP* 4.0b10 (Swofford, 2003). In order to obtain the most parsimonious tree, the heuristic search was chosen, with tree bisection-reconnection (TBR), branch-swapping algorithm, and MulTrees option in effect with the zero length-branches option switched on. MP was performed with the Maxtrees option set on; auto increase by 100, and a stepwise addition option was used to obtain start trees for branch swapping. For ML analyses, the heuristic search strategy was used as described by Harrison and Langdale (2006) with few modifications.

The best-fit model of nucleotide evolution was estimated using ModelTest 3.8 (Posada and Crandall, 1998; Posada, 2006) prior to ML and Bayesian searches. The model selected under Akaike's (1974) information criterion (AIC) was GTR+G+I for both *Morchella* and *Macrolepiota* datasets. The 1,000 bootstrap replicates (Felsenstein, 1985) were performed to evaluate tree branch robustness in maximum parsimony (MP) and maximum likelihood (ML).

Bayesian analysis

A. *Morchella* dataset

Bayesian Markov Chain Monte Carlo (MCMC) analyses were carried out using MrBayes v 3.1 (Ronquist et al., 2005). Bayesian MCMC analyses were performed with two simultaneous chains, with 500,000 generations for the ITS region, and 200,000 generations for the EF-1α fragment. The nucleotide substitution models were GTR+I+G for the ITS region and GTR+G for the EF-1α fragment. Generations were added until standard deviation of split frequencies fall below 0.01 (Ronquist et al., 2005). For every 100th generation, the tree with the best likelihood score was saved, resulting in 5,000 trees (ITS) and 2,000 (EF-1α). The burn in value was set to 25%.

B. *Macrolepiota* dataset

Bayesian Markov Chain Monte Carlo analyses were performed using MrBayes v3.1 (Ronquist et al., 2005). Bayesian MCMC analyses were performed with two simultaneous chains, and 1,000,000 generations were performed. The generations were added until standard deviation of split frequencies fell below 0.01 (Ronquist et al., 2005). For every 1,000th generation, the tree with the best likelihood score was saved, resulting in 10,000 trees. The burn in value was set to 25%.

C. Combined matrices

Mesquite Software v2.6 (Maddison and Maddison, 2007) was used to build a combined ITS and EF-1α dataset for *Morchella* and for *Macrolepiota* datasets. The congruency of a combined dataset (ITS and EF-1α for *Morchella*; ITS and EF-1α for *Macrolepiota*) was tested with the partition homogeneity test (Farris et al., 1994, 1995), as implemented in PAUP* 4.0b10. Trees were viewed using Tree View (Page, 1996).

RESULTS AND DISCUSSION

Phylogenetic analysis within *Morchella* Dill. ex Pers. (Ascomycota, Pezizales) species inferred from nrDNA ITS and EF-1α gene sequences

The relationships of *Morchella* species were investigated using both molecular (nrDNA ITS and EF-1α sequences) data, and compared with morphological characteristics. Based on molecular data, the genus as a whole is monophyletic. However, several ingroups within the phylogram appeared polyphyletic. The results showed that the genus *Morchella* united several species complexes which can be arranged into two main distinct groups: "yellow morels" (*Morchella crassipes*, *M. esculenta* s. str. and *M. spongiola*), and "black morels" (*M. elata* s. str., *M. costata*, *M. gigas*, and *M. vulgaris* (= *M. conica*)). All these confirm the earlier publications of Royse and May (1990), and Wipf et al. (1999). Results supported evidence that *M. elata* and *M. angusticeps* are identical taxa, an idea proposed in morphology-based publications. Morphological data together with molecular ones did not support the proposals that *M. esculenta* and *M. crassipes* are a single species (Volk, 1990). After phylogenetic studies, the list of *Morchella* species from Israel was re-evaluated. It was shown that the genus is currently represented by *M. elata, M. esculenta, M. rufobrunnea*, and *M. vulgaris* (=*M. conica*).

The species relationships within the genus *Morchella* was investigated using both molecular (nrDNA ITS and EF-1α) and morphological characteristics, in order to re-evaluate the taxonomic position of several controversial species, for which the taxonomic status still remains unclear.

The full ITS region including ITS-1, 5.8S, ITS-2, and EF-1α were analyzed first as separate datasets, and later as combined dataset.

Results of ITS dataset analyses

The alignment with 50 sequences resulted in 1,587 total characters from which 534 characters were constant, 137 variable characters were parsimony-uninformative, and 916 of which were parsimony informative. The ITS data yielded 1,885 equally parsimonious trees (CI=0.641, RI=0.862, RC=0.552).

Parsimony and Bayesian analyses of the ITS sequences produced similar trees in which two main clades of the "yellow" and "black" morel groups were clearly separated. All clades were retrieved in both analyses with strong or moderate bootstrap support and posterior probability values (80-100%). As shown in Fig. 7.2, at least three different sequence types were detected within analyzed specimens of the yellow morel group with high (100%) bootstrap and posterior probability support values (Fig. 7.2; Clade 2). Following Wipf et al. (1999), these could be named *M. esculenta, M. crassipes*, and *M. spongiola*, respectively. *M. rufobrunnea* the red-brown blushing species (Guzmán and Tapia, 1998) (Clade 1), comprising the yellow morel group formed the sister group (bootstrap and posterior probability values 100%).

The ITS data showed that clade 3 "black morels group" includes several ingroups: *M. gigas*, which is supported with 100% bootstrap and posterior probability values and well-defined species; *M. elata* nested within *M. vulgaris* group; besides, *M. elata* and *M. angustipes* appeared to be identical and supported with high bootstrap and posterior probability values (100%).

The outgroups, *Distiotis venosa, Gyromitra slonevskii*, and *Verpa bohemica* were very distant from all species of *Morchella*, as expected.

Results of EF-1α dataset analyses

The alignment with 37 sequences resulted in 1,308 total characters including exons, introns, and gaps. The introns were excluded from the analyses. A total

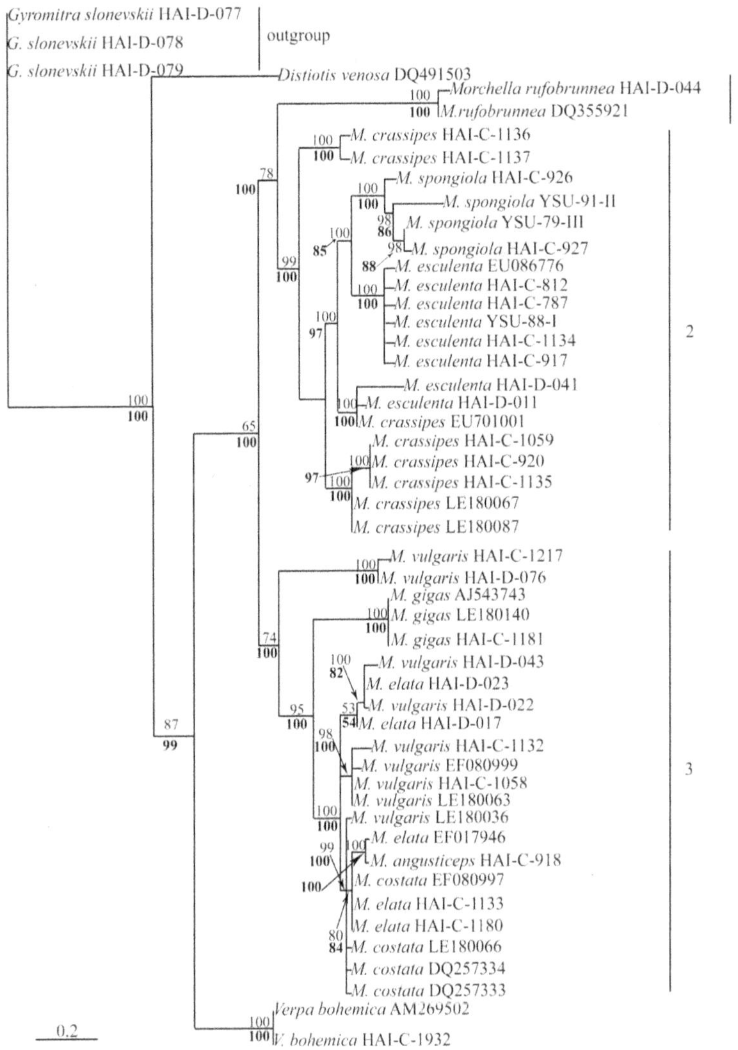

Fig. 7.2 Phylogram generated by Bayesian analyses based on ITS sequences data. Bootstrap values and posterior probability values higher than 50% are shown below respective branches. Topology is rooted with *Gyromitra slonevskii*. The lines with numbers indicate clades. The scale bar indicates the nucleotide substitution.

1,012 characters were examined for these analyses. Out of these, 206 characters were parsimony-informative, 685 characters were constant, and 121 variable characters were parsimony-uninformative. The score of the best trees found was 1963 (CI=0.706, RI=0.846, RC=0.597).

Topologies of the strict consensus MP and strict consensus ML trees were very similar. From the resulting above mentioned trees, *Morchella* species form two distinct groups: yellow (*M. crassipes, M. esculenta,* and *M. spongiola*) (Fig. 7.3; Clade 2), and black morels [(*M. angusticeps, M. elata, M. gigas,* and *M. vulgaris* (=*M.conica*)] with bootstrap (58 and 53%) and posterior probability (63 and 90%) supports respectively (Fig. 7.3; Clade 1 and Clade 2). These data confirm the data of Royse and May (1990) and Wipf et al. (1999), which have shown that the genus could be comprised of the two main groups ("black" and "yellow morels"). ITS data analysis also confirmed this fact with the highest bootstrap support. More significantly, results show that *M. elata* and *M. angusticeps* are identical taxa, and can be synonymized. These results confirm the classification by Heim (1936), but disagree with the Smith-Weber (1988) concept.

The outgroups, *Distiotis venosa, Gyromitra slonevskii,* and *Verpa bohemica* were very distant from all species of *Morchella,* as expected.

Results of combined ITS and EF-1α dataset analyses

The combined dataset was tested for incongruence with the partition homogeneity test (Farris et al., 1994, 1995). Two datasets were congruent (P=0.500) and were used for further analyses. The combined dataset of ITS and EF-1α sequences yielded 2,599 total characteristics from which 1,122 characters were parsimony-informative, 1,219 characters were constant, and 258 variable characters were parsimony-uninformative. Parsimony analyses yielded 2,000 equally parsimonious trees with the following characteristics: CI=0.654, RI=0.860 and RC=0.560.

All trees generated from the combined ITS and EF-1α datasets showed the same clades as revealed from the ITS-dataset without differences. The bootstrap and posterior probabilities values differed slightly.

Comparative analyses of molecular and morphological data

For fungi, the polymorphism of the ITS region has proven sufficient to delimit intraspecific groups in many taxa (Gardes et al., 1991; Lee and Taylor, 1992; Gardes and Bruns, 1993; Erland et al., 1994; Persson et al., 1996). The data based on EF-1α secuences were used for the first time to infer phylogenetic structure within the genus *Morchella.*

The phylogenetic tree based on ITS data presented here suggests that the "black morels" and the "yellow morels" are separate taxonomic groups. These data confirm results of Wipf et al. (1999), therefore, it seems justified to distinguish three different species within the *M. esculenta* group ("yellow morels"): *Morchella esculenta* s. str., *M. crassipes,* and *M. spongiola.* In the ITS topology *M. esculenta* and *M. crassipes* and replaced within several ingroups. In EF-1α phylogram

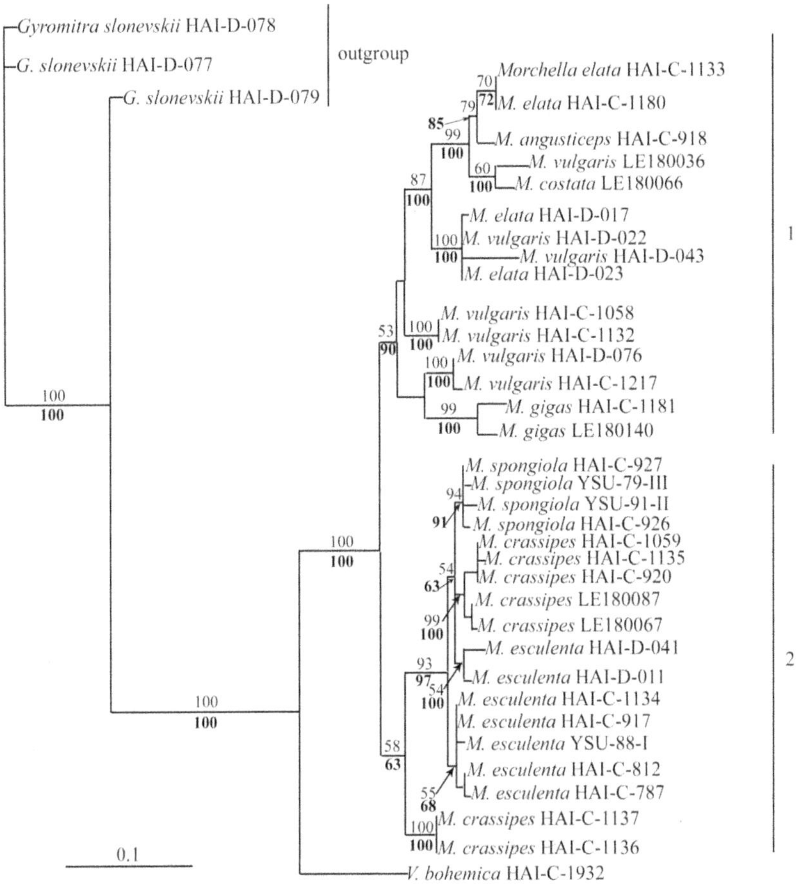

Fig. 7.3 Phylogram generated by Bayesian analyses based on EF-1α sequences data. Bootstrap values and posterior probability values higher than 50% are shown below respective branches. Topology is rooted with *Gyromitra slonevskii*. The lines with numbers indicate clades. The scale bar indicates the nucleotide substitution.

M. esculenta and *M. elata* appear in clearly separated ingroups. These results challenge the information published by Volk and Leonard (1989a,b) and Jung et al. (1993), which considers *M. esculenta* and *M. crassipes* as conspecific. However, in the ITS tree they are mixed through all clade.

Weber (1988) has discussed the descriptions of *M. crassipes* by Ventenat (1798), Persoon (1801), and Fries (1822) and considers the name invalid. She suggested that the large tan forms are just large forms of *M. esculenta*. Current results did not support her classification of the large forms as *M. esculenta*. *M. crassipes* was defined over 200 years ago in Europe as a yellow morel with a large, inflated stem. The name *M. crassipes* has been applied to a genetically distinct morel in several European studies based on ITS variation and RFLP analyses of nrDNA (Wipf et al., 1999; Kellner et al., 2005). Even before phylogenetic analyses of the group, however, mycologists began to suspect that "*Morchella crassipes*" merely represented mature, older versions of *M. esculenta* that can be found in late spring and early summer, when they have been growing for several weeks and have developed inflated stems. In the current study, all samples of the large tan forms, *M. esculenta*, grouped together with *M. crassipes* as genetically distinct taxa with the highest bootstrap and posterior probability support value 100%. *M. spongiola* and *M. esculenta* appeared to be distinct species and were supported with 100% bootstrap and 85% posterior probability values. Morphologically *M. spongiola* is similar in shape with *M. esculenta*, but slightly darker and much smaller.

M. rufobrunnea differs from *M. esculenta, M. crassipes,* and *M. spongiola* (the yellow morel group) in form and color of the ascoma, as well as in the type of blushing. There are no significant differences in the microscopic features of the hymenium except in width and form of paraphyses. The red-brown blushing morel, *M. rufobrunnea,* should be considered not only as a distinct species, but as a distinct group closer to "yellow morels" than to "black morels" with the highest bootstrap and posterior probability values 100%.

Within examined putative taxa of black morels, four different species are clearly separated: *M. elata* s. str., *M. costata*, *M. gigas*, and *M. vulgaris* (= *M. conica*). It is apparent that all the black *Morchella* species are grouped closely together. *M. angusticeps* can be seen as a synonym of *M. elata* (Fig. 7.2, Clade 3), based on identity displayed by *M. elata* sequences with *M. angusticeps.* This is in accordance with the classification proposed by Heim (1936), but challenges the concept of Smith-Weber (1988) that *M. angusticeps* and *M. vulgaris* (= *M. conica*) could be synonymous but different from *M. elata. M. costata* is distinguished as a different taxa from *M. elata* and *M. vulgaris* (=*M. conica*). From the current data, it is clear that the ITS polymorphism in black morels is much more reduced than in yellow morels, suggesting that more taxa could exist in yellow morels.

MP, ML, and Bayesian strict consensus trees, based on the EF-1α region dataset, have few differences compared with the trees gained from the ITS dataset. Dataset of EF-1α showed that *M. vulgaris* is identical with *M. costata* (bootstrap value 60% and posterior probability 100%). However, in ITS data, *M. vulgaris (=M. conica)* appeared to be an independent species.

A small amount of variation exists among species based on geographic differences. As shown in phylograms, there is no exact correlation between molecular and geographic data. For instance, specimens from India and Canada, as well as Armenia, USA, etc. are grouped together. However, the Israeli specimens are clustered together, and few geographic differences between Israeli specimens and others are apparent.

Kellner et al. (2005), based on his study of species diversity within the *M. esculenta* group in Germany and France, revealed that the occasional practice of separating morel species by ecological features may have led to an incorrect classification. Abiotic factors easily influence morphological characters in morels, which do not reflect their generic distance.

Based on phylograms obtained from ITS, EF-1α, and combined datasets, *Morchella* as a genus appears to be monophyletic, however, contains several polyphyletic ingroups.

Phylogeny of genus *Macrolepiota* (Basidiomycota, Agaricales) inferred from rDNA ITS and EF-1α sequences

The relationship of *Macrolepiota* species to one another within the genus was investigated using both molecular (nrDNA ITS and EF-1α sequences) and morphological characteristics, in order to re-evaluate the taxonomic position of several controversial species, in which the taxonomic status still remain unclear.

Based on molecular data, the genus *Macrolepiota* appeared to be monophyletic, however, several ingroups are polyphyletic (clade with *M. mastoidea* and its close related species, clade with *Chlorophyllum* species). Results supported the placement of *M. rachodes* into the genus *Chlorophyllum*, which was proposed in earlier publications. However, current molecular data together with morphological data did not support the proposals to treat *M. mastoidea* and its close related species as one taxon.

The full ITS region including ITS-1, 5.8S, ITS-2, and EF-1α were analyzed first as separate datasets, then as a combined dataset.

Results of analyses of ITS dataset

The aligned data was 847 bp long; 269 characters were parsimony-informative; 508 characters were constant and 70 variable characters were parsimony-uninformative. Parsimony analyses yielded 729 equally parsimonious trees with the following characteristics: CI=0.683, RI=0.830 and RC=0.567.

The results of our phylogenetic reconstructions are shown in Fig. 7.4. Topologies of the strict consensus MP and strict consensus ML trees were very similar.

The clades represented in the same topology in all three types of analyses: Bayesian, ML and MP. Trees generated by ML and MP analysis are not shown.

Five main clades were recorded: First clade comprised *M. psammophila*, *M. prominens* and *M. rickenni*. It has very short branches, it nests within clade 5, and is not supported by bootstrap value over than 50%, however, posterior probability value was 56%. The second clade contains *M. excoriata*, and is supported by 100% bootstrap and posterior probability values. The third clade consisted of several subclades, such as *M. procera*, *M. dolichaula* with relatively long branches, and subclade with *M. fuliginosa* and *M. rhodosperma*. Surprisingly, *M. puellaris* appears to be a sister taxon for *M. dolichaula*, and is supported with 100% bootstrap and 97% posterior probability values.

The fourth clade is the *Chlorophylum* clade. *Macrolepiota rachodes* belongs here. The clade has 97% bootstrap and 100% posterior probability value support. The clade has the longest branches in the entire phylogram.

The fifth clade appears to be quite variable. It combines closely related species such as *M. konradii*, *M. mastoidea*, *M. affinis*, *M. gracilenta*, *M. subsquarrosa*, *M. fuligineosquarrosa*, and *M. zeyheri*. However, this clade has no bootstrap support.

Results of analyses of EF-1α dataset

A total of 44 sequences were analyzed for EF-1α region. The aligned dataset showed a total of 1,184 characters including exons, introns and, gaps. The introns were non informative and were excluded from the analyses. A total of 1,011 characters were used for this analysis. The number of parsimony-informative characters was 167, 681 characters were constant, and 163 variable characters were parsimony-uninformative. The score of the best trees found is 455 (CI=0.846, RI=0.895, RC=0.757).

The ML and MP analyses separated the trees into the same clades (not shown). The tree separated into four clades (Fig. 7.5). The second clade with the relatively long branches comprised *M. procera* species, and was supported with a 72% bootstrap and 100% posterior probability. The second and fourth clades comprised *M. mastoidea* and all related species, such as *M. konradii*, *M. rickenii*, *M. affinis*, *M. subsquarrosa*, *M. fuligineosquarrosa*, *M. gracilenta*, *M. prominens*, *M. psammophila*, and *M. zeyheri*. These clades were not supported with bootstrap values. However clade 4 has 61% posterior probability support. The third clade contained *M. excoriata* with 90% bootstrap and 79% posterior probability support, and *M. rachodes* was represented as a separate sister node to *Leucoagaricus leucothites*.

Results of analyses of the combined ITS and EF-1α dataset

The combined dataset was tested for incongruence with the partition homogeneity test. Two datasets were congruent (P=0.500) and were used for further analyses. The combined dataset of ITS and EF-1α sequences yielded

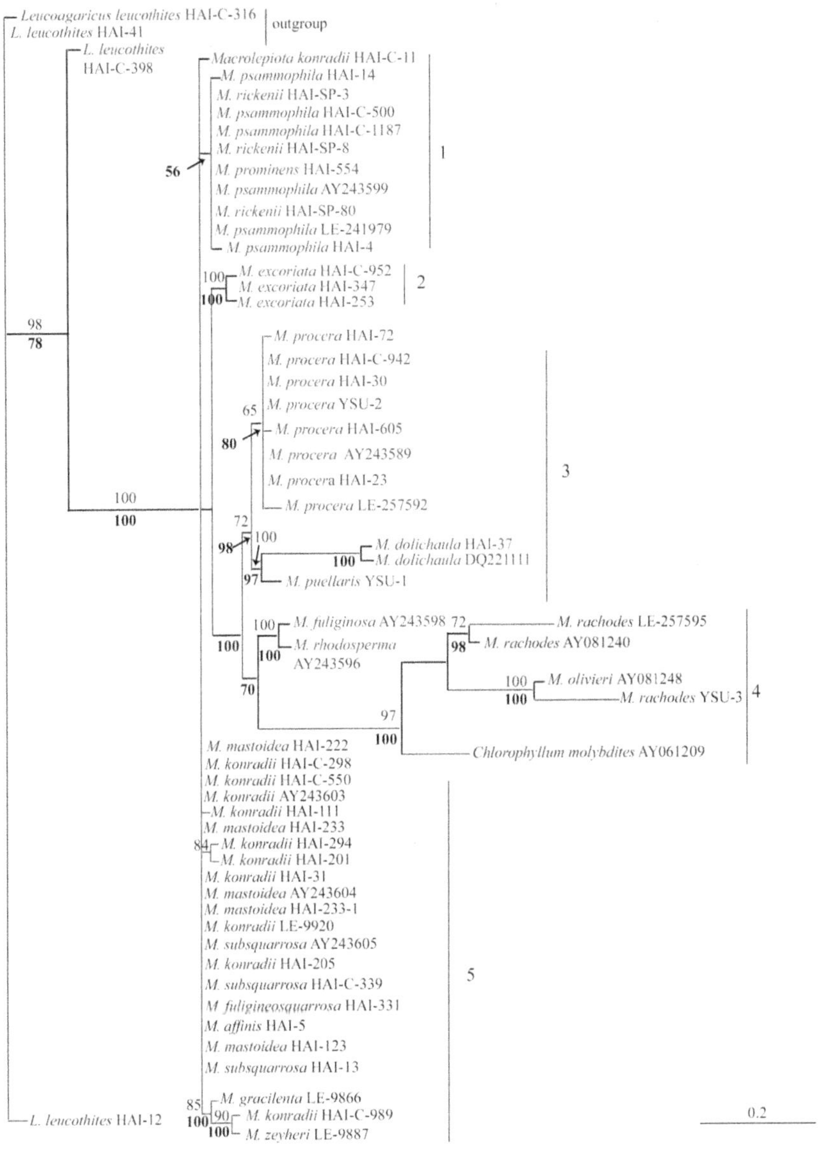

Fig. 7.4 Phylogram generated by Bayesian analysis based on ITS sequences data. Bootstrap values and posterior probability values higher than 50% are shown below the respective branches. Topology is rooted with *Leucoagaricus leucothites*. Lines with numbers indicate clades. The scale bar indicates the nucleotide substitution.

1,858 total characters from which 430 characters were parsimony-informative; 1,299 characters were constant, and 129 variable characters were parsimony-

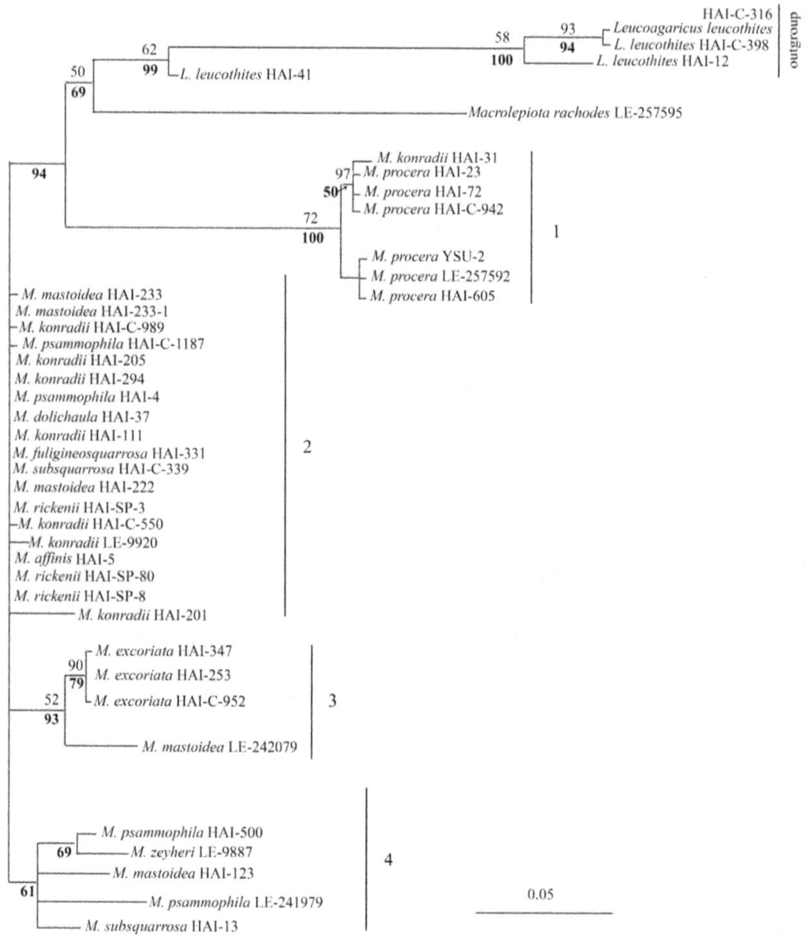

Fig. 7.5 Phylogram generated by Bayesian analysis based on EF-1α sequences data. Bootstrap values and posterior probability values higher than 50% are shown below the respective branches. Topology is rooted with *Leucoagaricus leucothites*. Lines with numbers indicate clades. The scale bar indicates the nucleotide substitution.

uninformative. Parsimony analyses yielded 1,032 equally parsimonious trees with the following characteristics: CI=0.716, RI=0.854, and RC=0.611.

All the trees generated from the combined ITS and EF-1α datasets showed the same clades as it was revealed from ITS-dataset, with few differences. *M. mastoidea* was nested within the *M. excoriata* clade. The bootsrap and posterior probability values differed slightly.

Comparative analyses of molecular and morphological data

All the trees generated from ITS, EF-1α, and ITS-EF-1α combined datasets separate genera into the same main clades: *M. mastoidea* and its close related species without any bootstrap support. *M. excoriata* group, *M. procera* and *Chlorophyllum* groups had a high bootstrap support. However, the position of several species was unresolved in the trees based on EF-1α.

The phylogram based on the ITS sequences, showed five distinct clades within *Macrolepiota* species. Several morphological characteristics supported the identified clades. Current results partially support earlier data (Vellinga, 2003 a, b, c).

M. rachodes belonged to clade 4, and as shown previously by Vellinga et al. (2003 a, b), it shared the following morphological features with the genus *Chlorophyllum*: the veil structure consists of clavate to laginiform terminal elements; stipe covering absent, spores lack germ pore or have truncate to rounded apex, without a hyaline cap over germ pore in *Chlorophyllum*. In *Macrolepiota* the veil structure consists of cylindrical to elongated elements, stipe is mainly banded, spores always have rounded apex, and germ pore is always covered by a hyaline cap.

Clade 2 included *M. excoriata*, which is a morphologically well defined species, and has 100% bootstrap support.

Clade 4 included several subclades. The first subclade included *M. procera*, which is a well-defined species. It was characterized by a brown concentrically squamulose pileus, with large separating scales, a clearly banded stipe, with double crown ring, with ellipsoid-amygdaliform spores 13-16 × 7.5-11 μm. *M. procera* subclade was supported with a 65% bootstrap value.

M. fuliginosa and *M. rhodosperma* represent another subclade within clade 4. These two species are very closely related with *M. procera*. *M. fuliginosa* differs mainly with a larger squamulose and fibrillous pileus, with more gray brown color, discoloration to reddish when scratched, and color of spore print (Candussso and Lanzoni, 1990). *M. rhodosperma* shares the same morphological features with *M. fuliginosa*, with differences in spore print color: cream colored in *M. fuliginosa* and pale salmon pink in *M. rhodosperma* (Breitenbach and Kränzlin, 1995). The phylogenetic analyses showed that these two taxa can be treated as identical.

Surprisingly, another subclade with the species *M. dolichaula* and *M. puellaris* was nested in this clade. Bon (1981) replaced *M. puellaris* in the section *Laevistipes*. It is mostly related to *M. rhacodes*, from which it differs in smaller spore size, pileus coloration and flesh coloration on exposure. Migliozii and Bizzi (1994), based on morphology, suggested that *M. puellaris* (as *M. nympharum*) is an intermediate between *Macrolepiota* and *Leucoagaricus*. Vellinga (2003c) showed that *M. nympharum* does not belong to *Macrolepiota* lineage, but it is a sister taxon of *Leucoagaricus leucothites*. However, current molecular data placed *M. puellaris* within clade 4 with *M. procera* and allies. *M. dolichaula* is paraphyletic to the *M. procera* group throughout the phylogram.

Clade 5 appears to be the most variable of all. It combines closely related species, such as *M. konradii*, *M. mastoidea*, *M. subsquarrosa*, *M. fuligineosquarrosa*, *M. affinis*, *M. gracilenta*, and *M. zeyheri*.

The position of these species has always remained unresolved. *M. mastoidea* is characterized with an umbonate pileus, covered with fine, thin, addressed scales or naked with a stellary-distrupted surface layer (Wasser, 1993). The pileipellis trichoderm consists of cylindrical elements with non encrusted walls. Stipe has small brown bands, with a small bulb, and a ring movable with age, with a double crown.

M. konradii differs from *M. mastoidea* mainly by pigmentations of velar elements. *M. konradii* is characterized by incrusted velar elements, which is not common for the other paler species of this clade. However, Vellinga (2001) reports that incrustations were found in typical pale *M. mastoidea* collections as well.

M. gracilenta is closely related to *M. mastoidea*, and differs from the latter mainly in the smaller size of spores.

M. affinis is very closely related to *M. mastoidea* as well. It is a relatively pale species, with acute umbo, with brownish-pink scales on the disk, with cracked cream margins (Candusso and Lanzoni, 1990).

M. fuligineosquarrosa was described in 1979 by Malençon from Morocco. It has a relatively smaller pileus, umbo obtuse, cuticule breaks to form a central patch surrounded by small concentric squamules on a light brown ground. Flesh turns pink and later is red-brown under exposure. Stipe has a relatively bigger bulb compared with the other species from this clade. Spores are almond shaped compared with other species in this clade, which have ellipsoid spores.

M. subsquarrosa is another closely related species of this group. It is closely related to *M. fuligineosquarrosa*, and differs from it by ellipsoid spores and flash reactions on exposure.

M. zeyheri is a relatively large parasol mushroom, about 10-25 cm nested in this clade as well. It is recognized by the very broad pinkish lamellae, the flocculose, pale brown pileal squamules on a chalky white background, and a stipe length that rarely exceeds the pileal diameter (Pegler, 1982). It has 85% bootstrap support in this clade.

Morever, besides clade 5, the phylogram separates clade 1 species that are closely related to *M. mastoidea*. These species are *M. psammophila*, *M. rickenii* and *M. prominens*.

M. rickenii and *M. prominens* were characterized by a brown granulose pileus and are closely related to *M. mastoidea* (Candusso and Lanzoni, 1990). *M. psammophila* was described in 1996 by Guinberteau, from dunes on the Atlantic Coast of France. *M. psammophila* was considered an ecological form of *M. fuligineosquarrosa* by Migliozzi and Ortega-Diaz (2001). It differs from type form by the root-like ending of the bulb, and bigger basidia. However it was considered as an independent species closely related to *M. mastoidea.*

The main differences between *M. mastoidea* and its closely related species are presented in Table 7.4. Information was collected from Pegler (1982), Bon (1996), Candusso and Lanzoni (1990), Wasser (1993), Vellinga (2001) and Didukh (2004), combined with personal observations.

The tree based on EF-1α data showed almost the same results, with few differences. The position of some species were unresolved within EF-1α phylogram: *M. konradii* was nested within *M. procera* clade and the clade was supported with 72% bootstrap and 100% posterior probability values; *M. mastoidea* was nested within *M. excoriata* clade, again with higher bootstrap and posterior probability values (52% and 93% respectively); *M. rachodes* appeared as a sister node for *Leucoagaricus* outgroup.

Based on the phylograms obtained from ITS, EF-1α and combined datasets *Macrolepiota* as a genus appeared to be monophyletic. However, on a species-specific level, few clades within the genus (e.g. clades 1 and 5 in the trees reveled from ITS and ITS-EF-1α combined datasets; clades 2 and 4 in the tree reveled from EF-1α dataset) can be considered as polyphyletic.

Molecular data combined with morphological data do not fully support the proposals set by De Kok and Vellinga (1998), Vellinga (2001), and Vellinga et al. (2003) in treating *M. mastoidea* and its related species as one taxon.

Comparison of numbers of observed substitutions versus inferred substitutions within ITS indicates that this region is much more informative than EF-1α.

CONCLUSIONS

Genus *Morchella* Dill. ex Pers.

1. Molecular data combined with micromorphological characters provide a powerful tool for the delimitation of natural groups within the genus *Morchella.*
2. Phylogenetic data and morphological data separate the genus into two main distinct clades: yellow (*M. crassipes*, *M. esculenta* s. str. and *M. spongiola*

Table 7.4 Synoptic keys to *Macrolepiota mastoidea* and its closely related species

Species	*Pileus* [*in* cm]	*Pileipelis - trichoderm elements* [*in* µm]	*Stipe* [*in* cm]	*Spores* [µm]	*Cheilo-cystidia* [µm]
M. affinis	5-12 umbo acute, brownish-pink scales on disc, margins creamy cracked	Long cylindrical (6-12 diam.), pigments intercellular, non-encrusting	7-13 × 0.6-1 (small bulb at base)	10-16 (20) × 7-9.5 (11) ellipsoid	20-45 × 8-15 ventricose-fusiform
M. gracilenta	5-12 obtuse-conic, umbo pale brown, the rest of surface whitish with yellowish-ochraceous adpressed scales	Cylindrical 5-10 (12) in diam., walls with/ without pale encrustations	10-15 × 0.8-2, enlarging toward the base	10-13 × 7-8, ellipsoid-ovate	30-35 × 12-15, clavate, bottle-shaped
M. fuligineo-squarrosa	6-8 umbo obtuse, cuticle breaks to form central patch surrounded by small concentric stumbles on light brown ground	Disc elements terminal (113-200 × 15-23) and palisade (118-125 × 13-18); margin elements cylindrical	8-10 × 1-1.5, (<2.7 diam. basal bulb)	12.5-17 × 7.5-10, ovoid, sub elliptic, almond-shaped	45-50 × 10-15 slightly clavate, fusiform
M. konradii	5-11 umbo low, disc umbrous, to black brownish margins whitish to gray, surface disrupted in asteroid fashion	Long (50-300 × 9-15), cylindrical, intercellular pigments brownish, incrusted toward base	7-15.5 × 0.5-1.5 (2.5-3 at base)	11.4-15.5 (20) × 7-10, ellipsoid, ellipsoid-ovate	17-40 × 10-13 clavate, subfusiform
M. mastoidea	8-12 umbo sharply, protruding, dark disc scales, appressed, fine floccose, margins whitish, cracked	Long (80-180 × 6-12), cylindrical, walls with/ without pale encrusations	11-16 × 0.3-1.1 (1.5-2 at base)	11-16.5 (19) × 7.5-10.5, ellipsoid	20-50 × 12-15, fusiform

Table 7.4 Contd. ...

M. psammophila	Disc 6-8, with small concentric brown squamules on light ground	20-60 (80) × 10-21 (28)	8-10 × 1.5 (3 at base, with root-like ending of the bulb)	12-17 × 7-10, ovoid, elliptic	40-50 × 10-15, clavate, subfusiform
M. prominens	8-12 umbo protruding, grayish-brown in the center, edge floccose-fibrillose	Long (100-200 × 7-12), cylindrical, walls with ochraceous encrustations	12-18 × 1-2.5, with bulb	9-11 × 7-8, ovate-elliptical	30-36 × 12-15, polymorphic
M. rickenii	6-12 disc, umbonate and with light brown scales	Cylindrical (8-14 diam), walls with/without pale encrustations	10-16 × 1-2 (2-2.5 at base)	12-17 × 7.5-10 ellipsoid	15-33 × 6-15 ventricose-subfusiform
M. subsquarrosa	6-8 (10), slightly umbonate, brownish, with concentric brownish-reddish squamules	Long cylindrical (200-250 × 8-15), walls with ochraceous incrustations	6-8 (10) × 0.8-1 (2 at base)	10-15 (20) × 7-11, ellipsoid-ovate	20-40 × 8-13 claviform to ventricose
M. zeyheri	10-25 cm, white cap, scaled light brown	–	10-20 (25) × 1.5-2, with broaden base	11.5-21 × 7.5-15.5	–

and black morels (*M. elata* s. str., *M. costata*, *M. gigas*, and *M. vulgaris* (= *M. conica*). These results confirm the earliest publications of Royse and May (1990), and Wipf et al. (1999). The results supported the evidence that *Morchella elata* and *M. angusticeps* are identical taxa as proposed by earlier workers. These data did not confirm data obtained from the cultural studies. Morphological data together with molecular data did not support the proposals that *M. esculenta* and *M. crassipes* are a single species (Volk, 1990).

3. *M. spongiola* and *M. esculenta* appeared to be distinct species, morphologically *M. spongiola* is similar in shape to *M. esculenta*, but slightly darker and much smaller (Smickaya, 1980).
4. *M. rufobrunnea* differs from *M. esculenta, M. crassipes* and *M. spongiola* (yellow morel group) in the form and color of the ascoma, as well as in the type of blushing. The red-brown blushing morel, *M. rufobrunnea,* should be considered not only as a distinct species, but as a distinct group closer to yellow morels than to black morels.

5. *M. angusticeps* can be seen as a synonym of *M. elata* based on identity displayed by *M. elata* sequences with *M. angusticeps*. This is in accordance with the classification proposed by Heim (1936), but challenges the concept of Smith-Weber (1988) that *M. angusticeps* and *M. vulgaris* (=*M. conica*) could be synonymous but different from *M. elata*. *M. costata* is distinguished as a different taxa from *M. elata* and *M. vulgaris* (=*M. conica*). From current data, it is clear, that the ITS polymorphism in black morels is much more reduced than in yellow morels, suggesting that more taxa could exist in yellow morels.
6. The dataset of EF-1α showed that *M. vulgaris* (=*M. conica*) is identical with *M. castata*. However, in ITS data *M. costata* appeared to be an independent species.
7. A small amount of variation exists among species based on geographic differences. As noted in phylograms, there is no exact correlation between molecular and geographic data. For instance, specimens from India and Canada, as well as Armenia, USA, etc. are grouped together. However, the Israeli specimens are clustered together and a slight geographical distance between Israeli specimens and others can be noted.
8. After the phylogenetic study, the list of *Morchella* species of Israel was re-evaluated, which is currently represented by *M. crassipes*, *M. elata*, *M. esculenta*, *M. rufobrunnea*, and *M. vulgaris* (=*M. conica*).

Genus *Macrolepiota* Singer

1. According to the phylogenetic data *Macrolepiota rachodes* belonged to the *Chlorophyllum* clade and was supported by high bootstrap value and morphological data.
2. According to the molecular and morphological data *Macrolepiota rhodosperma* and *M. fuliginosa* appeared to be identical.
3. Variable clade 5, comprising *M. mastoidea* and allies together with clade 1, which is composed of closely related species *M. rickenii*, *M. psammophila*, and *M. prominens*. Also, taking into consideration the morphological characteristics, the known data is not enough to support the proposals to treat *M. mastoidea* and its related species as one taxon.
4. The position of *M. puellaries* and *M. dolichaula* is not supported by morphological data and remains unresolved within the phylogram.
5. Result showed that for *Morchella*, EF-1α is informative, whereas for *Macrolepiota* – hardly so. With the topology presented for *Macrolepiota* based on EF-1α, it is logical to conclude that this marker is not as suitable for resolution of species complexes that are close in this particular taxon, genus *Macrolepiota*. Yet saying that the complexes should be treated as the same species is quite premature.

6. This study helped to accomplish the extensive revision of the putative *Macrolepiota* species in Israel; and re-evaluate the previous list of Israeli *Macrolepiota* species, which is currently presented by the following species: *M. affinis*, *M. excoriata*, *M. excoriata* var. *rubescens*, *M. fuligineosquarrosa*, *M. konradii*, *M. mastoidea*, *M. procera*, *M. procera* var. *mediterranea*, *M. prominens*, *M. psammophila*, *M. rickenii*, and *M. subsquarrosa*. *Macrolepiota oliveri* and *M. rachodes* were transferred to the genus *Chlorophyllum*.

ACKNOWLEDGEMENTS

Professor Sophia Menache, Dean of Graduate Studies, Haifa University and Dr. Moshe Inbar, Dean of Department of Evolutionary & Environmental Biology, Faculty of Science & Science Education of Haifa University, are deeply appreciated for supporting our research that resulted in this publication. We also thank Prof. Nanagulyan G. Siranush (Armenia) and Dr. I. Zmitrovich (Russia) for providing the herbarium material.

REFERENCES

Agerer, R. 2002. Rhizomorph structures confirm the relationship between Lycoperdales and Agaricaceae (Hymenomycetes, Basidiomycota). Nova Hedwigia, 75: 367-385.

Akaike, H. 1974. A new look at the statistical model identification. IEEE Transactions on Automatic Control, 19: 716-723.

Alexopoulos, C. J., Mims, C. W. and Blackwell, M. 1996. Introductory Mycology. 4th edn. John Wiley & Sons Inc., New York, USA.

Arora, D. 1986. Mushrooms Demystified. 2nd edn. Ten Speed Press, Berkeley, CA.

Baldauf, S. L., Roger, A. J., Wenk-Siefert, I. and Doolittle, W. F. 2000. A kingdom-level phylogeny of eukaryotes based on combined protein data. Science, 290: 972-977.

Baldwin, B. G. 1992. Phylogenetic utility of the internal transcribed spacers of nuclear ribosomal DNA in plants: An example from the Compositaogy. Molecular Phylogenetics and Evolution, 1: 3-16.

Baldwin, B. G., Sanderson, M. J., Porter, J. M., Wojciechowski, M. F., Campbell, C. S. and Donoghue, M. J. 1995. The ITS region of nuclear ribosomal DNA: a valuable source of evidence on angiosperm phylogeny. Annals of the Missouri Botanical Garden, 82: 247-277.

Berbee, M. L. and Taylor, J. W. 1992. Convergence in ascospore discharge mechanism among Pyrenomycete fungi based on 18S ribosomal RNA gene sequence. Molecular Phylogeny and Evolution, 1: 59-71.

Berbee, M. L. and Taylor, J. W. 2001. Fungal molecular evolution: gene trees and geologic time, pp. 229-245. In: The Mycota, Vol. 7B. Systematics and Evolution. D. J. McLaughlin, E. G. McLaughlin and P. A. Lemke (eds.). Springer-Verlag, Berlin, Germany.

Binder, M., Besl, H. and Bresinsky, A. 1997. Agaricales oder Boletales? Molekularbiologische Befunde zur Zuordnung einiger umstrittener Taxa. Zeitschrift für Mykologie, 63: 189-196.

Binder, M., Hibbett, D. S., Larsson, K. H., Larsson, E., Langer, E. and Langer, G. 2005. The phylogenetic distribution of resupinate forms across the major clades of mushroom-forming fungi (Homobasidiomycetes). Systematic Biodiversity, 3: 113-157.

Binder, M., Hibbett, D. S., Wang, Z. and Farnham, W. F. 2006. Evolutionary relationships of *Mycaureola dilseae* (Agaricales), a basidiomycete pathogen of a subtidal rhodophyte. American Journal of Botany, 93: 547-556.

Bodensteiner, P., Binder, M., Moncalvo, J. M., Agerer, R. and Hibbett, D. S. 2004. Phylogenetic relationships of cyphelloid homobasidiomycetes. Molecular Phylogenetics and Evolution, 33: 501-515.

Bon, M. 1981. Novitates. Combinaisons et noms nouveaus. Document Mycologia, 11: 28.

Bon, M. 1996. Die Grosspilzflora von Europa. Lepiotaceae, Vol. 3. IHW-Verlag, Germany.

Boudier, J. L. É. 1897. Diagnoses nouvelles de quelques espèces critiques de champignons. Bulletin de la Sociétè Botanique de France, 26: 228-236.

Bowman, E. J. and Knock, T. E. 1992. Structures of the genes encoding the α and β subunits of the *Neurospora crassa* mitochondrial ATP synthase. Gene, 114: 157-163.

Bresinsky, A. and Stangl, J. 1961. Zur Artabgrenzung innerhalb der Gattung *Morchella*. Z. Pilzkd, 27: 102-110.

Bresinsky, A., Glaser, W. and Stangl, J. 1972. Untersuchungen zur Sippenstruktur der Morchellaceen. Ber. Bayer. Bot. Ges., 43: 127-143.

Breitenbach, J. and Kränzlin, F. 1984. Fungi of Switzerland. Vol. 1. Ascomycetes. Mycological Society of Lucerne. Lucerne.

Breitenbach, J. and Kranzlin, F. 1995. Fungi from Switzerland. Agarics. 2nd part. IV. Mykologia Lucerne. Lucerne.

Brummelen, J. van. 1994. Problems in the Systematics of Pezizales. In: Ascomycete Systematics: problems and perspectives in the nineties. D. L. Hawksworth (ed.). NATO ASI Series 269. Plenum Press, New York, USA.

Bruns, T. D., Vilgalys, R., Barns, S. M., Gonzalez, D., Hibbett, D. S., Simon, D. J., Stickel, S., Szaro, T. M., Weisburg, W. G. and Sogin, M. L. 1992. Evolutionary relationships within the fungi: analysis of nuclear small subunit rRNA sequences. Molecular Phylogeny and Evolution, 1: 231-241.

Bruns, T. D., White, T. J. and Taylor, J. W. 1991. Fungal molecular systematics. Annual Review of Ecology and Systematics, 22: 525-564.

Bunyard, B. A., Nicholson, M. S. and Royse, D. J. 1994. A systematic assessment of *Morchella* using RFLP analysis of the 28S ribosomal RNA gene. Mycologia, 86: 762-772.

Candusso, M. and Lanzoni, G. 1990. *Lepiota* s.l. Fungi Europeai 4. Libreria editrice Giovanna Biella, Italy.

Cho, S., Mitchell, A., Regier, J. C., Mitter, C., Poole, R. W., Friedlander, T. P. and Zhao, S. 1995. A highly conserved nuclear gene for low-level phylogenetics: elongation factor

1-alpha recovers morphology-based tree for heliothine moths. Molecular Biology and Evolution, 12: 650-656.

Cubeta, M. A., Echandi, E., Abernathy, T. and Vilgalys, R. 1991. Characterization of anastomosis group of binucleate *Rhizoctonia* species using restriction analysis of an amplified ribosomal RNA gene. Phytopathology, 81: 1395-1400.

de Kok, R. P. J. and Vellinga, E. C. 1998. Notulae ad floram agaricinam neerlandicam - XXXII, *Macrolepiota*. Persoonia, 17: 69-79.

Dennis, R. W. G. 1981. [First published 1978]: British Ascomycetes. A.R. Gantner Verlag K.G., Vaduz.

Didukh, M. Ya. 2004. Taxonomy, Biology, and Phylogeny of Some Taxa of the Family Agaricaceae Chevall. With Special Attention to Biotechnologically Important Species. Ph.D. Thesis. University of Haifa, Israel.

Dubouzet, J. G. and Shinoda, K. 1999. ITS DNA sequence relationships between *Lilium concolor* Salisb., *L. dauricum* Ker-Gawl. and their putative hybrid, L. maculatum Thunb. TAG Theoretical and Applied Genetics, 98: 213-218.

Edgar, R. C. 2004. MUSCLE: multiple sequence alignment with high accuracy and high throughput. Nucleic Acids Research, 35: 1792-1797.

Erland, S., Henrion, B., Martin, F., Glover, L. A. and Alexander, I. J. 1994. Identification of the ectomycorrhizal basidiomycete *Tylospora fibrillose* Donk by RFLP analysis of the PCR-amplified ITS and IGS regions of ribosomal DNA. New Phytologist, 126: 525-532.

Farris, J. S., Kallersjo, M., Kluge, A. G. and Bult, C. 1994. Testing the significance of incongruence. Cladistics, 10: 315-319.

Farris, J. S., Kallersjo, M., Kluge, A. G. and Bult, C. 1995. Constructing a significance test for incongruence. Systematic Biology, 44: 570-572.

Fayod, V. 1889. Prodrome d'une histoire naturelle des Agaricinées. Annls. Sci. Nat. Bot. Sér.7, 9: 181-411.

Felsenstein, J. 1985. Confidence limits on phylogenies: An approach using the bootstrap. Evolution, 39: 783-791.

Franco-Molano, A. E. 1995. Observations on *Rugosospora* in the Neotropics. Mycologia, 87: 574-578.

Fries, E. M. 1822. System Mycologicum. Vol. 2. Lund, Sweden and Greifswald, Lund & Greifswald. Germany.

Gardes, M. and Bruns, T. D. 1993. ITS primers with enhanced specificity for basidiomycetes-application to the identification of mycorrhizae and rusts. Molecular Ecology, 2: 113-118

Gardes, M., White, T. J., Fortin, J. A., Bruns, T. D. and Taylor, J. W. 1991. Identification of indigenous and introduced symbiotic fungi in ectomycorrhizae by amplification of nuclear and mitochondrial ribosomal DNA. Canadian Journal of Botany, 69: 180-190.

Gargas, A. and Taylor, J. 1995. Phylogeny of Discomycetes and early radiations of the apothecial Ascomycotina inferred from SSU rDNA sequence data. Experimental Mycology, 19: 7-15.

Guzmán, G. and Tapia, F. 1998. The known morels in Mexico, a description of a new blushing species, *Morchella rufobrunnea*, and new data *M. guatemalensis*. Mycologia, 90: 705-714.

Guzmán, G., Bandala, V. M., Montoya, L. and Saldarriaga, Y. 1989. Nuevas evidências sobre las relaciones micoflorísticas entre África y el neotrópico. El género *Rugosospora* Heinem. (Fungi, Agaricales). Brenesia, 32: 107-112.

Hall, T. A. 1999. BioEdit: a user-friendly biological sequence alignment editor and analysis program for Windows 95/98NT. Nucleic Acids Symposium Series, 41: 95.

Hallen, H., Volk, T. and Adams, G. 2001. May is Morel Month in Michigan. Michigan State University Extention, East Lansing, MI.

Hallen, H. E., Watling, R. and Adams, G. C. 2003. Taxonomy and toxicity of *Conocybe lactea* and related species. Mycological Research, 107: 969-979.

Hanlin, R. T. and Hahn, C. G. 1990. Illustrated genera of Ascomycetes. The American Pathological Society St. Paul, MN.

Hansen, K. and Pfister, D. H. 2006. Systematics of the Pezizomycetes – the operculate discomycetes. Mycologia, 98: 1029-1040.

Harrington, F. A., Pfister, D. H., Potter, D. and Donoghue, M. J. 1999. Phylogenetic studies within the Pezizales. I. 18S rRNA sequence data and classification. Mycologia, 91: 41-50.

Harrison, C. J. and Langdale, J. A. 2006. A step by step guide to phylogeny reconstruction. Plant Journal, 45: 561-572.

Hashimoto, T., Nakamura, Y., Kamaishi, T., Adachi, J., Nakamura, F., Okamoto, K-I. and Hasegawa, M. 1995. Phylogenetic place of kinetoplastid protozoa inferred from a protein phylogeny of elongation factor 1α. Molecular and Biochemical Parasitology, 70: 181-185.

Hawksworth, D. L. 1991. The fungal dimension of biodiversity: magnitude, significance, and conservation. Mycological Research, 95: 641-655.

Hawksworth, D. L. 2001. The magnitude of fungal diversity: the 1.5 million species estimate revisited. Mycological Research, 105: 1422-1432.

Heim, R. 1936. La culture des Morilles. Review Mycologia (Suppl. 1): 10-11 and (Suppl. 2): 19-25.

Heim, R. 1971. The interrelationships between the Agaricales and Gasteromycetes, pp. 505-534. In: Evolution in the Higher Basidiomycetes. R. H. Petersen (ed.). Knoxville, University of Tennessee Press, USA.

Hershkovitz, M. A. and Lewis, L. A. 1996. Deep-level diagnostic value of the rDNA-ITS region. Molecular Biology and Evolution, 13: 1276-1295.

Hibbett, D. S. and Vilgalys, R. 1991. Evolutionary relationships of *Lentinus* to the Polyporaceae: evidence from restriction analysis of enzymatically amplified ribosomal DNA. Mycologia, 83: 423-439.

Hibbett, D. S. and Thorn, R. G. 2001. Basidiomycota: Homobasidiomycetes, pp. 121-168. In: The Mycota. VIIB. Systematics and Evolution. D. J. McLaughlin, E. G. McLaughlin and P. A. Lemke (eds.). Springer-Verlag, Berlin.

Hibbett, D. S., Pine, E. M., Langer, E., Langer, G. and Donoughue, M. J. 1997. Evolution of gilled mushrooms and puffballs inferred from ribosomal DNA sequences. Proceedings of the National Academy of Sciences, USA, 94: 12002-12006.

Hibbett, D. S., Binder, M., Bischoff, J. F., Blackwell, M., Cannon, P. F., Eriksson, O. E., Huhndorf, S., James, T., Kirk, P. M., Lücking, R., Thorsten Lumbsch, H., Lutzoni, F., Matheny, P. B., McLaughlin, D. J., Powell, M. J., Redhead, S., Schoch, C. L., Spatafora, J. W., Stalpers, J. A., Vilgalys, R., Aime, M. C., Aptroot, A., Bauer, R., Begerow, D., Benny, G. L., Castlebury, L. A., Crous, P. W., Dai, Y. C., Gams, W., Geiser, D. M., Griffith, G. W., Gueidan, C., Hawksworth, D. L., Hestmark, G., Hosaka, K., Humber, R. A., Hyde, K. D., Ironside, J. E., Kõljalg, U., Kurtzman, C. P., Larsson, K. H., Lichtwardt, R., Longcore, J., Miadlikowska, J., Miller, A., Moncalvo, J. M., Mozley-Standridge, S., Oberwinkler, F., Parmasto, E., Reeb, V., Rogers, J. D., Roux, C., Ryvarden, L., Sampaio, J. P., Schüssler, A., Sugiyama, J., Thorn, R. G., Tibell, L., Untereiner, W. A., Walker, C., Wang, Z., Weir, A., Weiss, M., White, M. M., Winka, K., Yao, Y.-J. and Zhang, N. 2007. A higher-level phylogenetic classification of the Fungi. Mycological Research, 111: 509-547.

Hopple, J. S. and Vilgalys, R. 1999. Phylogenetic relationship of mushroom genus *Coprinus* and dark-spored allies based on sequence data from the nuclear gene coding for the large ribosomal subunit RNA: divergent domains, outgroups and monophyly. Molecular Phylogentics Evolution, 13: 1-19.

Horak, E. 1968. Synopsis generum Agaricalium (Die Gattungstypen der Agaricales). Beiträge zur Kryptogamenflora der Schweiz, 13: 1-741.

Jacquetant, E. 1984. Les morilles. [The morels]. Piantanida. Lausanne, Switzerland.

James, T. Y., Kauff, F., Schoch, C. L., Matheny, P. B., Hofstetter, V., Cox, C. J., Celio, G., Gueidan, C., Fraker, E., Miadlikowska, J., Lumbsch, H. T., Rauhut, A., Reeb, V., Arnold, E. A., Amtoft, A., Stajich, J. E., Hosaka, K., Sung, G-H., Johnson, D., O'Rourke, B., Crockett, M., Binder, M., Curtis, J. M., Slot, J. C., Wang, Z., Wilson, A. W., Schüßler, A., Longcore, J. E., O'Donnell, K., Mozley-Standridge, S., Porter, D., Letcher, P. M., Powell, M. J., Taylor, J. W., White, M. M., Griffith, G. W., Davies, D. R., Humber, R. A., Morton, J. B., Sugiyama, J., Rossman, A. Y., Rogers, J. D., Pfister, D. H., Hewitt, D., Hansen, K., Hambleton, S., Shoemaker, R. A., Kohlmeyer, J., Volkmann-Kohlmeyer, B., Spotts, R. A., Serdani, M., Crous, P. W., Hughes, K. W., Matsuura, K., Langer, E., Langer, G., Untereiner, W. A., Lücking, R., Büdel, B., Geiser, D. M., Aptroot, A., Diederich, P., Schmitt, I., Schultz, M., Yahr, R., Hibbett, D. S., Lutzoni, F., McLaughlin, D. J., Spatafora, J. W. and Vilgalys, R. 2006. Reconstructing the early evolution of the fungi using a six-gene phylogeny. Nature, 443: 818-822.

Johnson, J. 1999. Phylogenetic relationship within *Lepiota sensu lato* based on morphological and molecular data. Mycologia, 91: 443-458.

Johnson, J. and Vilgalys, R. 1998. Phylogenetic systematics of *Lepiota sensu lato* based on nuclear large subunit rDNA evidence. Mycologia, 90: 971-979.

Jülich, W. 1981. Higher taxa of basidiomycetes. Bibliothec. Mycologia, 85: 5-485.

Jung, S., Gessner, R., Keudell, K. and Romano, M. 1993. Systematics of *Morchella esculenta* complex using enzyme-linked immunosorbent assay. Mycologia, 85: 677-684.

Kellner, H., Renker, C. and Buscot, F. 2005. Species diversity within the *Morchella esculenta* group (Ascomycota: Morchellaceae) in Germany and France. Organisms, Diversity & Evolution, 5: 101-107.

Kim, S. and Jung, H. 2000. Phylogenetic relationships of the Aphyllophorales inferred from sequence analysis of nuclear small subunit ribosomal DNA. The Journal of Microbiology, 38: 122-131.

Kimbrough, J. W. 1970. Current trends in classification of Discomycetes. Botanical Review, 36: 91-161.

Kirk, P., Cannon, P., David, J. and Stalpers, J. 2008. Dictionary of the Fungi. 10th edn. CABI Publishing, Wallingford, United Kingdom.

Kohn, L. M., Petsche, D. M., Bailey, S. R., Novak, D. A. and Anderson, J. B. 1988. Restriction fragment length polymorphisms in nuclear and mitochondrial DNA of *Sclerotia* species. Phytopathology, 78: 1047-1051.

Korf, R. P. 1973. Discomycetes and Tuberales. In: The Fungi. Vol. IVA. G. C. Ainsworth, F. K. Sparrow and A. S. Sussman (eds.). Academic Press, New York.

Krieglsteiner, G. J. 1993. Verbreitungsatlas der Grospilze Deutschlands (West). Band 2: Schlauchpilze. Ulmer, Stuttgart.

Kühner, R. 1980. Les Hymènomycétes agaricoides (Agaricales, Tricholomatales, Pluteales, Russulales). Etude générale et classification. Bulletin Mensuel de la Société Linnéenne de Lyon, 49: 1027.

Landvik, S. 1996. *Neolecta*, a fruit-body-producing genus of the basal ascomycetes, as shown by SSU and LSU rDNA sequences. Mycological Research, 100: 199-202.

Landvik, S., Egger, K. and Schumacher, T. 1997. Towards a subordinal classification of the Pezizales (Ascomycota): phylogenetic analyses of SSU rDNA sequences. Nordic Journal of Botany, 17: 403-418.

Larsson, K. H., Larsson, E. and Kõljalg, U. 2004. High phylogenetic diversity among corticioid basidiomycetes. Mycological Research, 108: 983-1002.

Lee, S. B. and Taylor, J. W. 1992. Phylogeny of five fungus like protoctistan *Phytophthora* species, inferred from the internal transcribed spacers of ribosomal DNA. Molecular Biology and Evolution, 9: 636-653.

Lumbsch, H. T. and Huhndorf, S. M. 2007. Outline of Ascomycota - 2007. Myconet, 13: 1-58.

Maddison, W. P. and Maddison, D. R. 2007. Mesquite: A modular system for evolutionary analysis. Version 2.6.

Magee, B. B., D'Souza, T. M. and Magee, P. T. 1987. Strain and species identification by restriction fragment length polymorphisms in the ribosomal DNA repeat of *Candida* species. Journal of Bacteriology, 169: 1639-1643.

Matheny, P. B. and Bougher, N. L. 2005. A new violet species of *Inocybe* (Agaricales) from urban and rural landscapes in Western Australia. Australasian Mycologist, 24: 7-12.

Matheny, P. B., Curtis, M. J., Hofstetter, M. C., Moncalvo, J. M., Ge, Z. W., Yang, Z. L., Slot, J. C., Ammirati, J. F., Baroni, T. J., Bougher, N. L., Karen, W. Hughes, K. W., Lodge, D. J., Kerrigan, R. W., Seidl, M. T., Aanen, D. K., DeNitis, M., Daniele, G. M., Desjardin, D. E., Kropp, B. R., Norvell, L. L., Parker, A., Vellinga, E. C., Vilgalys, R. and Hibbett, D. S. 2006. Major clades of Agaricales: a multilocus phylogenetic overview. Mycologia, 98: 982-995.

Medlin, L., Elwood, H. J., Stickel, S. and Sogin, M. L. 1988. The characterization of enzymatically amplified eukaryotic 16S-like rRNA-coding regions. Gene, 71: 491-499.

Meyer, C. P. and Paulay, G. 2005. DNA barcoding: error rates based on comprehensive sampling. PLoS BIOLOGY, 3: 2229-2238.

Migliozzi, V. and Bizzi, A. 1994. Considerazioni su una raccolta e studio di Macrolepiota puellaris (Fries) Moser. Funghi Ambiente, 66: 9-16

Migliozzi, V. and Ortega-Diaz, A. 2001. Alcune *Macrolepiota* della sezione *Macrosporae* in ambiente mediterraneo. Bollettino Gruppo Micologico G. Bresadola, 44: 5-30.

Mitchell, A., Cho, S., Regier, J. C., Mitter, C., Poole, R. W. and Mathews, M. 1997. Phylogenetic utility of elongation factor 1α in Noctuoidea (Insecta: Lepidoptera): the limits of synonymous substitution. Molecular Biology and Evolution, 14: 381-390.

Moncalvo, J. M., Lutzoni, F., Rehner, S. A., Johnson, J. and Vilgalys, R. 2000. Phylogenetic relationship of agaric fungi based on nuclear large subunit ribosomal DNA sequences. Systems Biology, 49: 278-305.

Moncalvo, J. M., Vilgalys, R., Redhead, S. A., Johnson, J. E., James, T. Y., Aime, M. C., Hofstetter, V., Verduiin, S. J. W., Larsson, E., Baroni, T. J., Thorn, R. G., Jacobsson, S., Clemençon, H. and Miller, O. K. 2002. One hundred and seventeen clades of euagarics. Molecular Phylogenetics and Evolution, 23: 357-400.

Moser, M. 1967. Die Röhrlinge und Blätterpilze (Agaricales). 3 Aufl. Gustav Fisher, Germany.

Moser, M. 1978. Die Röhrlinge und Blätterpilze (Polyporales, Boletales, Agaricales, Russulales). 5 Aufl. Gustav Fisher, Germany, United States.

Moser, M. 1983. Keys to Agarics and Boleti. Phillips, Germany.

Nordnes, S., Krauss, S. and Johansen, T. 1994. rDNA sequence of zebrafish (*Brachydanio rerio*) translation elongation factor-1α: molecular phylogeny of eukaryotes based on elongation factor-1α protein sequences. Biochimica et Biophysica Acta, 1219: 529-532.

Norvell, L. L. 2001. An informal poll for MSA members: should the type of *Coprinus* be changed? Inoculum, 52: 5.

O'Donnell, K., Cigelnik, E., Weber, N. E. and Trappe, J. M. 1997. Phylogenetic relationships among ascomycetous truffles and the true and false morels from 18S and 28S ribosomal DNA sequence analyses. Mycologia, 89: 48-65.

O'Donnell, K., Weber, N. S., Rehner, S. and Milles, G. 2003. Phylogeny and biogeography of *Morchella*. The 22nd Fungal Genetics Conference. Kansas City, MO: Fungal Genetics Stock Center. Fung. Gen. News letter, 50 (Supplement): Abstract 443.

Overholts, L. O. 1934. The morels of Pennsylvania. Proceedings Pennsylvania Academy Science, 8: 108-114.

Pacioni, G. 1981. Guide to Mushrooms. Simon & Schuster, New York.

Page, R. D. M. 1996. Treeview: An application to display phylogenetic trees on personal computers. Computer Applications in the Biological Sciences, 12: 357-358.

Pegler, D. N. 1982. Agaricoid and boletoid fungi (Basidiomycota) from Malawi and Zambia. Kew Bull., 37: 255-271.

Pegler, D. N. and Young, T. W. K. 1969. Ultrastructure of basidiospores in Agaricales in relation to taxonomy and spore discharge. Transactions of the British Mycological Society, 52: 491-513.

Peintner, U., Rainer, J. and Pöder, R. 2001. Biodiversity and concentration of airborne fungi in a hospital environment. Mycopathologia, 149: 87-97.

Persoon, C. H. 1801. Synopsis methodica fungorum, systems enumerationem omnium hucusque detectarum specierum cum brevibus descriptionibus nec non synonymis et observationibus selectis. Gottingae.

Persson, Y., Erland, S. and Jansson, H. B. 1996. Identification of nematode-trapping fungi using RFLP analysis of the PCR-amplified ITS region of ribosomal DNA. Mycological Research, 100: 531-534.

Petersen, R. H. 1971. Evolution in the Higher Basidiomycetes. University of Tennessee Press, USA.

Pilz, D., McLain, R., Alexander, S., Villarreal-Ruiz, L., Berch, S., Wurtz, T., Parks, C., McFarlane, E., Baker, B., Molina, R. and Smith, J. 2007. Ecology and management of morels harvested from the forests of Western North America. USDA. Forest Serv. Pacific Northwest Research Station, General Technical Report, (PNW-GTR-710).

Posada, D. 2006. ModelTest Server: a web-based tool for the statistical selection of models of nucleotide substitution online. Nucleic Acids Research, 34: 700-703.

Posada, D. and Crandall, K. A. 1998. Modeltest: testing the model of DNA substitution. Bioinformatics, 14: 817-818.

Redhead, S. A., Vilgalys, R., Moncalvo, J. M., Johnson, J. and Hopple, J. 2001. *Coprinus* Pers. and the disposition of *Coprinus* species *sensu lato*. Taxon, 50: 203-241.

Regier, J. and Shultz, J. W. 1997. Molecular phylogeny of the major arthropod groups indicates polyphyly of the crustaceans and a new hypothesis for the origin of hexapods. Molecular Biology and Evolution, 14: 902-913.

Rehner, S. A. and Buckley, E. P. 2005. A Beauveria phylogeny inferred from nuclear ITS and EF1-alpha sequences: evidence for cryptic diversification and links to *Cordyceps* teleomorphs. Mycologia, 97: 84-98.

Roger, A. J., Sandblom, O., Doolittle, W. F. and Philippe, H. 1999. An evaluation of elongation factor 1 alpha as a phylogenetic marker for eukaryotes. Molecular Biology and Evolution, 16: 218-233.

Ronquist, F., Huelsenbeck, J. P. and van der Mark, P. 2005. MrBayes 3.1 Manual.

Royse, D. J. and May, B. 1990. Interspecific allozyme variation among *Morchella* spp. and its inferences for systematic within the genus. Biochemical Systematics and Ecology, 18: 475-479.

Seaver, F. J. 1942. The North American Cup-fungi (Operculates). Hafner Publishing Co., New York.

Singer, R. 1946 (1948). New and interesting species of basidiomycetes. The Michigan Academy of Science Arts and Letters, 32: 103-105.

Singer, R. 1962. The Agaricales in Modern Taxonomy. J. Cramer, Liechtenstein.

Singer, R. 1975. The Agaricales in Modern Taxonomy. 3rd edn. J. Cramer, Liechtenstein.

Singer, R. 1986. The Agaricales in Modern Taxonomy. 4th edn. Koeltz Sci. Books, Germany.

Smickaya, M. 1980. Operculate Discomycetes. Mycophlora of Ukraine. Naukova Dumka Press, Kiev.

Smith, A. H. and Weber, N. S. 1980. The Mushroom Hunter's Field Guide. University of Michigan Press, Ann Arbor.

Smith, A. H., Smith, H.V. and Weber, N. S. 1981. How to Know the Non-gilled Mushrooms. Dubuque, Iowa: Wm. C. Brown, United States.

Smith-Weber, N. 1988. A Morel Hunters Companion. *Two*. Peninsula Press, Lansing, Michigan.

Sogin, M. L. 1990. Amplification of ribosomal RNA genes for molecular evolution studies. In: PCR Protocols, A Guide to Methods and Applications. M. A. Innis, D. H. Gelfand, J. J. Sninsky and T. J. White (eds.). Academic Press, New York.

Spatafora, J. W. 1995. Ascomal evolution among filamentous ascomycetes: evidence from molecular data. Canadian Journal of Botany, 73: 811-815.

Spatafora, J. W., Sung, G-H., Johnson, D., Hesse, C., O'Rourke, B., Serdani, M., Spotts, R., Lutzoni, F., Hofstetter, V., Miadlikowska, J., Reeb, V., Gueidan, C., Fraker, E., Lumbsch, T., Lücking, R., Schmitt, I., Hosaka, K., Aptroot, A., Roux, C., Miller, A. N., Geiser, D. M., Hafellner, J., Hestmark, G., Arnold, A. E., Büdel, B., Rauhut, A., Hewitt, D., Untereiner, W. A., Cole, M. S., Scheidegger, C., Schultz, M., Sipman, H. and Schoch, C. L. 2006. A five-gene phylogeny of Pezizomycotina. Mycologia, 98 (6): 1018-1028.

Sugiyama, J. 1998. Relatedness, phylogeny, and evolution of the fungi. Mycoscience, 39: 487-511.

Swofford, D. L. 2003. PAUP*: phylogenetic analysis using parsimony (*and other methods). Version 4 beta 10. Sinauer Associates, Sunderland, Mass.

Tehler, A., Farris, J. S., Lipscomb, D. L. and Kallersjo, M. 2000. Phylogenetic Analyses of the Fungi Based on Large rDNA Data Sets. Mycologia, 92: 459-474.

Van de Peer, Y., Baldauf, S. L., Doolittle, W. F. and Meyer, A. 2000. An updated and comprehensive rRNA phylogeny of (crown) eukaryotes based on rate calibrated evolutionary distances. Journal of Molecular Evolution, 51: 565-576.

Vellinga, E. C. 1999. Phylogeny and Taxonomy of the Lepiotaceous Fungi. Ph.D. Thesis. University of California, USA.

Vellinga, E. C. 2001. Genera *Macrolepiota, Leucocoprinus, Leucoagaricus, Lepiota*. In: Flora Agaricina Neerlandica. 5th edn., Publishers Lisse.

Vellinga, E. C. 2002. New combinations in *Chlorophyllum*. Mycotaxon, 83: 415-417.

Vellinga, E. C. 2003a. *Chlorophyllum* and *Macrolepiota* (Agaricaceae) in Australia. Australian Systematic of Botany, 16: 361-370.

Vellinga, E. C. 2003b. Phylogeny of *Lepiota* (Agaricaceae) - Evidence from nrITS and nrLSU sequences. Mycological Progress, 2: 305-322.

Vellinga, E. C. 2003c. *Chlorophyllum* and *Macrolepiota* in Nederland. Nieuwe inzichten door moleculair onderzoek. Coolia, 46: 177-188.

Vellinga, E. C. 2004a. Genera in the family Agaricaceae: evidence from nrITS and nrLSU sequences. Mycological Research, 10: 352-377.

Vellinga, E. C. 2004b. Ecology and distribution of lepiotaceous fungi (Agaricaceae) - A Review. Nova Hedwigia, 78: 273-299.

Vellinga, E. C. and de Kok, R. P. J. 2002. Proposal to conserve the name *Chlorophyllum* Masse against *Endoptychum* Czer. (Agaricaceae). Taxon, 51: 563-564.

Vellinga, E. C. and Yang, Z.-L. 2003. *Volvolepiota* and *Macrolepiota* - *Macrolepiota velosa*, a new species from China. Mycotaxon, 85: 183-186.

Vellinga, E. C., de Kok, R. P. J. and Bruns, T. D. 2003. Phylogeny and taxonomy of *Macrolepiota* (Agaricaceae). Mycologia, 95: 442-456.

Ventenat, E. P. 1798. Dissertation sur le genre *Phalus*. Mem. l'Inst. Nat. Sci. Arts. Sci. Math. Phys., 1: 503-523.

Volk, T. J. 1990. The current state of morel cultivation. Mushroom News, 38: 24-27.

Volk, T. and Leonard, T. 1989a. Experimental studies on the morel. I. Heterokaryon formation between monoascosporous strains of *Morchella*. Mycologia, 81: 523-531.

Volk, T. and Leonard, T. 1989b. Physiological and environmental studies of sclerotium formation and maturation in isolates of *Morchella crassipes*. Applied and Environmental Microbiology, 55: 3095-3100.

Wartchow, F., Putzke, J. and Queiroz Cavalcanti, M. A. 2008. Agaricaceae Fr. (Agaricales, Basidiomycota) from areas of Atlantic forest in Pernambueo, Brasil. Acta Botanica Brasilica, 22: 287-299.

Wasser, S. P. 1980. Flora Fungorum RSS Ucrainicae: Agaricaceae Cohn. Naukova Dumka Press, Ukraine.

Wasser, S. P. 1985a. Family Agaricaceae of the USSR. Naukova Dumka Press, Ukraine.

Wasser, S. P. 1985b. Results of studies on the family Agaricaceae (Fr.) Cohn of the Soviet Union. Agarica (Norway), 6: 285-301.

Wasser, S. P. 1993. Tribes Cystodermatae Singer and Leucocoprinae Singer of the CIS and Baltic States. IHW-Verlang, Eching, Germany.

Wasser, S. P. 2002. Family Agaricaceae (Fr.) Cohn of Israel Mycobiota I. Tribe Agariceae Pat. E Nevo and P. Volz (eds.). Ruggell A.R.A Gantner Verlag K.-G.

Weber, N. 1988. A Morel Hunter's Companion: a guide to the true and false morels of Michigan. Two Peninsula Press. Lansing, MI.

Weber, N. 1995. Overview of research pertinent to morels or *Morchella* spp. as special forest products, pp. 129-133. In: Dancing with the Elephant: proceedings, the business, and science of special forest products - conference and exposition. C. Schenepf (ed.). University of Idaho Extension Service, Moscow.

White, T. J., Bruns, T., Lee, S. and Taylor, J. 1990. Amplification and direct sequencing of fungal ribosomal RNA genes for phylogenetics. In: PCR Protocols, a Guide to Methods and Applications. M. A. Innis, H. D. Gelfand, J. J. Sninsky and T. J. White (eds.). Academic Press, New York, USA.

Wilmotte, A., Van der Auwera, G. and Wachter, R. D. 1993. Structure of the 16s ribosomal RNA of the thermophilic cyanobacterium *Chlorogloeopsis* HTF (*Mastigocladus laminosus* HTF) strain PCC7518, and phylogenetic analysis. FEBS Letters, 317: 96-100.

Wipf, D., Fribourg, A., Munch, J., Botton, B. and Buscot, F. 1999. Diversity of the internal transcribed spacer of rDNA in morels. Canadian Journal of Botany, 45: 769-778.

Wipf, D., Munch, J., Botton, B. and Buscot, F. 1996. DNA polymorphism in Morels: complete sequences of the internal transcribed spacer of genes coding for rDNA in "Morchella Esculenta" (Yellow Morel) and "Morchella Conica" (Black Morel). Applied and Environmental Microbiology, 62: 3541-3543.

Woese, C. R. and Olsen, G. J. 1986. Archaebacterial phylogeny: perspectives on the subkingdoms. Systematic and Applied Microbiology, 7: 161-177.

Zmitrovich, I. and Wasser, S. P. 2004. Modern view of the origin and phylogenetic reconstruction of Homobasidiomycetes fungi, pp. 231-263. In: Evolutionary Theory and Processes: Modern Horizons. S. P. Wasser (ed.). Kluwer Academic Publishers, Dordrecht, The Netherlands.

Zolan, M. E. and Pukkila, P. J. 1986. Inheritance of DNA methylation in *Coprinus cinereus*. Molecular and Cellular Biology, 6: 195-200.

Chapter 8

Phylogenetic Conundrum of the Mushroom-forming Fungi (Agaricomycetes)

Ivan V. Zmitrovich[1] and Solomon P. Wasser[2,3]

[1]V.L. Komarov Botanical Institute, Russian Academy of Sciences
2, Popova St., St. Petersburg, 197376, Russia

[2]Institute of Evolution and Department of Evolutionary and Environmental Biology, Faculty of Science and Science Education, University of Haifa
Mt. Carmel, 31905, Haifa, Israel

[3]N.G. Kholodny Institute of Botany, National Academy of Sciences of Ukraine
2,Tereshchenkovskaya St., Kiev, 01601, Ukraine
E-mail: iv_zmitrovich@mail.ru

ABSTRACT

The problem of the origin and phylogeny of Agaricomycetes remains controversial. However, various data accumulated in recent decades provide an opportunity to reach some definite conclusions. The conclusions are: 1) Final establishment of the circumscription of the class Agaricomycetes (Basidiomycetes with dolipore septa without nanopores, with perforated, or rarely unperforated, parenthosomata), and the proof of their evolutionary advancement compared with Tremellomycetes; 2) The loss of taxonomic status of such groups as Hymenomycetes/Gasteromycetes and Aphyllophorales; 3) The appearance, in the course of studies of rDNA, of some "molecular concepts", the most important of which are the rooting of Agaricomycetes within the Dacrymycetales/Auriculariales and the basal position of taxa with unperforated parentosomes; 4) Proof of evolutionary advancement of brown-rot fungi with considerable devaluation of the type of rot as a character in taxonomy of Basidiomycetes; 5) Proof of secondary origins of corticioid and gasteroid forms. Any phylogenetic reconstruction of Agaricomycetes should be based on the aforesaid concepts. However, some of the remaining difficulties and contradictions in the field of systematics are analyzed in the present chapter.

INTRODUCTION

The class Agaricomycetes contains all mushrooms, toadstools, tinder fungi, puffballs, club- and crust-fungi, and some representatives of jelly fungi, like *Sebacina*, or *Auricularia*. This new name is the result of typification of higher taxa (Dowled, 2001), whereas essentially this unit is more or less equivalent to Homobasidiomycetes, or Hymenomycetes in the traditional sense. The name Homobasidiomycetes weakly reflects any morphological evidence, because this group fixed all stages of homobasidium formation from some types of heterobasidia. The name Hymenomycetes is unsuccessful too, because the group contains some gasteroid forms, which were independently derived from hymenomycetous forms. The widely known genus *Agaricus* represents a rather good neutral representative of the group. It is characterized by a homobasidium and is a member of a rich order of the class, and this is why the group is frequently called "mushroom-forming fungi". The other two classes, Dacrymycetes and Tremellomycetes, are sisters to the Agaricomycetes, and all three groups form a separate subphylum Agaricomycotina, which together with Pucciniomycotina and Ustilagomycotina enter into the phylum Basidiomycota (Hibbett et al., 2007).

Despite the progress achieved during recent decades, the taxonomy of Agaricomycetes remains rather unstable. The "good times", when this group was "easily" subdivided into three large subgroups – Aphyllophorales, Agaricales, and Gasteromycetes, ended by the 1980s when this traditional taxonomic practice was questioned and doubted (Zerova and Wasser, 1974; Zerova et al., 1979; Kühner, 1980; Nuss, 1980; Wasser, 1980, 1982, 1985; Locquin, 1981, 1984; Jülich, 1982; Moser, 1983). The main shift of the classification paradigm, which is associated mainly with the work of Jülich (1982), turned the phylogenetic system of the mushroom-forming fungi into chaos; unfortunately, this disorder still prevails today. Multiple convergence [according to Donk's (1971) terminology] accompanying the huge evolutionary radiation of the group which covers not only epimorphology (so-called "life forms" sensu Bondartseva, 1993), but also microscopic structures of basidiomata are the main reasons that present enormous difficulties in solving the confusion in Agaricomycetes.

Application of molecular data to the taxonomy of this group is not likely to solve all problems. Certainly, these methods have helped to prove the monophyly of many compact groups, although the monophyletic status of the majority of these was already well tested by traditional morphological methods. Nevertheless, the reliability of any separation of large groups in various topologies based on single genes remains problematic and inconsistent, whereas the resulting multigene trees leave many open possibilities for future interpretations.

An approach using *a set of characters* in phylogenetic reconstruction, which would allow allocating and comparing the *types of organization* of existing groups of organisms, seems more promising. Thus, some differentiation of the morphological and molecular fields and data would be methodologically satisfactory. While "molecular taxa" are only partly justified, the traditionally understood taxa are mainly the objects of a "morphological" semantic field (Vasilyeva, 1999).

A "MORPHOLOGICAL" TREE

The basal radiations in Agaricomycetes

This aspect is connected with the problem of basidiomycete origin as a whole on the one hand, and the problem of interrelationships of heterobasidiomycetous fungi on the other.

The idea of polyphyletic origin of Basidiomycetes from various groups of Ascomycetes was expressed by Gäumann (1964) and supported by Petersen (1974). The reflection of this idea on classification schemes can be found in the systems of von Arx (1967) and Kreisel (1969), where ustilaginaceous (as well as exobasidial) fungi are united with Taphrinales and Hemiascomycetes as a separate class, Endomycetes.

In later studies, adherents of the idea of an isolated position of ustilaginaceous and exobasidial fungi were forced to reject the concept of their relationship with Endomycetes (representatives of this group have no dikaryophase, no chitin in the cell walls, and their GC-pairs content <50%). Hence a separate class, Ustomycetes was created (Moore, 1971; von Arx, 1979).

The splitting of basidial fungi into three classes – Ustomycetes (Ustilaginomycetes), Teliomycetes (Pucciniomycetes), and "true" Basidiomycetes – is maintained in manuals of 1980–90s, although it is impossible to understand from those whether an independent origin of these three groups of basidial fungi was postulated, or the elevation of their taxonomic status to the class rank was a result of some general trend of devaluation of taxonomic ranks.

An idea of monophyly of Basidiomycota is reflected in the system proposed by Wells (1994). This view is supported by basic common features as cell wall composition, GC-pairs contents, and mostly exogenous sporogenesis. The phylum is split into two subphyla, Teliomycotina (Uredinales s.l. + Septobasidiales +

Atractiellales etc. + Ustilaginales s.l. + Exobasidiales) and Basidiomycotina (classes Heterobasidiomycetes and Homobasidiomycetes).

Cytological data provide the basis for removing the ustilaginaceous and exobasidial fungi, as well as Uredinales, Septobasidiales, etc. from residual heterobasidiomycetous and homobasidiomycetous fungi. These groups are characterized by discoid spindle pole bodies (SPB) and simple (or, in some Ustilaginales, rather specific in their structure) septal pores, whereas Homobasidiomycetes and residual Heterobasidiomycetes are characterized by big lobular SPB and dolipore septa (Jones, 1973; Oberwinkler, 1977; Littlefield and Heath, 1979; Khan et al., 1981; Wells, 1994; Moore, 1996; Begerow et al., 1997; Cavalier-Smith, 1998; Zmitrovich, 2003; Weiss et al., 2004; Zmitrovich and Wasser, 2004). Yet these cytological data obviously testify against the concept of the possible polyphyly of Basidiomycetes. Ustilaginomycetes, the group most frequently separated from other Basidiomycetes is closely related cytologically to Uredinales. The linkage of the latter with doliporous Heterobasidiomycetes through Septobasidiales – Platygloeales is rather obvious (Moore, 1996). However, the question arises, as to which characters or the type of organization of Basidiomycetes can possibly be considered primitive in this group? We shall try to solve this problem with reference to two essential features of basidial fungi, namely their hyphal ultrastructure, and the type of basidium. It is rather obvious that simple ("ascomycetous") septal pores and discoid SPB should be considered primitive characters, since the doliporous state is an obvious apomorphy. The basal position of taxa with simple septal pores was also confirmed by molecular data (Swann and Taylor, 1993, 1995; Lutzoni et al., 2004; Hibbett et al., 2007).

Considering the issue of a primitive type of basidium, we should reject the thesis about the morphological simplicity of the holobasidium as its supposedly primitive feature (Raitviir, 1964; Oberwinkler, 1965). First, it is impossible to ignore the case of the obviously secondary origin of holobasidium in Tilletiaceae (this example was discussed in detail by Linder, 1940, see Fig. 8.1).

Second, the greatest "primitiveness" for fungi is the presence of a septate basidium (more precisely, a transversely septate phragmobasidium). The latter is connected, in turn, with a longitudinally septate tremelloid basidium (some intermediates are observed in several representatives of Auriculariales and Exidiales – see Bandoni, 1984; Boehm and McLaughlin, 1989).

The origin of holobasidium seems to be polyphyletic. The holobasidium of Ustilaginomycetes (including Exobasidiales) represents a structurally "rationalized" promycelium adapted to synchronous maturation of sporidia.

The holobasidium of Dacrymycetales is comparable with the phragmobasidium of some rust fungi, like *Coleosporium*, where the tetrakaryotic basidial body is formed inside the teleutospore, from where it sprouts by long sterigmata.

The holobasidia of *Tulasnella*- or *Ceratobasidium*-types in many respects are comparable with longitudinally septate cruciate basidia of the *Sebacina*-type.

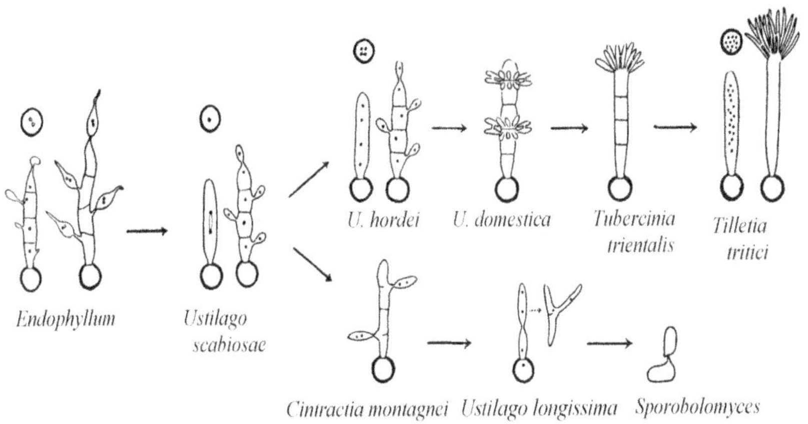

Fig. 8.1 The scheme of basidium (promycelium) morphogeny in the ustilaginaceous fungi according to Linder (1940). This scheme demonstrates an independent and secondary holobasidium origin in Basidiomycota.

Recently, it was assumed that the formation of holobasidium in *Oliveonia* progressed independently of similar processes in Tulasnellales–Ceratobasidiales (Roberts, 1998).

The holobasidium of Filobasidiales is also of independent origin: ultrastructure as well as life cycle features in this group testify to its close relationships with tremellaceous (Tremellales s. str.) dimorphic fungi in having cruciate phragmobasidia (Oberwinkler et al., 1984; Weiss et al., 2004).

All these types of holobasidia are at the same time heterobasidia, since they are subdivided into the hypo- and epibasidial segments, whereas during their ontogenesis the stages of probasidium and metabasidium can be easily observed. This character is also clearly correlated with the mode of germination of basidiospores: they are either divided into several segments, proliferating by hyphae or microconidia (Exobasidiales, Dacrymycetales), or they germinate as a series of secondary spores (Ceratobasidiales, Tulasnellales), or else begin to bud (Filobasidiales). Such features are not characteristic of the homobasidium in the majority of representatives of Agaricomycetes: the stage of probasidium is greatly reduced or not expressed at all, and the dividing of a mature basidium into hypo- and epibasidial segments is poorly expressed.

The problem of ancestral forms of Agaricomycetes is still far from being solved (as also a century ago). Many mycologists (in the 20th century and at present) specify the group of corticoid fungi as being a possible link between Tremellomycetes and Agaricomycetes (or vice versa).

Indeed, there are remarkable similarities in the formation of basidiome microstructures in some corticioid forms bearing homobasidia and those with heterobasidia (Oberwinkler, 1972).

Moreover, the presence of long-lasting probasidia is characteristic of some corticioid fungi (*Corticium*, *Coniophora*, etc.), whereas in other groups mature basidia are obviously differentiated into hypo- and epibasidial segments, although the probasidial stage in ontogenesis is not so clearly expressed. These are, first of all, urniform basidia (*Sistotrema*) characterized by ovoid (egg-like) hypobasidia and narrowed epibasidia; secondly, utriform basidia (*Hyphoderma*, *Punctularia*) having hypo- and epibasidial segments approximately equal in their width; thirdly, stalked basidia (podobasidia – *Radulomyces*, *Sphaerobasidium*) with narrow, twisting hypobasidial and extending epibasidial segments. All these types are interconnected by transitions. Such "aberrant" morphology of homobasidia is usually observed in wood-inhabiting fungi with poorly differentiated hymenophores. These types of basidia are most typical for Corticiaceae s. str., having a loose hymenial layer or a hyphidial hymenium (catahymenium, according to Lemke, 1964; the same type of hymenium is also characteristic for jelly fungi). A catahymenium is a good protection for developing basidia; in such a case, however, for effective sporulation they should protrude through a dense hyphidial layer. This is probably a secondary phenomenon connected with development of various strategies of effective sporulation. The same is most likely true for pleurobasidia of Xenasmataceae adhering to the substrate, or repetobasidia unprotected by sterile layers (in Repetobasidiaceae), which are comparable with microcyclic conidiophores of some Aspergilli. Stalked basidia are also characteristic for some wood-inhabiting heterobasidiomycetous Agaricomycetes (*Protodontia*, *Myxarium*). Interpretations of such original morphology and possible origin of "holo-heterobasidia" are discussed by some authors (Rogers, 1934; Martin, 1949; Donk, 1954a, b, 1956; Chadefaud, 1960, 1975; Oberwinkler, 1965, 1977, 1978, 1982; Parmasto, 1969; Bauer et al., 1997; Zmitrovich and Wasser, 2004; Malysheva, 2009).

Better success might result from a search for a relic basidium, considering data on the orientation of the microtubular spindle during the cell division process before the basidiospore formation. This character was first considered by Juel (1898), and later its importance was noted by Linder (1940) and Donk (1964). Unfortunately, these days it is hardly ever used in taxonomy; moreover, because of an unfortunate typo in Ainsworth & Bisby's Dictionary of the Fungi (2001, p. 62: the figures 9C and 9D are misnamed), the concept of modern researchers regarding this character has possibly become even more confused. According to Juel (1898), basidia (he discussed mainly holobasidia) are either stichic (the spindle is arranged longitudinally, nuclei are divided in a plane parallel to the longitudinal axis of the basidium) or chiastic (the nuclei are divided in a plane perpendicular to the basidial axis). Stichobasidia are characteristic only for a few representatives of homobasidiomycetous Agaricomycetes (Clavulinaceae, Cantharellaceae, *Sistotrema*, *Hydnum* s.str.) as well as for some Exobasidiales (Ustilaginomycetes), whereas chiastobasidia are characteristic for the majority of Homobasidiomycetes (see Donk, 1964).

In fact, Juel's terminology can be also easily extended to heterobasidia. Thus, cruciate tremelloid basidia (and related holobasidia of the *Oliveonia-* or *Ceratobasidium*-types) may be interpreted as chiastobasidia, while the transversely septate phragmobasidia, characteristic for Pucciniomycetes, Ustilaginomycetes, and some Agaricomycetes (Auriculariales), can be interpreted as stichobasidia. Most importantly, the two-spored (bisporous) basidium of Dacrymycetales is also stichic, and thus represents a kind of link between the phragmobasidium of Uredinales (Coleosporiaceae) and the stichic homobasidium of Clavulinaceae. Thus, Clavulinaceae and Sistotremataceae possessing stichobasidia are close to the Agaricomycetes progenitor. Also, the class Dacrymycetes that possesses stichic basidia links the group to the basal Basidiomycota lineages (rust fungi). The other groups of heterobasidiomycetous Agaricomycetes represent a series of septal reduction in the metabasidial body (Fig. 8.2).

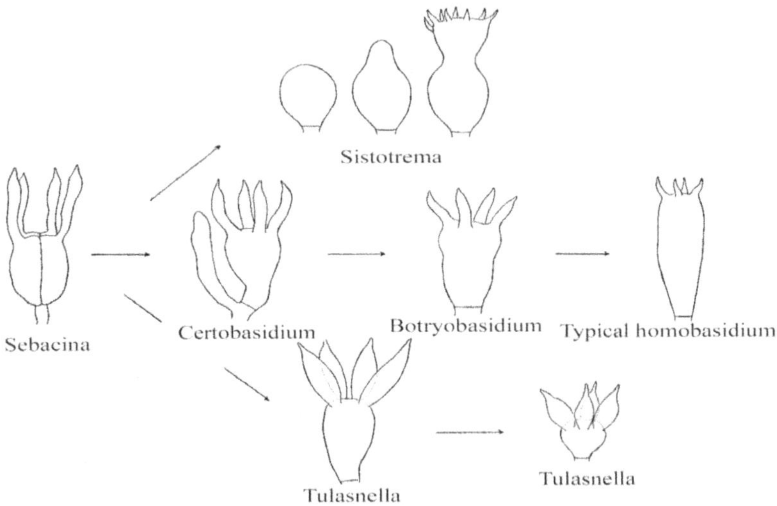

Fig. 8.2 The origin and morphogeny of holobasidia and homobasidium in the Agaricomycetes (after Linder, 1940; supported by modern data – see Matheny et al., 2007).

Groups like Auriculariales, Sebacinales, Ceratobasidiaceae, and Tulasnellaceae exhibit relationships with Cantharellales (Hibbett and Binder, 2002; Binder et al., 2005; Lawrey et al., 2007; Larsson, 2007; Matheny et al., 2007). As a result, we can more or less precisely outline an area of basal relationships of Agaricomycetes (Fig. 8.3).

The basal lineages of the class are Cantharellales with contiguous Auriculariales and Sebacinales groups.

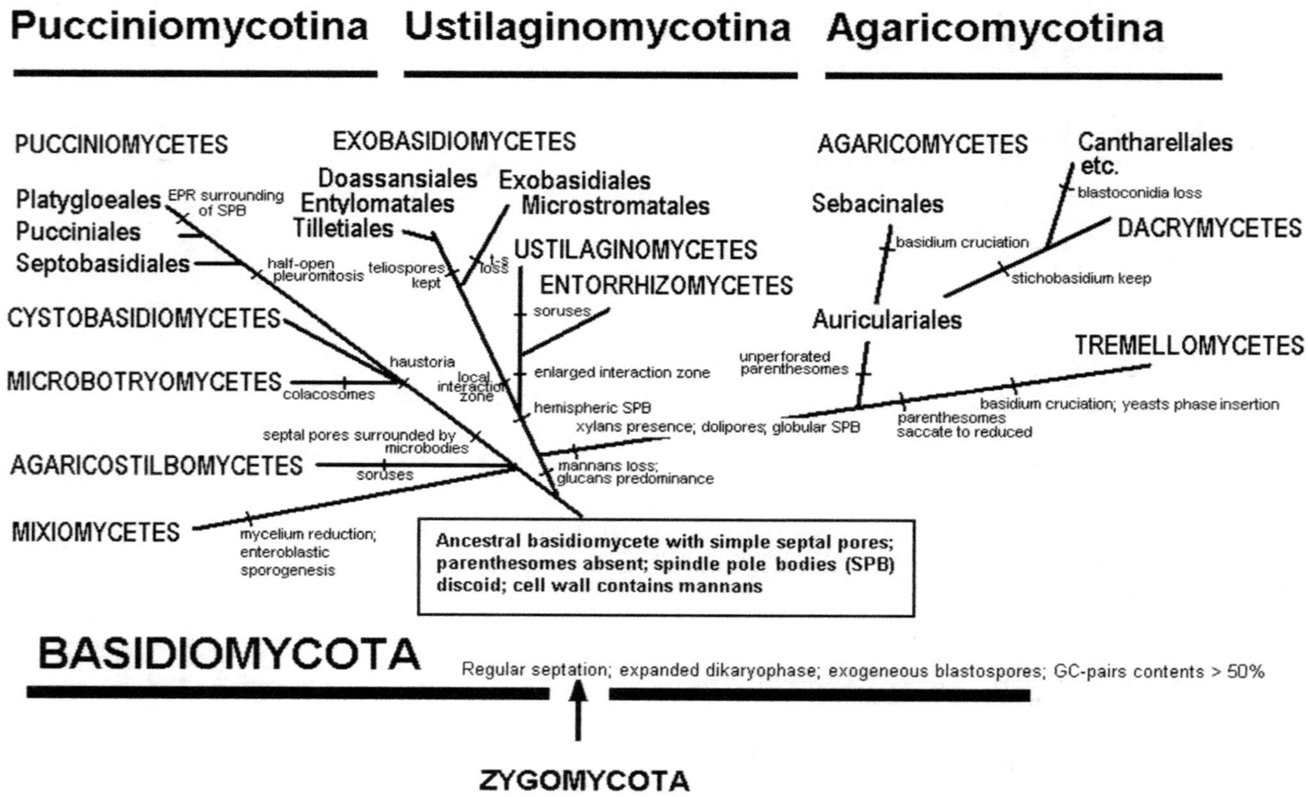

Fig. 8.3 Large-scale phylogeny of Basidiomycota based on divergence of key cytological markers.

Trends of morphogeny in Agaricomycetes

The formation of the homobasidium was a major evolutionary acquisition of the group, connected with its adaptation to conditions of terrestrial and independent (non-endobiotic) existence, which triggered its active speciation and wide taxonomic diversification. At present there are about 16,000 species of Agaricomycetes.

A wide diversification of the taxa was accompanied in the evolution of Agaricomycetes by a most impressive large-scale convergence phenomenon, probably comparable only to the patterns observed in Ascomycetes and red algae. No wonder, in all the three groups, the main evolving structures are apically growing filaments with centripetal septa, so the entire range of morphogenetic processes is determined by the "work" of apical cells.

Such a situation has created a rather complex problem for taxonomists. The problem of phylogeny of Agaricomycetes has been one of the most debated issues of phylogenetic taxonomy in general as judged by the available extensive literature. In delimiting large groups, the most essential role is probably played by characters of hyphal ultrastructure because these characters are most conservative (Fig. 8.4).

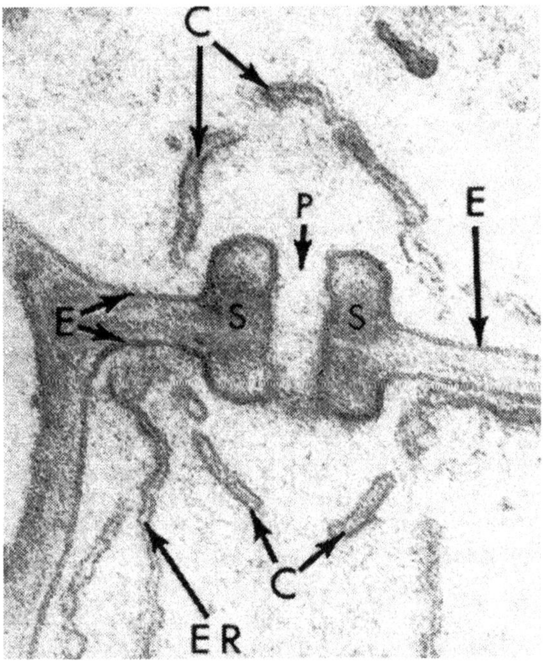

Fig. 8.4 Dolipore of *Schizophyllum commune* (after Wells, 1965): C – parenthosome (membrane cap), S – doliolum (septal thickening), E – ectoplast, ER – endoplasmic reticulum, P – septal pore.

With reference to the Agaricomycetes, we shall first consider the peculiarities of parenthosomata: continuous (unperforated) parenthosomata in taxa diverging from the basal area of the group, and perforated parenthosomata in the more derived taxa (see van Driel, 2007; Fig. 8.5).

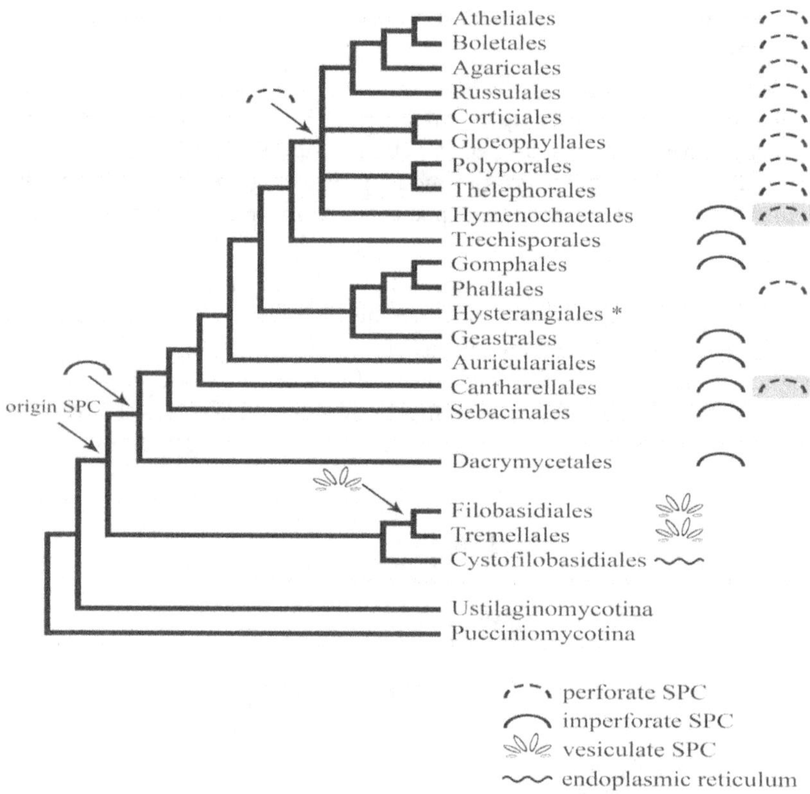

Fig. 8.5 The distribution of parenthosome types across the system of Agaricomycotina (after van Driel, 2007). The gray phone marks a possibility of a secondary loose arrangement of perforate structures in some Cantharellales and Hymenochaetales.

Mutual relationships between hypo- and epibasidial segments are also essential for taxonomy. The ratio between the width of the epibasidial segment and subhymenial hyphae is also important. This ratio remains stable in representatives of different orders (Zmitrovich et al., 2006). Both features are connected with modes of growth organization. In erect basidiomata with protected hymenophores and euhymenium, the basidial body tends to be of classical club-like shape without sufficient constrictions and sterigmata hypertrophy. In the case of the thread-like structure of prostrate basidiomata, the basidia tend to be with insufficient pleural outgrowths on horizontally oriented hyphae. In the case of the crustose appearance of prostrate basidiomata, the basidia originate in catahymenial palisade, therefore, epibasidia or sterigmata are narrow and elongated. Some constriction of the epibasidial segment is also characteristic for weakly protected hymenophores of some corticioid and cantharelloid forms which have urniform basidia. These and some other features require, therefore, a differential approach depending on the set of correlated characters.

As noted earlier, biochemical and molecular data require further accumulation and reassessment in light of new (not always finally generated) paradigms. However, one biochemical character, the type of rotting produced by fungi, has to be considered since many researchers, beginning with Nobles (1958), actively use this character in taxonomy although, nowadays, its taxonomic value has been somewhat revised.

After the discovery of the laccase gene-specific sequences in the genome of brown-rotters (D'Souza et al., 1996) and the oxidase-dependent (not hydrolytic!) mechanism of cellulose degradation (Hyde and Wood, 1997), it has become obvious that the brown-rot fungi represent a heterogeneous assemblage of the most advanced wood-rotters (Ander and Marzullo, 1997). Thus, phylogenetically brown-rot fungi are probably independent terminal branches of the phylum of Agaricomycetes (Gilbertson, 1980), but certainly neither the domain of its basal radiation nor the basal clade (Nobles, 1971).

We conclude that detecting evolutionary trends in Agaricomycetes is a rather complicated task with no straightforward solution, at least for now. Any further advance in this area is possible only by means of *whole sets* of characters reflecting the *types of organization*. Below we consider some possible trends in transformation of basic morphogenetic events in the evolution of Agaricomycetes.

Epimorphology. There are four basic versions of the basidiomata morphogeny in the Agaricomycetes: 1) from corticioid crustothecia to sessile and stipitate, and from these to gasteroid forms (Bessey, 1950; Raitviir, 1964; Parmasto, 1968, 1969; Zerov, 1972, among many others); 2) from gasteroid to pileate cantharelloid forms, then to sessile and resupinate aphyllophoraceous forms (Singer, 1962, 1971, 1975, 1986; Kreisel, 1969; Wasser, 1974); 3) from poorly differentiated "mycelial clusters" to terrestrial stipitate (+ gasteroid) on the one hand, and to wood-inhabiting sessile or resupinate on the other (Parmasto, 1986; Bondartseva, 1997; Zmitrovich, 1997, with some reservations); 4) from clavarioid to pileate forms, and from them to resupinate and gasteroid forms (Corner, 1954, 1964, 1966, 1968, 1970). Version 3 is the most popular at present. However, it does not take into account the possibility of heterochronous wood settling by various groups of terrestrial fungi.

We will demonstrate the validity of this possibility using one example. Practically all merulioid fungi, including those having basidiomata completely adhering to the substrate, demonstrate some anatomic features approaching those in agaricoid forms. These fungi have three-layered basidiomata consisting of a more or less gelatinized hymenophore, medullar tissue, and a superficial abhymenial layer constructed mainly according to the trichoderm type; the superficial layer also occurs in wholly prostrate forms. Merulioid fungi have as a rule lateral or dorsal attachment, and if basidiomata are adhering, they usually have free, inrolling margins. The presence of a characteristic abhymenial layer both in pileate and effused (prostrate) forms of merulioid fungi allows them to be considered not primary effused (prostrate), but resupinate (turned down), representing in fact dorsally attached pilei with their margins adhering to the substrate (Corner, 1971; Zmitrovich, 2001). The microscopic features of the hymenial layer make it possible to trace, without any reasonable doubt, the origin of merulioid fungi from agaricoid/boletoid (see Fig. 8.6).

Most of resupinate fungi with smooth (or almost smooth) hymenophores are easily traced evolutionarily from such transitional forms, assuming the most efficient adaptation to the positive geotropism as the basic line of evolution in that group (Corner, 1954). This assemblage is rather diverse in a morphogenetic aspect. Plagiotropic basidiomata can be distinguished into following types as.– (1) arachnoid type – when basal filaments are totally prostrate whereas pleuridia are loosely arranged; (2) crustose type – when basal filaments are totally prostrate, but pleuridia densely packed in uninflated hymenium; (3) resupinate type – when the basal structure is plate-like with more or less free margins, whereas pleuridia form hymenophoral outgrowths; (4) orbicular type – when the basal structure is prostrate and radially arranged, but pleuridia gave orthotropic hymenophoral outgrowths, too; and (5) rhizomoid type – when basal structures form mycelial strands of a anisotropic orientation, whereas secondary pleuridial structures overbuild above by a rhizomoid matrix on the manner of resupinate and orbicular types (Zmitrovich, 2007, 2008).

The remaining groups of corticioid fungi also can be connected either to reduced agarics – these lines are parallel to several "cyphelloid" lines (see Singer, 1945) or to less differentiated clavarioid and cantharelloid forms (Clavulinaceae – Sistotremataceae; *Ramaria* – *Ramaricium*, etc.).

Examples of evident secondary development of resupinate forms in Thelephoraceae are also well known (Corner, 1968; Kõljalg and Renvall, 2000).

Thus, we arrive at the "*Clavaria*-hypothesis" (version "4") proposed by Corner, which seems to be capable of explaining most successfully many inconsistent facts in the field of comparative morphology of Agaricomycetes (Petersen, 1971; Thiers, 1971; Watling and Largent, 1977; Pegler and Young, 1981; Jülich, 1982; Kovalenko, 1984; Gorovoy, 1989). It is most important that, in conformity with Cornerian hypothesis, we have to place clavarioid and cantharelloid taxa having a set of really primitive characters (hyphal ultrastructure, karyology) at the base of the evolutionary tree. The "*Clavaria*-hypothesis" concurs well with the idea of stichobasidial Clavulinaceae as the most primitive group of Agaricomycetes.

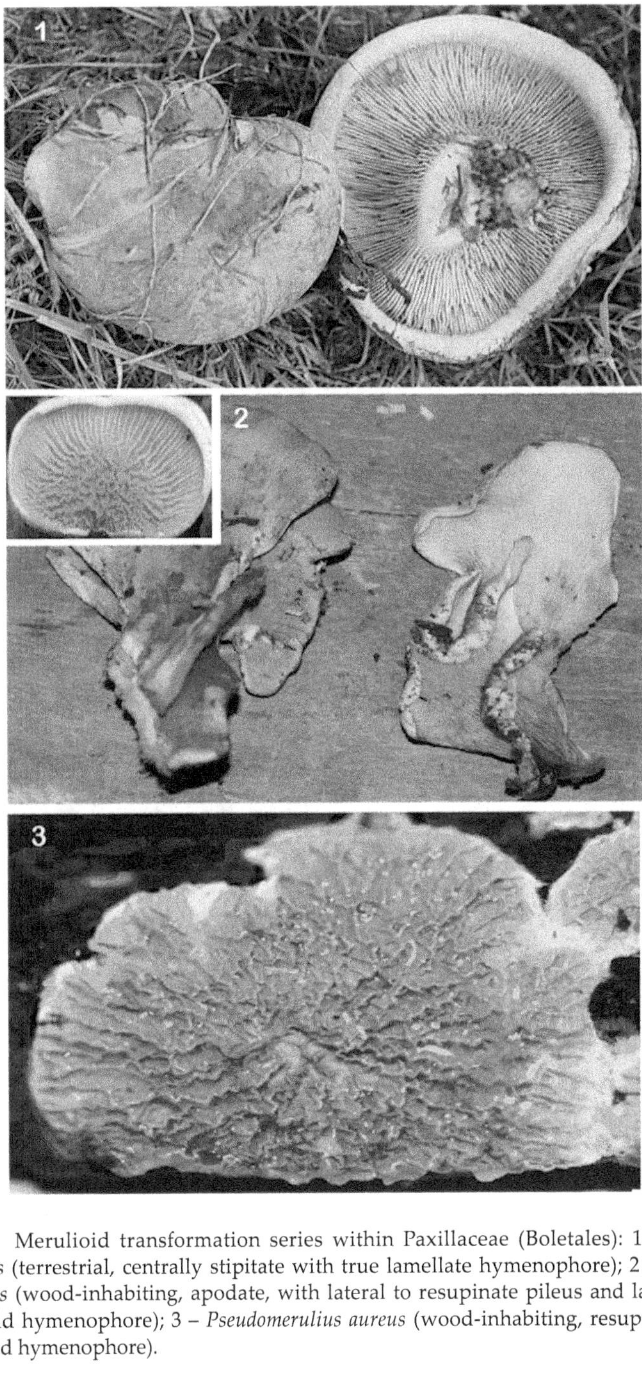

Fig. 8.6 Merulioid transformation series within Paxillaceae (Boletales): 1 – *Paxillus involutus* (terrestrial, centrally stipitate with true lamellate hymenophore); 2 – *Tapinella panuoides* (wood-inhabiting, apodate, with lateral to resupinate pileus and lamellate to merulioid hymenophore); 3 – *Pseudomerulius aureus* (wood-inhabiting, resupinate, with merulioid hymenophore).

If we accept the Cornerian hypothesis, the evolutionary tree of Homobasidiomycetes will look like a fountain, with "jets" formed by cantharelloid and agaricoid clades, and "splashes and drops" composed of various groups of aphyllophoroid resupinates. The "*Clavaria*-hypothesis" also postulates that Gasteromycetes are descendants of various groups of agaricoid fungi.

The hypothesis of the origin of secotiaceous and hymenogastraceous fungi from various groups of Agaricales s.l. is not new. It was actively developed by Gäumann (1964), Heim (1971), Zerov (1972), Wasser (1974, 1980) and others, as opposed to group of researchers, who considered Gasteromycetes as ancestors of basidial fungi (Buhgoltz, 1902 with some reservations; Singer and Smith, 1960; Singer, 1962, 1975; Demoulin, 1974). However, Heim (1971) and his followers tried not to emphasize the question of the origin of Endogasteromycetes, considering this group is rather primitive. Cornerian hypothesis in this respect is more radical, since it assumes the origin of Endogasteromycetes from secotioid fungi through their further loss of the initial agaricoid habit and transformation of sporogeneous tissues (Corner, 1964).

A brief summary of such a gasteromycetization process was provided by Thiers in his remarkable paper "The secotioid syndrome" (Thiers, 1984). According to the author, among agaricoid fungi adapting to open habitats, there was a selection of the forms in which the hymenium was protected against the loss of moisture. This protection proved most effective at fixing one of the juvenile stages of sporocarp development, namely, the stage with a closed pileus. At this stage, the function of active spore discharging was lost. This process developed independently in various groups of Agaricales s.l. affected by the "secotioid syndrome". When this active spore discharge ability was lost, the stipe also lost its selective advantages; it was preserved in some groups only as an enclosed columella. The loss of the stipe gradually led many fungi to the formation of underground carpophores. The completely angiocarpous development of the hymenophore also became possible due to stipe loss. Thus, a typical gasteroid basidiocarp was formed. The angiocarpous hymenophore development under the peridium resulted in its reduction to a system of lacunas.

The latest data show that in the morphologically fairly simple white-spored agaricoids (e.g., *Lentinus tigrinus*), the mutations can be induced that lead to the formation of secotioid basidiocarps (Hibbett et al., 1997). Sometimes such mutations occur in nature, however, under the current conditions they are not preserved by natural selection.

The basic trends of basidiome transformations in the evolution of Agaricomycetes according to the proposed concept can be reflected in the following scheme (Fig. 8.7).

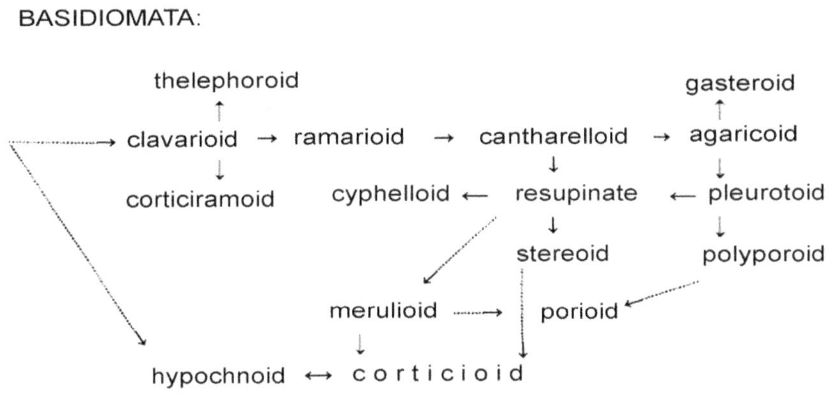

Fig. 8.7 The basic trends of basidiome transformation in the evolution of Agaricomycetes.

Shape and Structure of Hymenophore. Opinions of various authors on the patterns of evolution of hymenophore are also inconsistent. Undoubtedly, in the first Agaricomycetes, the hymenophore was formed by a smooth palisade layer. This type is observed, on the one hand, in clavarioids with negatively geotropic, vertically growing basidiomata, and on the other hand in positively geotropic resupinates, which were placed in the large family Corticiaceae (Herter, 1910; Bondarzew and Singer, 1941; Eriksson, 1958; Donk, 1964, 1971; Parmasto, 1968, 1986; Jülich, 1972, 1974; Hallenberg and Parmasto, 1998). It is remarkable that the latter group was most frequently considered initial in Agaricomycetes evolution, whereas the possibility of a secondary shift to a smooth hymenophoral surface during their evolution was mostly overlooked. However, gradually other facts have also become available which show a different side to the seemingly stable evolutionary concepts. First, systematists following "classical" opinions have begun to show some concern over the case of cyphelloid fungi, since the reduced nature of their hymenophore could hardly be denied. On the other hand, in many "hemiagaricoid" fungi (representatives of Tricholomataceae, Lentinacae, Paxillaceae, Gomphidiaceae, and other groups), as well as in some cantharelloids, the trend towards formation of gill anastomoses is also rather obvious. This tendency is especially clearly expressed in wood-inhabiting fungi. When basidiomata transform from laterally attached to resupinate (turned down, with dorsal attachment), the radial orientation of the gills (folds) becomes completely lost, and the anastomoses become perfectly "equal-in-rights" elements of the resulting merulioid (irregularly folded) hymenophore. This explanation is readily supported by the case of hymenophore variability in *Tapinella panuoides* and *Pseudomerulius curtisii*, depending on the basidiome attachment to the substrate (Zmitrovich, 2001). A merulioid hymenophore under certain conditions can be transformed into a smooth or nearly smooth form. In some species (*Serpulomyces borealis*, *Amylocorticiellum molle*, *Ceraceomyces* *microsporus*) the initially merulioid hymenophore becomes smooth upon drying (Zmitrovich and Spirin, 2002; Zmitrovich, 2008).

Thus, consideration of corticioid fungi as a homogeneous primitive family of Hymenomycetes should be abandoned. Attempts to subdivide this aggregate into more natural and smaller units have been made by Kreisel (1969). He was the first mycologist who stressed the need to use the priority name Peniophoraceae Lotsy, 1907 for a redefined group of Corticiaceae (Parmasto, 1968, 1986, 1995; Jülich, 1982). Despite these efforts, some difficulties still remain.

Let us consider other types of hymenophores which are derivatives of the merulioid type. The polyporoid type can be considered a specific "update" of the merulioid type. The main difference is that the folds (or tubes) become deeper, and their edges become sterile. Transitions from the merulioid to polyporoid hymenophore are evident in various species: *Gloeoporus pannocinctus*, *Ceriporia tarda*, *Oxyporus latemarginatus*, *Castanoporus castaneus*, etc. However, in some lines of development, the gilled hymenophore directly transforms into the polyporoid form through the merulioid stage. A number of examples can be found among the *Polyporus–Lentinus*-conglomerate in which there are species with plentiful gill anastomoses; for other representatives the cellular (favoloid) hymenophore is already a characteristic (Krüger and Gargas, 2004; Zmitrovich, 2010).

Labyrinthiform (daedaleoid) hymenophores are interpreted as polyporoid aberrations. Radially-foliaceous (*Daedaleopsis*, *Lenzites*, *Cerrena*, and *Gloeophyllum*) and irpicoid (*Irpex*, *Antrodiella*, *Trichaptum*, and *Cyclomyces*) forms are supposed to be the intermediate forms. The polyporoid nature of these deviations is obvious to specialists (Bondartsev, 1953), although with regard to morphogenetic parallelism with lentinaceous fungi, this fact, as far as we know, has not been discussed. At the same time, the secondary formation of gills by polyporoid fungi may be regarded as an atavistic phenomenon, evidence of the existence of "modular possibility" to re-building of radially arranged elements of the hymenophore.

The hymenophore of *Schizophyllum* presents a certain problem. It is regarded as a gilled derivative by some authors (of the *Lentinus*-type; Pegler and Young, 1971), as polycyphelloid by others (Cooke, 1961; Donk, 1962, 1964; Ginns, 1974; Singer, 1975), or as a modified merulioid form by still others (Stalpers, 1988; Nordic Macromycetes, 1997). Probably all these three ideas can be reasonably united. Since both the cyphelloid and merulioid types of hymenophore originated from the radially-gilled (-folded) form, it is logical to assume that the hymenophore of *Schizophyllum* represents a specialized, originally agaricoid or merulioid hymenophore adapted to sporulating under dry conditions. The morphogenesis of *S. commune* was actively studied (Yli-Mattila, 1990). However, the interpretations of experimental results by various authors are controversial. For example, Gorovoy (1989) assumed that the polycyphelloid structure of its fruit bodies is formed initially. Stalpers (1988), by contrast, specifies that the germinating basidiomata of *S. commune*, are in fact, single cupulate basidiomata, resembling those of *Auriculariopsis ampla*, whereas a more complex structure of hymenophore appears at subsequent stages, when its surface becomes ridged, and invaginations arise on ridge edges and it starts splitting further on its external and internal borders. In our opinion, *Schizophyllum* is one of the preserved links between hemiagaricoid (like *Panellus*) and merulioid (*Auriculariopsis*, *Plicaturopsis*, and *Merulius*) fungi. It stabilized in its secondary, xeromorphic state, which has developed from some disturbance at the initial stages of basidiome morphogenesis. Other polycyphelloids (*Fistulina*, *Stromatoscypha*, etc.) probably also represent some secondary results of the process of development of polyporoid and merulioid types of hymenophore.

We should also briefly consider the issue of origin of the toothed hymenophore. It seems incorrect to treat its origin as a result of some complication of the smooth hymenium. Under the initial "fountain" hyphal arrangement in negatively geotropic basidiomata, a more economical way would be expansion of the hymenial surface through formation of radial folds. Studies of variation of the hymenophore and its tramal structure in various representatives of *Phlebia*, *Byssomerulius*, *Phanerochaete*, *Hyphoderma*, and *Kneiffiella* lead us to conclude that the toothed hymenophore is more logically connected with the merulioid type via phlebioid and raduloid (scraper-like) types. The latter has evolved in two directions: through tuberculate to smooth and to odontoid (consisting of regular conical aculei) types. The odontoid hymenophore, in turn, had the potential for further progressive complicated development (formation of the pseudocystidial core, elongation of aculei), and for ephemerous imitations like grandinioid (spot-like), farinaceous, or finely uneven types. Essentially the same processes probably occurred in precursors of stipitate hydnums. The irpicoid hymenophore, as stated above, originated from the polyporoid type (*Trametes–Antrodiella–Steccherinum*; *Schizopora–Hyphodontia*).

Below we propose a scheme reflecting possible trends of transformation of the hymenophore during the process of evolution in Agaricomycetes (Fig. 8.8).

Summarizing our discussion of the evolutionary pathways of macro-structure in Agaricomycetes, we emphasize the convergent nature of their evolution. The ability to form both negative and positive geotropic (resupinate) basidiomata is observed practically in all phyla of basidial fungi and it is obviously adaptive; here again we see some parallels with red algae, where closely related coralloid and crustaceous forms occur. The above conclusions are also true in respect to the problem of evolution of the hymenophore (Fig. 8.9).

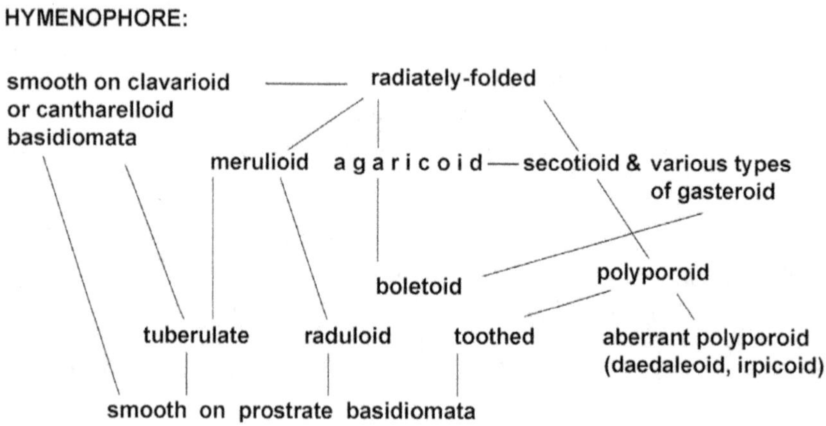

Fig. 8.8 The ways of morphogeny of the hymenophore of Agaricomycetes (the basic trend reads left to right and up to down).

Hyphal Structure. To clarify phylogenetic relationships between taxa, some information on the features of the hyphal structure of carpophores can be useful. The concept of a hyphal system (monomitic, dimitic, and trimitic) was introduced in basidiomycete morphology by Corner (Corner, 1932a, b, 1953). The type of hyphal system has been considered as a taxonomic character. However, it was soon realized that its successful use is possible only in combination with other data and characters. The loss of the plasmatic contents of the hyphae and the formation of sclerified elements are characteristics of practically all groups of Agaricomycetes - clavarioid, gasteroid, agaricoid (especially wood-inhabiting), polyporoid, and stereoid forms. Various groups are characterized by the specificity of their hyphal differentiation.

As was assumed by us (Zmitrovich et al., 2009), the hyphal systems of the Polyporales can be subdivided as follows according to their pattern of hyphal differentiation (Table 8.1).

1) Monomitic (basidiocarp composed of generative hyphae only)

2) Pseudodimitic (there is an additional type of hyphae, namely pseudoskeletal hyphae)

3) Subdimitic (there are generative, pseudoskeletal and skeletal hyphae)

4) Dimitic (there are only generative and skeletal hyphae)

5) Amphimitic (there are generative and binding hyphae)

6) Subtrimitic (there are generative, pseudoskeletal, skeletal, and binding hyphae)

7) Trimitic (there are generative, skeletal and binding hyphae).

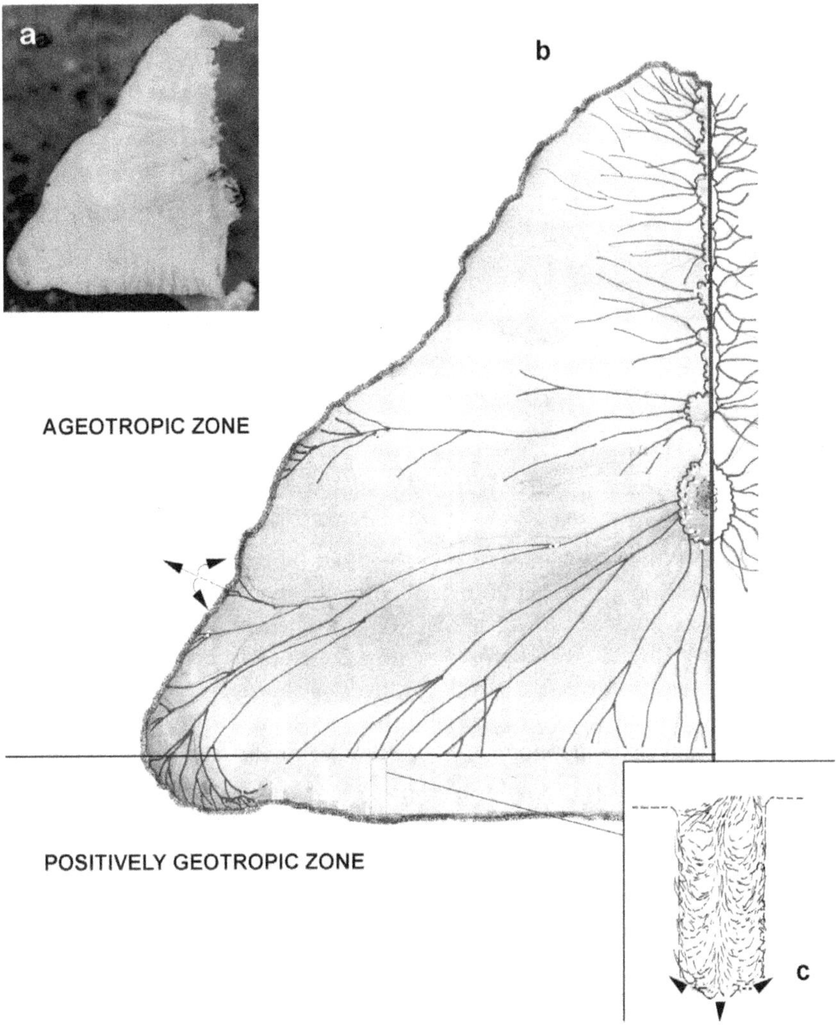

Fig. 8.9 An insolation determined model of the morphogenesis of the tinder fungus basidiome: **a** – Cross-section of annual basidiome of *Fomitopsis pinicola* (the directions of hyphal expansion corresponed with dimensions of fibre fascicles), **b** – The scheme which demonstrates insolation determined directions of hyphal expansion. The shadow zone, which has been formed during radial growth (so-called "positively geotropic zone") stays during a time by basic increase of the center, because the hyphae of the tubular core tends to form a light-covered surface, after which it achieves a change in growth direction (**c**). The tubes thus accrue in length.

Moreover, certain modifications have also been described, depending on the presence or absence of inflated hyphal elements, within the above mentioned

Table 8.1 The types of hyphal systems in the Agaricomycetes

Type of hyphal system	*Types of hyphae composing basidiomata*			
	1	2	3	4
1. Monomitic	+			
2. Pseudodimitic	+	+		
3. Subdimitic	+	+	+	
4. Dimitic	+		+	
5. Amphimitic	+			+
6. Subtrimitic	+	+	+	+
7. Trimitic	+		+	+

Note: *1* – generative hyphae, *2* – pseudoskeletal hyphae, *3* – skeletal hyphae, *4* – binding hyphae.

hyphal systems. They are named as polyphysaloid (sclerified hyphae composed of inflated cells), loculiphysaloid (sclerified hyphae with irregular swellings), and fibroid forms (sclerified hyphae that are fiber-like and uninflated). Corner's term "sarcodi(tri)mitic" is recognized as unsuccessful, the better term for this type seems to be polyphysaloid pseudodimitic hyphal system. It is observed that a number of agaricoid taxa, considered earlier as having a sarcodimitic system, really have a polyphysaloid monomitic one. Redhead's thesis of a relic nature of sclerohyphae in agaricoid fungi (Redhead, 1987) as well as Corner's "*Hericium*-hypothesis" of the origin of trametoids (Corner, 1984) are rejected. It appeared as a clear transformation ranging from thin-walled physalohyphae via meripiloid irregularly swelled hyphae to corioloid uninflated fiber-like hyphal elements.

A case of a very successful use of hyphal structure as a character in taxonomy and phylogeny of a group can be exemplified by comparative studies of lentinaceous gilled fungi and Polyporaceae. In fact, it is impossible to reveal any anatomic differences between the gilled genera *Lentinus*, *Pleurotus*, *Phyllotopsis* and the tubular genera *Polyporus* or *Dichomitus*. This had already become obvious to taxonomists by the middle of the last century (in many respects due to the work of Corner), which was reflected in taxonomic interpretations of the position of lentionoid agarics (we omit the extensive literature on this issue as it has been summarized by Stankovicova, 1973). However, numerous parallels with lentinaceous fungi in the microscopic structures are revealed not only in the stipitate polypores (Polyporaceae s.str.), but also in trametoid and fomitoid forms. The concept of Polyporaceae without trametoid and fomitoid elements is thus rather vulnerable because of the presence of some groups representing in their microstructure characters a continuous series of transitions:

(*Polyporus*–[*Microporus, Xerotus, Dichomitus*]–*Trametes*–[*Hexagonia, Daedaleopsis*]–*Fomes*) (see Wasser et al., 2006).

The pigment content of hyphal walls is also of some taxonomic importance (see Donk, 1948; Wasser, 1980, 1982, 1985, 2002; Besl et al., 1986; Singer, 1986; Clémençon, 2004).

The peculiarities of internal hyphal structures are usually not used in taxonomy, though based on these characters (the presence of secondary septation), for example, the genus *Pseudocraterellus* has been described (Corner, 1966; Petersen, 1971). The value of the presence/absence of clamp-connections in various taxa was discussed above. In some cases the pattern of arrangement of these structures is useful for taxonomists. For example, *Coniophora* and some representatives of *Phanerochaete* are characterized by multiple clamp connections on one septum, whereas in the genera *Serpula, Amylocorticium, Ceraceomyces*, or *Parmastomyces* the clamps have a central aperture (observed as a small "eye").

Some researchers consider the presence of differentiated gloeoplerous elements (gloeocystidia, pseudocystidia, macrocystidia, gloeoplerous hyphae, and vesicles) in the basidiocarp as an important taxonomic character. In particular, based on the presence of gloeocystidia and gloeoplerous hyphae in the basidiocarp, in combination with the character of amyloidity of spore walls, Stalpers (1996) united a number of diverse agaricoid, clavarioid, and corticioid taxa in the large order Hericiales (after some time it was merged into Russulales). In our opinion, the taxonomic value of gloeopleroid elements, which actually represents analogues of secretory formations (Corner, 1964, 1970), is greatly exaggerated. Besides "Hericiales", these elements can be expressed in various taxa that are not connected by direct phylogenetic relationships (*Clavulicium, Gomphus, Russula, Mycena, Albatrellus, Laricifomes, Fomitopsis, Inonotus, Antrodia, Oxyporus, Cylindrobasidium*, and *Punctularia*), whereas in the majority of Agaricomycetes, the gloeoplerous system is completely reduced. In fact, this character is not correlated in any way with other characters, in particular with amyloidity of the spore wall (Table 8.1). In this respect it is enough to analyze the pattern of distribution of these two characters in the following pairs of closest relatives: *Stereum* (amyloid)–*Peniophora* (inamyloid), *Aleurodiscus* (amyloid)–*Corticium* (inamyloid), *Dichostereum* (amyloid)–*Scytinostroma* (reaction varying). The reaction of the gloeocystidia content with sulfoaldehyde (Larsen and Burdsall, 1976) is a more reliable marker, thus sulfopositive Hericiaceae, Gloeocystidiellaceae, Aleurodiscaceae, Stereaceae and Peniophoraceae with the background of some other intermediate linkers, were merged into Russulales together with agaricoid (*Russula, Lactarius*) and gasteroid (*Martellia, Zelleromyces*) taxa.

As a whole, we can believe that in ancestral Agaricomycetes ("Proto-Clavulinaceae") the gloeoplerous hyphal system was well advanced, and during its evolution in different phyla, it was to various degrees, preserved or modified, or disappeared. However, sterile elements of the hymenophore (except hyphidia and pseudocystidia) could be considered as its derivatives. Studies of peculiarities of hymenial elements (and their transitional stages) in *Stereum, Peniophora, Metulodontia, Oxyporus, Hyphoderma*, and *Subulicystidium* clearly show that lamprocystidia and various grades of leptocystidia (guttating and those that have lost this ability), as well as various "cysts", are connected by intermediate stages with gloeocystidia and vesicles, whereas tubular cystidia, metuloids, and setae demonstrate a certain similarity with pseudocystidia.

The peculiarities of basidia, as basic hymenial elements, indeed play an important role in taxonomy. However, their phylogenetic value greatly varies depending on the "background" determined by the specificity of evolutionary tendencies of any concrete phylum. The shape of the homobasidium is a rather deceptive character. Thus, the karyologically similar basidia of Clavulinaceae, Cantharellaceae, and Sistotremataceae are very different in their habit. Basidia of Cantharellaceae are surprisingly similar to long chiastobasidia of some stereoid fungi (like *Veluticeps*); basidia of *Sistotrema* remind suburniform basidia of *Kneiffiella*, and basidia of *Clavulicium* (Clavulinaceae) are similar to podobasidia of *Radulomyces* (Pterulaceae). On the other hand, in many cases, the particular shape of basidia correlates with other groups of characters, and in such cases, we probably observe well-distinguished natural groups of Agaricomycetes.

In trying to distinguish the basic trends in morphological transformations of homobasidia, we must proceed from the obvious fact that elongate cylindrical basidia are characteristic for negatively geotropic fungi with smooth or venose hymenophores. Shorter clavate basidia are developed on gilled and tubular hymenophores of negative or positive geotropic basidiomata, and in many groups not connected by close relationships; they have a common tendency to widening and truncation. In positively geotropic fungi with smooth hymenophores, we see some regularity in development of basidia (as well as other elements of the hymenium). All of them, to a greater or lesser degree, are inclined to adhere to the substrate. This tendency is maximally realized in *Xenasma*-like fungi with pleurobasidia and multi-radicate cystidia. In this respect, we cannot agree with the theory developed by Oberwinkler (1977, 1978 and 1982) on the primary, initial position of pleurobasidia in the evolution of Homobasidiomycetes. The medial constriction, which is evident in many types of homobasidia, is thus better interpreted as a borderline between the pro- and metabasidial segments, but not as a truncated hyphal end, as supposed by the author.

The spore characters (shape, size, color, and ornamentation) are also extremely important in the taxonomy of the group; however, with the use of these characters it is even more hazardous to disregard the peculiarities of the "body plan" of the classified organisms. This is explained this with positive and negative examples.

It is widely known that *Polyporus* and *Lentinus* are currently regarded as related based on their hyphal systems, the shape and dimensions of basidia, the presence of characteristic metuloids and hyphal pegs, an irregular trama, and also the tendency for spore elongation (from ellipsoid to fusoid-cylindrical). In this example, the spore features are well correlated with a complex of other characters determining the specificity of this group. As a negative example, the proposal by Locquin (1984) is mentioned to recognize the order of agaricoid fungi, "Mycenales", which is in fact based on a single character – thin-walled non-ornamented spores.

The thickening, melanization, and ornamentation of the spore wall are a relatively ordinary phenomenon in the evolution of the plant and fungal kingdoms. Some degree of convergence is possible here even between the representatives of various phyla, for example, Zygomycota (Mucorales) and Basidiomycota ("Boletogastreae"). In that respect it is probably inappropriate in phylogenetic reconstructions to segregate as the primary (basal) clades the dark-spored and light-spored lines of development of agaricoid fungi (Miller and Watling, 1987), especially since the phylogenetic connections between representatives of dark-spored and light-spored agarics are evident (Wasser, 1982, 2002). It is also hardly justified to unite various groups of amyloid-spored aphyllophoraceous fungi (Ryvarden, 1991; Stalpers, 1996); the so-called "sporocentrical" approach to solving difficult problems of gasteromycete phylogeny (Locquin, 1984) is also questionable.

Concerning the evolutionary transformation of the form (shape) of basidiospores, even in onthogenesis, its obvious tendency to elongation is well expressed (Corner, 1947). In different groups of ballistosporous (as well as statismosporous) Agaricomycetes, certain stages of lengthening were registered depending upon ecological conditions. In gasteroid statismosporous fungi, the basidiospores are usually spherical or regularly ellipsoid. Regularly spherical spores are not characteristic for ballistosporous Agaricomycetes. Their spores are commonly more or less extended along their longitudinal axis. Most typical for all these groups are ellipsoid spores with a suprahilar depression. After excessive increase in the longitudinal length (usually $Q > 2.5–3$), tendencies of some incurving of spores (so-called allantoid or sigmoid shape), or a concentration of mass around the equatorial plane (fusoid), less often on one of the poles (pyriform or amygdaloid shape) have been observed. Regularity is a common correlation between the spore size and the length of basidia that normally exists. In many evolutionary lines, a trend toward some reduction of basidia and diminishing of spores is observed (*Merulius–Gloeoporus–Skeletocutis*; *Trametes–Diplomitoporus*; *Leucogyrphana–Parmastomyces*, etc.).

Concerning evolution of the spore surface of Agaricomycetes, this issue is controversial until now. Of course, it is very difficult to detect any leading factor affecting spore morphogeny. It is evident, that a presence of tubercles, crests, spines, or reticular formations must be regarded as an apomorphic condition because these structures improve aerodynamic and adhesive spore characteristics (Frazier and Glaser, 1979). However, whether these structures appeared from an originally thin sporoderm, or were produced by degraded thick-walled spores, is not clear.

After the study of hemitectospores and thick-walled non-perforated spores, we favor the second viewpoint. Evidently, an original – protective – function of the spores causes a development of its thickened impenetrable tectum. Thick-walled spores are characteristic for many archaic Homobasidiomycetes groups (Cantharellales, Gomphales, Boletales), and represent, probably, a plesiomorphic stage. At the same time, as noted by Coleman (1927), their main deficiency is "slowed reaction" to an environment change, i.e. the belated germination.

In our opinion, an appearance of hemitectospores was connected to an evolutionary tendency to increase their environmental information exchange. The perforations, filled with hydrophilic fibrillar-crystalline matter and protected from a direct influence by the perine, allow the external signals soon to appear on the heat-shock proteins of cytoplasm.

An analogous phenomenon is known in the evolution of gymnospermous microspores. According to Krassilov (2004), the appearance of a perforated tectum is a special *prophetic* feature, i. e. preadaptation to a sudden change of the spore surface that could be very helpful to the spore in its dispersal and rooting to the substrate.

This change is accompanied with degradation of perine structure carried out either by 1) dehydration and collapse of perisporium matrix, when perine (as thin or mostly invisible exosporium) covers a perforated tectum, or 2) when a perisporium is partly reduced, and IKI-reaction shows a remnant of the fibrillar-crystalline matrix.

So, we think that the hemitectospores represent an unstable transitional condition between the archetypic thick-walled spores and the thinner-walled ornamented forms. An origin of smooth thin-walled spores is not considered here; we only note that they may be produced due to "sprinter" strategy of sporulation and appear in both "smooth thick-walled spored" and "hemitectosporous" taxa. In this case, smooth amyloid spores represent a rudimentary phenomena of exosporium reduction (compare with the SEM of a fractured spore of *Ganoderma valesiacum* with a vanished perine, whose exposed exine strongly resembles the surface of the *Heterobasidion* spore, – see Pegler and Young, 1973, pl. 21, f. 4, 5 – or an analogous picture demonstrated in Adaskaveg and Gilbertson, 1986, f. 14, 15).

For example, the spore spines of *Strobilomyces strobilaceus* (Boletales) are characterized by an exceptionally peculiar shape. The same outline is characteristic for *Crepidotus wakefieldii* (Agaricales), as seen in SEM by

Pegler and Young (1972; see Fig. 8.10). This spore pattern indicates that the cap-like tubercles of *Strobilomyces strobilaceus* and *Crepidotus wakefieldii* are remnants of continuous perisporium of a hemitectosporous ancestor. So, Singer (1975) was correct when he designated the spores of *Crepidotaceae* as ganodermoid; here we may amend only - *ancestrally ganodermoid* (Spirin et al., 2005).

One deals with numerous independent phylogenetic lines in which different stages of a cycle "thick-walled spores - hemitectospores - ornamented spores" are fixed. Within one genus, the species may have a spore integument of different types (e.g., *Crepidotus*). In this connection, the hemitectosporous genus such as *Haploporus* and its crassitunicate smooth-spored counterpart *Perenniporia* draw our attention. In our opinion, *Haploporus suaveolens* shows "the first steps" of integument perforation and degradation, while the genera *Wrightoporia* and *Heterobasidion* demonstrate the "latter" forms.

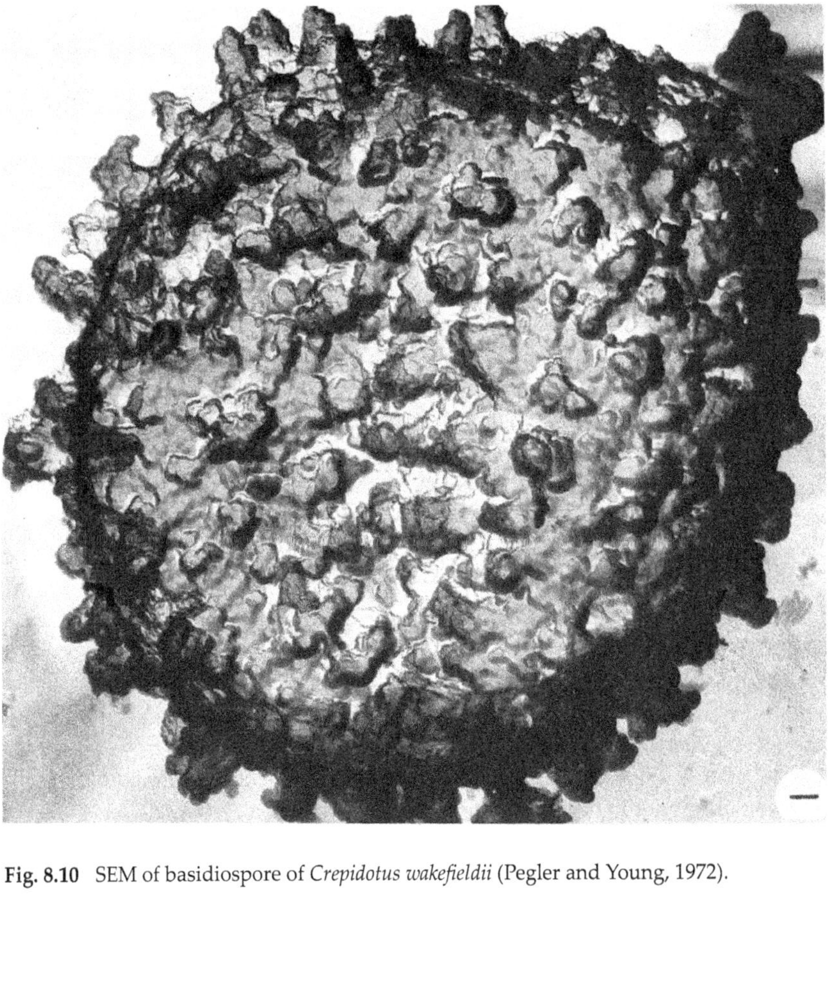

Fig. 8.10 SEM of basidiospore of *Crepidotus wakefieldii* (Pegler and Young, 1972).

A natural place of these genera in taxonomy of higher Basidiomycetes remains questionable. The aforegoing description of the spores of *Haploporus suaveolens* could give an impression that they are homologous to the spores of *Ganoderma*, and, on the other hand, to those of *Heterobasidion*. If some other features common to the Ganodermataceae, Perenniporiaceae and Wrightoporiaceae (incl. *Heterobasidion*), (Melzer-positive skeletal hyphae having a tendency to dichotomous branching, and apically widened clavate basidia) are emphasized, their regrouping into the Lachnocladiales (or the "large" Russulales) could be rather logical. However, known molecular data contradict this group, supporting a position of *Heterobasidion* within the Bondarzewiaceae (Russulales) (Hibbett et al., 1997; Larsson and Larsson, 2003), whereas *Ganoderma* and *Perenniporia* appear to belong to the "polyporoid clade" (Moncalvo, 2000). Undoubtedly, this problem may be decided when a large set of new data is analyzed.

Phylogenetic lines of Agaricomycetes

On the basis of the above discussion, we shall try to draw some conclusions on the possible phylogenetic lines in Agaricomycetes in general (Fig. 8.11).

Figure 8.11 presents a hypothetical phylogenetic tree of Agaricomycetes, which provides data on the origin and deviations from the common ancestor of large groups of this class. Circumscriptions and diagnostic characters are provided in Table 8.2.

As it can be seen from Table 8.2, most of the archetypal features are preserved in Clavulinales. This small group is characterized by basidia of a primitive structure (stichobasidia), having a small number of basidiospores (2, rarely 4) and continuous parenthosomata. Practically all representatives of the group demonstrate the tendency to form negatively geotropic fructifications.

Continuous parenthosomata is characteristic for all groups, derived from hypothetical evolutionary stock lower than the Thelephorales, but in some archaic groups (e.g. Cantharellales s.str.) the parenthosomes demonstrate perforated patterns. Probably, this event took place independently in the evolution of the phylum and was connected by a general tendency of integration of compartments in basidiomycete filaments, especially in the case of physalohyphae. In "higher" Cantharellales, the parenthosomata remain unperforated, as in the Hymenochaetales, the group, uniting xanthochroic polypores and stereoids. But, according to molecular data, non-xanthchroic also (*Hyphodontia, Oxyporus, Trichaptum)* bear unperforated parenthosomes (Langer, 1994, 1998; Keller, 1997), and minute agaricoid-cantharelloid fungi, however, are perforated like *Cantharellus* (van Driel et al., 2008).

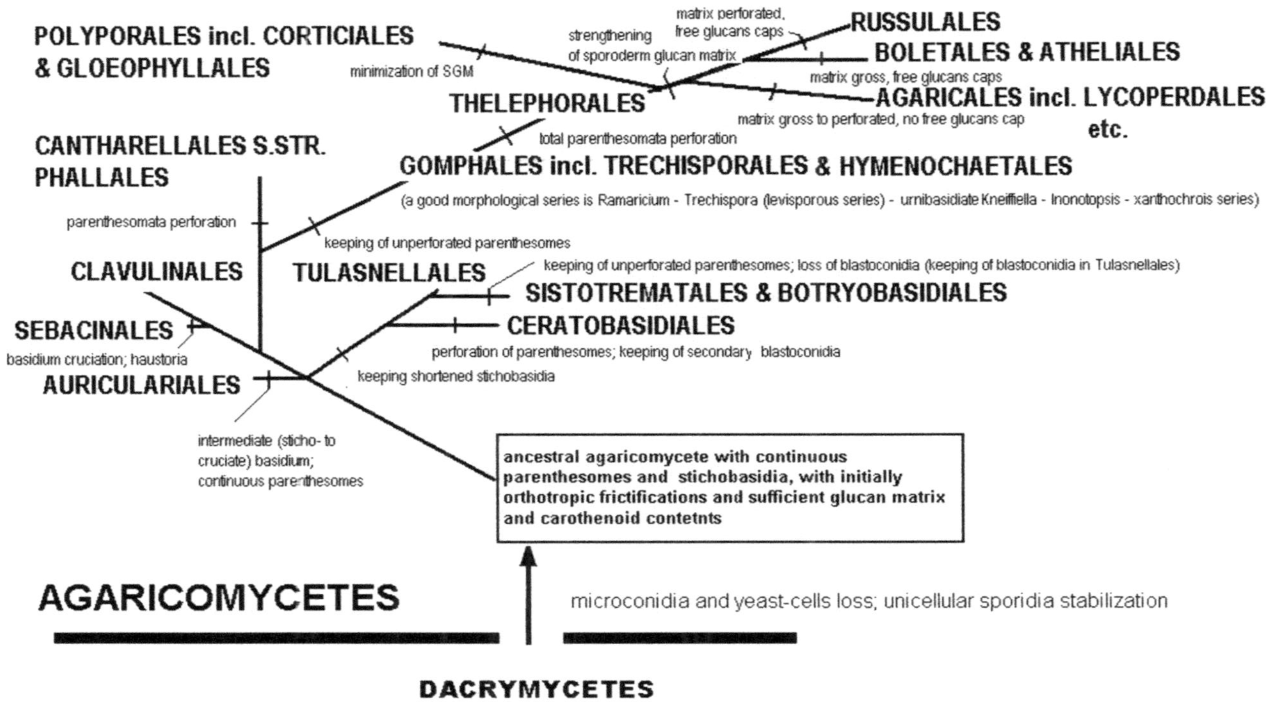

Fig. 8.11 Basic markers in the postulated phylogeny of Agaricomycetes. The events of parenthosomata perforation and stichobasidium cruciation is postulated as being in minimal numbers. The outer sporoderm layers degradation "higher" of Thelephorales demonstrates rather doomed trends corresponding with Agaricales-, Boletales-, Russulales-, and Polyporales-lines.

Table 8.2 Taxonomic subdivision of Agaricomycetes based on generalized morphological data

Orders	*Diagnostic characters*	*General families*
1. Clavulinales	Parenthosomata continuous; basidia stichic, utriform, 2–4-spored; hyphae/ epibasidium ratio ≥ 2; spores with distinct walls, smooth; basidiomata orthotropic to prostrate, monomitic; clamp-connections more or less inflated.	Clavulinaceae
2. Auriculariales (incl. Sebacinales)	Parenthosomata continuous; basidia stichic, intermediate, or chiastic, sepated; from tubular to stalked; hyphae/ epibasidium ratio ≥ 2; spores thin-walled, smooth; basidiomata orthotropic to prostrate, monomitic, mostly gelatinized; clamp-connections varying in shape.	Auriculariaceae Exidiaceae Hyaloriacaeae Patouillardinaceae Tremellodendro-psidaceae Sebacinaceae
3. Tulasnellales (incl. Sistotrematales and Botryobasidiales)	Parenthosomata continuous; basidia stichic to prolificated, 2–8-spored, without primary septation; basically head-like with pronounced sterigmata (metabasidia); hyphae/epibasidium ratio ≥ 2; spores thin-walled, smooth to spinose; basidiomata orthotropic to prostrate, monomitic, clamp-connections varying in shape and presence.	Tulasnellaceae Botryobasidiaceae Sistrotremataceae Hydnaceae s.str.
4. Cantharellales (incl. Phallales)	Parenthosomata perforated; basidia stichic, utriform, 2–6-spored; hyphae/ epibasidium ratio ≥ 2; spores with distinct walls, smooth or ornamented; basidiomata orthotropic to gasteroid, monomitic; clamp-connections more or less inflated.	Cantharellaceae Phallaceae Clathraceae Hysterangiaceae Geastraceae
5. Gomphales (incl. Trechisporales and Hymenochaetales)	Parenthosomata continuous; basidia chiastic, suburniform to clavate, 2–4 (6)-spored; sometimes repetobasidia; hyphae/ epibasidium ratio ≥ 2; spores thin- to thick-walled, rough of smooth; basidiomata clavarioid to sessile or resupinate, mono- to pseudodimitic or dimitic; clamp-connections, if present, of amullate to normal appearance.	Gomphaceae Lentariaceae Ramariaceae Trechisporaceae (?Phlebiellaceae) Rickenellaceae* Repetobasidiaceae Hymenochaetaceae Schizoporaceae
6. Thelephorales	Parenthosomata perforated; basidia chiastic, clearly utriform, 2–4-spored; hyphae/epibasidium ratio ≥ 2; spores commonly thick-walled, roughed;	Thelephoraceae Bankeraceae

Table 8.2 Contd. ...

	basidiomata clavarioid, cantharelloid, infundibuliform, mucronate, pileate, to dorsally attached and resupinate.	
7. Agaricales (incl. Lycoperdales, Tulostomatales, Podaxales)	Parenthosomata perforated; basidia chiastic, utriform to clavate, (2) 4-spored; hyphae/epibasidium ratio ≥ 2; spore wall basically gross to perforated, in some cases with gross glucans caps and Melzer's positive endosporial ougrowths; basidiomata agaricoid to resupinate or gasteroid, monomitic to trimitic; gloeoplerous hyphae present or not; clamp connections, when present, of normal appearance.	Tricholomataceae ?Fistulinaceae Clavariaceae Typhulaceae Hygrophoraceae Pterulaceae Catathelasmataceae Amanitaceae Pluteaceae Entolomataceae Agaricaceae (incl. Secotiaceae) Battareaceae Tulostomataceae Lycoperdaceae Nidulariaceae Coprinaceae (incl. Podaxaceae) Bolbitiaceae Strophariaceae Cortinariaceae Inocybaceae (incl. Crepidotaceae) Hymenogastraceae Gastrosporiaceae
8. Boletales (incl. Atheliales)	Parenthosomata perforated; basidia chiastic, utriform to clavate, (2)4–6-spored; hyphae/epibasidium ratio ≥ 2; spores with gross wall, commonly pigmented, if seemingly thin-walled, the free glucans Matrix is commonly detected by Melzer's media; basidiomata agaricoid to resupinate or gasteroid, monomitic to sarcotrimitic; gloeoplerous hyphae present or not; clamp connections, when present, slightly to grossly inflated.	Gomphidiaceae (incl. Rhizopogonaceae) Hygrophoro-psidaceae Paxillaceae (incl. Gyrodontaceae, Meiorganaceae syn. Serpulaceae and Coniophoraceae) Tapinellaceae (incl. Amylocorticiaceae syn. Plicaturaceae) Atheliaceae Boletaceae (incl. Xerocomaceae) Strobilomycetaceae Sclerodermataceae Astraeaceae

Table 8.2 Contd. ...

9. Russulales	Parenthosomata perforated; basidia chiastic, clavate, 4-spored; hyphae/epibasidium ratio ≥ 2; spore walls with gross mathrix, commonly perforated to perine wanished; ornamented exosporium mostly amyloid; sulfocystidia or lacticiferous hyphae commonly present; basidiomata clavarioid, sessile polyporoid, agaricoid, gasteroid, or resupinate, mono- to sarco(pseudo)dimitic.	Albatrellaceae Amylosporaceae Hericiaceae Gloeocystidiellaceae Peniophoraceae Stereaceae syn. Aleurodiscaceae ?Xenasmataceae Auriscalpiaceae Lentinellaceae Wrightoporiaceae Bondarzewiaceae Russulaceae (incl. Elasmomycetaceae, Hybogasteraceae)
10. Polyporales (incl. Gloeophyllales, Phanerochaetales, Meruliales, Hyphodermatales, Corticiales)	Parenthosomata perforated; basidia chiastic, clavate-cylindrical in catahymenial palisades imitating cystobasidial patterns of Pucciniomycetes, (2)4-spored; hyphae/epibasidium ratio commonly ≤ 2; spores inamyloid, smooth, mostly with distinct wall an a tendency to uniaxial elongation; mostly wood-inhabiting – pleurotoid or polyporoid to resupinate (proioid, meruliuid or corticioid) or sessile (tyromycetoid, trametoid, fomitoid), monomitic to sarcotrimitic, often pseudodimitic, predominantly di- to trimitic; sclerohyphae with rectangular or dendroid branching pattern; gloeoplerous hyphae rare to absent; clamp connectioins varying in shape, in some groups double or verticillate.	Polyporaceae (incl. Ganodermataceae, Lentinaceae, Trametaceae, Perenniporiaceae) Phanerochaetaceae (incl. Bjerkanderaceae, Ceriporiaceae, Irpicaceae, Climacodontaceae) Gloeophyllaceae Fomitopsidaceae (incl. Laetiporaceae, Phaeolaceae) Meripilaceae Meruliaeae (incl. Steccherinaceae, Hyphodermataceae, Punctulariaceae, ?Schizophyllaceae) Corticiaceae s.str. Vuilleminiaceae

* The family Rickenellaceae is fomally undescribed until now; moreover, this taxon is characterized by perforated parenthosomata (van Driel et al., 2008). In our opinion, despite of "rDNA-dictature", the position of the "Rickenellaceae" would be more natural as Cantharellaceae s. str.

"Higher" basidia in the Tulasnellales and Cantharellales are already chiastic, but in many representatives they preserve their suburniform shape, reminiscent of *Sistotrema*. A hypothetical archetype of the Hymenochaetales was probably similar to the tropical genus *Clavarichaete* characterized by clavarioid basidiomata, 6-spored basidia, and tramatic setae.

The order Gomphales, characterized already by perforated parenthosomata, utriform chiastobasidia, and more or less rough/thick-walled, often ornamented spores. It is obviously connected with Cantharellales, which can be seen in multiple parallels in the formation of cantharelloid, ramarioid and clavarioid forms.

Results of a comparative morphological analysis testify to a high probability that the basic phylogenetic radiation of Agaricomycetes (Agaricales–Boletales–Russulales) is connected with this group. All these groups, like Gomphales, have utriform to clavate chiastobasidia with more or less expressed medial constrictions, in many agaricoid and boletoid taxa with an inflated epibasidial part (see Oberwinkler, 1977, 1982; Stalpers, 1992). This pool is the foremost in its species diversity; its development proceeded mostly in the direction of colonizing mainly terrestrial soil habitats, although in various groups fairly successful attempts of adaptation to wood substrates are observed.

Representatives of Tricholomataceae have, in their basic features, preserved the *Cantharellus*-like organization, though in many evolutionary lines of the family tendencies are seen of transformations towards complication and reduction. The postulation of repeated and independent origin of this type of organization is considered by us as contradictory to the parsimony principle. Thus, we adhere to the basic building of a "stem", covering and connecting firstly all lamellulae-bearing taxa with subsequent independent deviations of aphyllophoraceous and gasteroid forms from this "core structure".

The problem of monophyly of Agaricales is still poorly resolved. If the monophyly concept is assumed, this group may be successfully traced phylogenetically from (Russulales) via taxa like Leucopaxillaceae. In this case, Hygrophoraceae, as well as various dark-spored taxa, will necessarily be phylogenetically deduced from Tricholomataceae. Some morphological lines would be tested more precisely, e.g. *Omphaliaster*-Hygrophoraceae, *Phaeolepiota*-Cortinariaceae, and *Catathelasma*-Agaricaceae. In the case of polyphyly, the Tricholomataceae seem to be rooted in the domain of Boletales/Russulales/Agaricales divergence, whereas the Hygrophoraceae and Cortinariaceae took root near the Gomphales (in this case, it would be worthwhile examining parallels in the pairs Hygrophoraceae–Clavariaceae, Entolomataceae–*Donkella*, and Cortinariaceae–Ramariaceae). In addition, the dark-spored taxa can also be rooted phylogenetically "lower" than Tricholomataceae (several possibilities exist to connect these lines with Cortinariaceae or Paxillaceae).

The various groups of Agaricales were involved in the process of gasteromycetization, and among these groups we shall look for the roots of such a large group of Gasteromycetes as Lycoperdaceae.

The order Polyporales is definitely evolutionarily connected with clitocyboid Tricholomataceae (*Clitocybe*–*Lentinus*–*Polyporus*-Trametoideae). Several sarcotrimitic polypores (such as *Abortiporus*, *Grifola*, *Osteina*, *Pilatoporus*, etc.), as well as some merulioid and lissed-spored corticioid fungi were recently transferred to this group (Zmitrovich et al., 2006). Other resupinate fungi from unnatural orders "Stereales" and "Lachnocladiales" are well connected with sulfocystidia-bearing Hericiaceae and as a whole can be deduced from lamellulae-bearing Russulales.

MOLECULAR TREE (MOLECULAR PHYLOGENY)

Many "evolutionary trees" of Agaricomycetes have been established during recent decades (Donoghue and Sanderson, 1991; Hibbett and Donoghue, 1995, 2001; Hibbett et al., 1997; Lee and Jung, 1997; Hallenberg and Parmasto, 1998; Boidin et al., 1998; Bruns et al., 1998; Langer, 1998, 2002; Bresinsky et al., 1999; Hibbett et al., 2000; Kim and Jung, 2000; Parmasto and Hallenberg, 2000; Hibbett and Thorn, 2001; Jarosch and Besl, 2001; Lim, 2001; Weiss and Oberwinkler, 2001; Wu et al., 2001; Binder et al., 2002; Binder and Hibbett, 2002; Hibbett and Binder, 2002; Redhead et al., 2002a, b; Larsso et al., 2002; Hsiau and Harrington, 2003; de Koker et al., 2003; Kottke et al., 2003; Larsson and Larsson, 2003; Moncalvo et al., 2003; Taylor et al., 2003; Yoon et al., 2003; Bodensteiner et al. 2004; Larsson et al. 2004; Wang et al., 2004; Binder et al., 2005; Hibbett et al., 2007; Larsson, 2007; Lawrey et al., 2007; Matheny et al., 2007), thus a global multigene tree of the class has left mainstream without sufficient deviations and rotations (Fig. 8.12).

The common traits of the SSU and LSU trees are their "roots" (sequences of representatives of Dacrymycetales) and their low resolved topology: the multifurcation "Auriculariales, Sebacinales, Cantharellales, Phallomycetidae" are basal for a poorly resolved crown, containing multufurcation "Hymenochaetales, Corticiales, Gloeophyllales, Polyporales – Thelephorales" and furnished by the trichotomy "Agaricales, Athelieles, Boletales". All these types of trees support the basal position of "heterobasidiomycete" and cantharelloid clusters.

Previous study of Moncalvo et al. (2003) is an example of elegant work on euagarics phylogeny using nLSU rDNA sequences. Their large-scale tree of Agaricomycetes supports monophyly of six of the above mentioned clades (the euagarics, bolete, hymenochaetoid, and thelephoroid, gomphoid-phalloid, and cantharelloid clades), the basal position of the cantharelliod clade, and the reciprocal monophyly of Polyporaceae and corticoid clades within the polyporoid clade. However, some difficulties arise in the resolution of the basal relationships in some major clades of the tree. For instance, the basal relationships of the euagarics clade are poorly resolved, although the maximum parsimony tree and bootstrap 50% majority rule consensus tree represent

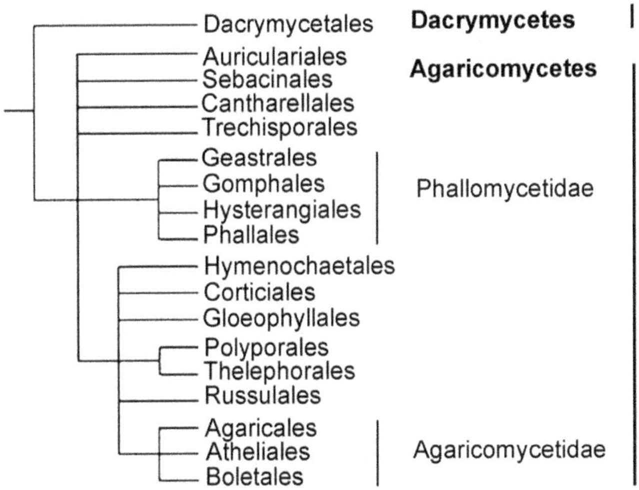

Fig. 8.12 Large-scale Agaricomycetes phylogeny based on analysis of SSU, LSU, *rpb1*, *rpb2*, and *tef1* sequence data (after Hibbett et al., 2007).

more than 100 distinct clades with considerable bootstrap support (exceeding 70% for most of them). The order of separating clades, starting from the base of the tree to the top, maintains the general pattern: cantharelloid – gomphoid-phalloid – thelephoroid – corticioid – hymenochaetoid – russuloid. The hymenochaetoid clade including a number of agaricoid taxa (such as *Omphalina brevibasidiata* or representatives of the genus *Rickenella*) is the neighbor of the bolete clade, the suggested sister group of the euagarics in Moncalvo's tree.

So trees based on the sequences of ribosomal genes have addressed several important issues regarding basic patterns of Agaricomycetes evolution—if there are monophyletic groups; what groups exist, and which groups are basal? All these questions are answered, but only in part, leaving many issues still open. For instance, the russuloid clade, the agaric-bolete-russuloid "triangle", and others, although included in studies of genes, vary in respect to rates of evolution (18S evolves slower than the nLSU), and, in general, should allow a better resolution of higher-level phylogenetic groupings (terminal as well as deeper levels of branching).

However, apart from these questions, the phylogenies confirm many groups previously outlined by classical methods (based on morphology, microstructures, and hyphal ultrastructure). It would be correct to address the following question: to what extent do both types of trees correspond to the morphological tree?

RECONSTRUCTING THE TRUE PHYLOGENETIC TREE

Despite dissimilarities, both groups of trees accord well with the most general features of morphological topologies. This accord is evident in patterns of placement and mutual distribution of groups of long and short branches, conditioned by the significant difference in age between them. The age difference is expressed in different degrees of rDNA polymorphisms.

Concerning taxa formation, first there are morphogenetical events of microadaptive character unattended. They are not connected immediately with evolution of basic cell functioning components, like RNAs and corresponding genes. In most cases, these events represent the result of regulatory genes mutations. Therefore, dissimilarities between "true phylogeny" and "molecular phylogeny" are determined by features of Evolution as transformation of *hierarchically organized* systems (Fig. 8.13).

The long branches appearing from the crown part of the Tremellomycetes thus produce cantharelloid and phalloid clades. It is also necessary to underline that traditional morphological as well as ultrastructural data confirm the basal position of these taxa.

As can be seen from Fig. 8.13, the "congruent rDNA–Morphology zone" takes place *near the basal part* of the cladogram, where the relationships are not overfilled by stochastic morphogenetic events of overbuilt adaptive radiations. It is obvious that major Agaricomycetes taxa separated and diverged during an

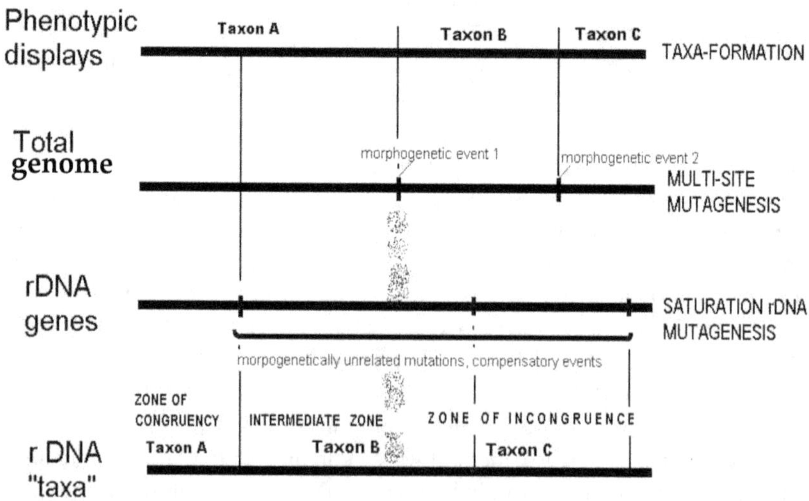

Fig. 8.13 The scheme displays incongruence between rDNA-phylogenies and phenotypic (morphologically-based) ones. The basic feature of rDNA-phylogenies concludes that there is no deep correlation with morphogenetic events in taxa formation.

insignificant period of time (probably middle Cretaceous), and later developed in parallel lineages.

If the gomphoid–phalloid clade is understood under this radiation region, it explains more or less why the bootstrap support of the nodes situated above significantly decreases (from 100% for the cantharelloid cluster to 57% of the node of large radiation in Larsson's tree); rDNA does not have enough resolution for reliable clusterization of the top polytomy.

Some discrepancies between morpohological evidence and molecular phylogeny were already debated (Zmitrovich et al., 2006). The most fundamental controversy remains the total disruption by rDNA phylogeny of morphological continuum of the generic appearance "*Panellus–Merulius–Schizophyllum*" (shuffled to Mycenaceae, Meruliaceae and Schizophyllaceae, respectively). A similar situation is the sharp molecular delimitation of *Pleurotus* (Agaricales) and *Lentinus* (Polyporales), whereas morphologically *Pleurotus–Lentinus* represent a network rather than two outlined objects. Figure 8.11 explains such discrepancies. In the work devoted to Phanerochaetineae, the following re-make of a classical sentence by Parmasto (1968) was given as: "An artificial system built on one (let even on molecular) feature is no more sophisticated than the last artificial systems created on the macromorphological ones" (Zmitrovich et al., 2006).

Thus, the general configuration of a molecular tree confirms the suggestion that the basic line tracing the Agaricomycetes is associated with the cantharelloid–gomphoid–phalloid area, and their further development was parallel. This agrees well with Corner's phylogeny, postulating a cantharelloid ancestor of the group (Corner, 1966) and further "gomphoid–agaricoid" radiation (Corner, 1972) accompanied by gasteromycetization and corticioid simplification, although it leaves open a number of other questions. In this respect, claims such as "gilled mushrooms evolved at least six times, from morphologically diverse precursors" (Hibbett et al., 1997) probably needs some amendment: more likely it is the existence of several (probably, six) lines, within which the initial cantharelloid or agaricoid ancestors gave rise to the formation of non-gilled forms.

Reconstruction of the phylogenies of Agaricomycetes is far from complete. No serious factual and methodological assumptions for advancing the process toward its end are seen. Molecular systematics, factors of which captured the minds of taxonomists several years ago, enters into a new and more constructive phase (in light of the previous experience). The appearance of numerous noncongruent molecular topologies has assisted the process to a great extent.

These days, despite the long-lasting debate on molecular versus morphological systematics, it is very important to realize that any reconstruction of phylogenies should be based on some analysis of molecular characters together with morphological (including anatomical and ultrastructural) traits. However, incorporation of such different data sets into a joint analysis entails a number of difficulties already widely discussed. The issues considered are the assigning of weight to molecular and nonmolecular characters, violation of weighting methods for various kinds of evolutionary changes on the incorporation of characters from different sources into one data set, etc. (Kluge, 1983; Miyamoto, 1983; Lavin and Doyle, 1991). Possible solutions are to analyze morphological and molecular characters separately and to create one hypothesis later on using agreed methods (Adams, 1972); to conduct a combined analysis (Hillis, 1987; Donoghue and Sanderson, 1991); to record recoding of terminal taxa of the molecular tree as states of one trait followed by its inclusion into a general (molecular-morphological) matrix (Schantser, 2003). In our opinion, a highly productive method is a separate use/combining of morphological vs molecular distance matrices (Prof. E. Kosman, pers. comm.).

Either way, progress in this field should be attributed to the accumulation of new data (both molecular and morphological) as well as the maintenance of one of the most important principles of systematics: *the totality of traits and combinations of characters is always more important than one trait or one character alone – no matter how important this sole character seems to be.*

REFERENCES

Adams, E. 1972. Consensus techniques and the comparison of taxonomic trees. Syst. Zool., 21: 390-397.

Adaskaveg, J. E. and Gilbertson, R. L. 1986. Cultural studies and genetics of sexuality of *Ganoderma lucidum* and *G. tsugae* in relation to the taxonomy of the *G. lucidum* complex. Mycologia, 78: 694-705.

Ander, P. and Marzullo, L. 1997. Sugar oxidoreductases and vetatril alcohol oxidase as related to lignin degradation. J. Biotechnol., 53: 115-131.

Arx, J. A. von. 1967. Pilzkunde. Lehre. Germany.

Arx, J. A. von. 1979. Propagation in the yeasts and yeast-like fungi, pp. 555-566. In: The Whole Fungus, Vol. 2. B. Kendrick (ed.). National Museum of Canada, Ottawa.

Bandoni, R. J. 1984. The Tremellales and Auriculariales: an alternative classification. Trans. Mycol. Soc. Japan, 25: 489-530.

Bauer, R., Oberwinkler, F. and Vánky, K. 1997. Ultrastructural markers and systematics in smut fungi and allied taxa. Can. J. Bot., 75: 1273-1314.

Begerow, D., Bauer, R. and Oberwinkler, F. 1997. Phylogenetic studies on large subunit ribosomal DNA sequences of smut fungi and related taxa. Can. J. Bot., 75: 2045-2056.

Besl, H., Bresinsky, A. and Kammerer, A. 1986. Chemosystematic der Coniophoraceae. Z Mykol., 52: 277-286.

Bessey, E. A. 1950. Morphology and Taxonomy of Fungi. 4th Print, Hafner Publishing Co., New York.

Binder, M. and Hibbett, D. S. 2002. Higher level phylogenetic relationships of homobasidiomycetes (mushroom-forming fungi) inferred from four rDNA regions. Molec. Phylog. Evol., 22: 76-90.

Binder, M., Hibbett, D. S. and Larsson, K.-H., Larsson, E. and Langer, G. 2005. The phylogenetic distribution of resupinate forms across the major clades of mushroom-forming fungi (Homobasidiomycetes). Systematics Biodiv., 3: 113-157.

Bodensteiner, P., Binder, M., Moncalvo, J.-M., Agerer, R. and Hibbett, D. S. 2004. Phylogenetic relationships of cyphelloid homobasidiomycetes. Molec. Phylog. Evol., 33: 501-515.

Boehm, E. W. A. and McLaughlin, D. J. 1989. Mitosis and ultrastructure in Eocronartium muscicola: meiosis and basidial development. Mycologia, 81: 98-114.

Boidin, J., Mugnier, J. and Canales, R. 1998. Taxonomie moleculaire des Aphyllophorales. Mycotaxon, 66: 445-491.

Bondartsev, A. S. 1953. The Polyporaceae of the European part of the USSR and the Caucasia. Acad. Sci. USSR, Moscow–Leningrad (in Russian).

Bondartseva, M. A. 1993. Life forms of higher fungi in European ecosystems, pp. 157-170. In: Fungi in Europe: Investigation, recording and conservation. D. N. Pegler, L. Boddy and P. M. Kirk (eds.). Royal Botanic Garden, Kew, U.K.

Bondartseva, M. A. 1997. Evolutionary relationships and taxonomic position of tinder fungi (Polyporaceae s. lato). Mikol Fitopatol., 31: 76-83 (in Russian).

Bondarzew, A. and Singer, R. 1941. Zur Systematik der Polyporaceae. Ann. Mycol., 39: 43-65.

Bresinsky, A., Jarosch, M., Fischer, M., Schönberger, I. and Wittmann-Bresinsky, B. 1999. Phylogenetic relationships within *Paxillus* s. l. (Basidiomycetes, Boletales): separation of a Southern Hemisphere genus. Plant Biology, 1: 327-433.

Bruns, T. D., Szaro, T. M., Gardes, M., Cullins, K. W., Pan, J. J., Horton, T. R., Kretzer, A., Garbelotto, M. and Li, Y. 1998. A sequence database for identification of ectomycorrhizal Basidiomycetes by phylogenetic analysis. Molec. Ecol., 7: 257-272.

Buchgoltz, F. V. 1902. Materials to the Morphology and Systematics of Hypogean Fungi (Tuberaceae and Gastromycetes pr. p.). Riga (in Russian).

Cavalier-Smith, T. 1998. A revised six-kingdom system of life. Biol. Rev., 73: 203-266.

Chadefaud, M. 1960. Les Végétaux non Vasculaires (Cryptogamie). In: Traite de Botanique Systématique. M. Chadefaud et L. Emberger L. (eds.). Masson et Cie Editeurs, Paris, t. 1, 1-1018.

Chadefaud, M. 1975. Sur le "basidies a répétition" des *Repetobasidium*. Rev. Mycol., 39: 173-179.

Clémençon, H. 2004. Cytology and Plectology of the Hymenomycetes/with assistance of V. Emmett and E. E. Emmett. Bibl. Mycol., 199: 1-488.

Coleman, L. C. 1927. Structure of spore wall in *Ganoderma*. Bot. Gaz., 83: 48-61.

Cooke, W. B. 1961. The genus *Schizophyllum*. Mycologia, 53: 575-599.

Corner, E. J. H. 1932a. The fruitbody of *Polystictus xanthopus* Fr. Ann. Bot., 46: 71-101.

Corner, E. J. H. 1932b. A *Fomes* with two systems of hyphae. Trans. Brit. Mycol. Soc., 17: 51-81.

Corner, E. J. H. 1947. Variation in the size and shape of spores, basidia, and cystidia in Basidiomycetes. New Phytol., 46: 195-228.

Corner, E. J. H. 1953. The construction of polypores. 1. Introduction. Phytomorph., 3: 152-167.

Corner, E. J. H. 1954. The classification of the higher fungi. Proc. Linn. Soc. Lond., 165: 4-6.

Corner, E. J. H. 1964. The Life of Plants. Clowes and Sons, London.

Corner, E. J. H. 1966. A Monograph of Cantharelloid Fungi. Oxford Univ. Press, Oxford.

Corner, E. J. H. 1968. A monograph of *Thelephora*. Beih Nova Hedwigia, 27: 1-110.

Corner, E. J. H. 1970. Supplement to "A monograph of *Clavaria* and allied genera". Beih Nova Hedwigia, 33: 1-299.

Corner, E. J. H. 1971. Merulioid fungi in Malaysia. Gdns'. Bull., Singapore, 25: 355-381.

Corner, E. J. H. 1972. *Boletus* in Malaysia. Gov. Printing Office, Singapore.

Corner, E. J. H. 1984. Ad *Polyporaceas* II & III. *Polyporus*, *Mycobonia*, and *Echinochaete*. *Piptoporus, Buglossoporus, Laetiporus, Meripilus*, and *Bondarzewia*. Beih. Nova Hedwigia, 78: 1-222.

Demoulin, V. 1974. The origin of Ascomycetes and Basidiomycetes. The case for a red algal ancestry. Bot. Rev., 40: 315-345.

Donk, M. A. 1948. Notes on Malesian fungi. Bull. Bot. Gdns. Buitenzorg, 17: 473-483.

Donk, M. A. 1954a. A note on sterigmata in general. Bothalia, 6: 301-302.

Donk, M. A. 1954b. Notes on resupinate Hymenomycetes. I. On *Pellicularia* Cooke. Reinwardtia, 2: 425-434.

Donk, M. A. 1956. Notes on resupinate Hymenomycetes. II. The tulasnelloid fungi. Reinwardtia, 3: 363-379.

Donk, M. A. 1962. Notes on Cyphellaceae – 2. Persoonia, 2: 331-348.

Donk, M. A. 1964. A conspectus of the families of Aphyllophorales. Persoonia, 3: 199-324.

Donk, M. A. 1971. Multiple convergence in the polyporaceous fungi, pp. 393-422. In: Evolution in the Higher Basidiomycetes. R. H. Petersen (ed.). Tennessee Press, Knoxville.

Donoghue, M. and Sanderson, M. 1991. The suitability of molecular and morphological evidence in reconstructing plant phylogeny, pp. 340-368. In: Molecular Systematics in Plants. D. Soltis, P. Soltis and J. Doyle. (eds.). Chapman & Hall, New York.

Doweld, A. 2001. *Prosyllabus Tracheophytorum, Tentamen systematis plantarum vascularium (Tracheophyta)*. Hdb. Moscow.

Driel, K. G. A. van. 2007. Septal Pore Caps in Basidiomycetes, Composition and Ultrastructure. Doctoral Thesis. Utrecht University, Utrecht.

Driel, K. G. A. van, Humbel, B. M., Verkleij, A. J., Stalpers, J., Müller, W. H. and Boekhout, T. 2008. Septal pore complex morphology in the *Agaricomycotina* (*Basidiomycota*) with emphasis on the *Cantharellales* and *Hymenochaetales*. Mycol. Res., 113: 559-576.

D'Souza, T. M., Boominathan, K. and Reddy, C. A. 1996. Isolation of laccase gene-specific sequences from white rot and brown rot fungi by PCR. Appl. Environ. Microbiol., 62: 3739-3744.

Eriksson, J. 1958. Studies in the Heterobasidiomycetes and Homobasidiomycetes–Aphyllophorales of Muddus National Park in North Sweden, Uppsala.

Frazier, W. and Glaser, L. 1979. Surface components and cell recognition. Ann. Rev. Biochem., 48: 491-523.

Gäumann, E. 1926. Vergleichende Morphologie der Pilze. Gustav Fischer, Jena.

Gäumann, E. 1964. Die Pilze. Birkäuser Verl., Basel/Stuttgart.

Gilbertson, R. L. 1980. Wood-rotting fungi of North America. Mycologia, 72: 1-49.

Ginns, J. 1974. *Schizophyllum commune*. Fungi Canadenses, 42.

Gorovoy, L. F. 1989. Morphogenesis in the Gilled Fungi. Naukova Dumka Press, Kiev (in Russian).

Hallenberg, N. and Parmasto, E. 1998. Phylogenetic studies in species of Corticiaceae growing on branches. Mycologia, 90: 640-654.

Hansen, L. and Knudsen, H. (eds.). 1992. Nordic Macromycetes, Vol. 2. Polyporales, Boletales, Agaricales, Russulales. Copenhagen, Denmark.

Hansen L. and Knudsen, H. (eds.) 1997. Nordic Macromycetes, Vol. 3. Heterobasidioid, Aphyllophoroid and Gastromycetoid Basidiomycetes. Copenhagen, Denmark.

Heim, R. 1971. The interrelationships between the Agaricales and Gasteromycetes, pp. 505-534. In: Evolution in the Higher Basidiomycetes. R. H. Petersen (ed.). Tennessee Press, Knoxville, USA.

Herter, W. 1910. Pilze. 2. Autobasidiomycetes. In: Kryptogamenflora der Mark Brandenburg, bd 6. S. 1-192.

Hibbett, D. S. and Binder, M. 2002. Evolution of complex fruiting-body morphologies in Homobasidiomycetes. Proc. Roy. Soc. London, 269: 1963-1969.

Hibbett, D. S. and Donoghue, M. J. 1995. Progress toward a phylogenetic classification of the Polyporaceae through parsimony analysis of mitochondrial ribosomal DNA sequences. Can. J. Bot., 73: suppl. 1: sect. E–H, 853-861.

Hibbett, D. S. and Donoghue, M. J. 2001. Analysis of character correlations among wood decay mechanisms, mating systems, and substrate ranges in homobasidiomycetes. Systematic Biology, 50: 215-242.

Hibbett, D. S. and Thorn, R. G. 2001. Basidiomycota: Homobasidiomycetes, pp. 121-168. In: The Mycota VII, Part B, Systematics and Evolution. D. J. McLaughlin, E. G. McLaughlin and P. A. Lemke (eds.). Springer-Verlag, Berlin, Heidelberg, N.Y.

Hibbett, D. S., Binder, M., Bischoff, J. F., Blackwell, M., Cannon, P. F., Eriksson, O. E., Huhndorf, S., James, T., Kirk, P. M., Lücking, R., Lumbsch, T., Lutzoni, F., Matheny, P. B., Mclaughlin, D. J., Powell, M. J., Redhead, S., Schoch, C. L., Spatafora, J. W., Stalpers, J. A., Vilgalys, R., Aime, M. C., Aptroot, A., Bauer, R., Begerow, D., Benny, G. L., Castlebury, L. A., Crous, P. W., Dai, Y.-C., Gams, W., Geiser, D. M., Griffith, G. W., Gueidan, C., Hawksworth, D. L., Hestmark, G., Hosaka, K., Humber, R. A., Hyde, K., Ironside, J. E., Kõljalg, U., Kurtzman, C. P., Larsson, K.-H., Lichtwardt, R., Longcore, J., Miądlikowska, J., Miller, A., Moncalvo, J.-M., Mozley-Standridge, S., Oberwinkler, F., Parmasto, E., Reeb, V., Rogers, J. D., Roux, C., Ryvarden, L., Sampaio, J. P., Schüßler, A., Sugiyama, J., Thorn, R. G., Tibell, L., Untereiner, W. A., Walker, C.,

Wang, Z., Weir, A., Weiß, M., White, M. M., Winka, K., Yao, Y.-J. and Zhang, N. 2007. A higher-level phylogenetic classification of the Fungi. Mycological Research, 111: 509-547.

Hibbett, D. S., Gilbert, L.-B. and Donoghue, M. J. 2000. Evolutionary instability of ectomycorrhizal symbioses in basidiomycetes. Nature, 407: 506-508.

Hibbett, D. S., Pine, E. M., Langer, E., Langer, G. and Donoghue, M. J. 1997. Evolution of gilled mushrooms and puffballs inferred from ribosomal DNA sequences. Proc. Natl. Acad. Sci., USA., 94: 12002-12006.

Hillism, D. 1987. Molecular versus morphological approaches in systematics. Ann. Rev. Ecol. Syst., 18: 23-42.

Hsiaum, P. T. W. and Harrington, T. C. 2003. Phylogenetics and adaptations of basidiomycetous fungi fed upon by bark beetles (Coleoptera: Scolytidae). Symbiosis, 34: 111-131.

Hyde, S. M. and Wood, P. M. 1997. A mechanism for production of hydroxyl radicals by brown rot fungus *Coniophora puteana*: Fe (III) reduction by cellobiose dehydrogenase and Fe (II) oxidation at a distance from the hyphae. Microbiol., 143: 259-266.

Jarosch, M. and Besl, H. 2001. *Leucogyrophana*, a polyphyletic genus of the order Boletales (Basidiomycetes). Plant Biology, 3: 443-448.

Jones, D. R. 1973. Ultrastructure of septal pore in *Uromyces dianthi*. Trans. Brit. Mycol. Soc., 60: 227-235.

Juel, H. O. 1898. Die Kernteilung in den Basidien und die Phylogenie der Basidiomyceten. Jahrb. Wiss. Bot., 32: 361-388.

Jülich, W. 1972. Monographe der Athelieae (Corticiaceae, Basidiomycetes). Willdenowia Bh., 7: 1-283.

Jülich, W. 1974. The genera of the Hyphodermoideae (Corticiaceae). Persoonia, 8: 59-97.

Jülich, W. 1982. Higher Taxa of Basidiomycetes. Bibl. Mycol., 85: 1-485.

Keller, J. 1997. Atlas des Basidiomycetes vus aux microscopes électroniques. Union des Societies Suisses de Mycologie, Neuchâtel.

Khan, S. R., Kimbrough, J. W. and Mims, C. W. 1981. Septal ultrastructure and the taxonomy of *Exobasidium*. Can. J. Bot., 59: 2450-2457.

Kim, S. Y. and Jung, H. S. 2000. Phylogenetic relationships of the *Aphyllophorales* inferred from sequence analysis of nuclear small subunit ribosomal DANN. J. Microbiol., 38: 122-131.

Kirk, P. M., Cannon, P. F., David, J. C. and Stalpers, J. A. (eds.) 2001. Dictionary of the Fungi, 9th edn., CABI, Wallingford, UK.

Kluge, A. 1983. Cladistics and the classification of the great apes, pp. 151-177. In: New Interpretations of Ape and Human Ancestry. R. Ciochon and R. Corrucini (eds.). Plenum Press, New York.

Koker de, T., Nakasone, K. K., Haarhof, J., Burdsall, H. H. Jr. and Janse, B. J. H. 2003. Phylogenetic relationships of the genus *Phanerochaete* inferred from the internal transcribed spacer region. Mycol. Res., 107: 1032-1040.

Kõljalg, U. and Renvall, P. 2000. *Hydnellum gracilipes* – a link between stipitate and resupinate Hymenomycetes. Karstenia, 40: 71-77.

Kottke, I., Beiter, A., Weiss, M., Haug, I., Oberwinkler, F. and Nebel, M. 2003. Heterobasidiomycetes form symbiotic associations with hepatics: Jungermanniales have sebacinoid mycobionts while *Aneura pinguis* (Metzgeriales) is associated with a *Tulasnella* species. Mycol. Res., 107: 957-968.

Kovalenko, A. E. 1984. The modern views on phylogenetic relationships and systematics of agaricoid fungi, pp. 118-136. In: Evolution and Systematics of the Fungi. N.A. Naumov (ed.). Leningrad (in Russian).

Kovalenko, A. E. 1989. Definitorium Fungorum URSS. Ordo Hygrophorales. Nauka, Leningrad (in Russian).

Krassilov, V. A. 2004. Macroevolutionary events and the origin of higher taxa, pp. 265-289. In: Evolutionary Theory and Processes: Modern Horizons. S. P. Wasser (ed.). Kluwer Academic Publishers, Dordrecht, Boston.

Kreisel, H. 1969. Grundzüge eines Natürlichen Systems der Pilze. G. Fischer Verlag, Jena. 245 S.

Krüger, D. and Gargas, A. 2004. The basidiomycete genus *Polyporus* - an emendation based on phylogeny and putative secondary structure of ribosomal RNA molecules. Feddes Repertorium, 115: 530-546.

Kühner, R. 1980. Hymenomycetes agaricoïdes, étude générale et classification. Bull. Soc. Linn. Lyon, 49: (No spec.), 1-1025.

Langer, E. 1998. Evolution of *Hyphodontia* (Corticiaceae, Basidiomycetes) and related Aphyllophorales inferred from ribosomal DNA sequences. Folia Cryptog Estonica, 33: 57-62.

Langer, E. 2002. Phylogeny of non-gilled and gilled Basidiomycetes: DNA sequence inference, ultrastructure and comparative morphology. Tübingen.

Langer, G. 1994. Die Gattung Botryobasidium Donk (Corticiaceae, Basidiomycetes). Bibl. Mycol., 159: 1-459.

Larsen, M. J. and Burdsall, H. H. 1976. A consideration of the term gloeocystidium. Mem. N.Y. Bot. Gard., 28: 123-130.

Larsson, E. and Larsson, K.-H. 2003. Phylogenetic relationships of russuloid Basidiomycetes with emphasison aphyllophoralean taxa. Mycologia, 95: 1037-1065.

Larsson, K.-H. 2007. Molecular phylogeny of *Hyphoderma* and the reinstatement of *Peniophorella*. Mycol. Res., 111: 186-195.

Larsson, K.-H., Larsson, E. and Kõljalg, U. 2002. Evolution of crust-like Homobasidiomycetes inferred from nuclear ribosomal DNA sequences. A manuscript.

Larsson, K.-H., Larsson, E. and Kõljalg, U. 2004. High phylogenetic diversity among corticioid Homobasidiomycetes. Mycol. Res., 108: 983-1002.

Lavin, M. and Doyle, J. 1991. Tribal relationships of Sphinctospermum (Leguminosae): integration of traditional and chloroplast DNA data. Syst. Bot., 16: 162-172.

Lawrey, J. D., Binder, M., Diederich, P., Molina, M. C., Sikaroodi, M. and Erttz, D. 2007. Phylogenetic diversity of lichen-associated Homobasidiomycetes. Molec. Phylog. Evol., 44: 778-789.

Lee, S.-S. and Jung, H. S. 1997. Phylogenetic analysis of the *Corticiaceae* based on gene sequences of nuclear 18S ribosomal DNAs. J. Microbiol., 35: 253-258.

Lemke, P. A. 1964. The genus *Aleurodiscus (sensu lato)* in North America. Can. J. Bot., 42: 723-768.

Lim, Y. W. 2001. Systematic Study of Corticioid Fungi Based on Molecular Sequence Analyses. Ph.D. Thesis, National University, Seoul.

Linder, D. H. 1940. Evolution of the Basidiomycetes and its relation to the terminology of the basidium. Mycologia, 32: 419-447.

Littlefield, L. J. and Heath, M. C. 1979. Ultrastructure of Rust Fungi. Academic Press, New York.

Locquin, M. 1984. Mycologie Générale et Structurale. Masson, Paris.

Locquin, M. V. 1981. Entaxie, taxotropie, néguentropie, valeur et qualité en taxonomie géneralisée. Ark'All Com., 2: 45-86.

Lutzoni, F., Kauff, F., Cox, C. J., McLaughlin, D., Celio, G., Dentinger, B., Padamsee, M., Hibbett, D., James, T. Y., Baloch, E., Grube, M., Reeb, V., Hofstetter, V., Schoch, C., Arnold, A. E., Miadlikowska, J., Spatafora, J., Johnson, D., Hambleton, S., Crockett, M., Shoemaker, R., Sung, G.-H., Lücking, R., Lumbsch, Th., O'Donnell, K., Binder, M., Diederich, P., Ertz, D., Gueidan, C., Hansen, K., Harris, R. C., Hosaka, K., Lim, Y.-W., Matheny, B., Nishida, H., Pfister, D., Rogers, J., Rossman, A., Schmitt, I., Sipman, H., Stone, J., Sugiyama, J., Yahr, R. and Vilgalys, R. 2004. Assembling the fungal tree of life: progress, classification, and evolution of subcellular traits. Am. J. Bot., 91: 1446-1480.

Malysheva, V. 2009. Heterobasidiomyctes, its position in the system of Basidiomycetes and the problems of taxonomy. Mikol. Fitopatol., 43: 20-32 (in Russian).

Martin, G.W. 1949. The genus *Ceracea* Cragin. Mycologia, 41: 77-86.

Matheny, P. B., Wang, Z., Binder, M., Curtis, J. M., Lim, Y. W., Nilsson, R. H., Hughes, K. W. and Hofstetter, V. 2007. Contributions of rpb2 and tefl to the phylogeny of mushrooms and allies (Basidiomycota, Fungi). Molec. Phylog. Evol., 43: 430-451.

Miller, O. K. Jr. and Watling, R. 1987. Whence cometh the agarics? A reappraisal, pp. 435-448. In: Evolutionary Biology of the Fungi. A. D. M. Rayner, C. M. Brasier and D. Moore (eds.). Cambridge Univ. Press, Cambridge.

Miyamoto, M. 1983. Frogs of the *Eleutherodactylus rugulosus* group: a clasdistic study of allozyme, morphological, and caryological data. Syst. Zool., 32: 109-124.

Moncalvo, J. M., Lutzoni, F. M., Rehner, S. A., Johnson, J. and Vilgalys, R. 2000. Phylogenetic relationships of agaric fungi based on nuclear large subunit ribosomal DNA sequences. Syst. Biol., 49: 278-305.

Moncalvo, J.-M., Vilgalys, R., Redhead, S. A., Johnson, J. E., James, T. Y., Aime, M. C., Hofstetter, V., Verduin, S. J. W., Baroni, T. J., Thorn, R. G., Jacobsson, S., Clémençon, H. and Miller, O. K. Jr. 2003. One hundred and seventeen clades of euagarics. Molecular Phylog. Evol., 23: 357-400.

Moore, R. T. 1971. An alternative concept of fungi based on their ultrastructure. Recent Adv. Microbiol., 10: 49-64.

Moore, R. T. 1996. An inventory of the phylum Ustomycota. Mycotaxon, 59: 1-31.

Moser, M. 1983. Die Röhrlinge and Blätterpilze (Polyporales, Boletales, Agaricales, Russulales). 5. Aufl. Gustav Fischer, Stuttgart; New York. 533 S. (Bd. Ii, b/2, Kleine Kryptogamenflora. Begründet von H. Gams).

Nobles, M. K. 1958. Cultural characters as a guide to the taxonomy and phylogeny of the Polyporaceae. Can. J. Bot., 36: 883-926.

Nobles, M. K. 1971. Cultural characters as a guide to the taxonomy of the Polyporaceae, pp. 169-195. In: Evolution in the Higher Basidiomycetes. R. H. Petersen (ed.). Tennessee Press, Knoxville, USA.

Nuss, I. 1980. Untersuchungen zur systematischen Stellung der Gattung *Polyporus*. Hoppea. 39: 127-198.

Oberwinkler, F. 1965. Primitive Basidiomyceten. Ann. Myc. Ser. II, 19: 17-72.

Oberwinkler, F. 1972. The relationships between the Tremellales and Aphyllophorales. Persoonia, 7: 1-16.

Oberwinkler, F. 1977. Das neue System der Basidiomyceten, pp. 59-105 In: Beiträge zur Biologie der niederen Pflanzen. W. Frey, H. Hurka and F. Oberwinkler (eds.). Fischer Verlag, Stuttgart, New York.

Oberwinkler, F. 1978. Was ist ein Basidiomycet? Z. Mykol., 44: 13-29.

Oberwinkler, F. 1982. The significance of the morphology of the basidium in the phylogeny of Basidiomycetes, pp. 9-35. In: Basidium and Basidiocarp. K. Wells and E. K. Wells (eds.). Springer-Verlag, New York, USA.

Oberwinkler, F., Bandoni, R. J., Bauer, R., Deml, G. and Kisimova-Horovitz, L. 1984. The life-history of *Christiansenia pallida*, a dimorphic, mycoparasitic heterobasidiomycete. Mycologia, 76: 9-22.

Parmasto, E. 1968. Conspectus Systematis Corticiacearum. Inst. Zool. Bot., Tartu.

Parmasto, E. 1969. *Paullicorticium curiosum* Parm. et Žukov sp. nov. and the phylogenetical development of the basidium of the corticiaceous fungi. Česká. Mykol., 23: 73-78.

Parmasto, E. 1986. On the origin of the Hymenomycetes (What are corticioid fungi?) Windahlia, 16: 3-19.

Parmasto, E. 1995. Corticioid fungi: a cladistic study of a paraphyletic group. Can. J. Bot., 73(suppl. 1): S843-S852.

Parmasto, E. and Hallenberg, N. 2000. A taxonomic study of phlebioid fungi. Nordic J. Bot., 20: 105-118.

Pegler, D. N. and Young, T. W. K. 1971. Basidiospore Morphology in the Agaricales. Beih Nova Hedwigia, 35: 1-210.

Pegler, D. N. and Young, T. W. K. 1972. Basidiospores form in the British species of *Crepidotus*. Kew Bull., 27: 311-323.

Pegler, D. N. and Young, T. W. K. 1973. Basidiospores form in the British species of *Ganoderma* Karst. Kew Bull., 28: 351-364.

Pegler, D. N. and Young, T. W. K. 1981. A natural arrangement of the Boletales, with reference to spore morphology. Trans. Brit. Mycol. Soc., 76: 103-146.

Petersen, R. H. 1971. Interfamilial relationships in the clavarioid and cantharelloid fungi, pp. 345-371. In: Evolution in the Higher Basidiomycetes. R. H. Petersen (ed.). Tennessee Press, Knoxville, USA.

Petersen, R. H. 1974. The rust fungus life cycle. Bot. Rev., 40: 453-513.

Raitviir, A. G. 1964. Review of Heterobasidial Fungi of USSR in Connection With Some Problems of Phylogeny and Mycogeography. Ph.D. Thesis. Tartu. (in Russian).

Redhead, S. A. 1987. The Xerulaceae (Basidiomycetes), a family with sarcodimitic tissues. Can. J. Bot., 65: 1551-1562.

Redhead, S. A., Lutzoni, F., Moucalvo, J.-M. and Vilgalys, R. 2002a. Phylogeny of Agarics: Partial systematics solutions for core omphalinoid genera in the Agaricales (euagarics). Mycotaxon, 83: 19-57.

Redhead, S. A., Moncalvo, J.-M., Vilgalys, R. and Lutzoni, F. 2002b. Phylogeny of agarics: partial systematics solutions for bryophilous omphalinoid agarics outside of the Agaricales. Mycotaxon, 82: 151-168.

Roberts, P. 1998. *Oliveonia* and the origin of the Holobasidiomycetes. Folia Crypt. Estonica, 33: 127-132.

Rogers, D. P. 1934. The Basidium. Univ. Iowa Studies Nat. Hist., 16: 160-183.

Ryvarden, L. 1991. Genera of polypores. Nomenclature and taxonomy. Fungiflora, Oslo.

Schantser, I. A. 2003. About Discrepancy Between "Phylogeny" of Molecules and True Phylogeny. Proceedings of the XIth International Symposium on Plant Phylogeny, Moscow, Russia, 109-111.

Singer, R. 1945. The *Laschia*-complex (Basidiomycetes). Lloydia, 8(3): 170-230.

Singer, R. 1962. The Agaricales in Modern Taxonomy. 2nd edn. J. Cramer, Weinheim.

Singer, R. 1971. A revision of the genus *Melanomphalia* as a basis of the phylogeny of the Crepidotaceae, pp. 475-480. In: Evolution in the Higher Basidiomycetes. R. H. Petersen (ed.). Tennessee Press, Knoxville, USA.

Singer, R. 1975. The Agaricales in Modern Taxonomy, 3rd edn. J. Cramer, Vaduz.

Singer, R. 1986. The Agaricales in Modern Taxonomy. 4th edn. Koeltz Scientific Books, Koenigstein, USA.

Singer, R. and Smith, A. H. 1960. Studies on secotiaceous fungi VII. *Secotium* and *Neosecotium*. Madroño, 15: 152-158.

Spirin, W. A., Zmitrovich, I. V. and Malysheva, V. F. 2005. Notes on Perenniporiaceae. All-Russian Institute of Plant Protection. St. Petersburg.

Stalpers, J. A. 1988. *Auriculariopsis* and the Schizophyllales. Persoonia, 13: 495-504.

Stalpers, J. A. 1992. *Albatrellus* and Hericiaceae. Persoonia, 14: 537-541.

Stalpers, J. A. 1996. The aphyllophoraceous fungi – II. Keys to the species of the Hericiales. Stud. Mycol., 40: 1-185.

Stankovicova, L. 1973. Hyphal structure in some pleurotoid species of Agaricales. Nova Hedwigia, 24: 61-85.

Swann, E. C. and Taylor, J. W. 1993. Higher taxa of Basidiomycetes: An 18S rRNA gene perspective. Mycologia, 85: 923-936.

Swann, E. C. and Taylor, J. W. 1995. Phylogenetic perspectives on Basidiomycetes systematics: Evidence from the 18S rRNA gene. Can. J. Bot., 73: S862-S868.

Taylor, D. L., Bruns, T. D., Szaro, T. M. and Hodges, S. A. 2003. Divergence in mycorrhizal specialization within Hexalectris spicata (Orchidaceae), a nonphotosynthetic desert orchid. Amer. J. Bot., 90: 1168-1179.

Thiers, H. D. 1971. Some ideas concerning the phylogeny and evolution of the boletes, pp. 423-436. In: Evolution in the Higher Basidiomycetes. R. H. Petersen (ed.). Tennessee Press, Knoxville, USA.

Thiers, H. D. 1984. The secotioid syndrome. Mycologia, 76: 1-8.

Vasilyeva, L. 1999. Systematics in Mycology. Bibliotheca Mycologica, Bd. 178. J. Cramer in der Gebr. Borntraeger Verlagsbuchhandlung, Berlin, Stuttgart.

Wang, Z., Binder, M., Dai, Y.-C. and Hibbett, D. S. 2004. Phylogenetic relationships of Sparassis inferred from nuclear and mitochondrial ribosomal DNA and RNA polymerase sequences. Mycologia, 96: 1015-1029.

Wasser, S. P. 1974. Problems of Systematics of Secotiaceous Mushrooms, pp. 62-65. In: Progress of Mycology and Lichenology in the Soviet Baltic Republics. Estonian Academic Science Press, Tartu.

Wasser, S. P. 1980. Flora Fungorum RSS Ucrainicae: Agaricaceae Cohn. Naukova Dumka Press, Kiew (in Russian).

Wasser, S. P. 1982. Family Agaricaceae (Fr.) Cohn of the Soviet Union (morphology, anatomy, ecology, geography, origin, evolution, and practical application). Dr. Sci. (Biol.) Thesis, Ukr. Acad. Sci., Kiev.

Wasser, S. P. 1985. Family Agaricaceae of the USSR. Naukova Dumka, Press Kiev. (in Russian).

Wasser, S. P. 2002. Family Agaricaceae (Fr.) Cohn (Basidiomycetes) of Israel Mycobiota. I. Tribe Agariceae Pat. E. Nevo and A. Volz (eds.). A. R. A. Gantner Verlag K.-G., Ruggell/Liechtenstein.

Wasser, S. P., Zmitrovich, I. V., Didukh, M. Ya., Spirin, W. A. and Malysheva, V. F. 2006. Morphological traits of *Ganoderma lucidum* complex Highlighting *G. tsugae* var. *jannieae*: The Current Generalization. Ruggell: A. R. A. Gantner Verlag K.-G.

Watling, R. and Largent, D. 1977. Macro- and microscopic analysis of the cortical zones of basidiocarps of selected agaric families. Nova Hedwigia, 28: 569-617.

Weiss, M. and Oberwinkler, F. 2001. Phylogenetic relationships in Auriculariales and related groups—hypotheses derived from nuclear ribosomal DNA sequences. Mycol. Res., 105: 403-415.

Weiss, M., Bauer, R. and Begerow, D. 2004. Spotlights on Heterobasidiomycetes, pp. 7-48. In: Frontiers in Basidiomycote Mycology. R. Agerer, M. Piepenbring, and P. Blanz (eds.). IHW-Verlag Eching, Munich.

Wells, K. 1965. Ultrastructural features of developing and mature basidia and basidiospores of *Schizophyllum commune*. Mycologia, 57: 236-261.

Wells, K. 1994. Jelly fungi, then and now! Mycologia, 86: 18-48.

Wu, S. H., Hibbett, D. S. and Binder, M. 2001. Phylogenetic analyses of *Aleurodiscus* s.l. and allied genera. Mycologia, 93: 720-731.

Yli-Mattila, T. 1990. Photobiology of Fruit Body Formation in the Basidiomycete *Schizophyllum commune*. Reps. Dep. Biol. Univ. Turku. No 27.

Yoon, S. I., Kim, S. E., Lim, Y. W. and Jung, H. S. 2003. Phylogenetic evaluation of stereoid fungi. J. Microbiol. Biotechnol., 13: 406-414.

Zerov, D. K. 1972. Outlines of Non-Vascular Plant Phylogeny. Naukova Dumka Press, Kiev (in Russian).

Zerova, M. Ya. and Wasser, S. P. 1974. Current problems of taxonomy of Agricales s.l, pp. 22-24. In: Progress of Mycology and Lichenology in the Soviet Baltic Republics. Estonian Acad. Sci. Press, Tartu.

Zerova, M. Ya., Sosin, P. E. and Rozhenko, G. I. 1979. Identification key of the fungi of Ukraine, Vol. 5: Basidiomycetes. Book 2. Naukova Dumka Press, Kiev (in Ukranian).

Zmitrovich, I. V. 1997. Corticioid fungi: modern systematics and problems of phylogenetics. Mikol. Fitopatol., 31: 79-91 (in Russian).

Zmitrovich, I. V. 2001. On the systematics of genus *Serpula* s. lato. Nov. Syst. P1. non-Vasc., 35: 70-89.

Zmitrovich, I. V. 2003. A revised eukaryote tree: the case for a euglenozoan root. Intern. J. Algae, 5: 1-38.

Zmitrovich, I. V. 2007. Modes of growth organization in corticioid basidiocarps. XV Congress of European Mycologists. Saint Petersburg, Russia, September 16–21, pp. 180-181 (Abstract).

Zmitrovich, I. V. 2008. Definitorium fungorum Rossicum. Aphyllophorales. 3. Familia Atheliaceae et Amylocorticiaceae. St. Petersburg, KMK. (in Russian).

Zmitrovich, I. V. 2010. The taxonomical and nomenclatural situation around medicinal mushrooms-containing genera of Polyporaceae (with possible solutions). Int. J. Medicinal Mushrooms (in press).

Zmitrovich, I. V. and Spirin, W. A. 2002. On the taxonomy of corticioid fungi. II. The genera *Serpula* (Pers.) Gray, *Serpulomyces* gen. nov., *Amylocorticiellum* gen. nov. Mikol. Fitopatol., 36: 11-26 (in Russian).

Zmitrovich, I. V. and Wasser, S. P. 2004. Modern view on the origin and phylogenetics reconstruction of Homobasidiomycetes fungi, pp. 230-263. In: Evolutionary Theory and Processes: Modern Horizons. S. P. Wasser. (ed.). Kluwer Academic Publishers, Dordrecht, Boston.

Zmitrovich, I. V., Malysheva, V. F. and Malysheva, E. F. 2009. Hyphal systems of polyporoid and pleurotoid fungi. 1. Types of hyphae. Ukr. Bot. J., 66: 71-87 (in Russian).

Zmitrovich, I. V., Malysheva, V. F. and Spirin, W. A. 2006. A new morphological arrangement of the Polyporales. I. Phanerochaetineae. Mycena, 6: 4-56.

Chapter 9

Phylogenetic Analyses of *Phellinus s. l.* and *Inonotus s. l.* (Hymenochaetales) Inferred from rDNA ITS Sequences and Morphological Data

Daniel Tura[1,*], Ivan V. Zmitrovich[2,], Solomon P. Wasser[1,3], Dina Raats[1] and Eviatar Nevo[1]**

[1]Department of Evolutionary and Environmental Biology, Institute of Evolution Faculty of Natural Sciences, University of Haifa, Mt. Carmel Haifa 31905, Israel

[2]Lab. of the Systematics and Geography of the Fungi, Komarov Botanical Institute, Russian Academy of Sciences, 2 Prof. Popova St. Petersburg 197376, Russia

[3]N.G. Kholodny Institute of Botany, National Academy of Sciences of Ukraine 2 Tereshchenkivska, Kiev 01001, Ukraine

E-mail: *turadaniel21@yahoo.com, **iv_zmitrovich@mail.ru

ABSTRACT

The phylogenetic status of several taxonomically important wood-decaying species belonging to *Phellinus s. l.* and *Inonotus s. l.* was investigated by analyzing sequence data from ribosomal DNA ITS regions of 26 species. As expected, an evaluation of the results confirmed *Phellinus s. l.* and *Inonotus s. l.* as being polyphyletic. The generic concept of *Inonotus* s. str. was revised from a molecular and morphological point of view. Further analyses are still needed for a more comprehensive delimitation between *Phellinus s. l.* and *Inonotus s. l.* A comparison of results found in several publications might elucidate the phylogenetic identity of taxonomically problematic taxa belonging to these groups. Two new combinations in the genus *Xanthoporia* Murrill were made.

INTRODUCTION

The genera *Phellinus* Quél. and *Inonotus* P. Karst., as other members of Hymenochaetales (Agaricomycotina), comprise wood-inhabiting saprotrophic and/or parasitic species characterized by permanently darkening tissue in KOH (xanthochroic reaction), the absence of clamps, presence of a continuous parenthesome in the dolipore apparatus, frequent occurrence of setae, and the production of white rot wood (Parmasto and Parmasto, 1979; Moore, 1980, 1985; Corner, 1991; Müller et al., 2000; Wagner and Fischer, 2001, 2002; Fischer and Binder, 2004; Larsson et al., 2007; Spirin et al., 2007).

All recent molecular analyses demonstrated that certain species of the former *Corticiaceae* (*Basidioradulum*, *Hyphodontia*), *Polyporaceae* (*Oxyporus*, *Schizopora*, and *Trichaptum*), *Stereaceae* (*Cotylidia*), and *Agaricaceae* (*Cantharellopsis*, *Omphalina*, and *Rickenella*) are related to Hymenochaetales (Hibbett and Donoghue, 1995, Ko et al., 1997; Langer, 1998; Jeong, 1999; Hibbett et al., 2000; Paulus et al., 2000; Hibbett and Thorn, 2001; Larsson, 2001; Larsson et al., 2007). Although many members of these groups exhibit a different morphology than Hymenochaetales, they are also related by the presence of an imperforate parenthesome.

The genus *Phellinus s. l.* is the largest group of species amongst the Hymenochaetales, with 150–180 species known worldwide, while the number of species in *Inonotus s. l.* ranges from 58–100 (Larsen and Cobb-Poulle, 1990; Spirin et al. 2007; Kirk et al., 2008). Both "genera" occur on living or dead hardwoods and conifers. Basidiocarps in *Phellinus* are perennial, hard-woody in consistency, often covered by a thick cracked crust, and with a tubular distinctly/indistinctly stratified hymenophore. The hyphal system is pseudodimitic to dimitic. The setae are abundant to rare or absent, while the setal hyphae are absent in all representatives. The hymenial cystidia are sometimes present, and the basidiospores range in shape from ellipsoid to nearly globose and are thick-walled, hyaline/pigmented, cyanophilous/acyanophilous, and dextrinoid/indextrinoid.

The genus *Inonotus s. l.* is characterized by annual soft-suberose basidiocarps, a monomitic hyphal system, setae with frequent occurrence in some species or absence in others, and the presence of setal hyphae in some species. Basidiospores are broadly ellipsoid to cylindric, thin- to slightly thick-walled, hyaline/pigmented, acyanophylous and Melzer's negative.

Recent studies have revealed that the traditional generic concept of *Phellinus s. l.* and *Inonotus s. l.* based on morphological features only (annual or perennial basidiocarp; monomitic or dimitic hyphal system) does not follow the evolutionary history of the species (Fiasson and Niemelä, 1984; Fischer, 1996; Hansen and Knudsen, 1997; Jeong, 1999; Niemelä et al., 2001; Wagner and Fischer, 2001, 2002; Rizzo et al., 2003; Jeong et al., 2005; Larsson et al., 2007; Spirin et al., 2007). As a consequence, the two groups were recognized as being polyphyletic and were subdivided into several smaller subunits, each one of which was considered as a distinct genus. Wagner and Fischer (2002) revised the classification of *Phellinus s. l.* and *Inonotus s. l.*, by using boreal and subtropical fungal strains, and reconfirmed both *Phellinus* and *Inonotus* as polyphyletic.

These groups were subdivided into the genera *Phellinus* s. str. [type species: *Ph. igniarius* (L.) Quél.], *Porodaedalea* Murrill [type *P. pini* (Brot.) Murrill], *Fomitiporia* Murrill (type *F. langloisii* Murrill), *Fomitiporella* Murrill [type *F. umbrinella* (Bres.) Murrill], *Fulvifomes* Murrill [type *F. robiniae* (Murrill) Murrill], *Fuscoporia* Murrill [type *F. ferruginosa* (Schrad.) Murrill], *Phellinidium* (Kotl.) Fiasson & Niemelä [type *Ph. ferrugineofuscum* (P. Karst.) Fiasson & Niemelä], *Phellopilus* Niemelä, T. Wagner & M. Fisch. [type *Ph. nigrolimitatus* (Romell) Niemelä, T. Wagner & M. Fisch.], *Aurificaria* D.A. Reid [type *A. indica* (Massee) D.A. Reid], *Onnia* P. Karst. [type *O. circinata* (Fr.) P. Karst.], and *Phylloporia* Murrill (type *Ph. parasitica* Murrill). The latter group was subdivided into the genera *Inonotus* s. str. [type *I. hispidus* (Bull.) P. Karst.], *Inocutis* Fiasson & Niemelä [type *I. rheades* (Pers.) Fiasson & Niemelä], *Mensularia* Lázaro Ibiza [type *M. radiata* (Sowerby) Lázaro Ibiza], *Pseudoinonotus* T. Wagner & M. Fisch. [type *P. dryadeus* (Pers.) T. Wagner & M. Fisch.], *Inonotopsis* Parmasto [type *I. subiculosa* (Peck) Parmasto], *Coltricia* Gray [type *C. connata* Gray], and *Coltriciella* Murrill [type *C. dependens* (Berk. & M.A. Curtis) Murrill]. However, the inclusion of more tropical and subtropical taxa is necessary in phylogenetic representations for a more natural classification of *Phellinus s. l.* and *Inonotus s. l.* (Fischer, 1996; Wagner and Fischer, 2001; Jeong et al., 2005).

The main problem in the classification of the *Inonotus–Phellinus* conglomerate is that morphological characters were "weighed" differently by various authors, whereas characteristic combinations in this study are distributed gradually from unit to unit. Concerning the hyphal system and fruitbody consistency as border markers between *Phellinus s. l.* and *Inonotus s. l.*, it is necessary to note that these characteristics are more complicated than previously considered. Many tropical "*Phellinus*" species are characterized by pseudodimitic hyphal construction, whereas cover elements of several "*Inonotus*" species are composed of setae, i.e., these species can be considered as setodimitic.

Recent molecular approaches have proved to be better tools for phylogenetic representation although the application of different methods may result into different conclusions. Mainly due to this, the taxonomic positions of some species are still unclear.

This chapter aims to re-evaluate the systematic status of some important taxa (Table 9.1) presented in some recent publications. Some original phylogenetic analyses were conducted by the authors by using nucleotide sequence data from the rDNA internal transcribed spacer (ITS1), 5.8S RNA gene and ITS2 region. Some of the chosen taxa names correspond to the current concepts of genera and have a well known position in taxonomy, whereas most of the included taxa were classified as taxonomically problematic.

Table 9.1 The taxonomic status of some critical *Phellinus* s. l. and *Inonotus* s. l. taxa according to recent publications

Species	*Authors*						
	Dai (1999)	*Jeong (1999)*	*Wagner & Fischer (2001)*	*Wagner & Fischer (2002)*	*Sutton et al. (2005)*	*Jeong et al. (2005)*	*Larsson et al. (2007)*
Inonotus andersonii	—	Not clear	—	*Inonotus* s. str.	—	Not clear	*Inonotus* s. str.
I. cuticularis	—	*Ph. rimosus*-gr.	*Inonotus* s.str.**	*Inonotus* s. str.**	*Inonotus* s. l.**	*Inocutis*	*Inonotus* s. str.
I. dryadeus	*Fomitiporia*	Not clear	*Pseudoinonotus**	*Pseudoinonotus**	—	*Pseudoinonotus**	*Pseudoinonotus*
I. levis	—	—	—	—	—	—	—
Phellinus baumii	*Fulvifomes*	*Ph. linteus*-gr.**	—	*Inonotus* s. str.**	Not clear	*Ph. linteus*-gr.**	*Inonotus* s. str.
Ph. linteus	—	*Ph. linteus*-gr.	—	*Inonotus* s. str.**	Not clear	*Ph. linteus*-gr.**	*Inonotus* s. str.
Ph. pachyphloeus	*Phellinidiopsis*	Not clear	—	*Inonotus* s. str.**	—	*Phellopilus* *	*Inonotus* s. str.
Ph. tropicalis	—	*Ph. linteus*-gr.	—	*Inonotus* s. str.	Not clear	*Ph. linteus*-gr.**	*Inonotus* s. str.
Ph. weirianus	—	*Ph. linteus*-gr.**	—	*Inonotus* s. str.**	—	*Ph. linteus*-gr.**	*Inonotus* s. str.
Ph. weirii	—	—	—	*Phellinidium* **	—	Not clear	*Phellinidium*

Note: '—' taxa not present in the given system or missing information; 'gr.' group; '**' high bootstrap support (above 80%); '*' low bootstrap support (under 50%).

MATERIALS AND METHODS

Fungal material and mycelial cultures. Fifteen taxa collected from various Israeli forests and other geographical regions of the world were deposited in the Culture Collection of the Institute of Evolution, University of Haifa, Israel (HAI) and used in this study. Together with 25 previously published sequences from GenBank (listed in Table 9.2), a total of 26 taxa were represented by 40 strains. Mycelial cultures were grown on malt-extract agar medium (15 g malt extract, 15 g agar in 1,000 ml dH_2O, pH 5.5) for several days at 27 °C in the dark and then transferred to liquid medium flasks containing 0.4 g NaH_2PO_4, 0.8 g KH_2PO_4, 0.5 g $MgSO_4$, 2.0 g NH_4NO_3, 2.0 g yeast extract, and 10 g glucose. After 7–14 days of growth, the mycelia were ground under liquid nitrogen and then processed for further analyses.

DNA isolation and PCR amplification. Total genomic rDNA was extracted from liquid-cultured mycelia with the DNeasy Plant Mini Kit (cat. No. 69104, Qiagen, Israel) according to the manufacturer's protocol. Quantity and quality of DNA samples were examined on 1% agarose gels containing the nucleic acid dye GelRed™ (Biotium, Israel), followed by DNA concentration measurements with a NanoDrop 3300 Spectrophotometer (NanoDrop Technologies, Israel).

The internal transcribed spacer (ITS) of the nuclear ribosomal DNA, including 5.8S RNA gene, was amplified using the primers ITS-1F (5'-CTTGGTCATTTAGAGGAAGTAA-3') and ITS4 (5'-TCCTCCGCTTATT GATATGC-3') under the following conditions: denaturation step for 3 min at 94°C, annealing step for 30 sec at 55°C, extension step for 1 min at 72°C, and with a final extension step of 10 min at 72°C. The cycle was repeated 35 times. The PCR was performed in 25 µl reaction tubes containing an optimal concentration of extracted DNA solution, 2.50 µl 10× PCR buffer, 0.50 µl 10 mM dNTPS mix, 1.00 µl 25 mM $MgCl_2$, 2.50 µl 5 uM ITS-1F and ITS4 primer mix, 0.25 µl Taq DNA polymerase, and variable amounts of dH_2O. The PCR master-mix preparation was performed with the Taq PCR core Kit (cat. No. 201223, Qiagen, Israel). The amplified PCR products were analyzed on agarose gels and then purified with the Qiaquick PCR purification Kit (cat. No. 28104, Qiagen, Israel) following the manufacturer's instructions.

Sequencing. Cycle sequencing reactions were conducted using the ABI PRISM BigDye Terminator Cycle Sequencing Ready Reaction Kit (Applied Biosystems, California) and analyzed on a 3130 xl Genetic Analyzer (Applied Biosystems). Approximately 650 b to 750 b of the ITS rDNA region were sequenced.

Alignment and phylogenetic analysis. The sequences resulting from this study were automatically aligned together with sequences obtained from GenBank database (NCBI) using the Clustal W package integrated in BIOEDIT sequence alignment editor (Hall, 1999). Manual adjustments were made for alignment optimization. Regions of ambiguous alignment were excluded from the analyses. The obtained sequences were deposited in GenBank database (GQ253451–GQ253465).

Table 9.2 List of analyzed species, sources, strains, and GenBank accession numbers of taxa used in this study

Species	*Sources*	*Isolate*	*GenBank accession no.*
Fomitiporia mediterranea M. Fisch.	Israel	HAI-1213	GQ253458
	Israel	HAI-0772	GQ253457
	Israel	HAI-1212	GQ253459
Inonotus andersonii (Ellis & Everh.) Černý	GenBank	SFCC-50025	AY558599
I. cuticularis (Bulliard) P. Karst.	GenBank	QFB-888	AF237730
I. dryadeus (Pers.) Murrill	GenBank	FP-105836-4	AM269808
I. dryophilus (Berk.) Murrill	Russia	HAI-1248	GQ253464
I. levis P. Karst.	Israel	HAI-0796	GQ253462
I. radiatus (Sowerby) P. Karst.	Germany	HAI-1247	GQ253460
I. rheades (Pers.) Bondartsev & Singer	Lithuania	HAI-1246	GQ253454
I. rickii (Pat.) D. A. Reid	Israel	HAI-1240	GQ253461
I. tamaricis (Pat.) Maire	Israel	HAI-1287	GQ253451
	Israel	HAI-1289	GQ253452
	Israel	HAI-1291	GQ253453
I. tropicalis (M.J. Larsen & Lombard) T. Wagner & M. Fisch.	GenBank	CBS-617.89	AF534077
	GenBank	UTHSC-02-617	AY641432
	GenBank	UAMH-10376	AY599487
I. obliquus (Ach. ex Pers.) Pilát	Ukraine	HAI-1253	GQ253463
	GenBank	IFO-8681	AY558593
I. pachyphloeus (Pat.) T. Wagner & M. Fisch.	GenBank	CBS-193.37	AY558635
Phellinus baumii Pilát	GenBank	SA-05	EF694975
	GenBank	MPNU-7006	AF200231
Ph. cancriformans (M. J. Larsen, Lombard & Aho) M. J. Larsen & Lombard	GenBank	IMSNU-32031	AF200242
Ph. chrysoloma (Fr.) Donk	GenBank	CFMR-5406	AY558612
Ph. igniarius (L.) Quél.	Turkey	HAI-0785	GQ253465
	Ukraine	HAI-0590	GQ253456
Ph. linteus (Berk. & M. A. Curtis) Teng	GenBank	MPNU-7002	AF200228
	GenBank	IFO-6980	AF200226

Table 9.2 Contd. ...

Ph. punctatus (Fr.) Pilát	GenBank	CBS-386.66	AY558640
Ph. ribis (Schumach.) Quél.	GenBank	IMSNU-32011	AF200237
	GenBank	CBS-579.50	AY558644
Ph. robustus (P. Karst.) Bourdot & Galzin	Belarus	HAI-0264	GQ253455
Ph. spiculosus (W.A. Campb. & R.W. Davidson) Niemelä	GenBank	KTCC-6658	AY558648
Ph. sulphurascens Pilát	GenBank	TYPE-11	EF527216
Ph. torulosus (Pers.) Bourdot & Galzin	GenBank	PT-6	EF068238
Ph. weirii (Murrill) Gilb.	GenBank	CNU-6017	AF251438
	GenBank	CFS-586	AY829342
Ph. weirianus (Bres.) Gilb.	GenBank	CBS-618.89	AY558654
Hyphodontia alutaria (Burt) J. Erikss.	GenBank	GEL-3183	DQ340318
H. pruni (Lasch) Svrček	GenBank	3007B	DQ873624

Phylogenetic trees were obtained by using parsimony and likelihood methods conducted with PAUP* version 4.0b10 (Swofford, 2003). In order to obtain the most parsimonious trees, the heuristic search strategy was chosen with tree bisection-reconnection (TBR), branch-swapping algorithm, and MulTrees option in effect with the zero length-branches option switched on. Characteristics were treated as unordered and equally weighted, and gaps were treated as missing data. Maximum parsimony (MP) was performed with the Maxtrees option set on; auto increase by 100, and the stepwise addition option was used to obtain start trees for branch-swapping. For maximum likelihood analyses (ML) the heuristic search strategy was used, as described by Harrison and Langdale (2006) with few modifications. Likelihood scores were calculated for 56 models of evolution by PAUP* 4.0b10.

The best-fit maximum likelihood model was estimated using MODELTEST 3.8 (Posada and Crandall, 1998; Posada, 2006), which is available at the following address: http://darwin.uvigo.es/software/modeltest_server.html.

The model selected (GTR+G+I) was produced by Akaike's (1974) information criterion (AIC). Based on this, a ML neighbor joining (NJ) tree was generated. To compare tree topologies of alternative phylogenetic hypotheses for the dataset, several constrained analyses were conducted.

The bootstrap analyses (Felsenstein, 1985) by 1000 replicates were performed to evaluate tree-branch robustness in MP strict-consensus and ML neighbor-joining trees. Trees were viewed using TreeView (Page, 1996).

RESULTS

Sequences of the ITS region resulting from our data ranged from 650 b to 750 b, while the length of selected GenBank DNA fragments ranged between 560 b and 870 b. The alignment of 40 sequences resulted in 902 characteristics with 129 parsimony uninformative variable characteristics and 215 constant characteristics, 558 of which were parsimony informative. Parsimony analyses yielded 2,635 equally parsimonious trees (l = 2640, CI = 0.537, RI = 0.725, RC = 0.390). Analyses of the first-derived strict consensus MP tree topology reflected the data used to generate it, indicating alignment data as having a sufficient phylogenetic signal.

The results of our phylogenetic reconstructions are shown in Figs. 9.1-9.3. Topologies of the MP strict-consensus, neighbor-joining, and ML neighbor-joining trees were closely similar under the aspect of taxa relationship, but somewhat different concerning the distribution of groupings within trees.

The strict-consensus MP tree topology depicted in Fig. 9.3 shows that *Phellinus s. l.* and *Inonotus s. l.* are polyphyletic; both genera are clearly separated into smaller groups supported by high bootstrap values.

The *Xanthoporia*-clade comprising *Inonotus andersonii* as a sister taxon to *I. radiatus* (bootstrap value 100%) is well represented as a monophyletic group. Although, *Porodaedalea* formed by *Phellinus cancriformans* and *Ph. chrysoloma* (bootstrap value 100%) exhibits poor morphological congruency to *Xanthoporia,* appears to be a sister group but with low support value in both NJ trees (Figs. 9.1, 9.2). *Xanthoporia* and *Porodaedalea* maintain morpho-anatomical and ecological differences, i.e., in substrate occurrence, fruitbody consistency, wall thickness of basidiospores, and reaction in Melzer's solution (Table 9.3).

Inonotus s. str. clade (bootstrap value 74%) seems to be a polyphyletic group that unites taxa from *Inonotus s. l.* and *Phellinus s. l.*, many of which show intermediate morphological features of the two genera. In all the three representations its place is well-defined. Within the *Inonotus*-clade, two subclades are distinguishable as follows: *Inonotus cuticularis*-subclade (bootstrap value 96%) composed of *I. cuticularis, I. rickii, I. levis,* and *I. obliquus* having *I. tropicalis* as a sister taxon (bootstrap value 100%) and *Phellinus linteus*-subclade (bootstrap value 100%) including three morphologically similar species, *Ph. linteus, Ph. baumii,* and *Ph. weirianus.*

Inocutis-clade (bootstrap 100%) consisting of *Inonotus dryophilus, I. rheades,* and *I. tamaricis* formed a monophyletic unit. The result is also supported by the similar morphological characteristics shared by these species. *Inocutis* in our phylogenetic representations has a stable position because it is closely related to *Inonotus* s. str.

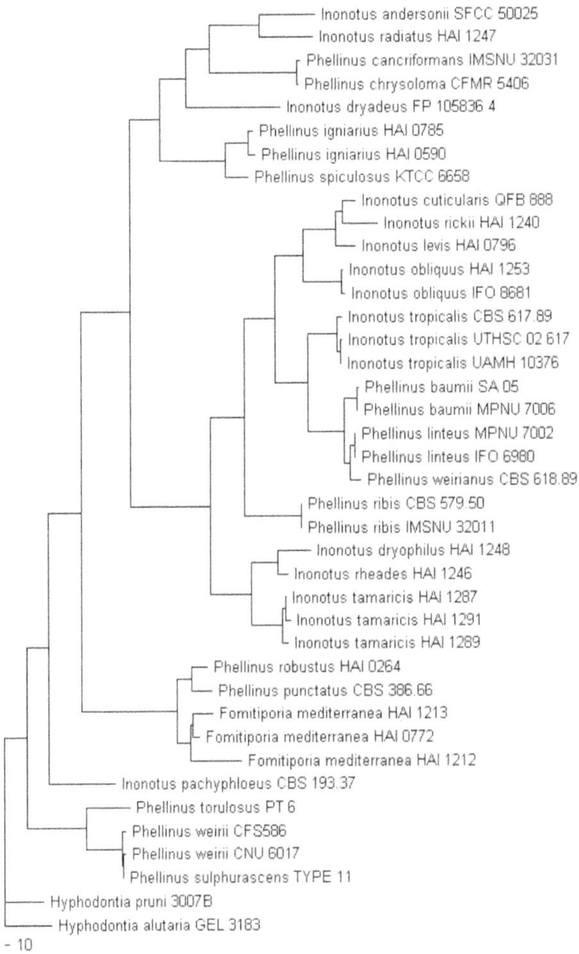

Fig. 9.1 Neighbor-joining maximum-parsimony tree derived from ITS rDNA sequence dataset.

The inclusion of *Inonotus dryadeus* within our analyses resulted in a distinct lineage in the strict consensus tree placed between *Inocutis* and *Porodaedalea*. Its position in our representations showed this species to have affinity towards *Porodaedalea* and *Phellinus* s. str. (Figs. 9.2, 9.3), however, the species currently belongs to *Pseudoinonotus*.

The *Phellinus* s. str.-clade (bootstrap 100%) comprising *Ph. spiculosus* and *Ph. igniarius* is linked as a sister group to *Fomitiporia*-clade (bootstrap 100%) including *Phellinus robusutus*, *Ph. punctatus*, and *Fomitiporia mediterranea*. The strong relationship between the two groups is well supported also by their morphological traits.

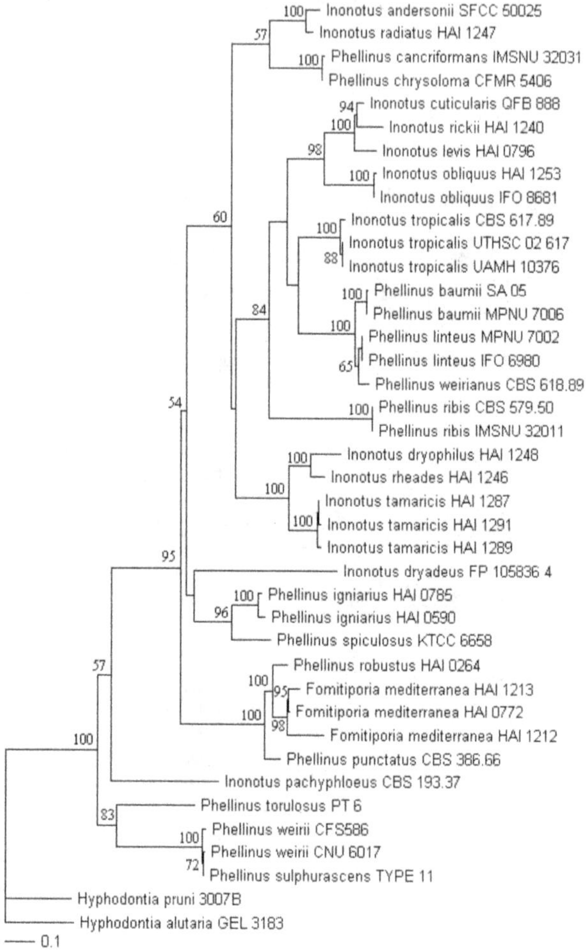

Fig. 9.2 Neighbor-joining maximum-likelihood tree derived from ITS rDNA sequence dataset. Bootstrap values above 50%, based on 1000 replicates, are indicated next to branches.

The *Fomitiporia*-group appeared to be a sister clade to *Phylloporia* represented by *Phellinus ribis.* In both NJ trees *Ph. ribis* clustered as a basal independent lineage as a sister taxon to the *Inonotus* s. str. group with low branch support. Morphologically, the species position is confirmed due to intermediate features between *Inonotus* and *Phellinus,* however, in the strict-consensus tree, it appeared to be placed next to *Fomitiporia.*

Inonotus pachyphloeus and *Phellinus torulosus* formed two distinct lineages closely related to *Phellinidium* in the strict-consensus tree. *Ph. torulosus,* in

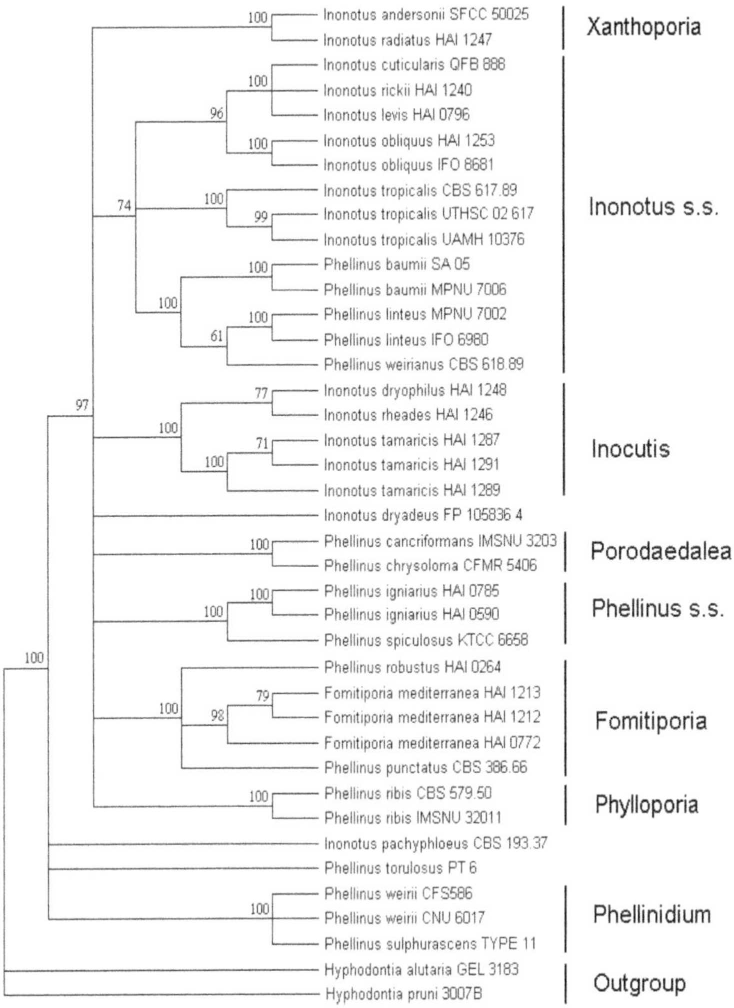

Fig. 9.3 Strict-consensus tree of 20 equally parsimonious trees, resulting from parsimony analyses based on 40 isolates from *Phellinus* and *Inonotus* taxa. Bootstrap values above 50%, based on 1000 replicates, are indicated next to branches.

our opinion, belonging to *Mucronoporus* clustered with representatives of the *Phellidinium*-clade in both NJ trees. Its position highlights the strong morphological and anatomical similarities between the two groups.

Phellinidium-clade (bootstrap 100%), comprising *Ph. weirii* and *Ph. sulphurascens,* formed a divergent group having a basal position within *Inonotus s. l.* and *Phellinus s. l.* assemblage.

The basal lineages of Hymenochaetales, represented here by *Hyphodontia alutaria* and *H. pruni* (of which the latter is currently considered as a member of *Xylodon*), represent a distinct lineage and are therefore used here as an outgroup.

DISCUSSION

The species "*Inonotus*" *radiatus* (= *Mensularia radiata* sensu auct. contemp.) is characterized by annual, effused-reflexed to resupinate, rather small fruitbodies of fibrous-fleshy consistency with pores ranging from 2 to 5 per mm, monomitic (to pseudodimitic) hyphal system, small and clavate basidia, short subhymenial thick-walled acute or hooked setae, and ellipsoid, slightly thickened, hyaline to yellowish-brown when mature, and strong dextrinoid basidiospores. Species of this genus produce a white rot of fallen or living deciduous wood and an obvious specificity for *Fagaceae*. "*Inonotus*" *andersonii,* which is a true pathogen of living trees with a pronounced specificity for oak, exhibits annual, widely effused fruitbodies, monomitic hyphal system, and broadly ellipsoid, yellowish, thick-walled basidiospores. Except that this species has non-dextrinoid basidiospores, the link of "*I*". *andersonii* with "*I*". *radiatus* is well supported here by molecular data and similar morphological characteristics. The generic unit which includes both species under discussion needs to be renamed according to the nomenclatural priority rule (McNeill et al., 2006). There are three previously published names based on *Polyporus radiatus* Sowerby, namely, *Inoderma* P. Karst., *Inodermus* Quél., and *Mensularia* Lázaro Ibiza. The name *Inoderma* P. Karst. 1879 is not legitimate because it is a previously known homonym of *Inoderma* (Ach.) Gray 1821, which is applied to lichenized Ascomycota. The genus *Inodermus* Quél. 1886 is a superfluous orthographical variant of *Inoderma* (Donk, 1974; Ryvarden, 1991). The third name, *Mensularia* Lázaro Ibiza 1916, is therefore valid and was re-introduced by Wagner and Fischer (2001). However, our introduction of "*I*". *andersonii* in this unit should change the correct generic name in the nomenclature because *Mucronoporus andersonii* Ellis & Everh. in the nomenclature appears as a type of genus *Xanthoporia* Murrill 1916. Both names, *Mensularia* and *Xanthoporia,* were described in 1916, but *Mensularia* was published in late 1916 (Shaun Pennycook, pers. comm.), whereas *Xanthoporia* was published on 21 January 1916, therefore, has priority over *Mensularia.*

Some authors (e.g., Wagner and Fischer, 2002) proposed that such taxa as *Phellinus baumii* Pilát, *Ph. linteus* (Berk. & M.A. Curtis) Teng, *Ph. pachyphloeus* (Pat.) Pat., *Ph. tropicalis* M.J. Larsen & Lombard, *Ph. vaninii* Ljub., and *Ph. weirianus* (Bres.) Gilb. should be transferred to the genus *Inonotus.* However, the consequent taxonomical study of Jeong et al. (2005), based on combined ITS and mitochondrial SSUrDNA sequence data, showed incongruent results with those of Wagner and Fischer (2002). Their phylogenetic analyses subdivided

Inonotus into two groups (*Ph. linteus*-complex and *Inonotus*-subclade) according to high bootstrap values. Moreover, the authors suggested keeping the complex within the traditional boundaries of *Phellinus*. No significant conclusion was drawn from the phylogenetic relationship from four taxa i.e. *Phellinus occidentalis*, *Ph. weirii*, *Ph. pachyphloeus*, and *Ph. nigrolimitatus*. In both the aforementioned papers, the inclusion of *Ph. spiculosus*, *Ph. bicuspidatus*, *Ph. conchatus*, and *Ph. occidentalis* into *Phellinus* s. str. was supported by low bootstrap values. In our findings, *Inonotus* formed a polyphyletic group (bootstrap value 74%) that combined intermediate forms between *Inonotus s. l.* and *Phellinus s. l.*

From this point of view, our conclusions are congruent with those of Wagner and Fischer (2002) according to which this genus should be treated in a wider sense. Therefore, the genus comprises taxa with an annual and monomitic hyphal system as well as taxa with (sub) dimitic and perennial basidiocarps.

The *Phellinus linteus*-subclade (bootstrap value 100%) comprises three morphologically similar species here i.e. *Ph. linteus*, *Ph. baumii*, and *Ph. weirianus*. Several morphological and molecular analyses demonstrated the close phylogenetic relationship between these taxa with robust bootstrap support; in addition, *Inonotus tropicalis* also proved to be closely related (Jeong, 1999; Parmasto and Parmasto, 2001; Jeong et al., 2005; Sutton et al., 2005; Larsson et al., 2007). The four taxa from *Phellinus s. l.* were transferred back to *Inonotus* by Larsson et al. (2007), while their position remained uncertain according to Sutton et al. (2005). Members of the *Ph. linteus*-subclade have common morphological characteristics including the presence of setae that are ovoid to subglobose, have hyaline or slightly pigmented basidiospores, and react negatively in Melzer's reagent. *Ph. baumii*, *Ph. linteus*, and *Ph. weirianus*, united under similar morphological features, such as perennial fruitbody and dimitic hyphal system, have a certain specificity for inhabiting hardwood genera, while *Inonotus tropicalis* occurs exclusively on coniferous wood and represents an intermediate form between *Inonotus s. l.* and *Phellinus s. l.* with annual to perennial fruitbodies and a monomitic hyphal system; morphological differences from our results place this species apart from the other members of the *Ph. linteus*-subclade.

In our investigation members of *Inonotus cuticularis*-subclade (bootstrap value 96%) are annual, have a monomitic hyphal system, and are similar in basidiospore shapes. Basidiospores are more or less variable in color, while setae or setal hyphae are present except *I. levis*. However, all these species differ from one another by distinct peculiarities as branched setal hyphae (*I. cuticularis*), growth habit, and development of sterile conks (*I. obliquus*), chlamydospore presence, anamorph and teleomorph stages of development (*I. rickii*), granular core presence, and cyanophilous basidiospores (*I. levis*).

Dai (2000) accepted *Inonotus levis* within *Inocutis*. *I. levis* was considered the only species belonging to *Inocutis*, having large and thick-walled cyanophilous basidiospores (Ghobad-Nejhad and Kotiranta, 2008). Phylogenetic data on *I. levis* is missing in the literature (Table 9.1), however, this species in our investigation was included in the *I. cuticularis*-subclade supported by high bootstrap value (100%). In spite of this, further molecular and morphological analyses are needed to provide additional evidence concerning the phylogenetic position of *I. levis*.

Inonotus cuticularis was considered as a member of *Inonotus* s. str. by Wagner and Fischer (2001, 2002) and later supported by Sutton et al. (2005) and Larsson et al. (2007). This species resulted as a part of *Inocutis* in the study of Jeong et al. (2005), and within the *Phellinus rimosus*-complex in the study of Jeong (1999). In our opinion this species should be kept within *Inonotus* s. str.

The phylogenetic position of *Inocutis* (bootstrap value 100%) is well positioned in our tree because it is closely related to *Inonotus* s. str. *Inocutis*, originally proposed by Fiasson and Niemelä (1984), and was accepted by mycologists worldwide. *Inocutis* is characterized by annual fruitbodies with a monomitic hyphal system, ellipsoid, yellowish-brown, non-dextrinoid basidiospores, and lack of setae. *I. rheades*, *I. tamaricis*, and *I. dryophilus* are well known species of this group. All occur on a deciduous substrate, have a mycelial core, and exhibit sclerid structures under a microscope.

Considered as a member of *Inonotus s. l.* by Ryvarden and Gilbertson (1993), *I. dryadeus* was confirmed as belonging to *Fomitiporia* by Dai (1999), based on morphological similarities, and was recognized as basal in the *Phellinus* s. str. group showing a strong phylogenetic relationship with *Ph. occidentalis* and *Ph. conchatus* (Jeong, 1999; Nunez and Ryvarden, 2000; Wagner and Fischer, 2002; Jeong et al., 2005). Later, this species was transferred to the new genus *Pseudoinonotus* (bootstrap value under 50%) as suggested by Wagner and Fischer (2001), which was based on inconclusive molecular data. In spite of this, the position of *I. dryadeus* within *Pseudoinonotus* was recognized by further publications such as Wagner and Fischer (2002), Wagner and Ryvarden (2002), Jeong et al. (2005), and Larsson et al. (2007). Our data confirms a separate taxonomic position for this species supporting the previous studies of Wagner and Fischer (2001, 2002).

Porodaedalea is characterized by perennial, pseudodimitic to subdimitic, effused-reflexed to pileate fruitbodies, with a crustose abhymenial surface, rounded to labyrinthine pores, duplex context, hymenial setae in all species, and ellipsoid to subglobose hyaline to slightly yellowish non-dextrinoid, acyanophilous basidiospores, without a central vacuole (Spirin et al., 2007); produces a white pocket rot of living and dead conifers. *Phellinus cancriformans* was first treated as a variety of *Ph. pini*, and later recognized as a new species by Larsen and Cobb-Poulle (1990) and placed in *Porodaedalea*. Currently, *Ph. cancriformans* and *Ph. crysoloma* are well established as members of *Porodaedalea* based on the reports of Fiasson and Niemelä (1984), Wagner and Fischer (2002), Jeong et al. (2005), and Larsson et al. (2007).

Phellinus s. str. and *Fomitiporia* appear as unrelated clades in our strict-consensus tree, both units have common features as substrate occurrence, fruitbody consistency and shape, basidiospore shape, color, and wall thickness.

Phellinus s. str. is a well represented group characterized by perennial and hard consistency of fruitbodies with the surface of the pileus covered with thick-cracked crust of dark to almost black color; context hard-woody, often with streaks of white mycelium at the base; margin dull and sterile below; hymenial surface porose, pores with thick dissepiments; tube layer distinctly stratified; hyphal system dimitic; setae rare to abundant or absent of hymenial or subhymenial origin; setal hyphae absent in all representatives; basidiospores ellipsoid to nearly globose, thick-walled, hyaline-yellowish to brown, non-dextrinoid, cyanophilous or acyanophylous; produces a white rot of both dead and living broad-leaved trees. According to Gilbertson and Ryvarden (1986), *Ph. igniarius*-complex represents another difficult taxonomic problem. *Ph. igniarius* was treated as a member of *Ochroporus* by Fiasson and Niemelä (1984) under the synonym *Ochroporus alni* (Bondartsev) Fiasson & Niemelä. Currently, it is considered that *Ochroporus* is an obligate synonym of *Phellinus* (Spirin et al., 2007). *Ph. igniarius* and *Ph. spiculosus* were clustered as sister taxa (bootstrap support 100%), fact that supports other reports (e.g., Jeong, 1999; Wagner and Fischer, 2002; Jeong et al., 2005; Larsson et al., 2007).

The genus *Fomitiporia* resembles characteristics of *Phellinus* s. str., like perennial fruitbodies inhabiting broad-leaved or coniferous wood. Basidiospores are globose, hyaline, and (in contrast to *Phellinus* s. str.) strongly dextrinoid and cyanophylous. The hyphal system is pseudodimitic, the setae are rare to absent, and ampulla-shaped cystidioles are present in the hymenium.

Our results concerning taxa clustering into *Fomitiporia*-clade are congruent with those of Jeong (1999), Wagner and Fischer (2001, 2002), Fischer (2002), Jeong et al. (2005), and Larsson et al. (2007). Investigations on mating behavior in species of *Fomitiporia* are impeded by the fact that, for unknown reasons, germination rates of basidiospores are very low under best conditions in a laboratory (Fischer, 1987). On the basis of mycelia isolated from infected wood of *Vitis*, *Fomitiporia* (*Phellinus*) *punctata* had been misidentified as *Phellinus igniarius* (Fischer, 2002). *Fomitiporia mediterranea*, recently described as a new species by Fischer (2002), morphologically and structurally resembles *Phellinus punctatus* as both the species possess characters like perennial, resupinate, hard-woody fruitbodies, dimitic hyphal system, absent or very rare setae, broadly ellipsoid to subglobose, hyaline, thick-walled, dextrinoid and cyanophilous basidiospores. Furthermore, as stated by Fischer, "because of identical morphological and anatomical features, "*Phellinus punctatus*" growing on *Vitis vinifera* probably is assignable to "*Fomitiporia mediterranea*". However, examination of both species, in culture under the same laboratory conditions, proved *F. mediterranea* as being faster growing than *F. punctata*, while molecular data differentiated the two species into two separate groups.

The genus *Phylloporia*, according to Gilbertson and Ryvarden (1986), is closely related to *Inonotus* (*s. l.*) and *Cyclomyces* Kunze ex Fr. and is separated from the former by having a thick, persistent tomentum delimited from context by a thin dark line and small yellowish basidiospores, while from *Cyclomyces* it is separated by long subulate setae and hyaline basidiospores. Based on partial sequences of n-LSUrDNA and the morphological examination of 28 taxa, Wagner and Ryvarden (2002) showed the genus to be monophyletic. Additionally, the phylogenetic relationship of this genus to *Fulvifomes* was highlighted. *Phylloporia* was treated as a distinct genus by Ryvarden (1978). With a higher diversity in the tropics, this genus is known to comprise species resistant to drought; in such conditions, the pileal surface, covered with a fine tomentose thick layer of spongy consistency, has a moisture retaining role. This group often includes species inhabiting living branches of bushes and trees. Frequently used in molecular representations, the taxonomic position of *Phylloporia ribis* was thoroughly discussed by many authors. The species is distinguishable by perennial fruitbodies, small pores (6–7 per mm), indistinct stratified tube layer, lack of setae, small pale-yellowish non-dextrinoid, and acyanophilous basidiospores. Positioned separately from any taxon of *Phellinus*, *Ph. ribis* appeared to be an intermediate between *Inonotus s. l.* and *Phellinus s. l.*, as reported by Wagner and Fischer (2001). Jeong et al. (2005) reported this species as a sister taxon to *Fulvifomes*, and Larsson et al. (2007) reported it between *Fulvifomes* and *Inocutis*.

In our representation the taxonomic position of *Phylloporia* is somewhat questionable. In both NJ trees the genus proved to be basal within the *Inonotus* s. str. group, represented by a distinct lineage, while in the MP strict-consensus it appears to be an independent clade. The consensus result supports an independent status of *Phylloporia*, as concluded by Wagner and Ryvarden (2002).

Characteristics for *Mucronoporus* Ellis & Everh. [type *M. gilvus* (Schwein.) Ellis & Everh. – see Spirin et al., 2007] are the ellipsoid to nearly globose basidiospores, pseudodimitic to subdimitic hyphal system, and homogeneous context. The phylogenetic status of *Mucronoporus*, *Phellinidium*, *Phellopilus*, and *Fuscoporia* exhibits similar morpho-anatomical characteristics; differences are revealed in basidiospore characteristics in some taxa (pale, fusiform in *Fuscoporia* and *Phellopilus* vs. ellipsoid pigmented in *Mucronoporus*) and hyphal system features (pseudodimitic in *Mucronoporus* and *Fuscoporia*, trimitic in *Phellopilus* and macrosetal-bearing in *Phellinidium*) (see Table 9.3). *Mucronoporus torulosus* was treated by some authors as belonging to *Fuscoporia* (Jeong, 1999; Wagner and Fischer, 2001, 2002; Jeong et al., 2005; Larsson et al., 2007).

Inhabiting dead and living broadleaved or coniferous trees, *Inonotus pachyphloeus* resembles perennial firm to woody hard consistency fruitbodies, and a pseudodimitic hyphal system. Basidiospores are ellipsoid to globose, thick-walled, pale-yellowish in age, and negative in Melzer's reagent. Because

Table 9.3 Synoptic keys to genera of *Phellinus s. l.*, and *Inonotus s. l.*

Genus	Substrate	Consistency	Hyphal	Setae	Basidiospores				
					Shape	*Wall*	*Color*	M^+	CB^+
Fomitiporia	D/C	P	PSD	–/+*	S/G	THK	–/+	+	+
Inonotus s.str.	D/C	A/P	M/PSD/SB	+/-	E/S	THI/THK	–/+	–/+	–/+
Inocutis	D	A	M	–	E	THK	+	–	–
Xanthoporia	D	A	M/(PSD)	+	E	THK	–/+	+	–
Mucronoporus	D/C	A/P	PSD	+	E/S	THI	+	–	–
Phellinidium	C	A/P	PSD	+	E/A/CL	THI	–	–	–
Phylloporia	D	A/P	M/PSD/SB	+/–	E/S	THI/THK	–/+	–/+	–/+
Phellinus s.str.	D/C	P	DM	+/–	E/S	THK	–/+	–/+	–/+
Porodaedalea	C	P	PSD/SB	+	E/S	THI	–/+	–	–

Note: D = deciduous wood; C = coniferous wood; A = annual; P = perennial; M = monomitic; PSD = pseudodimitic; DM = dimitic; SB = subdimitic; +* = rare; CL = cylindrical; E = ellipsoid; S = subglobose; G = globose; THK = thick-wall; THI = thin-wall; M^+ = reaction in Melzer's solution; CB^+ = cyanophily.

this species shares common features with *Phellinidium*, except in basidiospore wall thickness and color, Dai (1999) suggested the subgenus *Phellinidiopsis* for this species; however, the species was classified as belonging to *Inonotus* s. str. in the study of Wagner and Fischer (2002) and was inconclusive in the study of Jeong (1999); therefore, it was placed into *Phellopilus* by Jeong et al. (2005) and considered as a member of *Inonotus* by Larsson et al. (2007). *I. pachyphloeus* formed a distinct lineage in our strict consensus tree because of its close relatedness to *Muconoporus torulosus*. Further analyses are needed for a more comprehensive overview on the phylogenetic position of this species.

Phellinidium was introduced as a subgenus of *Phellinus* by Kotlaba (1968), and later raised to the rank of genus by Fiasson and Niemelä (1984). *Phellinidium* is currently characterized by annual to perennial resupinate, soft-coriaceous to suberose fruitbodies, pseudodimitic hyphal system, no true skeletal hyphae, macrosetae presence, hyphoid to ampullate leptocystidia, and globose to allantoid, hyaline, thin-walled basidiospores, acyanophilous, and negative in Melzer's reagent. The taxonomic status of *Ph. sulphurascens* and *Ph. weirii* has been controversial. Considered as member of *Inonotus* by Larsen et al. (1994), *Ph. sulphurascens* was transferred to *Phellinidium* by Dai (1995) whose position was supported by Larsson et al. (2007). The hyaline ovoid basidiospores and the conspicuous setal hyphae were stated as being important characteristics of *Ph. weirii*, linking it to *Ph. ferrugineofuscum* (Gilbertson and Ryvarden, 1986).

In this phylogenetic study we demonstrated that the generic concept of *Inonotus* s. str. should be considered in a wider sense and accepted. This fact is also highlighted by previous studies that are in conformity with our results. Unfortunately, there has been a controversy in considering taxonomical characteristics and their importance for classification. According to our opinion, morphological characters such as consistency in hyphal system or presence and absence of setae should have less importance in phylogenetic analyses, while traits such as shape of basidiospores and basidia, basidiospore wall-thickness, ornamentation, color, reaction in chemical solutions should be considered as being more important.

Additional taxonomically problematic species need to be included in future studies in order to provide clarification on their phylogenetic status.

Nomenclatural novelties

Xanthoporia radiata (Sowerby) Ţura, Zmitr., Wasser, Raats & Nevo comb. nova. – Basionymum: *Boletus radiatus* Sowerby, Col. fig. Engl. Fung. Mushr. 2: 83, 1799.

Xanthoporia nodulosa (Fr.) Ţura, Zmitr., Wasser, Raats & Nevo comb. nova (ad interim). – Basionymum: *Polyporus nodulosus* Fr., Epicr. syst. mycol.: 474, 1838.

ACKNOWLEDGEMENTS

The authors extend special thanks to Krugman, T., Zamorzeva, I., Jun, Y., Barshegyan, G., Koshakyan, A., and Sela, H. for their valuable suggestions and technical support. Hadid Y. deserves special thanks for providing the much needed help for sequencing procedures. Also, we thank Dr. Sh. Pennycook, Prof. G. Moreno, Dr. W.A. Spirin, Dr. E.O. Yurchenko and Prof. G. Moreno for critical reviews of the chapter.

REFERENCES

Akaike, H. 1974. A new look at the statistical model identification. IEEE Transactions on Automatic Control, 19: 716-723.

Corner, E. J. H. 1991. Ad Polyporaceas VII: The xanthochroic polypores. Beih Nova Hedwigia, 101: 1-173.

Dai, Y. C. 1995. Changbai wood-rotting fungi 3. The genus *Phellinidium* (Basidiomycetes) and a new species, *P. aciferum*. Ann. Bot. Fennica, 32: 63-73.

Dai, Y. C. 1999. *Phellinus* sensu lato (Aphyllophorales, *Hymenochaetaceae*) in East Asia. Acta Bot. Fenn., 166: 1-115.

Dai, Y. C. 2000. A checklist of polypores from Northeast China. Karstenia, 40: 23-29.

Donk, M. A. 1974. Checklist of European polypores. Verh. Koninkl. Ned. Akad. Wetenschappen. Natuurk. D. 62. P. 1-469.

Felsenstein, J. 1985. Confidence limits of phylogenies: An approach using bootstrap. Evolution, 39: 783-791.

Fiasson, J. L. and Niemelä, T. 1984. The Hymenochaetales: A revision of the European poroid taxa. Karstenia, 24: 14-28.

Fischer, M. 1987. Biosystematische Untersuchungen an den Porlingsgattungen *Phellinus* Quél. und *Inonotus* Karst. Bibliotheca Mycologica, 107: 1-133.

Fischer, M. 1996. Molecular and microscopical studies in the *Phellinus pini* group. Mycologia, 88: 230-238.

Fischer, M. 2002. A new wood-decaying basidiomycete species associated with esca of grapevine: *Fomitiporia mediterranea* (Hymenochaetales). Mycol. Prog., 1: 315-324.

Fischer, M. and Binder, M. 2004. Species recognition, geographic distribution and host-pathogen relationships: a case study in a group of lignicolous Basidiomycetes, *Phellinus s. l.* Mycologia, 96: 799-811.

Ghobad-Nejhad, M. and Kotiranta, H. 2008. The genus *Inonotus* sensu lato in Iran, with keys to *Inocutis* and *Mensularia* worldwide. Ann. Bot. Fennici, 45: 465-476.

Gilbertson, R. L. and Ryvarden, L. 1986. North American polypores. Vol. 1. Fungiflora, Oslo.

Hall, T. A. 1999. BioEdit: a user-friendly biological sequence alignment editor and analysis program for Windows 95/98/NT. Nucl. Acids. Symp. Ser., 41: 95-98.

Hansen, L. and Knudsen, H. 1997. Nordic Macromycetes. Vol. 3. Heterobasidioid, aphyllophorales and gastromycetoid Basidiomycetes. Nordvamp, Copenhagen.

Harrison, C. J. and Langdale, J. A. 2006. A step by step guide to phylogeny reconstruction. The Plant Journal, 45: 561-572.

Hibbett, D. S. and Donoghue, M. J. 1995. Progress toward a phylogenetic classification of the *Polyporaceae* through parsimony analyses of mitochondrial ribosomal DNA sequences. Can. J. Bot., 73: S853-S861.

Hibbett, D. S. and Thorn, R. G. 2001. Homobasidiomycetes, pp. 121-168. In: The Mycota VII, Part B, Systematics and Evolution. D. J. McLaughlin, E. G. McLaughlin and P. A. Lemke (eds.). Springer-Verlag: Berlin, Heidelberg.

Hibbett, D. S., Gilbert, L. B. and Donoghue, M. J. 2000. Evolutionary instability of ectomycorrhizal symbioses in Basidiomycetes. Nature, 407: 506-508.

Jeong, W. J. 1999. Phylogenetic Analysis of *Phellinus* Based on Internal Transcribed Spacer and Mitochondrial Small Subunit Ribosomal Sequences. Ph.D. Thesis. Seoul National University, Seoul, Korea [http://www.plaza1.snu.ac.kr/~minervas/pdf/jwj.pdf].

Jeong, J. W., Lim, Y. W., Lee, S. J. and Jung, H. S. 2005. Phylogeny of *Phellinus* and related genera inferred from combined data of ITS and mitochondrial SSU rDNA sequences. J. Microbiol. Biotechnol., 15: 1028-1038.

Kirk, P. M., Cannon, P. F., Minter, D. W. and Stalpers, J. A. 2008. Ainsworth and Bisby's Dictionary of the Fungi. 10th edn. CAB International, Wallingford, UK.

Ko, K. S., Hong, S. G. and Jung, H. S. 1997. Phylogenetic analysis of *Trichaptum* based on nuclear 18S, 5.8S and ITS ribosomal DNA sequences. Mycologia, 89: 727-734.

Kotlaba, F. 1968. *Phellinus pouzarii* sp. nov. Česká Mykol., 22(1): 24-31.

Langer, E. 1998. Evolution of *Hyphodontia* (*Corticiaceae*, Basidiomycetes) and related Aphyllophorales inferred from ribosomal DNA sequences. Folia Cryptog. Estonica, 33: 57-62.

Larsen, M. J. and Cobb-Poulle, L. A. 1990. *Phellinus* (*Hymenochaetaceae*). A survey of the world taxa. Fungiflora, Oslo.

Larsen, M. J., Lombard, F. F. and Clark, J. W. 1994. *Phellinus sulphurascens* and the closely related *P. weirii* in North America. Mycologia, 86: 121-130.

Larsson, K-H. 2001. The position of *Poria mucida* inferred from nuclear ribosomal DNA sequences. Harv. Papers Bot., 6: 131-138.

Larsson, K-H., Parmasto, E., Fischer, M., Langer, E., Nakasone, K. K. and Redhead, S. A. 2007. Hymenochaetales: a phylogeny of the hymenochaetoid clade. Mycologia, 98: 926-936.

McNeill, J., Barrie, F. R., Burdet, H. M., Demoulin, V., Hawksworth, D. L., Marhold, K., Nicolson, D. H., Prado, J., Silva, P. C., Skog, J. E., Wiersema, J. H. and Turland, N. J. (eds.). 2006. International Code of Botanical Nomenclature (Vienna Code) adopted by the Seventeenth International Botanical Congress Vienna, Austria, July 2005. Gantner Verlag, Ruggell, Liechtenstein.

Moore, R. T. 1980. Taxonomic significance of septal ultrastructure in the genus *Onnia* Karsten (*Polyporineae*/*Hymenochaetaceae*). Bot. Not., 133: 169-175.

Moore, R. T. 1985. The challenge of the dolipore/parenthosome septum, pp. 175-212. In: Developmental Biology of Higher Fungi. D. Moore, L. A. Casselton, D. A. Wood and J. C. Frankland (eds.). Cambridge Univ. Press, UK.

Müller, W. H., Stalpers, J. A., Van Aelst, A. C., de Jong, M. D. M., van der Krift, T. P. and Boekhout, T. 2000. The taxonomic position of *Asterodon*, *Asterostroma* and *Coltricia* inferred from the septal pore cap ultrastructure. Mycol. Res., 104: 1485-1492.

Niemelä, T., Wagner, T., Fischer, M. and Dai, Y. C. 2001. *Phellopilus* gen. nov. and its affinities within *Phellinus* s. lato and *Inonotus* s. lato (Basidiomycetes). Ann. Bot. Fennici, 38: 51-62.

Nuñez, M. and Ryvarden, L. 2000. East Asian polypores. Synop. Fung., 13: 1-164.

Page, R. D. M. 1996. TREEVIEW: an application to display trees on personal computers. Computer Applications in the Biosciences, 12: 357-358.

Parmasto, E. and Parmasto, I. 1979. The xanthochroic reaction in Aphyllophorales. Mycotaxon, 8: 202-232.

Parmasto, E. and Parmasto, I. 2001. *Phellinus baumii* and related species of the *Phellinus linteus* group (*Hymenochaetaceae*, Hymenomycetes). Folia Cryptog. Estonica, 38: 53-61.

Paulus, B. N., Hallenberg, P. K. and Chambers, G. K. 2000. A phylogenetic study of the genus *Schizopora* (Basidiomycota) based on ITS DNA sequences. Mycol. Res., 104: 1155-1163.

Posada, D. 2006. ModelTest Server: a web-based tool for the statistical selection of models of nucleotide substitution online. Nucleic Acids Research, 34: W700-W703.

Posada, D. and Crandall, K. A. 1998. Modeltest: testing the model of DNA substitution. Bioinformatics, 14: 817-818.

Rizzo, D. M., Gieser, P. T. and Burdsall, H. H. 2003. *Phellinus coronadensis*: a new species from southern Arizona, USA. Mycologia, 95: 74-79.

Ryvarden, L. 1978. The *Polyporaceae* of North Europe. Vol. 1–2. Fungiflora, Oslo.

Ryvarden, L. 1991. Genera of polypores. Nomenclature and taxonomy. Synopsis Fungorum Vol. 5. Oslo.

Ryvarden, L. and Gilbertson, R. L. 1993. European Polypores. Fungiflora, Oslo.

Spirin, W. A., Zmitrovich, I. V. and Malysheva, V. F. 2007. To the systematics of *Phellinus s. l.* and *Inonotus s. l.* (*Mucronoporaceae*, Hymenochaetales). Novitates Systematicae Plantarum Non-vascularium, 40: 153-188.

Sutton, D. A., Thomson, E. H., Rinaldi, M. G., Iwen, P. C., Nakasone, K. K., Jung, H. S., Rosenblatt, H. M. and Paul, M. E. 2005. Identification and first report of *Inonotus* (*Phellinus*) *tropicalis* as an etiologic agent in a patient with chronic granulomatous disease. J. Clin. Microbiol., 43: 982-987.

Swofford, D. L. 2003. PAUP*: phylogenetic analyses using parsimony (*and other methods). Version 4. Sinauer Associates, Sunderland, MA.

Wagner, T. and Fischer, M. 2001. Natural groups and a revised system for the European poroid Hymenochaetales (Basidiomycota) supported by nLSU rDNA sequence data. Mycol. Res., 105: 773-782.

Wagner, T. and Fischer, M. 2002. Proceedings towards a natural classification of the wordwide taxa *Phellinus s. l.* and *Inonotus s. l.*, and phylogenetic relationships of allied genera. Mycologia, 94: 998-1016.

Wagner, T. and Ryvarden, L. 2002. Phylogeny and taxonomy of the genus *Phylloporia* (Hymenochaetales). Mycol. Prog., 1: 105-116.

Chapter 10

Morphological and Molecular Taxonomy of Highly Toxigenic *Fusarium* Species from Small Cereal Grains in Northern Europe and Asia

Tapani Yli-Mattila

Molecular Plant Biology, Department of Biochemistry and Food Chemistry,
University of Turku, FI-20014 Turku, Finland
E-mail: tymat@utu.fi

ABSTRACT

Several *Fusarium* species are involved in Fusarium Head Blight (FHB), which reduces both crop yield and quality in cereals. FHB is among the most important fungal diseases of cereals worldwide.

The genus *Fusarium* was established by Link in 1809. Traditional identifications and classifications of *Fusarium* species have been used for grouping isolates to species and grouping species according to shared morphological characteristics. During the last years researchers have started to use alternative ways for species identification and classification based on molecular data and phylogenetic analyses. This has totally revised fungal taxonomy. The best way to identify and classify *Fusarium* isolates is the polyphasic approach by using all available characters.

The aim of this chapter is to update the molecular and morphological taxonomy of problematic *Fusarium* species of sections *Discolor*, *Sporotrichiella* and *Roseum* in northern Europe and Asia.

INTRODUCTION

Fusarium, a common group of fungi in many parts of the world, is widely distributed in soil and is associated with many important diseases of plants and animals. This cosmopolitan fungus plays an important role in ecosystem functions and has a broad impact on human activities and therefore, has been an object of intensive investigations.

The genus *Fusarium* was established by Link in 1809. Traditional identifications and classifications of *Fusarium* species have been used for grouping isolates to species and grouping species according to shared morphological characteristics. Currently, researchers have started to use alternative ways for species identification and classification based on molecular data and phylogenetic analyses. Most of the phylogenetic characters are based on DNA sequences, which has dramatically increased the number of characters to be used for the identification. This has resulted in a total revision of the systematics of this group of fungi (Nirenberg and O'Donnell, 1998; O'Donnell et al., 1998, 2000, 2008, 2009a,b; Taylor et al., 2000).

Several *Fusarium* species are involved in Fusarium Head Blight (FHB), which reduces both crop yield and quality of cereals. FHB is one of the most important fungal diseases of cereals worldwide (Parry et al., 1995; O'Donnell, 1996; McMullen et al., 1997; Bottalico and Perrone, 2002) including Europe and Asia (Langseth et al., 1999; Eskola et al., 2001; Levitin 2001, 2004; Kosiak et al., 2003; Yli-Mattila et al., 2004b,c; Zhang et al., 2007; Qu et al., 2008; Suga et al., 2008; Yang et al., 2008; Lee et al., 2009; Yli-Mattila and Gagkaeva, 2010).

FHB was first discovered in England in 1884. Since then FHB has increased worldwide (Goswami and Kistler, 2004). In Russia, FHB caused by *F. graminearum* was first reported at the end of 19th century in the Far East region, where the environmental conditions are favourable for FHB (Voronin, 1890). This fungus has not been detected in Siberia and the Ural regions, while in the southern Russia region (North Caucasus), where the first outbreak was reported in 1933 (Pronicheva, 1935), *F. graminearum* is the most common causal agent of FHB. Since it was first detected in the North Caucasus region, FHB has reached epidemic levels several times in southern Russia during 1960s, '80s and '90s and caused reduction in cereal yields (Ivanchenko, 1960; Kirienkova, 1992). In central Russia, where *F. avenaceum* tends to be the predominant cause of FHB, *F. graminearum* isolates have been found in the central-chernozem territory since 1980s (Selivanova et al., 1991).

Recently, *F. graminearum* has been spreading northward in Europe (e.g. Nicholson et al., 2003; Waalwijk et al., 2003) and has been replacing the closely related *F. culmorum.* In the north-western region of Russia, *F. graminearum* was not found until in 2003, but since then this pathogen has become more common in this region (Gagkaeva et al., 2009). In Belorussia this pathogen was also not detected before 2003 (Shashko and Gagkaeva, 2010), but in the Ukraine both *F. graminearum* and *F.culmorum* are common pathogens of cereals (Kryuchkova et al., 2001). In Finland, *F. graminearum* was already reported in cereals in the 1960s (Ylimäki, 1981), and it has been present in Finland since then [(Uoti and Ylimäki, 1974; from 1976-1977 (Ylimäki et al., 1979); from 1982-1984; (Rizzo, 1993); and in 1998 (Eskola et al., 2001; Yli-Mattila et al., 2002)].

In northern Europe, the most common *Fusarium* species in cereals are *F. avenaceum* (including *F. arthrosporioides*), *F. tricinctum, F. poae, F. culmorum, F. graminearum, F. sporotrichioides* and the new species *F. langsethiae* (Kosiak et al., 2003; Levitin, 2004; Yli-Mattila et al., 2004a, 2008; Stepien et al., 2008; Gagkaeva et al., 2009; Nicolaisen et al., 2009). In western Siberia, *F. avenaceum, F. sporotrichioides* and *F. poae* predominate, while in eastern Siberia, *F. avenaceum* and *F. acuminatum* are the most common *Fusarium* species. The greatest diversity of *Fusarium* species in cereals has been found in the Russian Far East, where *F. graminearum* and *F. poae* are the most common species (Levitin, 2004).

Fusarium species produce mycotoxins dangerous to humans and animals (Bottalico and Perrone, 2002; Logrieco et al., 2002). Trichothecenes, which are the most important *Fusarium* mycotoxins according to chemical features, can be divided in type A, including T-2 toxin, HT-2 toxin, diacetoxyscirpenol (DAS), monoacetoxyscirpenol (MAS), and type B, including deoxynivalenol (DON, also called vomitoxin) and nivalenol (NIV) and their mono- and di-acetylated derivatives (Bottalico, 1998).

This chapter embodies an account of the molecular and morphological taxonomy of the most toxigenically significant *Fusarium* species in northern Europe and Asia.

Taxonomic status of *Fusarium* species

Fusarium (Link, 1809) comprises anamorphic filamentous fungi (*Hypocreales, Ascomycota*) in the genera *Gibberella* and *Haematonectria*. This genus is characterized by having fusoid, curved (banana-shaped), hyaline septate macroconidia, often gathered in slimy masses (sporodochia) on branched conidiophores.

The taxonomical status of *Fusarium* species has been the subject of considerable debate for 200 years. The existing classifications with widely differing species concepts have been based on morphological and cultural characteristics. This heterogeneous feature of *Fusarium* species has resulted into several classification systems. One of them was embodied in a landmark publication – *Die Fusarien* by Wollenweber and Reinking (1935). They grouped species into 16 sections, 65 species, 22 forms and varieties, and provided an excellent starting point for understanding this genus. Many of the species described by Wollenweber and

Reinking (1935) based on rigorous morphometric analyses have been validated even today by molecular analyses, but many of them, such as *F. graminearum*, *F. moniliforme*, *F. oxysporum*, *F. solani*, *F. avenaceum*, *F. sporotrichioides* and *F. equiseti*, actually represent species complexes based on phylogenetic analyses (O'Donnell et al., 1998, 2000, 2008, 2009a, b; Paavanen-Huhtala et al., 1999; Yli-Mattila et al., 2002, 2004a, 2009b, c).

After Wollenweber and Reinking (1935), taxonomical schemes that are based on morphological characters were developed by Booth (1971), Gerlach and Nirenberg (1982) and Nelson et al. (1983). The concepts of these taxonomic systems are very similar, but there are some differences in approaches to nomenclature.

Hughes (1953) emphasized the importance of the conidiogenous events to the systematics of the genus. Booth (1971) introduced the morphology of conidiogenous cells in differentiation of *Fusarium* species. According to his system, the genus *Fusarium* consists of 44 species, 7 varieties and 12 sections. Gerlach and Nirenberg (1982) have differentiated nearly 100 species based on minimal morphological characters and limited number of studied isolates. The greatest contribution in *Fusarium* systematics was made by Nelson et al. (1983), by describing 30 *Fusarium* species in 12 sections. In their next publication (Marasas et al., 1984), the toxin production ability was added in the species description. The book by Leslie and Summerell (2006) has dealt with the present morphological and molecular systematics along with mycotoxins produced by *Fusarium* species.

The most important *Fusarium* species associated with damaged cereal grains and mycotoxins fall into three *Fusarium* sections: *Discolor* (or *Fusarium* by Gerlach and Nirenberg (1982)), *Sporotrichiella* and *Roseum*.

SECTION DISCOLOR: *F. graminearum* Schwabe (1839), *F. culmorum* (W. G. Sm.) Sacc. (1892), and *F. cerealis* (Cooke) Sacc. (1886)

Fusarium species in section Discolor produce the classical form of banana-shaped mostly 3-5 septate macroconidia. These species do not produce microconidia.

Colony on high carbon media forms floccose aerial mycelium which is whitish, turning brownish to rose, and with vivid red and reddish-brown pigment. Conidiophores are usually branched and aggregated in orange-red-brown sporodochia. Conidiogenous cells are monophialides. Chlamydospores are present.

Among these *Fusarium* species, only *F. graminearum* has the sexual state *Gibberella zeae*. According to morphological characters, *F. cerealis* is between *F. graminearum* and *F. culmorum* considering the size and form of macroconidia and the color of pigmentation (Table 10.1, Figs. 10.1, 10.2). The macroconidia of *F. cerealis* are notably thicker in the middle as compared to *F. graminearum* and larger as compared to *F. culmorum*. *F. cerealis* was recently detected for the first time in the Russian Far East, the South-European part of Russia and Belorussia (Gagkaeva, 2009; Shashko and Gagkaeva, 2010; Yli-Mattila and Gagkaeva, 2010). The reason for this is that in the most popular literature of systematics in the former Soviet Union which was written by Bilai (1977), *F. cerealis* was treated as a synonym of *F. culmorum*. This was based on species concepts of Booth (1971). Raillo (1950) and Gerlach and Nirenberg (1982) considered this fungus as a synonym with *F. sambucinum* Fuck. var. *cereale* (Cke.) Wr. These three species are also macroscopically similar (Fig. 10.1).

F. graminearum, F. culmorum and *F. cerealis* are the main types B trichothecene producers, but *F. culmorum* produces less DON than *F. graminearum* (Langseth et al., 2001; Jennings et al., 2004; Jestoi et al., 2004a, 2008). All known *F. cerealis* isolates are typically NIV producers, (Miller et al., 1991; Chandler et al., 2003; Yli-Mattila and Gagkaeva, 2010).

F. graminearum and *F. culmorum* isolates can be subdivided based on mycotoxin production into two main groups depending on production profiles. One group produces DON and its acetylated derivatives and the other group produces NIV and its acetylated derivatives. The DON group of *F. graminearum* can be further divided into 3ADON producers and 15ADON producers, while all DON producers of *F. culmorum* belong to 3ADON producers. These molecular chemotypes are not species-specific and they can also be detected in other closely related *Fusarium* species of the *F. graminearum* species complex. The genetic basis of DON versus NIV production results from differences in *TRI13* gene. Chemotype-specific PCR primers (Ward et al., 2002; Jennings et al.,

Table 10.1 Measurements of the macroconidia of *F. graminearum, F. culmorum* and *F. cerealis* (CLA, two weeks of growth in dark, 23°C)

Species	*No. of septate*	*Median, μm*	*Range, μm*	*Length and width ratio*
F. graminearum	3	41.9 x 5.5	40.5 – 42.6 x 4.7 – 6.5	7.5
F. cerealis		30.0 x 5.5	23.5 – 33.6 x 4.3 – 7.0	5.2
F. culmorum		28.7 x 6.0	21.5 – 34.8 x 4.2 – 7.6	5.6
F. graminearum	4	44.2 x 5.4	37.4 – 50.0 x 4.7 – 6.5	8.2
F. cerealis		38.2 x 5.9	25.4 – 48.1 x 4.9 – 7.0	6.3
F. culmorum		34.7 x 6.2	28.4 – 39.4 x 5.0 – 7.6	5.6
F. graminearum	5	47.2 x 5.7	39.4 – 57.5 x 4.6 – 6.7	8.3
F. cerealis		41.3 x 5.9	31.6 – 58.1 x 4.8 – 7.2	7.0
F. culmorum		36,7 x 6.5	30.2 – 44.9 x 5.1 – 7.5	5.5

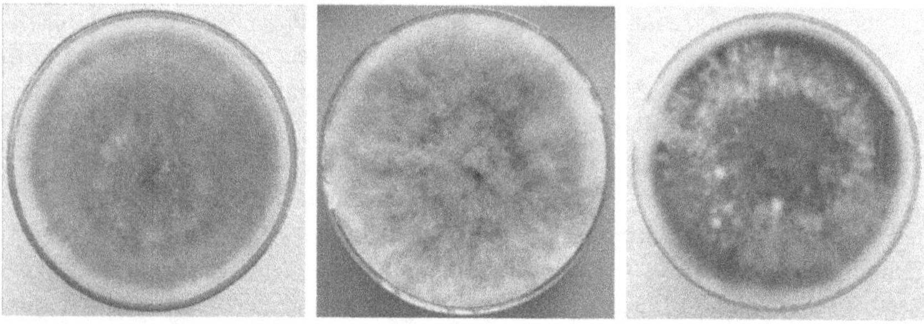

F. graminearum F. cerealis F. culmorum

Fig. 10.1 *Fusarium* species (PSA, 2 weeks in dark): *F. graminearum, F. cerealis, F. culmorum. Color image of this figure appears in the color plate section at the end of the book.*

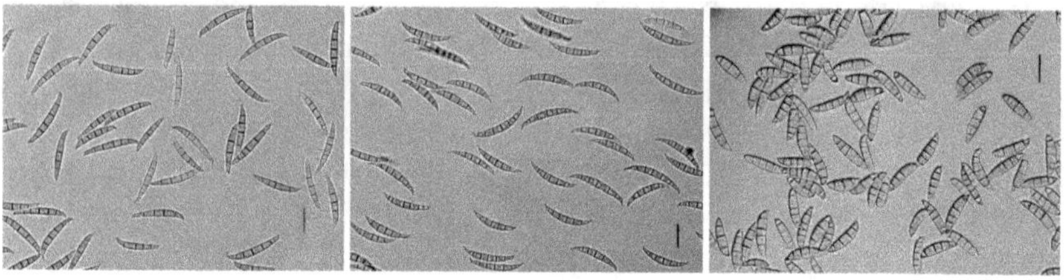

F. graminearum F. cerealis F. culmorum

Fig. 10.2 Macroconidia of *F. graminearum, F. cerealis and F culmorum* (SNA, 2 weeks).

2004; Starkey et al., 2007) and multilocus genotyping that utilizes a suspension microarray (Ward et al., 2008) have been developed to predict the chemical phenotype (chemotype) of *F. graminearum* isolates and other closely related main type B trichothecene producers.

There are three distinct sub-populations of *F. graminearum* in the European part of Russia (southern, central and north-western), which is adjacent to the Finnish population. Even though isolates of *F. graminearum* are segregated for 3-acetyldeoxynivalenol (3ADON) and 15-acetyldeoxynivalenol (15ADON) chemotypes in nearly equal frequencies in the regions sampled in Europe and Asia, significant differences in the geographic distribution of isolates producing these acetyl ester derivatives of deoxynivalenol (DON) have been observed in Europe. The *F. graminearum* population in north-western Russia is probably more closely related to the Finnish and Scandinavian populations of *F. graminearum* than to those in southern Russia, because the majority of the isolates belong to 3ADON chemotype unlike in southern and central Russia, where >90% of the isolates possess the 15ADON chemotype (Langseth et al., 2001; Elen et al., 2007; Yli-Mattila et al., 2009b; Yli-Mattila and Gagkaeva, 2010).

Multilocus molecular phylogenetic analyses have resolved the *F. graminearum sensu lato* as 13 phylogenetically distinct species (O'Donnell et al., 2000, 2004, 2008; Starkey et al., 2007; Yli-Mattila et al., 2009b). The latest new phylogenetic species, *F. ussurianum* was found in the Russian Far East. Results of these analyses resolved the isolates of *F. ussurianum* as phylogenetically distinct, reciprocally monophyletic sister lineage of *F. asiaticum*, which together with *F. vorosii*, form a newly discovered Asian clade within the *F. graminearum* species complex. None of the 13 species of the *F. graminearum* species complex can be identified by morphology (Yli-Mattila et al., 2009b). *F. culmorum* and *F. cerealis* are closely related to *F. graminearum sensu lato*, but they can be identified both by morphology and molecular markers.

SECTION SPOROTRICHIELLA: *F. sporotrichioides* Sherb. (1915), *F. poae* (Peck) Wollenw. (1913), *F. langsethiae* Torp et Nirenberg (2004), and *F. tricinctum* (Corda) Sacc. (1886)

The teleomorph stage in section Sporotrichiella has been described only for *F. tricinctum* as *Gibberella tricincta* El-Gholl, McRitchie, Schoult. & Ridings, 1978, which according to molecular and phylogenetic analyses is not closely related to the rest of the species of this section (Yli-Mattila et al., 2002; Kristensen et al., 2005; Leslie and Summerell, 2006; Harrow et al., 2010). Species of section Sporotrichiella are relatively weak pathogens as compared to *F. graminearum* and *F. culmorum* (Parry et al., 1995).

F. sporotrichioides and *F. poae* isolates produce abundant aerial mycelium which is floccose to cottony, whitish, pink-red and peach (Fig. 10.3). Conidiophores in *F. sporotrichioides* are both unbranched and branched with monophialides and polyphialides. Macroconidia, mesoconidia and different types of microconidia are abundant. Sporodochia are rare and red-brown. Chlamydospores are intercalary, mostly in chains, light brown and smooth-walled (Fig. 10.4). Macroconidia in *F. poae* are not usually formed, but they are only produced sparsely under UV light (Torp and Nirenberg, 2004). Microconidia are abundant, hyaline, globose, apiculate, pyriform, mostly one-celled and formed in unstable false heads. Phialides are only monophialidic (Van Wyk et al., 1991). Conidiophores with cylindric cells at first unbranched, ater branched with short and broad monophialides. In some strains a strong sweetish odour is present.

The differences of *F. langsethiae* in cultural characters as compared to *F. poae* include less abundant aerial mycelium without odour (Table 10.2, Fig. 10.3). The colour of aerial mycelia varies from creamy white to pinkish white. Pigmentation in reverse is creamy-white, pale red and pale violet, in the centre often with a

Fig. 10.3 *Fusarium* species (PSA, 2 weeks in dark): *F. poae, F. langsethiae, F. sporotrichioides.* *Color image of this figure appears in the color plate section at the end of the book.*

more bluish tinge Microconidia are abundant, hyaline, globose, apiculate, very rarely pyriform, mostly one-celled and formed in stable false heads. Ampulliform and bent cylindrical phialides, sometimes with polyphialides with 2 loci are typical for *F. langsethiae*. Macroconidia, sporodochia, chlamydospores, are absent (Fig. 10.4).

F. langsethiae is a European species, which was first found in Norway (Torp and Langseth, 1999) and later characterized and named as a new species (Knutsen et al., 2004; Torp and Nirenberg, 2004; Yli-Mattila et al., 2004a). Previously, *F. langsethiae* strains were usually identified as *F. poae.*

F. poae, F. sporotrichioides and *F. langsethiae* are type A trichothecene producers, but only a few *F. poae* isolates can produce small amounts of T-2 and HT-2, while NIV is the main mycotoxin produced by *F. poae* (Jestoi et al., 2004a and b, 2008; Thrane et al., 2004; Vogelgsang et al., 2008). In *F. poae* the TRI16 gene, which is needed for T-2 production, is truncated and therefore most likely nonfunctional, which probably explains, why it can not produce T-2 toxin (Proctor et al., 2009).

Morphologically *F. langsethiae* (Fig. 10.4a) is more similar to *F. poae* (Fig. 10.4b), but its metabolite profile is closer to that of *F. sporotrihichioides* and it is capable of producing high levels of HT-2 and T-2 toxins unlike *F. poae* (Torp and Langseth, 1999; Thrane et al., 2004; Torp and Nirenberg, 2004; Yli-Mattila et al., 2004a; Wilson et al., 2004). A morphologically intermediate isolate between *F. langsethiae* and *F. sporotrichioides* was found in Norway. According to molecular and toxin data it was, however, more closely related to *F. sporotrichioides* than to other European *F. langsethiae* isolates (Thrane et al., 2004; Torp and Nirenberg, 2004; Yli-Mattila et al., 2004a; Fig. 10.5). Two subgroups of *F. langsethiae* were found based on ribosomal IGS (intergenic spacer) sequences and RFLP patterns (Konstantinova and Yli-Mattila, 2004). More *Fusarium* isolates similar to *F. poae* and *F. langsethiae* have recently been found in Siberia and the Russian Far East

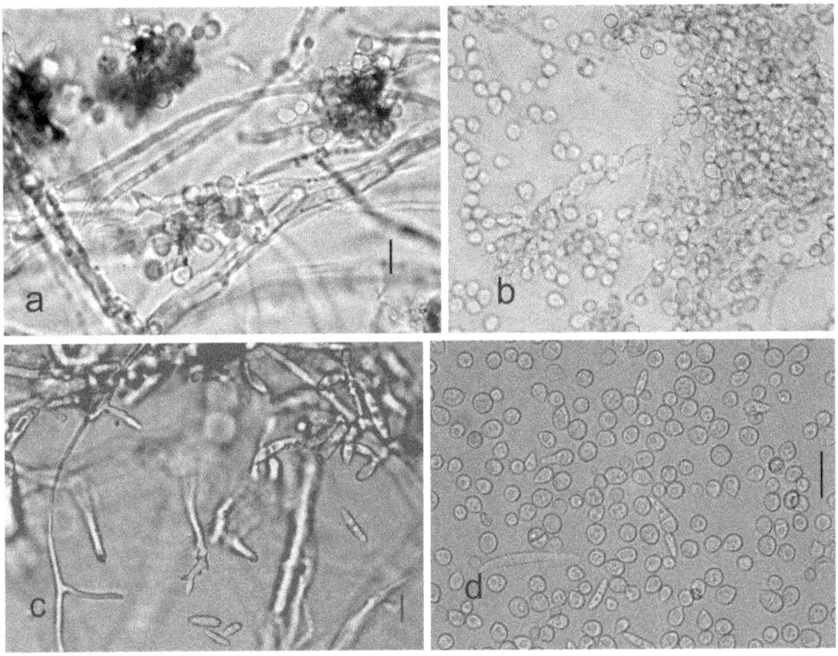

Fig. 10.4a *F. langsethiae (*SNA, 2 weeks). Microconidia formed in stable false heads, phialides mono- and polyphialides with two loci.

10.4b *F. poae* (SNA, 2 weeks). Branched conidiophores with monophialides formed in false heads.

10.4c *F. sporotrichioides* (PSA, 2 weeks). Mono- and polyphialides.

10.4d *F. sporotrichioides* (PSA, 2 weeks). Macroconidia and different types of microconidia.

(Gagkaeva et al., 2007a; Burkin et al., 2008; Yli-Mattila et al., 2009c, 2011). The atypical *F. poae* isolates produce much more T-2 mycotoxin than *F. poae* isolates (Burkin et al., 2008). According to preliminary results (Yli-Mattila et al., 2011) these Asian isolates are similar to isolate IBT9959, which was found earlier in Norway (Yli-Mattila et al., 2004a).

Based on the current morphological taxonomic systems, *F. tricinctum* (Corda) Sacc. (1886) belongs to section *Sporotrichiella* due to its ability to produce pyriform macroconidia but the toxin profile of this species is more similar to *F. avenaceum* than to *F. sporotrichioides* and *F. poae* (Burmeister and Plattner, 1987; Chelkowski et al., 1989; Jestoi et al., 2004a, 2008). There is no *TRI5* gene in *F. tricinctum* (Logrieco et al., 1990; Tan and Niessen, 2003; Kristensen et al., 2005). Moreover, it has been shown that there are cross-reactions with species-specific primers for *F. tricinctum* and *F. avenaceum* (Turner et al., 1998; Yli-Mattila et al., 2002). These findings together with phylogenetic investigations suggest that *F. tricinctum* should be moved from section *Sporotrichiella* to

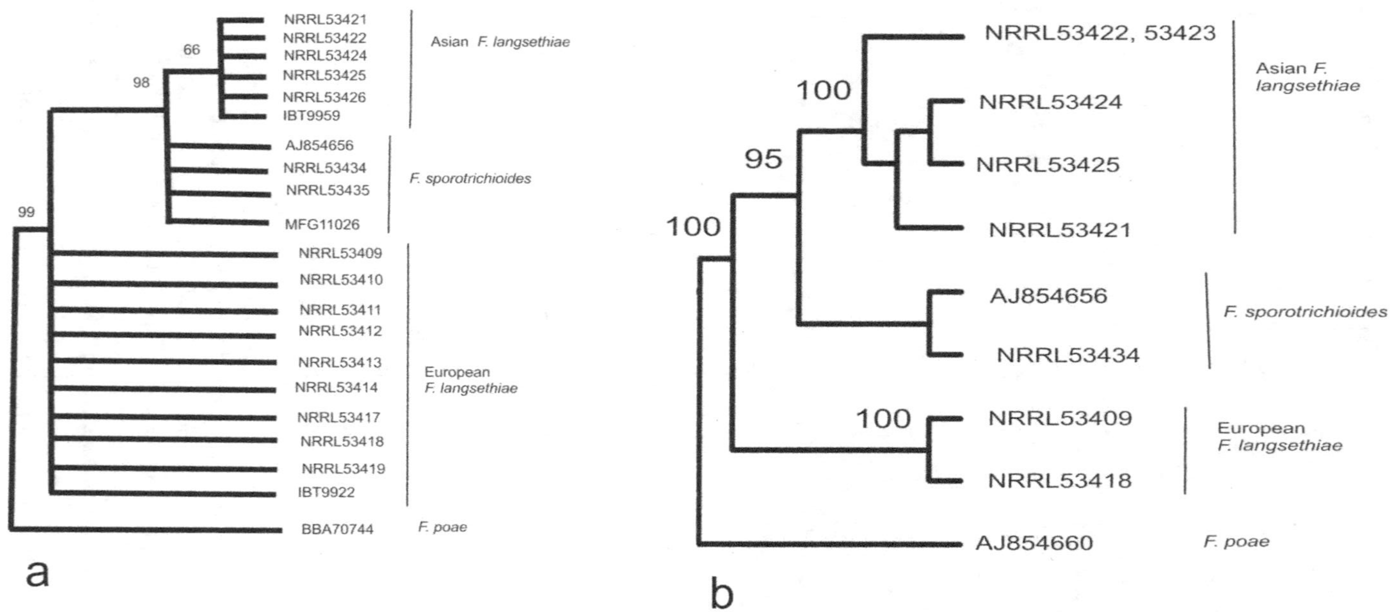

Fig. 10.5 Majority rule and strict NJ consensus tree for partial (a) and complete (b) IGS sequences. Only branches with a bootstrap value higher than 50% are shown. Gaps longer than 3 bp were excluded.

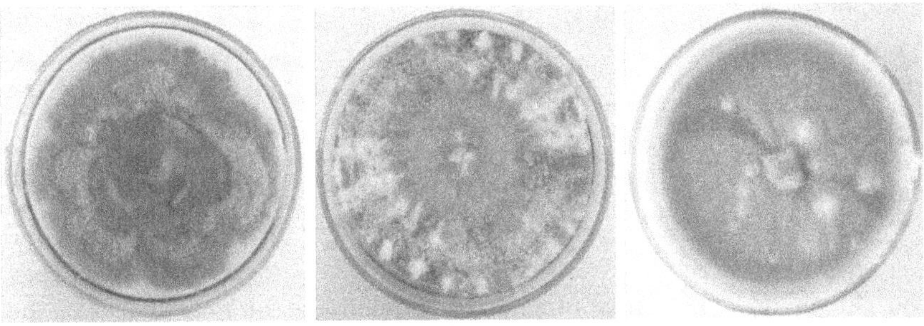

Fig. 10.6 *Fusarium* species (PSA, 2 weeks in dark) *F. tricinctum, F. avenaceum, F. arthrosporioides.*

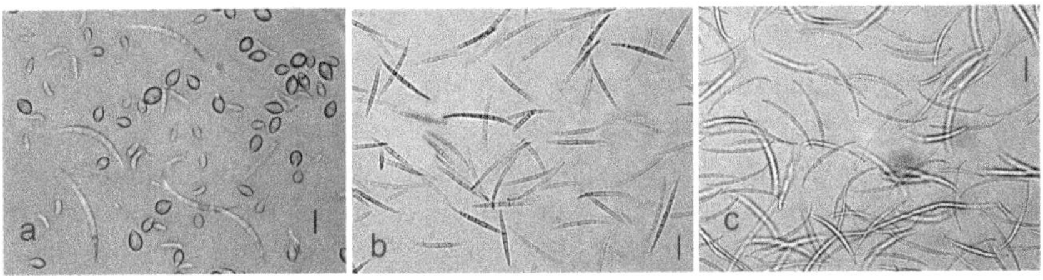

Fig. 10.7a *F. tricinctum.* (PSA, 2 weeks). Macroconidia and different types of microconidia
10.7b *F. arthrosporioides* (SNA, 2 weeks). Macroconidia
10.7c *F. avenaceum* (SNA, 2 weeks). Macroconidia.

section Roseum (Marasas et al., 1984; Turner et al., 1998; Yli-Mattila et al., 2002, 2006; Kristensen et al., 2005; Harrow et al., 2010). The cultures of *F. tricinctum* are similar to *F. avenaceum*, but they usually produce carmin pigment in agar (Fig. 10.6) and their branched conidiophores, bear slender monophialides. The most distinguished character in *F. tricinctum* is the ability to form marked lemon-shaped microconidia together with fusiform, puriform, ellipsoidal ones (Fig. 10.7). Chlamydospores occur rarely.

SECTION ROSEUM: *F. avenaceum* (Fr.) Sacc. (1886), and *F. arthrosporioides* Sherb. (1915)

The teleomorph stage in section *Roseum* has been described only for *F. avenaceum* – *Gibberella avenacea* R.J. Cook (1967). Colonies of *F. avenaceum* are reddish-yellow with whitish overgrowth of aerial mycelium. Later the cultures become yellow-brown in the centre. *F. avenaceum* produces slender macroconidia which are sickle-shaped to straight, narrowing at both ends, 4-7 septate with elongated, elegantly curved apical cell and a distinct foot cell. They are produced in sporodochia which are orange coloured (Fig. 10.6). Microconidia rarely occur in the aerial mycelium and are borne on polyblastic conidiogenous cells. The microconidia are fusiform, 0–3 septate, and variable in size. Conidiophores are simple arising from the aerial mycelium and later branch in the sporodochia which have monophialides. Chlamydospores are absent (Fig. 10.7).

The cultures of *F. avenaceum* are heterogeneous. *F. avenaceum* is considered a species morphologically very closely related to *F. arthrosporioides* (Gerlach and Nirenberg, 1982). These fungi are often confused with each other as the latter species is distinguished from *F. avenaceum* by only a few traits (Gerlach and Nirenberg, 1982). Reddish-pink cottony mycelium, characteristic of *F. arthrosporioides,* never forms the distinctive orange sporodochia as are formed in *F. avenaceum* (Fig. 10.6). Macroconidia are the same as in *F. avenaceum.* Conidiogenous cells are blastic (with 2–3 loci) and phialidic. Often loosely branching conidiophores form on the surface of cultures in dry beige-cream aggregations. Its distinctive feature is the formation of more or less abundant fusiform and comma shape 0-3 septate microconidia (Fig. 10.7). Several reports consider these species as synonymous conspecific (Uhlig et al., 2007).

F. avenaceum is a well known component of the species complex implicated in *Fusarium* head blight of cereals worldwide and it is a particularly dominant species in cool cereal-growing regions. The monitoring of the occurrence of *Fusarium* species on cereals in different geographical areas of Russia has shown that the population of *F. avenaceum/F. arthrosporioides* is common among cereals across the whole of Russia, although these two species were isolated with high frequency from kernel samples mainly from the north-western regions (Gagkaeva, 2003; Gagkaeva et al., 2009). *F. arthrosporioides* also occurs as a contaminant of cereals worldwide (Yli-Mattila et al., 2002, 2004b; Kosiak et al., 2004; Uhlig, 2005; Uhlig et al., 2007).

These two species are considered to be chiefly responsible for the accumulation of cyclic hexadepsipeptides (enniatins including beauvericin) and moniliformin in cereals in northern Europe (Logrieco et al., 2002; Jestoi et al., 2004b, 2008; Gagkaeva et al., 2007a; Uhlig et al., 2007). Toxin profiles have also been shown to be similar for both species and also with *F. tricinctum* (Langseth et al., 1999; Jestoi et al., 2004b, Uhlig et al., 2007). Although, toxin profiles of all these species are similar, they are not able to produce trichothecenes (Jestoi et al., 2004b, 2008; Uhlig et al. 2007).

Molecular data such as IGS, ITS and β-tubulin sequences have indicated a close relationship between these species (Yli-Mattila et al., 2002). It is also observed that several *F. arthrosporioides* isolates from different parts of Europe are closely related to a population of *F. avenaceum* called main group II by Yli-Mattila et al. (1998). This population was clearly separated from other

F. avenaceum isolates based on RAPD-PCR (Yli-Mattila et al., 1996) and UP-PCR (Yli-Mattila et al., 1997). Moreover, further phylogenetic analysis of combined β-tubulin, IGS and ITS sequences of both species showed that *F. arthrosporioides* and *F. avenaceum* did not resolve into clearly separated clades, confirming the genetic similarity between the isolates of both species and the difficulties in obtaining a clear separation of the two species (Yli- Mattila et al., 2004, 2006). However, some subgroups of the *F. arthrosporioides* and *F. avenaceum* isolates were supported both by DNA sequences and chemical data (Yli-Mattila et al., 2006). It is, therefore, possible to distinguish most of the strains of the three species, including degenerated strains by using a combination of three primer pairs designed for the main group of *F. avenaceum, F. arthrosporioides* and *F. tricinctum* (Turner et al., 1998; Yli-Mattila et al., 2002, 2004c).

Differences in ITS, IGS, mtSSU and β-tubulin sequences, UP-PCR hybridization and morphology have been analyzed between mainly northern European and international *F. avenaceum, F. anguioides, F. arthrosporioides, F. tricinctum, F. graminum* and *F. acuminatum* strains (Yli-Mattila et al., 2002), while Harrow et al. (2010) used mtSSU, β-tubulin and elongation factor 1 alpha sequences for analyzing *F. avenaceum, F. tricinctum* and *F. acuminatum* strains. In both cases quite a lot of variation was found within *F. avenaceum* and *F. tricinctum* isolates. Thus, further phylogenetic work is required in this group of *Fusarium* species.

Morphological and biological species concepts

Macroscopic and microscopic features, such as, colour of the colony, shape and length of the macroconidia, existence, shape and arrangement of microconidia, and presence or absence of chlamydospores are key features for the differentiation of *Fusarium* species. Identification by using the morphologial characters has some problems. The most important among them are the high variability of species-specific characters and overlapping of data. A lot of experience is needed for distinguishing closely related species, and often different investigators use different standards of species concepts. The stability of morphological characters in culture collections depends on conditions, and sometimes studied cultures of fungi are degenerated. Often among identified isolates, some isolates are difficult to differentiate because the macroconidia are not produced or the sum of acceptable characters are intermediate and are not typical of either species. Morphological species identification is also time consuming.

Biological species concept (Leslie and Summerell, 2006) has been useful in some cases (especially in *Gibberella fujikoroi* species complex), but it also has limitations as a tool for identification of unknown *Fusarium* isolates due to the lack of sexual stage in many *Fusarium* species. This concept has been developed only for a few species having a teleomorph and it is a slow method to use and sometimes the results are not clear.

Molecular methods and species concept

DNA extraction and molecular markers

The basis for molecular DNA analyses of plant pathogenic *Fusarium* fungi is DNA extraction and purification. In fungi and plants, the cell walls should be broken and different PCR inhibitors should be removed. Younger plants and fungi usually contain fewer inhibitory compounds than older ones. RNA can be degraded by RNAase. Different commercial kits are now available for DNA extraction and should be optimized for the fungal DNA from plant material and soil. The quality and quantity of DNA can be estimated by gel electrophoresis (Fig. 10.8) and spectrophotometry (e.g. Yli-Mattila et al., 1996, 2008) or by fluorescence of specific dyes (e.g. Yli-Mattila et al., 2010).

Species-specific primers can be designed by comparing known *Fusarium* DNA sequences from different species. Alternatively, primers can be designed based for species-specific RAPD-PCR (random amplified polymorphic DNA), UP-PCR (universally primed PCR), RAMS (random amplified microsatellites), etc. products (Yli-Mattila and Paavanen-Huhtala, 2007; Niessen, 2008). The use of species-specific primers (e.g. Doohan et al., 1998, Waalwijk et al., 2004; Yli-Mattila et al., 2004c; Jurado et al., 2006), microarrays and SNP (single nucleotide polymorphisms) analysis has made it possible to quickly detect, identify and quantify isolates of *Fusarium* species based on molecular data (Kristensen, 2007a, b). The advantages of TaqMan qPCR analyses have been discussed by Waalwijk et al. (2004), Sarlin et al. (2006) and Yli-Mattila et al. (2008, 2009a). Molecular data is also useful material for phylogenetic studies (O'Donnell et al., 2000, 2004; Yli-Mattila et al., 2004a; Kristensen et al., 2005). It is important that enough well-identified isolates and/or DNA sequences from GenBank, of the species for which the primers or probes are designed and of closely related species, are available for testing species-specific primers and probes before they are used for identification; the isolates should be single-spore isolates, identified by both morphological and molecular characters.

For AFLPs (amplified fragment length polymorphism) the overall similarity of the bands is usually more important than individual bands. Within the Liseola section, strains that share 60% of the bands usually belong to the same species, those that share <40% of the bands belong to different species and those that share 40-60% of their bands need additional evaluation by different molecular and morphological methods (Leslie et al., 2007).

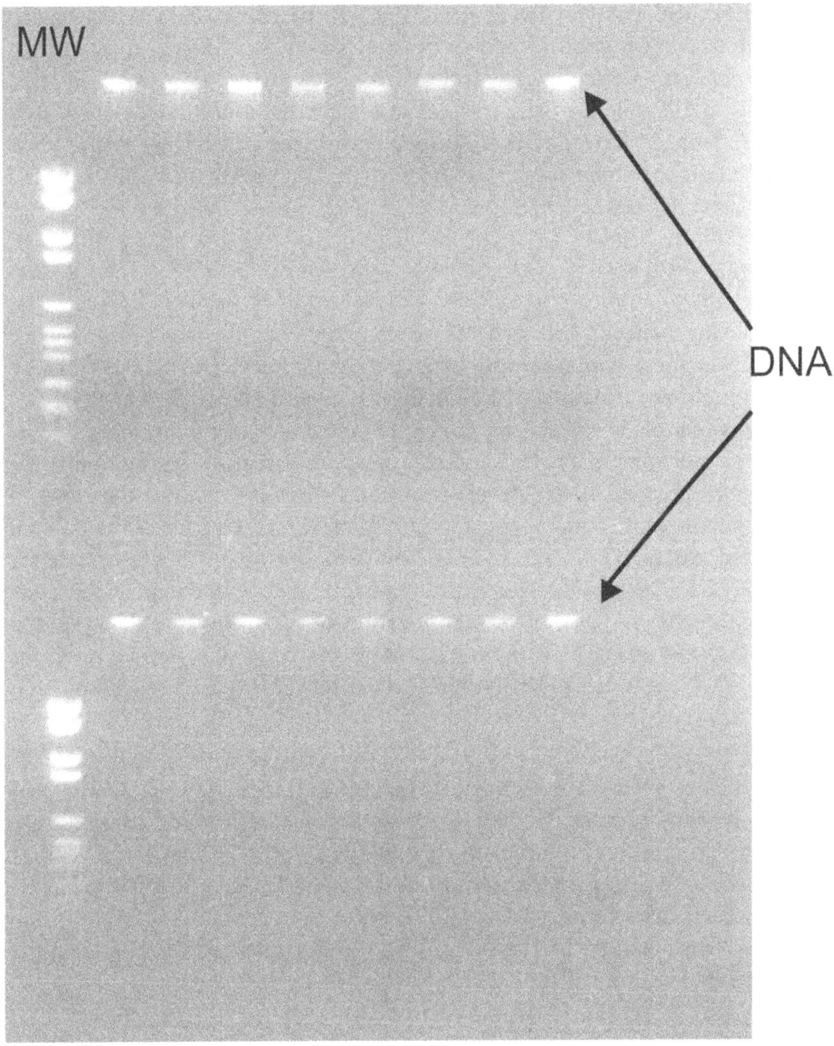

Fig. 10.8 DNA from 16 grain samples (on two lines) as compared to the bands of molecular weight marker (first wells on the left). The DNA amount in the bands of molecular weight marker varied between 3 and 30 ng and the amount of DNA in grain samples varied between 6-15 ng.

The advantage of primers based on ribosomal DNA is that there are numerous ribosomal DNA copies in the genome, making the reaction more sensitive. Internal transcribed spacer (ITS) and intergenic spacer (IGS) region sequences of ribosomal DNA are also highly variable. On the other hand, there may be different ITS sequences present in the same species, which in certain cases makes ITS sequences unreliable in species identification and especially in phylogenetic analyses (O'Donnell et al., 1997; Leslie et al., 2007). According to Jurado et al. (2005), species-specific primers based on IGS sequences can be used for separating *F. culmorum* isolates from *F. graminearum* isolates, while according to O'Donnell et al. (2004) *F. culmorum* is nonmonophyletic based on IGS sequences. Thus, further studies with more isolates are required to determine how well these species can be separated on the basis of IGS sequences (Yli-Mattila et al., 2004a, 2007).

DNA sequencing and phylogenetic analyses

Several phylogenetic analyses of DNA sequences during the last few years have revealed that most commonly reported *Fusarium* species actually represent species complexes (O'Donnell et al., 2000; Kvas et al., 2009; Migheli et al., 2010). This means that many species can not be identified by using morphology alone. Intron-rich portions of protein-coding genes are generally considered to be favourable markers for species-level phylogenetics in fungi, because they tend to evolve at a higher rate than more commonly applied markers such as the internal transcribed spacer (ITS) regions of the nuclear ribosomal RNA gene (Geiser et al., 2004). Elongation factor 1 alpha sequences are at the moment the most widely used DNA sequences for identification of *Fusarium* species. A special FUSARIUM-ID database (http://isolate.fusariumdb.org/) has been formed for *Fusarium* DNA sequences. Kristensen et al. (2007a, b) have used Elongation factor 1 alpha sequences for developing an microarray for identification of trichothecene and moniliformin producing *Fusarium* species.

Ribosomal ITS (internal transcribed spacer) sequences have been also used for species identification. But in several cases it has been found that they are not species-specific, which means that closely related species may have identical ITS sequences (e.g. in *F. graminearum* complex and between *F. sporotrichioides* and *F. langsethiae* isolates) or several ITS types may be found within a group of species, which can lead to incorrect phylogenetic inferences (O'Donnell and Cigelnik, 1997; O'Donnell et al., 2000; Yli-Mattila et al., 2004a).

In the genus *Fusarium*, ribosomal IGS sequences have also often been useful in species identification (Appel and Gordon, 1996; Yli-Mattila et al., 2004a, 2008; Konstantinova and Yli-Mattila, 2004; Jurado et al., 2006; O'Donnell et al., 2008, 2009), but in some cases IGS sequences have also been nonmonophyletic (O'Donnell et al., 2004) as was already mentioned above. The phylogenetic analysis by combining sequences of several loci (multilocus sequencing) is called Genealocical Concordance Phylogenetic Species Recognition (GCPSR, Taylor et al., 2000).

It has been demonstrated that in some cases certain genes may have different evolutionary histories from the majority of genes. For example trichothecene gene cluster is trans-specific and appears to originate from the common ancestor of type B trichothecene producers (Ward et al., 2002), which means that these genes are not suitable for GCPSR. So, it is best first to look at the genes and their phylogenetic trees individually to find out, how well they fit with each other. The sequences can also be tested by partition homogeneity test (Swofford, 2001). If the trees are very different and the partition homogeneity test shows incongruence in the combined data set, it is better to try by different combinations of gene sequences and to exclude the most different sequences in order to find a combined data set with a congruent partition test value.

Luminex system

In MLGT (multilocus genotyping) analyses e.g. SNPs of several loci are used for genotyping and species identification. The species and trichothecene chemotype composition of *F. graminearum* species complex and closely related species, such as *F. culmorum* and *F. cerealis*, can be analyzed by using multiplex PCR with several primer pairs followed by probes of the gene sequences produced by PCR in the multilocus genotyping (MLGT) assay (Fig. 10.9).The products of the first PCR were used as templates in the multilocus genotyping assay. Multiplex amplifications, allele-specific primer extensions, microsphere hybridization and detection are performed as described by Ward et al. (2008). Similar MLGT analyses should also be developed for other *Fusarium* species complexes.

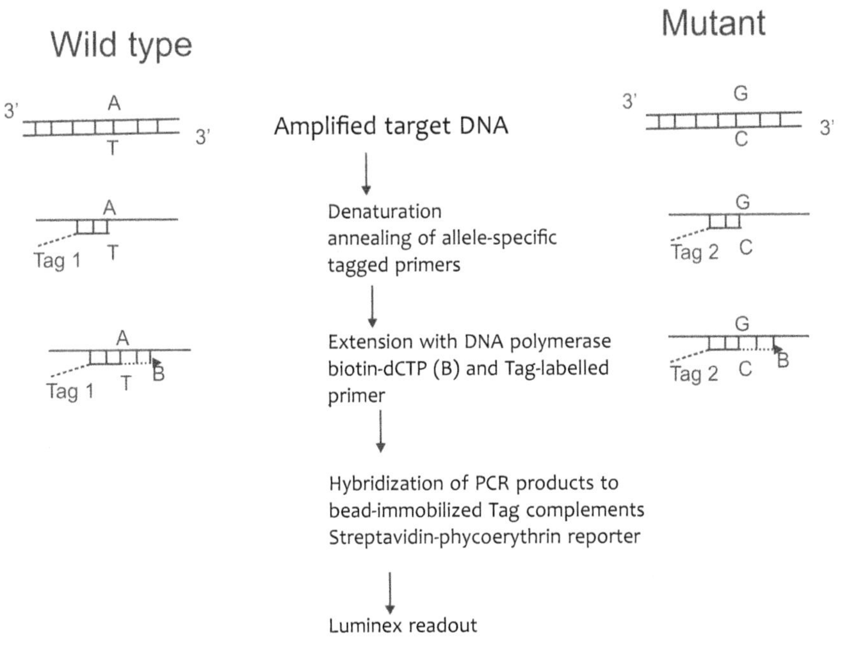

Fig. 10.9 A typical primer extension reaction.

CONCLUSIONS

In the most reliable species definitions morphological, biological and phylogenetic species concepts are consistent, but the concept of biological species cannot be used for anamorphic species, while the subjective nature of the concept of morphological species may also cause problems, especially with closely related species. The use of phylogenetic species concept alone may also cause problems, because it may yield results, which are not biologically meaningful. The best way to identify and classify *Fusarium* isolates is the polyphasic approach by using all available characters.

FUTURE PERSPECTIVES

In future, it will probably be possible to find more lineages and phylogenetic species within most *Fusarium* species. The current molecular methods will be improved, validated and automated, making it possible to use them in routine identification, detection and quantification of *Fusarium* species. The most promising techniques for the large-scale identification of different fungal species and chemotypes are microarray-based methods and other hybridization methods (e.g. Luminex) detecting single nucleotide polymorphisms. These methods will reveal the scientific classification of genus *Fusarium* for not only the researchers relying on traditional microscopic and macroscopic characters and classification, but also for those, who are investigating the real nature of organisms based on all polyphasic characters.

The currently available quick systems for PCR-based detection, identification and quantification in cereal grains should be used in food and feed quality control and even better and more reliable molecular methods should be developed, because they will also tell about the mycotoxigenic potential of analyzed samples. This information can be used in order to quickly decide, which samples should be analyzed by more expensive and slower chemical analyses of mycotoxins. The PCR-based methods will also be useful in scoring FHB in resistance breeding against *Fusarium* species, because it is difficult to detect and score FHB in infected heads or grains visually in barley and oats.

ACKNOWLEDGEMENTS

This work was supported by grants from the Academy of Finland (no. 126917 and 131957) and the Nordic network project New Emerging Mycotoxins and Secondary Metabolites in Toxigenic Fungi of Northern Europe (project 090014), which is funded by the Nordic Research Board. The author is grateful to Dr. Tatiana Gagkaeva from the All-Russian Plant Protection Institute in St. Petersburg in Russia for photographs, tables, valuable comments and references for the manuscript, and to Dr. Olga Gavrilova from the same institute for checking the text.

REFERENCES

Appel, D. J. and Gordon, T. R. 1996. Relationships among pathogenic and nonpathogenic isolates of *Fusarium oxysporum* based on the partial sequence of the intergenic spacer region of the ribosomal DNA. Molecular Plant–Microbe Interactions, 9: 125-238.

Bilai, V. I. 1977. *Fusarium* genus (biology and systematic). Kiev, Nauk. Dumka (in Russian).

Booth, C. 1971. The genus *Fusarium*. Commonwealth Mycological Institute, Kew, Surrey, England.

Bottalico, A. 1998. *Fusarium* diseases of cereals: species complex and related mycotoxin profiles in Europe. Journal of Plant Pathology, 80: 85-103.

Bottalico, A. and Perrone, G. 2002. Toxigenic *Fusarium* species and mycotoxins associated with head blight in small-grain cereals in Europe. European Journal of Plant Pathology, 108: 611-624.

Burkin, A. A., Soboleva, N. A. and Kononenko, G. P. 2008. T-2 toxin production ability of *Fusarium poae* isolates originated from East Siberian and the Far Eastern region of Russia. Mycology and Phytopathology, 42: 354-358 (in Russian).

Burmeister, H. R. and Plattner, R. D. 1987. Enniatin production by *Fusarium tricinctum* and effect on germinating wheat seeds. Phytopathology, 77: 1483-1487.

Chandler, E. A., Simpson, D. R., Sthonsett, M. A. and Nicholson, P. 2003. Development of PCR assays to *Tri7* and *Tri13* trichothecene biosynthetic genes, and characterisation of chemotypes of *Fusarium graminearum* and *Fusarium cerealis*. Physiological and Molecular Plant Pathology, 62: 355-367.

Chelkowski, J., Manka, M., Kwasna, H., Visconti, A. and Golinski, P. 1989. *Fusarium sporotrichioides* (Sherb.), *Fusarium tricinctum* (Corda) Sacc. and *Fusarium poae* Wollenw.: cultural characters, toxigenicity and pathogenicity towards cereals. Journal of Phytopathology, 124: 155-161.

Doohan, F. M., Parry, D. W., Jenkinson, P. and Nicholson, P. 1998. The use of species-specific PCR-based assays to analyse *Fusarium* ear blight of wheat. Plant Pathology, 47: 197-205.

Elen, O. N., Klemsdal, S. S., Clasen, P-E. and Razzaghian, M. J. 2007. *Fusarium* spp. and mycotoxins in Norwegian wheat grain (2001-2004). XV Congress of European Mycologists, St. Petersburg, Russia, September 16-21, 2007, 277-278 (Abstract).

Eskola, M., Parikka, P. and Rizzo, A. 2001. Trichothecenes, Ochratoxin A and Zearalenone contamination and *Fusarium* infection in Finnish cereal samples in 1998. Food Additives and Contaminants, 18: 707-718.

Fredlund, E., Gidlund, A., Olsen, M., Börjesson, T., Spliid, N. H. H. and Simonsson, M. 2008. Method evaluation of *Fusarium* DNA extraction from mycelia and wheat for down-stream real-time PCR quantification and correlation to mycotoxin levels. Journal of Microbiological Methods, 73: 33-40.

Gagkaeva, T. 2003. Importance of Fusarium Head Blight in Russia and the search for new sources of genetic resistance in wheat and barley, pp. 219-222. In: Proceeding of the National Fusarium Head Blight Forum, S. M. Canty, J. Lewis and R. W. Ward (eds.). Dec. 13-15, 2003, MN, USA.

Gagkaeva, T. 2009. Phytopathogenic fungus *Fusarium cerealis* on the territory of Russia. Mycology and Phytopathology, 43: 331-342.

Gagkaeva, T. and Yli-Mattila, T. 2004. Genetic diversity of populations of *Fusarium graminearum* in Europe and Asia. European Journal of Plant Pathology, 110: 551-562.

Gagkaeva, T., Levitin, M. and Yli-Mattila, T. 2002. Vegetative compatibility and incompatibility of nitrate nonutilizing mutants in *Fusarium avenaceum*. Journal of Applied Genetics, 43A: 45-54.

Gagkaeva, T., Gavrilova, O., Kononenko, G., Burkin, A. and Yli-Mattila, T. 2007a. Distribution, cultural characters and toxin production of *Fusarium poae*, *F. sporotrichioides* and *F. langsethiae* isolates. XV Congress of European Mycologists, St. Petersburg, Russia, September 16-21, 2007: 277-278 (Abstract).

Gagkaeva, T., Somma, S., Mule, G., Logrieco, A., Ritieni, A., Sumikova, T., Nicholson, P. and Moretti, A. 2007b. Genetic diversity among *Fusarium avenaceum* strains originated from Russia. XV Congress of European Mycologists, St. Petersburg, Russia, September 16-21, 2007: 276-277 (Abstract).

Gagkaeva, T. Yu., Levitin, M. M., Sanin, S. S. and Nazarova, L. N. 2009. *Fusarium* head blight of small cereals and causal organisms associated with this disease on the territory of Russia during 2004-2006. Agro, 21: 4-6 (in Russian).

Geiser, D. M., Jimenez-Gasco, M. N., Kang, S. M. I., Veeraghavan, N., Ward, T. J., Zhang, N., Kuldau, G. A. and O'Donnell, K. 2004. Fusarium-ID v.1.0: a DNA sequence database for identificatifying *Fusarium*. European Journal of Plant Pathology, 110: 473-479.

Gerlach, W. and Nirenberg, H. 1982. The Genus *Fusarium* – A Pictorial Atlas. P. Parey, Berlin.

Goswami, R. S. and Kistler, C. 2004. Heading for disaster: *Fusarium graminearum* on cereal crops. Molecular Plant Pathology, 5: 515-525.

Harrow, S. A., Farrokhi-Nejad, R., Pitman, A. R., Scott, I. A. W., Bentley, A., Hide, C. and Cromey, M. G. 2010. Characterisation of New Zealand *Fusarium* populations using a polyphasic approach differentiates the *F. avenaceum/F. acuminatum/F. tricinctum* species complex in cereal and grassland systems. Fungal Biology, 114: 293-311.

Hughes, S. J. 1953. Conidiophores, conidia, and classification. Canadian Journal of Botany, 31: 577-659.

Ivanchenko, M. Y. 1960. Causal organisms of Fusarium head blight and their teleomorph stage in North Ossetia. Plant Protection, 12: 26-28 (in Russian).

Jennings, P., Coates, M. E., Walsh, K., Turner, J. A. and Nicholson, P. 2004. Determination of deoxynivalenol- and nivalenol-producing chemotypes of *Fusarium graminearum* isolates from wheat crops in England and Wales. Plant Pathology, 53: 643-652.

Jestoi, M., Paavanen-Huhtala, S., Uhlig, S., Rizzo, A. and Yli-Mattila, T. 2004a. Mycotoxins and cytotoxicity of Finnish *Fusarium* strains grown on rice cultures, pp. 405-409.

In: Proceedings of the 2nd International Symposium on Fusarium Head Blight; incorporating the 8th European Fusarium seminar; 2004, 11-15; Orlando, Fl, USA. S. M. Canty, T. Boring, J. Wardwell and R. W. Ward (eds.). Michigan State University, East Lansing, MI., USA.

Jestoi, M., Rokka, M., Yli-Mattila, T., Parikka, P., Rizzo, A. and Peltonen, K. 2004b. The presence and concentrations of the *Fusarium*-related mycotoxins Beauvericin, Enniatins and Moniliformin in Finnish grain samples. Food Additives and Contaminants, 21: 794-802.

Jestoi, M., Paavanen-Huhtala, S., Parikka, S. and Yli-Mattila, T. 2008. *In vitro* and *in vivo* mycotoxin production of *Fusarium* species isolated from Finnish grains. Archives of Phytopathology and Plant Protection, 41: 545-558.

Jurado, M., Vazquez, C., Marin, S., Patino, B. and Gonzales-Jaen, M. T. 2006. PCR detection assays for the trichothecene-producing species *Fusarium graminearum*, *Fusarium culmorum*, *Fusarium poae*, *Fusarium equiseti* and *Fusarium sporotrichioides*. Systematic and Applied Microbiology, 28: 562-568.

Kirienkova, A. E . 1992. Fusarium head blight of cereals in Krasnodar krai. In: Fusarium Head Blight of Cereals. V. T. Goncharov (ed.). Krasnodar. 27: 28 (in Russian).

Knutsen, A. K., Torp, M. and Holst-Jensen, A. 2004. Phylogenetic analyses of the *Fusarium poae, Fusarium sporotrichioides* and *Fusarium langsethiae* species complex based on partial sequences of the translation elongation factor-1 alpha gene. International Journal of Food Microbiology, 95: 287-295.

Konstantinova, P. and Yli-Mattila, T. 2004. IGS-RFLP analysis and development of molecular markers for identification of *F. poae, F. langsethiae, F. sporotrichioides* and *F. kyushuense*. International Journal of Food Microbiology, 95: 321-331.

Kosiak, B., Torp, M. and Skjerve, E. 2003. The prevalence and distribution of *Fusarium* species in Norwegian cereals: a survey. Acta Agriculturae Scandinavica Section B-Soil and Plant Science, 53: 168-176.

Kosiak, B., Torp, M., Skjerve, E. and Andersen, B. 2004. *Alternaria* and *Fusarium* in Norwegian grains of reduced quality – a matched pair sample study. International Journal of Food Microbiology, 93: 51-62.

Kristensen, R., Gauthier, G., Berdal, K. G., Hamels, S., Remacle, J. and Holst-Jensen, A. 2007a. DNA microarray to detect and identify trichothecene- and moniliformin-producing *Fusarium* species. Journal of Applied Microbiology, 102: 1060-1070.

Kristensen, R., Berdal, K. G. and Holst-Jensen, A. 2007b. Simultaneous detection and identification of trichothecene- and moniliformin-producing DNA microarray to detect and identify trichothecene- and moniliformin-producing *Fusarium* species based on multiplex SNP analysis. Journal of Applied Microbiology, 102: 1071-1081.

Kristensen, R., Torp, M., Kosiak, B. and Holst-Jensen, A. 2005. Phylogeny and toxigenic potential is correlated in *Fusarium* species as revealed by partial translation factor 1 alpha gene sequences. Mycological Research, 109: 173-186.

Kryuchkova, L. O., Raichuk, L. B. and Michailenko, S. V. 2001. The pathogen of *Fusarium* head blight. Zernovi, 3: 12-13 (in Ukrainian)

Kvas, M., Marasas, W. F. O., Wingfield, B. D., Wingfield, M. J. and Steenkamp, E. T. 2009. Diversity and evolution of *Fusarium* species in the *Gibberella fujikuroi* complex. Fungal Diversity, 34: 1-21.

Langseth, W., Bernhoft, A., Rundberget, T., Kosiak, B. and Gareis, M. 1999. Mycotoxin production and cytotoxicity of *Fusarium* strains isolated from Norwegian cereals. Mycopathologia, 144: 103-113.

Langseth, W., Olen, O. and Rundberget, T. 2001. Occurrence of mycotoxins in Norwegian cereals 1988-1999, pp. 105-110. In: Occurrence of Toxigenic Fungi Mycotoxins in Plants, Food and Feed in Europe. A. Logrieco (ed.). COST Action 835, European Commission.

Lee, J., Chang, I. Y., Kim, H., Yun, S. H., Leslie, J. F. and Lee, Y.-W. 2009. Genetic diversity and fitness of *Fusarium graminearum* populations from rice in Korea. Applied and Environmental Microbiology, 75: 3289-3295.

Leslie, J. F. and Summerel, B. A. 2006. The Fusarium Laboratory Manual. Blackwell Publishing Professional, Ames, IA. USA.

Leslie, J. F., Anderson, L. L., Bowden, R. L. and Lee, Y.-W. 2007. Inter- and intra-specific genetic variation in *Fusarium.* International Journal of Microbiology, 119: 25-32.

Levitin, M. 2001. Distribution and toxicology *Fusarium* species on cereals in Russia, pp. 151-157. In: Occurrence of Toxigenic Fungi Mycotoxins in Plants, Food and Feed in Europe. A. Logrieco (ed.). COST Action 835, European Commission.

Levitin, M. 2004. Toxigenic fungi and mycotoxins in cereals grain and food in Russia. Toxigenic fungi and mycotoxins in Finnish cereals, pp. 195-199. In: An Overview on Toxigenic Fungi and Mycotoxins in Europe. A. Logrieco and A. Visconti (eds.). Kluwer Academic Publishers, The Netherlands.

Link, H. F. 1809. Observations in ordines plantarum naturals, Dissetatio I. Gesellschaft Naturforschender Freunde, Berlin, 3: 3-42.

Logrieco, A., Chelkowski, J., Botalico, A. and Visconti, A. 1990. Further data on specific trichothecene production by *Fusarium* section Sporotrichiella. Mycological Research, 94: 587-589.

Logrieco, A., Rizzo, A., Ferracane, R. and Ritieni, A. 2002. Occurrence of beauvericin and enniatins in wheat affected by *Fusarium avenaceum* head blight. Applied and Environmental Microbiology, 68: 82-85.

Marasas, W. F. O., Nelson, P. E. and Toussoun, T. A. 1984. Toxigenic *Fusarium* species. Identity and Mycotoxicology. Pennsylvania State University Press, University Park, USA.

McMullen, M., Jones, R. and Gallenberg, D. 1997. Scab of Wheat and Barley: A Re-emerging Disease of Devastating Impact. Plant Disease, 81: 1340-1346.

Migheli, Q., Balmas, V., Harak, H., Sanna, S., Schern, B., Aoki, T. and O'Donnell, K. 2010. Molecular phylogenetic diversity of dermatologic and other human pathogenic fusarial isolates from hospitals in northern and central Italy. Journal of Clinical Microbiology, 48: 1076-1084.

Miller, J. D. R, Greenhalgh, R., Wang, Y. and Lu, M. 1991. Trichothecene chemotypes of three *Fusarium* species. Mycologia, 83: 121-130.

Nelson, P. E., Tousson, T. A. and Marasas, W. F. O. 1983. *Fusarium* species. An Illustrated Manual for Identification. Pennsylvania State University Press, University Park, USA.

Nicholson, P., Chandler, E., Draeger, R. C., Gosman, N. E., Simpson, D. R., Thomsett, M. and Wilson, A. H. 2003. Molecular tools to study epidemiology and toxicology of *Fusarium* head blight of cereals. European Journal of Plant Pathology, 109: 691-703.

Nicolaisen, M., Supronien, S., Nielsen, L. K., Lazzaro, I., Spliid, N. H. and Justesen, A. F. 2009. Real-time PCR for quantification of eleven individual *Fusarium* species. Journal of Microbiological Methods, 76: 234-240.

Niessen, L. 2008. PCR-based diagnosis and quantification of mycotoxin-producing fungi. Advances in Food and Nutrition Research, 54: 81-138.

Nirenberg, H. I. and O' Donnell, K. 1998. New *Fusarium* species and combinations within the *Gibberella fujikuroi* species complex. Mycologia, 90: 434-458.

O' Donnell, K. 1996. Progress towards a phylogenetic classification of *Fusarium*. Sydowia, 48: 57-70.

O' Donnell, K. and Cigelnik, E. 1997. Two divergent intragenomic rDNA ITS2 types within a monophyletic lineage of the fungus *Fusarium* are nonorthologous. Molecular Phylogenetics and Evolution, 7: 103-116.

O' Donnell, K., Cigelnik, E. and Nirenberg, H. I. 1998. Molecular systematics and phylogeography of the *Gibberella fujikuroi* species complex. Mycologia, 90: 465-493.

O' Donnell, K., Kistler, H. C., Tacke, B. K. and Casper, H. H. 2000. Gene genealogies reveal global phylogeographic structure and reproductive isolation among lineages of *Fusarium graminearum*, the fungus causing wheat scab. Proceedings of the National Academy of Sciences of the United States of America, 97: 7905-7910.

O' Donnell, K., Sutton, K. D. A., Fothergill, A., McCarthy, D., Rinaldi, M. G., Brandt, M. E., Zhang, N. and Geiser, D. M. 2008. Molecular phylogenetic diversity, multilocus haplotype nomenclature and in vitro antifungal resistance within the *Fusarium solani* species complex. Journal of Clinical Microbiology, 46: 2477-2490.

O' Donnell, K., Ward, T. J., Geiser, D. M., Kistler, H. C. and Aoki, T. 2004. Genealogical concordance between the mating type locus and seven other nuclear genes supports formal recognition of nine phylogenetically distinct species within the *Fusarium graminearum* clade. Fungal Genetics and Biology, 41: 600-623.

O' Donnell, K., Gueidan, C., Sink, S., Johnston, P. R., Crous, P. W., Glenn, A., Riley, R., Zitomer, N. C., Colyer, P., Waalwijk, C., van der Lee, T., Moretti, A., Kang, S., Kim, H. S., Geiser, D. M., Juba, J. H., Baayen, R. P., Cromey, M. G., Bithell, S., Sutton, D.A., Skovgaard, K., Ploetz, R., Kistler, H. C., Elliott, M., Davis, M. and Sarver, B. A. J. 2009a. A two-locus DNA sequence database for typing plant and human pathogens within the *Fusarium oxysporum* species complex. Fungal Genetics and Biology, 46: 936-948.

O' Donnell, K., Sutton, K. D. A., Rinaldi, M. G., Gueidan, C., Crous, P. W. and Geiser, D. M. 2009b. A novel MLST scheme reveals high genetic diversity of human pathogenic members of the *Fusarium incarnatum* – *F. equiseti* and *F. chlamydosporium* species complexes within the United States. Journal of Clinical Microbiology, 47: 3851-3861.

Paavanen-Huhtala, S., Hyvönen, J., Bulat, S. A. and Yli-Mattila, T. 1999. RAPD-PCR, isozyme, rDNA RFLP and rDNA sequence analyses in identification of Finnish *Fusarium oxysporum* isolates. Mycological Research, 103: 625-634.

Parry, D. W., Jenkins, P. and McLeod, L. 1995. *Fusarium* ear blight (scab) in small grain cereals – a review. Plant Pathology, 44: 207-238.

Proctor, R., McCormick, S., Alexander, N. and Desjardins, A. 2009. Evidence that a secondary metabolic biosynthetic gene cluster has grown by gene relocation during evolution of the filamentous fungus *Fusarium*. Molecular Microbiology, 74: 1128-1142.

Pronicheva, L. L. 1935. *Fusarium* head blight of wheat in Azovo-Chernomorskom krai in 1934 and effect on yield. Plant Protection, 9: 129-137 (in Russian).

Qu, B., Li, H. P., Zhang, J. B., Xu, Y. B., Huang, T., Wu, A. B., Zhao, C. S., Carter, J., Nicholson, P. and Liao, Y. C. 2008. Geographic distribution and genetic diversity of *Fusarium graminearum* and *F. asiaticum* on wheat spikes throughout China. Plant Pathology, 57: 15-24.

Railo, A. I. 1950. Fungi of genus *Fusarium*. Moscow (in Russian).

Rizzo, A. F. 1993. Determination of Major Naturally Occurring Fusarium Toxins in Finnish Grains and Feeds. The Hemolytic Activity of DON and T-2 Toxin, and the Lipid Peroxidation Induced in Experimental Animals. Ph.D. Thesis, University of Helsinki, Helsinki.

Sarlin, T., Yli-Mattila, T., Jestoi, M., Rizzo, A., Paavanen-Huhtala, S. and Haikara, A. 2006. Real-time PCR for quantification of toxigenic *Fusarium* species in barley and malt. European Journal of Plant Pathology, 114: 371-380.

Selivanova, T. N., Baikova, O. V. and Chernenko, V. Yu. 1991. *Fusarium* head blight in central-chernozem region. In: Problems of Protection of Cereals against Fusarium Head Blight and Others Diseases. L. N. Morosova (ed.). Minsk. 64: 67 (in Russian).

Shashko, M. N. and Gagkaeva, T. Yu. 2010. Detection of *Fusarium cerealis* (Cooke) Sacc. in the pathogen complex of *Fusarium* head blight in Belorussia. Agriculture and forestry (in press, in Russian).

Starkey, D. E., Ward, T. J., Aoki, T., Gale, L. R., Kistler, H. C., Geiser, D. M., Suga, H., Tóth, B., Varga, J. and O'Donnell, K. 2007. Global molecular surveillance reveals novel *Fusarium* head blight species and trichothecene toxin diversity. Fungal Genetics and Biology, 44: 1191-1204.

Stepien, L., Popiel, D., Koczyk, G. and Cehlkowsky, J. 2008. Wheat-infecting *Fusarium* species in Poland – their chemotypes and frequencies revealed by PCR assay. Journal of Applied Genetics, 49: 433-441.

Suga, H., Karugia, G. W., Ward, T., Gale, L. R., Tomimura, K., Nakajima, T., Miyasaka, A., Koizumi, S., Kageyama, K. and Hyakumachi, M. 2008. Molecular characterization of the *Fusarium graminearum* species complex in Japan. Phytopathology, 98: 159-166.

Swofford, D. L. 2001. PAUP*. Phylogenetic Analysis Using Parsimony (*and Other Methods). 4th edn. Sinauer Associates, Sunderland, MA.

Tan, M.-K. and Niessen, L. M. 2003. Analysis of rDNA ITS sequences to determine genetic relationships among, and provide a basis for simplified diagnosis of *Fusarium* species causing crown rot and head blight of cereals. Mycological Research, 107: 811-821.

Taylor, J. W., Jacobson, D. J., Kroken, S., Kasuga, T., Geiser, D. M., Hibbett, D. S. and Fisher, M. C. 2000. Phylogenetic species recognition and species concepts in Fungi. Fungal Genetics and Biology, 31: 21-31.

Thrane, U., Adler, A., Clasen, P., Langseth, W. and Nielsen, K. F. 2004. Chemical diversity of *Fusarium poae, F. sporotrichioides* and *F. langsethiae* sp. nov. International Journal of Food Microbiology, 95: 257-266.

Torp, M. and Langseth, W. 1999. Production of T-2 toxin by a *Fusarium* resembling *Fusarium poae*. Mycopathologia, 147: 89-96.

Torp, M. and Nirenberg, H. I. 2004. Description of *Fusarium langsethiae* sp. nov. International Journal of Food Microbiology, 95: 89-96.

Turner, A. S., Lees, A. K., Rezanoor, H. N. and Nicholson, P. 1998. Refinement of PCR-detection of *Fusarium avenaceum* and evidence from DNA marker studies for phenetic relatedness to *Fusarium tricinctum*. Plant Pathology, 47: 278-288.

Uhlig, S. 2005. *Fusarium avenaceum* Toxic Metabolites, their Occurrence and Biological Effects. Ph.D. Thesis, The Norwegian School of Veterinary Science, National Verinary Institute, Oslo, Norway.

Uhlig, S., Jestoi, M. and Parikka, P. 2007. *Fusarium avenaceum* - the North European situation. International Journal of Food Microbiology, 119: 17-24.

Uoti, J. and Ylimäki, A. 1974. The occurrence of *Fusarium* species in cereal grain in Finland. Annales Agriculturae Fenniae, 13: 5-17.

Van Wyk, P. S., Wingfield, M. J., Marasas, W. F. O. and Bosman, J. L. 1991. Development of microconicia in *Fusarium* section Sporotrichiella. Mycological Research, 93: 284-289.

Vogelgsang, S., Sulyok, M., Bänziger, I., Krska, R., Schuhmacher, R. and Forrer, H.-R. 2008. Effect of fungal strain and cereal substrate on *in vitro* mycotoxin production by *Fusarium poae* and *Fusarium avenaceum*. Food Additives and Contaminants, 25: 745-757.

Voronin, M. C. 1890. About tempulent corn in the South-Ussuriiskij region. Botanical notes. SPb, Russia. 3, 1: 13-21 (in Russian).

Waalwijk, C., Kastelein, P., de Vries, I., Kerenyi, Z., van der Lee, T., Hesselink, T. and Kema, G. 2003. Major changes in *Fusarium* spp. in wheat in the Netherlands. European Journal of Plant Pathology, 109: 743-754.

Waalwijk, C., van der Heide, R., de Vries, I., van der Lee, T., Schoen, C., Costrel-de Corainville, G., Häuser-Hahn, I., Kastelein, P., Köhl, J., Lonnet, P., Demarquet, T. and Kema, G. H. J. 2004. Quantitative detection of *Fusarium* species in wheat using TaqMan. European Journal of Plant Pathology, 110: 481-494.

Ward, T. J., Bielawski, J. P., Kistler, H. C., Sullivan, E. and O'Donnell, K. 2002. Ancestral polymorphism and adaptive evolution in the trichothecene mycotoxin gene cluster of phytopathogenic *Fusarium*. Proceedings of the National Academy of Sciences of the United States of America, 99: 9278-9283.

Ward, T. J., Clear, R., Rooney, A., O'Donnell, K., Gaba, D., Patrick, S., Starkey, D., Gilbert, J., Geiser, D. and Nowicki, T. 2008. An adaptive evolutionary shift in *Fusarium* head blight pathogen populations is driving the rapid spread of more toxigenic *Fusarium graminearum* in North America. Fungal Genetics and Biology, 45: 473-484.

Wilson, A., Simpson, D., Chandler, E., Jennings, P. and Nicholson, P. 2004. Development of PCR assays for the detection and differentiation of *Fusarium sporotrichioides* and *Fusarium langsethiae*. FEMS Microbiology Letters, 233: 69-76.

Wollenweber, H. W. and Reinking, O. A. 1935. Die *Fusarium*, ihre Beschreiburg, Schadwirkung und Bekampfung. Paul Parey, Berlin.

Yang, L., van der Lee, T., Yang, X., Yu, D. and Waalwijk, C. 2008. *Fusarium* populations on Chinese barley show a dramatic gradient in mycotoxin profiles. Phytopathology, 98: 719-727.

Ylimäki, A. 1981. The mycoflora of cereal seeds and some feedstuffs. Annales Agriculturae Fenniae, 20: 74-88.

Ylimäki, A., Koponen, H., Hintikka, E.-L., Nummi, M., Niku-Paavola, M.-L., Ilus, T., and Enari, T. M. 1979. Mycoflora and occurrence of *Fusarium* toxins in Finnish grain. Technical Research Centre of Finland, Materials and Processing Technology Publication, Valtion Painatuskeskus, Helsinki, 21: 1-28.

Yli-Mattila, T. and Gagkaeva, T. 2010. Molecular chemotyping of *Fusarium graminearum, F. culmorum* and *F. cerealis* isolates from Finland and Russia, pp. 159-177. In: Molecular Identification of Fungi. Y. Gherbawy and K. Voigt (eds.). Springer-Verlag, Berlin.

Yli-Mattila, T. and Paavanen-Huhtala, S. 2007. Molecular identification and detection of plant pathogenic and toxigenic *Fusarium* fungi, pp. 82-111. In: The Biodiverse World of Fungi. B. N. Ganguli and S. K. Deshmukh (eds.). Anamaya Publishers, New Delhi, India and UK.

Yli-Mattila, T., Mironenko, N. V., Alekhina, I. A., Hannukkala, A. and Bulat, S. A. 1997. Universally primed polymerase chain reaction analysis of *Fusarium avenaceum* isolated from wheat and barley in Finland. Agricultural and Food Science in Finland, 6: 25-36.

Yli-Mattila, T., Mironenko, N. V., Alekhina, I., Hannukkala, A., Hyvönen, J. and Bulat, S. A. 1998. Identification of *Fusarium avenaceum* isolates by RAPD-PCR and UP-PCR, pp. 114-123. In: Advances in the Detection of Plant Pathogens by Polymerase Chain Reaction. Proceedings of COST 823 Workshop Ceske Budejovice. C. Manceau and J. Spak (eds.). 20-21 June, 1996.

Yli-Mattila, T., Paavanen-Huhtala, S., Bulat, S. A., Alekhina, I. A. and Nirenberg, H. I. 2002. Molecular, morphological and phylogenetic analysis of the *Fusarium avenaceum/F. arthrosporioides/F. tricinctum* species complex – a polyphasic approach. Mycological Research, 106: 655-669.

Yli-Mattila, T., Paavanen, S., Hannukkala, A., Parikka, P., Tahvonen, R. and Karjalainen, R. 1996. Isozyme and RAPD-PCR analyses of *Fusarium* strains from Finland. Plant Pathology, 45: 126-134.

Yli-Mattila, T., Mach, R., Alekhina, I.A., Bulat, S. A., Koskinen, S., Kullnig-Gradinger, C. M., Kubicek, C. and Klemsdal, S. S. 2004a. Phylogenetic relationship of *Fusarium langsethiae* to *Fusarium poae* and *F. sporotrichioides* as inferred by IGS, ITS, β-tubulin sequence and UP-PCR hybridization analysis. International Journal of Food Microbiology, 95: 267-285.

Yli-Mattila, T., Paavanen-Huhtala, S., Parikka, P., Konstantinova, P. and Gagkaeva, T. 2004b. Molecular and morphological diversity of *Fusarium* species in Finland and northwestern Russia. European Journal of Plant Pathology, 110: 573-585.

Yli-Mattila, T., Paavanen-Huhtala, S., Parikka, P., Hietaniemi, V., Jestoi, M. and Rizzo, A. 2004c. Toxigenic fungi and mycotoxins in Finnish cereals, pp. 83-100. In: An Overview on Toxigenic Fungi and Mycotoxins in Europe. A. Logrieco and A. Visconti (eds.). Kluwer Academic Publishers, The Netherlands.

Yli-Mattila, T., Paavanen-Huhtala, S., Parikka, P., Hietaniemi, V., Jestoi, M., Gagkaeva, T., Sarlin, T., Haikara, A., Laaksonen, S. and Rizzo, A. 2008. Real-time PCR detection and quantification of *Fusarium poae, F. graminearum, F. sporotrichioides* and *F. langsethiae* as compared to mycotoxin production in grains in Finland and Russia. Archives of Phytopathology and Plant Protection, 41: 243-260.

Yli-Mattila, T., Paavanen-Huhtala, S., Parikka, P., Jestoi, M., Klemsdal, S. and Rizzo, A. 2006. Genetic variation, real-time PCR, metabolites and mycotoxins of *Fusarium avenaceum* and related species. Mycotoxin Research, 22: 79-86.

Yli-Mattila, T., Parikka, P., Lahtinen, T., Ramo, S., Kokkonen, M., Rizzo, A., Jestoi, M. and Hietaniemi, V. 2009a. *Fusarium* DNA levels in Finnish cereal grains, pp. 107-138. In: Current Advances in Molecular Mycology. Y. Gherbawy, L. Mach and M. Rai (eds.). Nova Science Publishers, Inc., New York.

Yli-Mattila, T., Gagkaeva, T., Ward, T. J., Aoki, T., Kistler, H. C. and O' Donnell, K. 2009b. A novel Asian clade within the *Fusarium graminearum* species complex includes a newly discovered cereal head blight pathogen from the Far East of Russia. Mycologia, 101: 841-852 .

Yli-Mattila, T., Gavrilova, O., O'Donnell, K., Ward, T., Proctor, R. and Gagkaeva, T. 2009c. Asian *Fusarium langsethiae* and the two subgroups of European *F. langsethiae*. Nordforsk mould and mycotoxin seminar 14-15 April, 2009, Uppsala, Sweden (http://fou02.planteforsk.no/NordforskNetworkMycotox/PDFs/PDFs%20Uppsala%20 2009/Nordforsk%20mould%20and%20mycotoxin%20seminar.htm).

Yli-Mattila, T., Rämö, S., Tanner, R., Loiveke, H. and Hietaniemi, V. 2010. *Fusarium* DNA levels as compared to mycotoxin levels in Finnish and Estonian grain samples. *Fusarium* Seminar in Warsaw, September 20-23, 2010 (Abstract, in press).

Yli-Mattila, T., Ward, T., O'Donnell, K., Proctor, R. H., Burkin, A., Kononenko, G., Gavrilova, O., Aoki, T., McCormick, S. P., and Gagkaeva, T. Yu. 2011. *Fusarium sibiricum* sp. nov., a novel type A trichothecene-producing *Fusarium* from northern Asia closely related to *F. sporotrichioides* and *F. langsethiae*. International journal of Food Microbiology, 147: 58-68.

Zhang, J-B., Li, H-P., Dang, F-J., Qu, B., Xu, Y-B., Zhao, C-S. and Lia, Y-C. 2007. Determination of the trichothecene mycotoxins chemotypes and associated geographical distribution and phylogenetic species of the *Fusarium graminearum* clade from China. Mycological Research, 111: 967-975.

Chapter 11

Understanding the Phylogeny of the Alternarioid Hyphomycetes: What can be the Consequences in Taxonomy?

Philipp B. Gannibal

All-Russian Institute of Plant Protection (VIZR), shosse Podbelskogo 3
Pushkin, St. Petersburg, 196608, Russia
E-mail: phbgannibal@yandex.ru

ABSTRACT

The current system of the genus *Alternaria* and allied (alternarioid) hyphomycetes is an invention of intuitive classification based solely on morphological characters. Scrupulous study of the morphology of this group of fungi describes 10 genera consisting of about 370 species. However, contemporary taxonomy does not reflect the phylogenesis sufficiently, and approaches to the recognition of species' borders are void of a population genetic component. Results of phylogenetic analysis revealed that the genus *Alternaria*, as currently circumscribed by morphology, is paraphyletic, while some allied genera, e.g. *Ulocladium* and *Embellisia*, are polyphyletic. At least 19 phylogenetic groups can be resolved well in alternarioid hyphomycetes by molecular markers. Such a division correlates with morphological, physiological and ecological traits. Surprisingly, phylogenetic species of *Alternaria* in some cases was discovered to include several morphospecies. When systematics of alternarioid hyphomycetes will be adjusted in accordance with genealogic patterns, inevitably generic and species boundaries will have to be changed. It is predicted that new taxa at the ranks of genera and sections will have to be established. However, a tendency in alteration of accepted species number is still unclear.

INTRODUCTION

The genus *Alternaria* and related hyphomycetes comprises a group of anamorphs of the family Pleosporaceae (Pleosporales, Ascomycota). The genus *Alternaria* was established by Nees in 1816 and since then its taxonomy has been disputed. So far more than 1100 species of *Alternaria* are recorded including many invalid epithets and synonyms. This large data array, that is the morphology-based taxonomy of the genus, has been critically reviewed by E.G. Simmons (Simmons, 1992, 2007). During the last few years, a number of articles have been published on the molecular phylogeny of *Alternaria* and allied genera. These publications are not in agreement with the current taxonomic system of the genus and the phylogenetic pattern. These aspects have been interpreted and discussed in this chapter.

The term "phaeodictyosporic hyphomycetes" is used to designate the group of fungi including the genus *Alternaria* and several other morphologically similar genera having pigmented or dark colored muriform spores. The "phaeodictyosporic hyphomycetes" usually include 5 genera, *Alternaria, Embellisia, Nimbya, Stemphylium* and *Ulocladium*. However, in describing phylogeny it is more appropriate to determine the group based on their relationships rather than on their morphology. Therefore, in this chapter the term "alternarioid hyphomycetes" is used keeping in mind a monophyletic group which includes all or majority of *Alternaria* species and some closely related taxa. On a cladogram, this group looks compact and has clear hiatuses between sister clades. The details of alternarioid hyphomycetes are discussed below. It is necessary to mention that alternarioid hyphomycetes differ from phaeodictyosporic hyphomycetes by inclusion of several additional genera as well as by the exclusion of the genus *Stemphylium*.

To date, perhaps all available morphological traits have been inventoried and utilized for alternarioid hyphomycetes taxonomy. E.G. Simmons has made major contributions in this area and is credited for studying *Alternaria* and other phaeodictyosporic hyphomycetes for 50 years and publishing a number of articles and an identification manual (Simmons, 2007). He has to his credit about 150 new *Alternaria* species and 25 new combinations.

Various classification approaches deal with different aspects of taxonomic reality, therefore, each (or at least many) of them has a certain portion of truth and consequently has a right of existence (Pavlinov, 2003). The system of alternarioid hyphomycetes is based on traditional principles of intuitive typology with no phylogenetic and population genetic premises. This system can be regarded as a stage of development. The step of primary accumulation of the knowledge about an object and species description without determination of their relationship was called by Mayr et al. (1953) as alpha systematics. Study of intraspecific diversity, compliance with evolution scenarios, and substantial classification hypothesis appealing to explanations of reasons of biodiversity are characters of the gamma systematics which is the highest level of maturity of the discipline. Beta systematics is an intermediate stage between alpha and gamma levels.

Obviously, E.G. Simmons' taxonomy of alternarioid hyphomycetes is close to completion of its first stage and proposes to reflect maximum morphological diversity to catalog biodiversity and as far as possible to make taxa identification possible. This results in a lack of link-up between morphological species and "real" biological (functional) entities. However, it is essential for ecologic, genetic and evolutionary studies to operate with biologically substantial entities, e.g. genospecies, biospecies or phylospecies that have not been described enough in alternarioid hyphomycetes.

This chapter is an attempt to analyze phylogenetic and other taxonomy-related data accumulated during the last decade and to assess the necessity and the means to bring the system of alternarioid hyphomycetes in correspondence with phylogeny.

THE CURRENT CONCEPT OF GENERA IN ALTERNARIOID HYPHOMYCETES

Taxonomy of the genus *Alternaria* and allied genera, since the time of their description (Nees, 1816), and still is a lively discussion. Taxonomy has undergone several serious revisions. The history of this subject matter was described several times (Tweedy and Powell, 1963; Simmons, 1992). Other publications that need to be mentioned here are those of Elliott (1917), Wiltshire (1933), Groves and Skolko (1944), Neergaard (1945), Joly (1964), and Simmons (1967, 1992, 2007) who worked according to the traditions of the classical school of intuitive typological systematics.

Until the 1970s all phaeodictyosporic hyphomycetes were divided into three big genera: *Alternaria* Nees (1816), *Macrosporium* Fries (1832) and *Stemphylium* Wallroth (1833). The fourth genus, *Ulocladium* Preuss described in 1851, was disregarded by taxonomists for a period of more than 100 years. The generic boundaries were formidably confused. Researchers had different notions about the characteristics of those genera. However, E.G. Simmons started to resolve the problem of genera delimitation by typification and providing clear re-descriptions of three genera: *Alternaria*, *Ulocladium* and *Stemphylium* (Simmons, 1967). The generic name *Macrosporium* was abolished because the type of the genus was a species, *M. cheiranthi* that was considered to be an *Alternaria* (Simmons, 1992).

Furthermore, the taxonomic study of *Alternaria* and related genera and the description of a number of new species resulted in understanding of phaeodictyosporic hyphomycetes as a big and very polymorphic group.

Therefore, five novel genera were introduced: *Prathoda* C.V. Subramanian (1956), *Embellisia* E.G. Simmons (1971), *Nimbya* E.G. Simmons (1989), *Alternariaster* E.G. Simmons (2007), *Chalastospora* E.G. Simmons (2007) and *Teretispora* E.G. Simmons (2007). Differentiation of the aforesaid taxa was based on a number of exclusively morphological traits, particularly the shape of their conidia (Simmons, 1967, 1992, 2007).

A comprehensive approach to the analysis of alternarioid hyphomycetes biodiversity that equally included both morphological and molecular data has been the basis for inclusion of two fungal forms regarded as *Dendryphion* species into the genus *Brachycladium* Corda (Inderbitzin et al., 2006). One species each of *Embellisia* and *Helminthosporium* were merged with a new genus *Undifilum* (no authority explicitly specified) (Pryor et al., 2009).

Thus, the system of all phaeodictyosporic and *Alternaria*-like hyphomycetes includes 11 genera, 7 of which exhibit close phylogenetic relationships. Species of the phaeodictyosporic genus *Stemphylium* form a sister clade for alternarioid hyphomycetes. No phylogenetic study has been done for the species of two genera – *Alternariaster* and *Teretispora.* The phylogenetic position of the genus *Prathoda* is not convincingly established. Taking into account the up to date systematics and phylogenetic data, it can be stated that alternarioid hyphomycetes comprise of following 10 genera with approximately 370 currently known, valid and distinguishable species, of which 75% are the species of *Alternaria.*

Alternaria – 276 species (teleomorph – *Lewia*, 14 species)
Alternariaster – 1 species
Brachycladium – 2 species (teleomorph – *Crivellia*, 1 species)
Chalastospora – 4 species
Embellisia – 23 species (teleomorph – *Allewia*, 2 species)
Nimbya – 17 species (teleomorph – *Macrospora*, 4 species)
Teretispora – 1 species
Ulocladium – 26 species
Undifilum – 2 species
(?)*Prathoda* – 2 species

THE CURRENT CONCEPT OF SPECIES IN ALTERNARIOID HYPHOMYCETES

The history of *Alternaria* and whole alternarioid hyphomycetes taxonomy is fairly dramatic. The number of specific names published for *Alternaria* is approximately twice as large as the number of "good" species. Unfortunately, descriptions of many species are not adequately detailed and their type specimens have become worthless or are not available anymore. As a result, there are more than 100 legally published names that have been determined as unidentifiable ("not distinguished as a unique taxon", "a small-spored *Alternaria* taxon that cannot be defined more closely", *inserta sedis*) and their use was suggested to be meaningless (Simmons, 2007). More than 150 names are synonyms, *nomen nudum* or invalidly published names.

The main problem lies in a group of small-spored *Alternaria* species or *A. alternata* in its former wide sense. Wide distribution, high frequency, considerable level of polymorphism and variability of morphology within the individual species has resulted into the description of a large number of species. On the other hand, there have been problems of having insufficiently described old species as new taxa when they were found on a new host plant. Such descriptions mainly contain and focus on the conidial shape and size. Although it has been indicated several times that the shape and size of conidia of alternarioid hyphomycetes depend on substratum, temperature, light and other factors to a great extent (Leach, 1970; Sobers and Doupnik, 1973; Misaghi et al., 1978; Simmons, 1998; Singh et al., 2000).

The work of E.G. Simmons differs from those of his predecessors. He used pure cultures and herbarium material equally. He established the standards of culture conditions for taxonomic studies of phaeodictyosporic hyphomycetes. He also paid greater attention to a new character i.e. three-dimensional sporulation pattern. By exploiting this trait a number of species were discriminated in the groups previously thought as being doubtful.

Simmons has focused on a proposition "need to know". "If the taxon under observation has stable colony development in axenic culture, distinctive sporulation pattern and features, and microscopic characters distinguishable from those of similar taxa, then the taxon requires a unique tag to hold its place for retrieval in the published literature. We choose to call this tag a species name" (Simmons, 2007).

The system of pathotypes is another approach to the classification of small-spored *Alternaria* strains having minor differences between them. Six forms of *A. alternata sensu lato* differing in their ability to produce host-specific toxins and induce disease symptoms in a few susceptible plant species or cultivars have been described. It was proposed by Nishimura and Kohmoto (1983) to name such forms as pathotypes. This discrimination between saprotrophic *A. alternata* and their pathotypes is related to physiological and biochemical traits but less to molecular features of these fungi. However, less regard has been paid to their morphological features which have caused a nomenclatural mess.

The significant example of the pathotype ideology is *A. alternata* apple pathotype that has been distinguished from other *A. alternata* variants by their ability to produce host-specific AM-toxins and infect apple trees. Gene associated with synthesis of AM-toxin is situated in a small conditionally dispensable (CD) chromosome (Johnson et al., 2001). Spontaneous loss of this chromosome does not affect growth of a strain but leads to the lack of toxin synthesis capability and thereby the pathogenicity. It was shown that several genes required for pathogenicity in fungi are located on CD chromosomes (Covert, 1998). However, there is no satisfactory answer to the question, where and how these chromosomes were acquired in the first place. To date a lot of evidence has been collected supporting the presence of horizontal gene transfer in phytopathogenic fungi (Khaldi et al., 2008) as well as spontaneous loss of pathogenicity. If this is the case, such transition from saprotroph to pathogen and *vice versa* in small-spored *Alternaria* looks rather simple in the genetic and evolutional scale to be considered as speciation. Thus, *A. alternata* apple pathotype is a variant of small-spored *Alternaria* species with one additional non-obligatory chromosome. It seems reasonable to mark such variations just as a pathotype but not as a species. However, it is still unclear as to which species these pathotypes may belong?

THE TAXA OF INTERMEDIATE RANK

Since the work of Elliott (1917) several attempts have been made to assign some informal subgeneric groups or taxa with ranks higher than species and lower than a genus within *Alternaria*. However, none of these groupings followed the rules of nomenclature and none became popular. All these groups were based on three characters: spore size, beak length and presence/length of conidial chains. The number of subgeneric groups varied from two to six. According to Neergaard (1945), *Alternaria* species were divided into three sections: *Longicatenatae* – species with long conidial chains, *Brevicatenatae* – species producing short chains of 3-5 conidia and *Noncatenatae* – species with no or rare conidial chains. However, Neergaard (1945) simultaneously separated the genus *Stemphylium* into two subgenera, one of which, *Pseudostemphylium*, was later considered as a part of the genus *Alternaria*.

Some informal subgeneric groups were also proposed by E.G. Simmons, keeping in mind the sporulation pattern as an important trait. He introduced a concept of species-group (complex) and defined species-group as a group of morphologically-describable species that have similar patterns of producing conidia (three-dimensional sporulation patterns). Now this term is relatively widely used in literature concerning *Alternaria* (Simmons, 1990a, 1995). He also underlined that this term should be used in relation to morphology only and its usage has no relationship to phylogeny or nomenclatural rules (E.G. Simmons, pers. comm.).

Simmons (1992) also described 14 other informal groups of species to facilitate the identification process of *Alternaria* species, generally based on the morphology of conidia. Further elaboration of this system was done when the genus was divided into two nameless sections and 14 groups, some of which contained several subgroups (Simmons, 2007). It should be mentioned once more that these units do not have taxonomic status or morphological descriptions, but are intended to amend the identification key. These enumerated groups have no meaning in classification.

Thus, no taxa of intermediate rank are currently maintained or in use in alternarioid hyphomycetes.

Accumulation of phylogenetic data

Various workers have added to the molecular data of alternarioid hyphomycetes. The most interesting data were obtained in the laboratories of B.M. Pryor and T.L. Peever. A total of almost 80 species belonging to 8 genera of alternarioid hyphomycetes were used for the phylogenetic studies that represent about 20% of their entire diversity (Pryor and Gilbertson, 2000; Chou and Wu, 2002; Pryor and Bigelow, 2003; Hong et al., 2005a, 2005b). A comparison of current taxonomy with that of the phylogenetic pattern resulted in some discrete clades within the group. The work of Huson et al. (2004) described an application of an algorithm for inferring a phylogenetic super-network (Z-closure method) using alternarioid hyphomycetes as a model. Phylogenetic studies mainly utilized molecular characters; however, a few investigators used phenotypic, morphological, biochemical and physiological traits as well for species differentiation.

The sequence data for alternarioid hyphomycetes in GenBank up to December 2009 consist of about 3200 entries referring to about 30 genomic loci. Twelve genomic loci can now be used for phylogenetic studies as they have a proper level of conservatism/variability and their sequences are obtained from sufficient number of species (Table 11.1). Since various investigators based their studies on different genomic loci, GenBank does not contain enough data at present to construct a common phylogenetic tree for all the species that have already been studied.

Most sufficient phylogenetic signals were found when the sequence data from genes of glyceraldehyde 3-phosphate dehydrogenase (gpd), *Alternaria* major allergen Alt a 1 and endopolygalacturonase (EndoPG) were used. Sequences of rDNA from the nuclear ITS regions and mitochondrial small subunit (mtSSU) were less useful in estimating phylogenies especially at the species level. Sequence analysis of beta-tubulin gene and translation elongation factor 1-alpha was fruitful for some species complexes while moderately useful for others. Some studied regions of the genome including genes of calmodulin (CAL), actin (ACT), 1,3,8-trihydroxynaphthalene (THN) and chitin synthase (CHS) were almost variable between species.

Results of phylogenetic analyses indicated alternarioid hyphomycetes as a large monophyletic clade with *Stemphylium* as a sister taxon. The genera such as *Embellisia* and *Ulocladium*, currently circumscribed, are polyphyletic,

Table 11.1 Genomic loci broadly used in phylogenetic studies of alternarioid hyphomycetes

Locus/gene name	*Locus/gene product*	*Number of sequences**	*Representation of alternarioid hyphomycetes in sequence data*
rDNA**	Nuclear ribosomal RNA	>1200	many species, different phylogroups***
Alt a 1	*Alternaria* major allergen Alt a 1	256	many species, different phylogroups
Gpd	Glyceraldehyde 3-phosphate dehydrogenase	269	many species, different phylogroups
EndoPG	Endopolygalacturonase	195	alternata phylogroup
OPA2-1	Unknown	187	alternata phylogroup
OPA1-3	Unknown	161	alternata phylogroup
OPA10-2	Unknown	138	alternata phylogroup
mtSSU	Mitochondrial small subunit ribosomal RNA	86	many species, different phylogroups
Btub1	Beta-tubulin	66	radicina phylogroup and a few species from other groups
Btub2	Beta-tubulin	62	some species, different phylogroups
H3	Histone 3	58	some species of alternata and infectoria phylogroups
EF-1α (tef1)	Translation elongation factor 1-alpha	49	radicina phylogroup and a few species from other groups

*Number of sequence entries in GenBank on December 2009

**Mainly 18S-28S ribosomal RNA gene and ITS1-5.8S-ITS2 region

***For description of the phylogroups see text below.

but the genus *Alternaria* is paraphyletic (Pryor and Bigelow, 2003). Within a group of species, examined with molecular phylogenetic approaches, several clades and a few single separate species were resolved. Most of the species were distributed on clades inspite of the genomic region used for analysis. Those clades were sometimes called species-groups that can cause confusion with morphologically defined species-groups *sensu* Simmons. I would prefer to use the term "phylogroups" to designate such clades (phylogenetic groups) appearing as clusters of alternarioid hyphomycetes relatively well separated on a phylogram.

DISCRIMINATION OF PHYLOGENETIC GROUPS

Assessment of phylogenetic signals from sequences of ITS, mtSSU and gpd resolved 8 clades (alternata, brassicicola, embellisia, infectoria, nimbya, porri, radicina and ulocladium) and one species (*U. alternariae*) was placed beyond any clade (Pryor and Gilbertson, 2000; Pryor and Bigelow, 2003). Data accumulated later, based on the analysis of larger number of species, allowed adding undifilum phylogroup and resolved species of the genus *Embellisia* as members of five clades (Pryor et al., 2009).

A similar cladification pattern was obtained as a result of the analysis of solely ITS regions (Chou and Wu, 2002; de Hoog and Horré, 2002). Additional facts that should be mentioned are as follows: (i) *A. brassicae* and *A. dianthi* (= *A. nobilis*) were not part of any clusters. (ii) Two new phylogroups were seen. They have been called photistica and dianthicola on the basis of the member species, *A. photistica* and *A. dianthicola*, respectively.

Further, genealogy of gpd and a newly studied gene, Alt a 1, as well as addition of some more species in phylogenetic studies allowed Hong et al. (2005b) to support earlier described clades and determine a new phylogroup, sonchi. However, six species were beyond any group.

Molecular phylogeny combined with morphological studies became a base for recognition of new phylogroups, brachycladium (Inderbitzin et al., 2006; Pryor et al., 2009) and chalastospora (Braun et al., 2003; Andersen et al., 2009; Crous et al., 2009), the last one has been described as a new genus.

Data presented by a number of other investigators clarified the content of some phylogroups: alternata (Peever et al., 2000, 2002, 2004, 2005; Kang et al., 2002; Akimitsu et al., 2003; Andrew et al., 2009), infectoria (Andersen et al., 2009), radicina (Pryor and Gilbertson, 2002; Park et al., 2008) and ulocladium (Wang et al., 2009, 2010).

In this chapter an attempt has been made to collect all information about subdivision of alternarioid hyphomycetes and describe all phylogroups more precisely. In doing so, all the molecular phylogenetic data needed were consolidated and analyzed. A total of 17 phylogroups were seen on several phylogenetic trees. To design a hypothetical synoptic consensus cladogram (Fig. 11.1), the most parsimonious (MP) trees resulting from analysis of combined ITS, gpd gene, and mtSSU rDNA region sequence data (Pryor and Gilbertson, 2000; Pryor and Bigelow, 2003; Pryor et al., 2009) or Alt a 1 and gpd exons sequences (Hong et al., 2005b) were used as a base because they represent the alternarioid hyphomycetes diversity more completely and resolve 13 clades. Also 4 clades were entered into the cladogram after analysis of other trees: nobilis and dianthicola – from a neighbor-joining (NJ) tree based on ITS sequences (de Hoog and Horré, 2002); chalastospora – from MP tree based on ITS sequences (Crous et al., 2009); and photistica – from NJ trees based on ITS sequences (de Hoog and Horré, 2002; Chou and Wu, 2002). Several species that were included into molecular phylogenetic research could not be referred to any clade listed, because data and methods used did not allow doing it. Probably, they are members of additional phylogroups scantly represented in the studied samples.

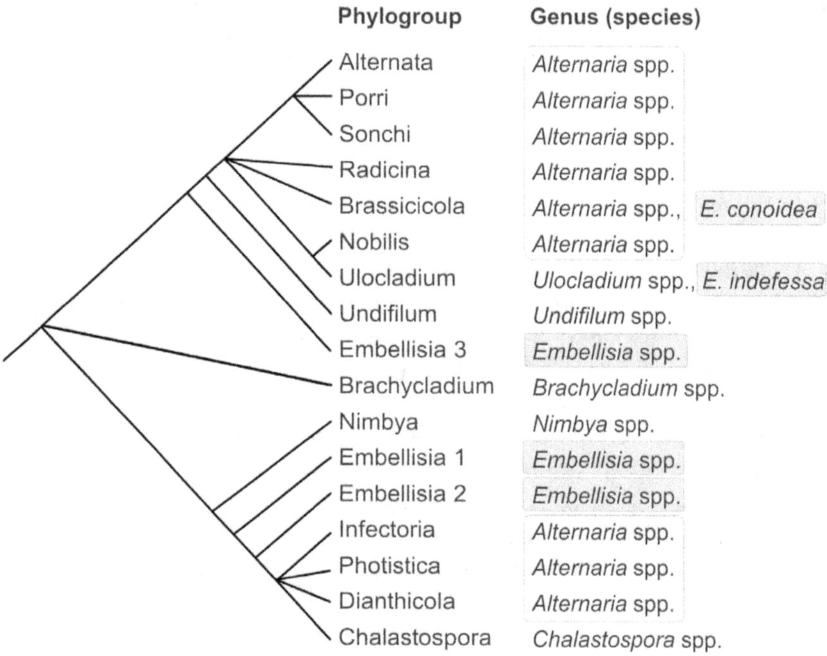

Fig. 11.1 Hypothetical consensus cladification derived from the combination of several phylogenetic trees based on analysis of nuclear and mitochondrial rDNA, gpd and Alt a 1 sequences (Pryor and Gilbertson, 2000; de Hoog and Horré, 2002; Pryor and Bigelow, 2003; Hong et al., 2005b; Crous et al., 2009; Pryor et al., 2009). Clear and shaded boxes represent clades containing species of *Alternaria* and *Embellisia*, respectively.

Examples of such species are *A. argyranthemi*, *A. brassicae*, *A. eryngii* and *U. alternariae*.

A special place is occupied by the genus *Prathoda* which contains 2 species, *P. longissima* (syn. *A. longissima*) and *P. saparva* (syn. *A. saparva*). The comparison of 18S rDNA sequences with those of other fungi indicated that *A. longissima* (=*P. longissima*) is phylogenetically distinct from *Alternaria* and *Ulocladium* species while it has a high degree of sequence identity with *Leptosphaeria* spp. (Pryor and Gilbertson, 2000). Further studies are needed to clarify phylogenetic and taxonomic positions of the species of *Prathoda*. *Prathoda* species are, therefore, regarded as a separate phylogroup of alternarioid hyphomycetes.

The phylogroups so defined were preliminarily described by morphology. Then rest of alternarioid hyphomycetes which were not used in phylogenetic revision were distributed among the phylogroups according to their morphological descriptions. A total of 367 distinguishable species that currently have legitimate taxonomic status, appropriately published descriptions, were used for this analysis. The following morphological reports were utilized: *Alternaria* - Simmons (2007), *Alternariaster* - Simmons (2007), *Brachycladium* - Inderbitzin et al. (2006), *Chalastospora* - Simmons (2007), Crous et al. (2009), *Embellisia* - Simmons (1971, 1983, 1990b, 2004), David et al. (2000), *Nimbya* - Simmons (1989, 1995, 1997, 2000, 2004), Johnson et al. (2002), Zhao and Zhang (2005), *Prathoda* - Simmons (2007), Ellis (1976), *Teretispora* - Simmons (2007), *Ulocladium* - Simmons (1967, 1990a, 1998, 2004), Wang et al. (2008, 2009, 2010), Zhang and Zhang (2002), *Undifilum* - Pryor et al. (2009).

The majority of species were distributed among the groups with sufficient level of confidence. However, many could not be related to any phylogroup on the basis of morphological analysis. About 25 species remained doubtful, many of which have intermediate characteristics. Morphology of 7 species sufficiently differed from that of any phylogroups allowing one to describe one extra group (morphogroup) i.e. molesta.

Similarly, two species (*Teretispora leucantemi* and *Alternariaster helianthi*) appeared morphologically unique and could be combined into a separate group. No special phylogenetic work has been done for those species, but some sequences for them are available in the GenBank. Analysis of rDNA sequences of *Teretispora leucanthemi* by BLUST with megablast algorithm (Altschul et al., 1990) has revealed its high similarity (99-100%) with *Alternariaster helianthi* while the level of similarity with other species of alternarioid hyphomycetes did not exceed 90%. Sporulation pattern and shape of conidia of those two species in general were also similar.

Thus, it appears reasonable to have about 20 separate groups within alternarioid hyphomycetes before more phylogenetic, genetic, biochemical and physiological data is obtained. Briefly, some characters of the phylogroups are tabulated (Table 11.2).

So far only a smaller number of alternarioid hyphomycetes have been used in phylogenetic studies. Phylogenetic analyses conducted for different groups utilized different datasets. When more species are assessed, new clades will more likely be revealed and the subdivisions described here will be corrected. However, species studied to date represent different groups of species keeping in mind the variability in morphology and ecology and, therefore, the correction should not be dramatic.

Superposition of functional phenotypic characters in a phylogenetic tree may be able to reveal some evolutionary trends and patterns of speciation in alternarioid hyphomycetes. Such analysis may reveal certain connections between molecular and ecological (level of parasitism, substrate adherence) or genetic (presence/absence of sexual reproduction and type of mating system) characters. One superclade (upper part on the cladogram, Fig. 11.1) includes the biggest phylogroups and most of *Alternaria* species. It is characterized by inclusion of saprotrophs and parasites; most important phytopathogens; the most common and widespread species; species with no recorded sexual

Table 11.2 Summary of Phylogroups

#	*Clade*	*Number of species*	*Sexual state*	*Mating system*	*Lifestyle*
1	Alternata	58	Unknown	Heterothallism	Saprotrophy, parasitism
2	Porri	120	Unknown		Parasitism
3	Sonchi	6	Unknown		Parasitism
4	Radicina	6	Unknown		Parasitism
5	Brassicicola	8	Unknown	Heterothallism	Parasitism
6	Nobilis	11	Unknown		Parasitism?
7	Ulocladium	30	Unknown		Saprotrophy
8	Undifilum	2	Unknown		Endophytism, parasitism
9	Embellisia-3	11	2 species	Homothallism	
10	Brachycladium	2	1 species	Hetero-, homothallism	
11	Nimbya	17	4 species	Homothallism	
12	Embellisia-1	2	Unknown		
13	Embellisia-2	4	Unknown		
14	Infectoria	43	9 species	Homothallism	Saprotrophy
15	Photistica	12	3 species	Homothallism	Saprotrophy
16	Dianthicola	3	Unknown		
17	Chalastospora	4	Unknown		
18	Prathoda	2	Unknown		
19	Teretispora	2	Unknown		Parasitism
20	Molesta (morphogroup)	7	Unknown		

Phylogroups are listed in the same order as in Fig. 11.1, except groups 18–20 which have no defined phylogenetic position. Clear and shaded boxes represent phylogroups containing species of *Alternaria* and *Embellisia*, respectively.

reproduction; and species with heterothallic ancestors (4 species were studied). Apparently this group is evolutionary advanced.

Interestingly, a close relationship was found between two of the richest groups diverging by ecological strategies: alternata consists primarily of saprotrophs forming a large number of small conidia (r-strategy) while porri includes necrotrophic parasites with big conidia and commonly weak sporulation (K-strategy). However, members of alternata group can synthesize a number of non-specific phytotoxins. Some species with acquired ability to produce host-specific toxins at once change into true pathogens called pathotypes.

The second superclade (lower part of the cladogram, Fig. 11.1) consists of mainly saprotrophs. Only species of infectoria group are wide spread and relatively abundant. Out of 85 species 16 have sexual states. Many or all of them are homothallic.

Phylogeny of alternarioid hyphomycetes in some cases contradicts their taxonomy. The bigger genera, e.g. *Alternaria*, *Embellisia* and *Ulocladium* are more problematic than recently described or revised small taxa, e.g. *Brachycladium*, *Chalastospora* and *Undifilum*. As it was previously mentioned those big genera are polyphyletic. On the contrary, several clades include species of two genera. For instance, the primary clade containing most *Ulocladium* spp. includes an *Embellisia* species and, thereby, is paraphyletic (Pryor and Bigelow, 2003).

Species of the genus *Embellisia* are distributed between five phylogroups. Moreover, recently, before the genus *Undifilum* was described (Pryor et al., 2009), the number of phylogroups where *Embellisia* was placed was bigger by one. A single *Embellisia* species in *Ulocladium* cluster as well as one species of *Embellisia* in brassicicola phylogroup should be transferred into appropriate genera. The rest of the genera must be split into two or three genera.

Species of the genus *Alternaria* on the cladogram (Fig. 11.1) are distributed among 9 phylogroups and it is seen that those clades are grouped in two parts of the cladogram. In terms of robust phylogeny-based taxonomy it makes sense to consider a set of three phylogroups (infectoria, photistica and dianthicola) as a separate genus. All 9 phylogroups may be described as sections which will help identification and will have predictive value due to rather higher correlation between phylogeny and ecologic, biochemical, physiological and economic characters.

Morphologic description as well as information about sexual reproduction and ecology of each phylogroup is presented below.

Alternata phylogroup (totally *ca.* 58 species, including 15 species grouped according to molecular cladistic analysis: *A. alternata*, *A. arborescens*, *A. citriarbusti*, *A. citrimacularis*, *A. dumosa*, *A. gaisen*, *A. interrupta*, *A. limoniasperae*, *A. longipes*, *A. mali*, *A. perangusta*, *A. tangelonis*, *A. tenuissima*, *A. toxicogenica* and *A. turkisafria*). Phylogenetically and morphologically this group is rather well separated from the other ones. This group includes the majority of species called small-spored. Many of them very often are mentioned in the literature under one collective name "*A. alternata*".

Primary conidiophores are of various shapes and sizes. Conidia are formed in long or moderately long simple or branched chains. Conidia are small obclavate, ovoid or long-ellipsoid, 20-50(-60)×8-12 μm in culture and somewhat bigger in nature. Conidia usually contain 3-7(-11) transverse and 1-2 longitudinal septa in one or a few of the transverse divisions. Apical secondary conidiophores are commonly present, short (usually not exceeding 5-10 μm or sometimes up to 35 μm). Lateral secondary conidiophores sometimes appear.

Teleomorph is not known. *A. alternata*, *A. tenuissima* or their ancestor was heterothallic (Arie et al., 2000; Berbee et al., 2003).

A big number of species are saprotrophs with no host/substrate specialization. There are also some species known as strong parasites, each of them is able to infect only a few genotypes (cultivars) of a host plant. Such species become pathogenic due to a limited number of genes associated with synthesis of host-selective toxins, but presumably they have all the adaptations necessary for saprotrophic growth as closely relative species. Wide occurrence in nature, fast growth, rich sporulation and small size of conidia indicate r-strategy.

Difference of *A. alternata*, *A. tenuissima* and *A. arborescens* from *A. infectoria* species-group (infectoria phylogroup) was convincingly shown by comparison of metabolite profiles (Andersen and Thrane, 1996; Andersen et al., 2002).

Brachycladium (2 species: *B. penicillatum* and *B. papaveris*). Analysis of several sequence data sets resulted into different genealogical positions of *Brachycladium*. In some cases it was included into brassicicola (Inderbitzin et al., 2006) or embellisia phylogroups (Pryor et al., 2009). Examination of combined ITS, gpd and mtSSU sequence data resolved two *Brachycladium* species as a separate clade (Pryor et al., 2009). Morphologically this group is well defined.

Primary conidiophores are 20-130×4-6 μm in length, sometimes branching; apex of each branch bears 2-4 conidiogenous sites (pores). Conidia remain solitary or form short branching chains, small cylindric or subcylindric, usually with no conspicuous constriction near the septa, 17-70×4-8 μm, with 3-6 transepta and no longisepta. Secondary conidiophores are very short or (mostly) secondary conidia are produced through the pores on the apex of the conidium body. *B. penicillatum* produces microsclerotia and reproduce sexually also (*Crivellia papaveracea*) under natural conditions and is proven to be heterothallic, while another one (*B. papaveris*) has no described teleomorph and possess mixed (out-of-order) mat-idiomorph that is characteristic rather for homothallic species (Inderbitzin et al., 2006).

Both species are common parasites of poppy.

Brassicicola (8 species, including *A. brassicicola*, *A. japonica*, *A. mimicula* and *E. conoidea*). Analysis of morphology allows including in this group *A. alstroemeriae*, *A. nepalensis A. septorioides* and *A. solidaccana*. Phylogenetic analysis placed *U. alternariae* in brassicicola clade when sequences of ITS-regions (Inderbitzin et al., 2006) were used and into separate terminal group when data set contained Alt a 1 exon and gpd exon sequences (Hong et al., 2005b).

Primary conidiophores are simple or branched, 15-150×5-6 μm bearing 1-3 conidiogenous loci on the apices. Conidia appear in chains of different length and level of branching. Conidia are small to medium-sized, 30-95×7-22 μm, subcylindrical or obclavate, with apical and lateral secondary conidiophores. In some species, a larger number of conidia remain small (20-40 μm in length). Conidia are constricted near the septa, and sides of conidial cells (sections) are frequently and obviously prominent and rounded. Conidia are 3-7 transeptate (1-3 septa in small conidia). In many species majority of conidia do not have longitudinal septa, but the oldest primary conidia can be divided by 1-2 longisepta in 1-4 sections.

Teleomorph is not known. *A. brassicicola* or its ancestor was heterothallic (Berbee et al., 2003).

Two widely distributed and well studied species are associated with a few (*A. japonica*) or several genera (*A. brassicicola*) of one family (Brassicaceae). Out of eight species three are described only on cruciferous plants.

Chalastospora (4 species of the genus *Chalastospora*, including *C. cetera*, *C. gossypii*, *C. ellipsoidea* and *C. obclavata*). Taxonomy and phylogeny of the genus *Chalastospora* are exhaustively described by Crous et al. (2009).

Primary conidiophores are relatively short, 10-30×3-4 µm with 1-3 conidiogenous loci. Chains of conidia usually are long (5-15 units in a rank) and branched. Conidia are small, up to 35×6 µm, ellipsoid or less frequently long-ovoid, with no septa or with 1-3 transverse septa only. Secondary conidiophores are usually absent, conidiogenous pores are formed at the apex of conidial body or on small protuberances.

Teleomorph has not been found.

Probably saprotrophs.

Dianthicola (3 species, including *A. dianthicola*). In molecular studies carried out by several authors, *A. dianthicola* was clustered in different groups (de Hoog and Horré, 2002; vs. Chou and Wu, 2002) because of the different strains sequenced. The work of de Hoog and Horré (2002) was followed here that allows revealing a separate group, which include *A. dianthicola* (authentic strain of Neergaard) also representative strains of *A. chlamydospora* and *A. mouchaccae*. The last two species were understood widely and seem to be frequently misidentified. Morphology of *A. chlamydospora* and *A. mouchaccae* is closer to *A. molesta* and respective group (see below) than to *A. dianthicola*. The Dianthicola group according to morphological as well as phylogenetic analyses share the most recent common ancestor with the infectoria group.

Primary conidiophores of *A. dianthicola* are simple, usually long, 200-350×5-6 µm. Conidia are solitary or in short simple or branching chains (up to 4 units in a rank). Apices of most conidia have conical, long-cylindrical cells or secondary conidiophores with 1-3 conidiogenous sites. Conidia are medium-sized with 80-125×11-18 µm bodies and 8-80 µm beaks or apical secondary conidiophores. Conidia have 4-10 transverse septa in the body and a few in secondary conidiophores. Longisepta are not produced or appear as 1 in 1 or 2 transverse segments.

Teleomorph is not known.

Embellisia (2+ species, including *E. allii* and *E. tellustris*). This group contains type species of the genus *Embellisia*, i.e. *E. allii*. Probably other *Embellisia* species can be defined as members of the group.

Primary conidiophores are 30-100×5-7 µm, simple or branched with 3-10 conidiogenous loci on each apex. Conidia are solitary, ellipsoidal, subcylindric,

30-56×10-17 µm, with 4-7(-10) transepta and 0-2 longisepta. Transverse septa are as much as 2 µm thick. Sometimes microsclerotia are formed.

Teleomorph is not known.

Embellisia-2 (4+ species, including *E. didymospora* and *E. phragmospora*). Most likely the group comprises *E. oxalidicola* and *E. chlamydospora*.

Primary conidiophores are simple or branched, up to 75×4 µm long and with 1-10 conidiogenous sites. Conidia are solitary or in some species (e.g. *E. phragmospora*) can form short chains. Conidia are spheroid, ovoid, cylindrical or obclavate, 10-17 or 25-40(-50) µm in length, 8-10(-12) µm in width with 1-2 or 3-5(-9) thick transepta and 0(1-2) longisepta. Secondary conidiophores, if present, are basal, lateral and apical. Chlamydospores and microsclerotia are abundant, terminal or intercalary, dark, thick-walled, and sometimes have knob-like excrescences.

Teleomorph is not known.

Embellisia-3 (11+ species, including *E. hyacinthi, E. leptinellae, E. lolii, E. novae-zelandiae, E. planifunda, E. proteae* and *E. tumida*). The group probably includes *E. verruculosa, E. abundans, E. dennisii* and *E. eureka*.

Conidiophores are simple or branched, up to 50×5 µm or longer, with 1-10+ conidiogenous loci. Conidia can form short simple chains, but most often do not and are located on conidiophores as fascicles. Conidia are cylindrical or oval, 20-65×8-18 µm. Mature conidia have 3-7 to 8-10 thick transverse and 0-5 longitudinal septa. Some species are characterized by the presence of mainly lateral or apical secondary conidiophores having 1-5 conidiogenous loci. Chlamydospores and microsclerotia may be abundant or rare.

Two species (*E. proteae* and *E. eureka*) have teleomorphs reported in cultures.

Infectoria (43 species, including *A. conjuncta, A. ethzedia, A. infectoria, A. metachromatica, A. oregonensis* and *A. triticina*). Many other similar species that are mentioned by Simmons (2007) as members of *A. infectoria* species-group can be included into infectoria phylogroup.

Primary conidiophores are simple or branched, 20-150×3-4 µm, with (1-)2-6 conidiogenous sites at each apex. Conidia are obclavate or in some species subcylindrical conidia are common, 30-60×6-15 µm; sometimes a large number of conidia remain smaller (12-30×5-10 µm). Several species (e.g. *A. triticina*) can produce relatively larger conidia, 25 µm wide. Usually many conidia have apical secondary conidiophores, some of which (5-30%) are long (10-60 µm or more) with multiple (2-3(-8)) additional conidiogenous geniculations. Lateral secondary conidiophores are rare. Due to secondary sporulation and polytretic conidiogenous cells, chains of conidia are usually branched. Chains are either short having a few conidia or (often) may appear bushy containing a few to hundreds of conidia. Conidial body has (1-)4-7(-10) transverse and 1-2(-3) longitudinal and oblique septa in 1-4 transverse divisions.

There are species where pseudothecia (approx. 9 species) were found on natural substrata. Some strains are able to produce proascomata in axenic culture that suggests that this group, at least in part, is homothallic (Andersen et al., 2009).

Common saprotrophs having wide preference to substrates.

Morphogroup **Molesta** (7 species: *A. anigozanthi*, *A. armoraciae*, *A. broccoli-italicae*, *A. geniostomatis*, *A. molesta*, *A. nucis* and *A. soliaridae*). *A. mouchaccae* morphologically seems to be a member of this group, but phylogenetically is attributed to the dianthicola group on the basis of a study of a representative strain. Phylogeny of listed species is still unknown, but I hypothesize that this group has a close relationship with the infectoria group, however it can also be a relative of the alternata, brassicicola or embellisia groups.

Primary conidiophores are mainly simple and short (no longer than 100 μm) with several conidiogenous loci, but in some species conidiophores may have other characteristics. Conidia often are *Stemphylium*-like, oval, long-ellipsoid, cylindrical; solitary or in short chains of 2-4 units. Conidia are constricted near the septa, conidial cells are obviously prominent. Conidia are small 15-50×8-20 μm, having 3-7 to 7-11 transverse septa, a part of them is thick. Longitudinal septa appear usually in several or sometimes in one or in all transverse segments. Secondary conidiophores are absent in some species or if present may be short (up to 5-10 μm), apical (*A. armoraciae*, *A. broccoli-italicae*, *A. geniostomatis*) or lateral (*A. broccoli-italicae*).

Teleomorph is not known.

Nimbya (17 *Nimbya* species, including *N. caricis* and *N. scirpicola*). The uniqueness of this phylogroup is evident by the study of Hong et al. (2005b), however, within the group both species have very big genetic differences. Morphology of *N. rapontici* fits better in the porri group with the exception of the presence of *Drechslera*-like transepta. *N. rapontici* is associated with a plant from *Asteraceae* (*Raponticum carthamoides*) as many other members of the porri group. Thus, inclusion of the species in the nimbya group is doubtful.

Primary conidiophores are mainly simple, 20-150 μm long, having 1-4 conidiogenous loci. Conidia are solitary, sometimes in simple or branched chains of 2-8 units. Intercalary conidia usually are shorter than basal and apical ones. Basal conidia may have short geniculate lateral and apical secondary conidiophores forming 1-3 conidiogenous pores. All conidia or at least some of them are extended obclavate, or subcylindrical, and with 30-130×6-22 μm body and (0-)20-100 μm beak. In some species a few conidia are long-ovoid, subcylindrical with extended cone-shaped apical cell, and 260 μm long. Transverse septa are 6-13 per conidia and thin *Alternaria*- or thick *Drechslera*-like. Longitudinal septa are absent or present (*N. major*) as 1 in 1-4 transverse divisions. *N. scirpinfestans* produces microsclerotia.

Four species (*N. juncicola, N. scirpicola, N. scirpivora* and *N. scirpinfestans*) presumably are homothallic (Johnson et al., 2002) and undergo sexual sporulation under natural and artificial conditions.

Six species of the group are described from monocotyledonous plants (families *Juncaceae* and *Cyperaceae*). All species are known to have teleomorphs, including the species *N. caricis* and *N. scirpicola*, for which molecular phylogeny was studied. Six species are found on *Amarantaceae*.

Nobilis (approximately 12 species, including *A. nobilis* and *A. cheiranthi*). Simmons (2002, 2007) stated morphological similarity of *A. diversispora, A. ellipsoidea, A. ellisii, A. guaranitica, A. gypsophilae, A. hibiscina, A. nobilis, A. saponariae* and *A. vaccariae*. Probably, *A. subelliptica* also should be included in this group, however, this species as well as *A. ellisii* are known only as herbarium samples hence information about their sporulation patterns etc. are incomplete. In the opinion of the author this group can by expanded by the addition of *A. juxtiseptata*. The species has smaller conidia with lesser number of septa in comparison with other species of the group, but *A. juxtiseptata, A. vaccariae* and *A. nobilis* have similar sporulation patterns and represent pathogens of caryophyllaceous plants. Species similar to *A. cheiranthi* (e.g. *A. interna, A. papaveris* and *A. shaanxiensis*) may also turn out to be members of the group.

Primary conidiophores are mainly short or sometimes long, up to 250× 5-6 μm, with 1-2 conidiogenous loci. Conidia form simple or branched chains of 3-7 units. Apical secondary conidiophores have 1-2(-4) conidiogenous loci. Lateral secondary conidiophores sometimes appear and reach about half of the length of conidial body. Conidia are obclavate, from 40-70 to 125-176× 15-30 μm with 10-15(-19) transepta. Transverse segments of the basal conidia have 1-8 lateral, oblique and secondary transverse distosepta, hence, the central segments are long. A few eusepta in conidial body are wider than others.

Teleomorph is not known. Five species are associated with host plants of the family *Caryophyllaceae*.

Photistica (approximately 12 species, including *A. panax* and *A. photistica*). Molecular phylogeny and fingerprinting support inclusion of two other species into this group in addition to those mentioned above. *A. avenicola* was found to be closely related to *A. photistica* (Gannibal, unpubl.), while *A. calycipyricola* is allied to *A. avenicola* (Roberts, 2007). Sporulation pattern, size of conidia and character of septation of many members of photistica group resembles that of infectoria complex (Simmons, 2007). Phylogenetic data confirm their close relationship, but allow resolving them as two diverged branches. Besides the four species mentioned, it is the opinion of the author that *A. prasonis* and *A. cumini* have morphological traits that can make them potential members of photistica group. Several similar species (*A. asphodeli, A. catharanthicola, A. ornatissima, A. putrefaciens, A. triticicola* and *A. ulmi*) are known only by herbarium specimens, that makes their morphological assessment difficult and not appreciably reliable.

Primary conidiophores may be short simple (*A. avenicola*) or long branched (*A. photistica*). Conidiogenous loci usually are solitary or in some species 1-3 per conidiophore apex. Chains of conidia are simple or with a few branches. Basal conidia in old cultures often have a pronounced ability to produce secondary conidiophores, either at the tips of conidium beaks or laterally (2-6 or more) from conidium body cells. Conidia are wide oval, obclavate, wide ellipsoid or long ellipsoid, 40-160×10-25 μm. Conidia have 5-8 or more transverse septa and 1-2 longisepta in a few transverse segments or 1-8 longisepta in many segments (*A. triticicola*). In culture, cells of conidia may become greatly protuberant and spheroid.

Three species (*A. avenicola, A. calycipyricola* and *A. photistica*) have teleomorphs described in culture (Simmons, 1986; Kwaśna and Kosiak, 2003; Roberts, 2007). At least two of them are homothallic (Gannibal and Kazartsev, unpubl.).

A. avenicola has a relatively wider distribution and was discovered on plants of different families (Gannibal, unpubl.), however such information is not available for other species.

Porri (approximately 120 large-spored species, including *A. blumeae, A. capsici, A. crassa, A. cucumerina, A. dauci, A. euphorbiicola, A. limicola, A. linicola, A. macrospora, A. porri, A. pseudorostrata, A. solani* and *A. tagetica*). One large-spored species, *A. brassicae,* has closer connection with the porri group than with others, but usually it is placed on dendrogramms as a separated lineage.

Primary conidiophores are mainly short, simple, rarely long (up to 250 μm) and/or branched, usually with 1-3 conidiogenous loci on the top. Conidia are solitary, sometimes in short chains of 2-3 units, relatively larger (due to which they are called as large-spored), obclavate, and subcylindrical. At least some mature conidia have one or a few long, sometimes branching, filiform beaks. Occasionally the beak apex is conversed into secondary conidiophores. Basal conidia of some species can form apical and 1-2 lateral secondary conidiophores. Conidium body is 40-110×10-25(-30) μm; beak can be up to 80-300 (-400) μm long. The majority of species have 5-9 to 8-13 transepta in the conidium body and no or several septa in the beak. Longitudinal septa appear as one or sometimes a few in two or larger number of transverse divisions.

Teleomorph is not known.

Specialized parasites of many different plants, a lot of species are known on *Asteraceae*. Several species have been discovered on *Apiaceae, Cucurbitaceae, Euphorbiaceae, Solanaceae* and other families.

Prathoda (2 species of the genus *Prathoda, P. longissima* and *P. saparva*). Probably this group is phylogenetically distinct from alternarioid hyphomycetes and has a closer relationship with *Leptosphaeria* spp. (Pryor and Gilbertson, 2000). However, further investigation is required.

Primary conidiophores are simple and unbranched. They become approximately 250-300 μm long through geniculate extension and produce conidia at 1-6 loci. A very high percentage of conidia are narrow-ovoid with a simple, very long, gradually narrowing, and whip-like beak. Sometimes such conidia are *Cercospora*-like. The narrow beak frequently becomes slightly swollen and conidiogenous at its tip, initiating a branching chain of 3-5+ long-beaked conidia. The overall length of rostrate conidia reach a range of 150-300(-500+) μm, with the narrow-ovoid base contributing ca 70-90 μm of the length. A number of conidia remain broadly ovoid, beakless, and 45-55× 8-14 μm. The conidium body has 4-9 transverse septa and usually no longisepta, or occasionally with one in one transverse segment.

Teleomorph is not known.

Both species were found on graminicolous as well as other plants.

Radicina (6 species, including *A. carotiincultae*, *A. petroselini*, *A. radicina*, *A. selini* and *A. smyrnii*). Besides the species mentioned, *A. atrocariis* fits into the morphological criteria of this group.

Primary conidiophores are of different types and lengths, with 1-4 conidiogenous loci at the apex. Conidia are relatively small oval, wide elliptical, subcylindric or rarely subspheroid, and solitary or occasionally in chains of 2-3; 30-80(-96)×15-38 μm. Conidia usually have 3-12 transepta and 1-3 longisepta in several transverse segments. Two species (*A. atrocariis* and *A. selini*) produce lateral secondary conidiophores. *A. smyrnii* forms microsclerotia.

Teleomorph is not known.

All 6 species are strongly specialized pathogens of umbellifers.

Sonchi (6 species, including *A. cinerariae* and *A. sonchi*). According to Simmons (2007) there exists a similarity between *A. aliena* and *A. sonchi*. *A. dianthicola* and *A. euphorbiicola* are also similar according to him. However in the author's opinion, *A. dianthicola* and *A. euphorbiicola* most likely belong to other phylogenetic groups. Also, one can find a similarity between *A. cinerariae* and *A. panax* or *A. sonchi*, *A. rudbeckiae* and *A. cinerariae*. Nevertheless, according to morphological analysis, the author believes this group includes *A. aliena*, *A. liriodendri*, *A. rdbeckiae*, and *A. triglochinicola* besides *A. cinerariae* and *A. sonchi*.

Primary conidiophores are simple, short (less than 100 μm) or long (up to 250 μm in culture), in some species rarely branched, and with 1-3 conidiogenous loci. Conidia are subcylindrical, sometimes obclavate, solitary or in short chains, and 70-150×15-30 μm. Apical cells of conidia are often conical. Mature conidia usually are sufficiently constricted near the septa (especially in *A. cinerariae*). Transepta are 3-7 to 8-12 per conidium. Conidia usually have 1-3 longisepta in (1-)2-4(-7) transverse divisions. Conidial cell walls and septa usually are thin and pale.

Teleomorph is not known.

Presumably specialized pathogens, 3 species are known from asteraceous plants.

Teretispora (2 species, *Alternariaster helianthi* and *Teretispora leucanthemi*). No special phylogenetic work has been done for these species, but rDNA sequence analysis supports high similarity of these species and their separation from other phylogroups (data not shown).

Primary conidiophores are short or long with 1-3 conidiogenous loci. Conidia are solitary or sometimes in chains of two conidia when conidia form lateral secondary conidiophores. Conidia are cylindrical, 80-130(-160)×17-23 (-30) μm and with 7-14(-17) transepta. Longisepta present in 3-7 wide transverse segments (*T. leucanthemi*) or are very uncommon (*A. helianthi*).

Teleomorph is not known.

Specialized pathogens of asteraceous plants.

Ulocladium (approximately 30 species, including *E. indefessa*, *U. atrum*, *U. botrytis*, *U. cantlous*, *U. capsicuma*, *U. chartarum*, *U. consortiale*, *U. cucurbitae*, *U. dauci*, *U. multiforme*, *U. obovoideum*, *U. oudemansii*, *U. septosporum*, *U. solani*, *U. subcucurbitae*). The group also includes other species of the genus *Ulocladium* except *U. alternariae*. This phylogroup can be expanded by species similar to *U. atrum* (e.g. *A. levis* and *A. trachelospermi*) or *A. cheiranthi* and similar species (see nobilis phylogroup).

Primary conidiophores are simple or branched, as much as 5-200× 3-5 μm, and have at least 2-3, but usually 6-10 conidiogenous loci. Conidia appear in clusters when conidiophore development is closely geniculate, or conidia are distant when the geniculations are well-separated. Short chains of conidia are characteristic for some species, but in many species all or almost all conidia remain solitary. Conidia are small 14-40×7-21 μm fundamentally obovoid, sometimes subspherical or short cylindrical. Commonly basal cell of the conidium terminates into a small and rounded apiculus. Conidia have 1-7 transverse septa and 1-3 longitudinal septa in 1-4 transverse segments. Conidium walls may be smooth but more commonly are variously roughened. Secondary conidiophores when present are short, simple or geniculate bearing several conidiogenous loci.

Teleomorph is not known.

Several species are widely distributed, saprotrophs or secondary plant parasites. All species have small or moderate population size in spite of wider area of distribution.

Undifilum (2 species of the genus *Undifilum*: *U. bornmuelleri* and *U. oxytropis*).

Primary conidiophores are simple or sometimes branched, short, 10-60 (-130)×6-8 μm with 1-2 conidiogenous loci. Conidia are solitary or in chains of

2 units, cylindrical or long ellipsoid, 30-90×8-12(-15) μm with (2-)3-8 transepta which often are thick. Longitudinal septa occur very rarely.

Teleomorph is not known.

Both species are associated with legumes as endophytes or parasites (Pryor et al., 2009).

Most of the morphological traits are plastic; therefore several states of each character may appear in one phylogroup. On the other hand, a character state may usually be present in several phylogroups indicating convergent or parallel evolution of many morphological traits. The best example of such convergence is the "morphological" genus *Embellisia*. Thus, many qualitative and all quantitative traits are not useful for phylogenetic analysis. The author's attempts to determine the accessible simple morphological traits and use them for phylogenetic work failed. A phylogenetic tree designed had too many of unresolved polytomy.

Only a few, rather complex traits, can be used for satisfactory characterization of phylogroups, for example (i) the shape of conidia including type and length of apical and lateral rostra; (ii) conidial size and number of transverse septa; and (iii) number of longitudinal septa. Sporulation pattern is also a valuable trait though it is a derivative of appearance and length of secondary conidiophores and number of conidiogenous loci on them.

Probably, it could be more interesting to superimpose biochemical and physiological data on the phylogenetic tree. Structure and ecological niche of micromycetes affect the biochemical and physiological features and play a greater role in environmental interactions than morphological traits. Changes in enzymatic apparatus, primary and secondary metabolites influence the evolution of fungi and determine their fitness for a specific ecological niche. Unfortunately, data on biochemical and physiological features of alternarioid hyphomycetes are still meager and, therefore, cannot be adequately used in phylogenetic analysis.

PHYLOGENETIC SPECIES RECOGNITION IN ALTERNARIOID HYPHOMYCETES

The issue of the concept of species has been discussed since a very long time, but still remains unsolved. The lack of the sexual reproduction in many fungi causes difficulties of philosophical and methodological nature. Phylogenetic approaches, utilizing molecular data, seem to be more convenient than other alternatives. However, they also have their limitations.

Several attempts have been made to compare *Alternaria* species using phylogenetic approaches and sequence data. One or two genomic loci, commonly ITS regions, were sequenced for this purpose (Jasalavich et al.,

1995; de Hoog and Horré, 2002; Kang et al., 2002; Serdani et al., 2002). But it was found that ITS regions are too conservative to ensure high resolution phylograms for alternarioid hyphomycetes. Moreover, genealogy of a single locus cannot serve as an appropriate base for species delimitation using ideology of genealogic concordance phylogenetic species recognition (GCPSR; Taylor et al., 2000). Only a few detailed studies relating to species boundaries in alternarioid hyphomycetes based on the sizable datasets and proper phylogenetic approaches as GCPSR have been published (Taylor et al., 2000). Comparison of morphospecies and phylospecies in different alternarioid hyphomycetes gave fairly different results.

Analysis of two morphologically similar umbelliferous pathogens from radicina phylogroup resulted into an entirely morphologically and phylogenetically matching recognized species. Some authors were of the view that the morphological differences between *A. radicina* and *A. carotiincultae* are partly due to the toxin production (radicinin) and that these fungi are conspecific (Pryor and Gilbertson, 2002). But further studies convincingly indicated the differences between the species (Park et al., 2008). The phylogenetic relationship between *A. radicina* and *A. carotiincultae* was re-examined based on morphology, sequence analysis of rDNA, five protein coding genes, and RAPD and ISSR analysis of total genomic DNA. The results suggested that *A. carotiincultae* and *A. radicina* are closely related, but they have distinct lineages and, therefore, they are separate species.

Several studies (see reviews of Taylor et al., 2000 and O'Donnell et al., 2004) showed an increase in numbers of taxa, when one switches from morphological species recognition to GCPSR in fungi. In some studies with alternarioid hyphomycetes the authors have found the opposite. Since the morphology of alternarioid hyphomycetes is studied very rigorously and many minor morphological characters have been endowed with taxonomic value, these fungi are a big group of "small" (I mean species that are a result of intensive splitting) morphologically similar species. Not all of these species can be recognized with phylogenetic approaches.

A lot has been done for the small-spored *Alternaria* species (alternata phylogroup) associated with citrus (Peever at al., 2000, 2002, 2004, 2005). Twelve morphospecies are separated across eight clades. Three clades contain more than one morphospecies. While these lineages have no diagnostic morphological traits and no correlation with the host, investigators categorically advocate merging all small-spored citrus-associated isolates of *Alternaria* into a single species, *A. alternata*. The investigations regarding small-spored *Alternaria* species in general were furthered by Andrew and co-workers (Andrew et al., 2009). No strict congruence between morphology and phylogenetic lineage was found among isolates identified as *A. alternata* and *A. tenuissima* using morphology. In contrast isolates grouped morphologically with *A. arborescens* fell into discrete clades for all datasets. Although, 5-9 well supported clades were evident among isolates, it remains unclear if these clades should be considered phylogenetic species or emerging evolutionary lineages.

Sequence analyses of the ITS region and two protein coding genes of 39 in *A. infectoria* species-group strains resulted in a tree of unclear polytomy (Andersen et al., 2009). Therefore, according to GCPSR, all studied strains are one phylogenetic species while morphological analysis gave at least 10 different species. Parallel chemical classification gave two different clusters and one single strain. Very similar results were obtained from UP-PCR analysis, where the question remained whether the 52 isolates are one species or 52? (Gannibal and Yli-Mattila, 2007). Applying GCPSR to the *A. infectoria* species-group would lead to synonymizing of all the morphological species into the *A. infectoria* species-group under one name: *A. infectoria*. However, merging of morphologically, chemically and ecologically diverse taxa into one appears impudent. Moreover, according to Taylor et al. (2000) GCPSR cannot be applied to truly clonal fungi. Morphological species recognition, an alternative approach is also not workable. If it is applied, almost each strain in the *A. infectoria* species-group would represent a new "emerging" species that would require a unique name and a formal description (Andersen et al., 2009).

Thus, the number of phylogenetic species in alternata and infectoria phylogroups is more than one and obviously all small-spored species cannot be combined into *A. alternata*. However, some recently described morphologically very similar species are not supported by phylogeny and the number of species in these two phylogroups is probably overestimated.

CONCLUSIONS

It is necessary to underline once more that current taxonomy of alternarioid hyphomycetes does not completely coincide with their phylogeny. To make the system conform to the phylogenetic pattern, it will be necessary to make the changes in genera boundaries. As a result, probably several new taxa of subgeneric ranks will emerge. The genus *Embellisia* is obviously polyphyletic and should be divided into 2 or 3 genera. Three phylogroups of *Alternaria* (infectoria, photistica, dianthicola) are more closely related to *Nimbya* and some *Embellisia* species than to other Alternarias and, therefore, they should be assigned the status of separate genera. The monotypic genera *Alternariaster* and *Teretispora,* after precise analysis of molecular data, could be combined into one.

Several phylogroups have been revealed within *Alternaria*. They may be formally described in future as sections or subgenera.

The species boundaries still remain an issue. Very likely in a number of cases phylogenetic species of alternarioid hyphomycetes include several morphospecies. While, in many other fungi it is not the case. Neither morphological nor phylogenetic species recognition approach works well for alternarioid hyphomycetes, at least for asexual saprotrophic *Alternaria* species.

In future studies of alternarioid hyphomycetes full consideration should be given to the affiliation of species with the phylogroup. In order to construct the most complete phylogram of alternarioid hyphomycetes, it is necessary to include in analysis, species from all phylogroups described above. Also, phylogenetic research may be continued separately at the same time for each group.

To avoid a terminological mess it is proposed to use the term "phaeodictyosporic hyphomycetes" for a morphologically defined group while the term "alternarioid hyphomycetes" should be used for a monophyletic group of fungi related to the genus *Alternaria*. Usage of "species-group" should be restricted to reflect a morphologic meaning in order to name a group of species with similar sporulation patterns, while for phylogenetically defined groups, the author proposes naming them 'phylogroups'.

ACKNOWLEDGEMENT

The research was supported by the Ministry of Education and Science of the Russian Federation (State contract 16.518.11.7068) and RFBR (grant no. 09-04-13753).

REFERENCES

Akimitsu, K., Peever, T. L. and Timmer, L. W. 2003. Molecular, ecological and evolutionary approaches to understanding *Alternaria* diseases of citrus. Mol. Plant Pathol., 4: 435-446.

Altschul, S. F., Gish, W., Miller, W., Myers, E. W. and Lipman, D. J. 1990. Basic local alignment search tool. J. Mol. Biol., 215: 403-410.

Andersen, B. and Thrane, U. 1996. Differentiation of *Alternaria infectoria* and *Alternaria alternata* based on morphology, metabolite profiles, and cultural characteristics. Can. J. Microbiol., 42: 685-689.

Andersen, B., Krøger, E. and Roberts, R. G. 2002. Chemical and morphological segregation of *Alternaria arborescens*, *A. infectoria* and *A. tenuissima* species-group. Mycol. Res., 106: 170-182.

Andersen, B., Sorensen, J. L., Nielsen, K. F., van den Ende, B. G. and de Hoog, S. 2009. A polyphasic approach to the taxonomy of the *Alternaria infectoria* species-group. Fungal Genet. Biol., 46: 642-656.

Andrew, M., Peever, T. L. and Pryor, B. M. 2009. An expanded multilocus phylogeny does not resolve morphological species within the small-spored *Alternaria* species complex. Mycologia, 101: 95-109.

Arie, T., Kaneko, I., Yoshida, T., Noguchi, M. and Nomura, Y. 2000. Mating-type genes from asexual phytopathogenic ascomycetes *Fusarium oxysporum* and *Alternaria alternata*. Mol. Plant-Microbe Interact., 13: 1330-1339.

Berbee, M. L., Payne, B. P., Zhang, G., Roberts, R. G. and Turgeon, B. G. 2003. Shared ITS DNA substitutions in isolates of opposite mating type reveal a recombining history for three presumed asexual species in the filamentous ascomycete genus *Alternaria*. Mycol. Res., 107: 169-182.

Braun, U., Crous, P. W., Dugan, F., Groenwald, J. Z. and de Hoog, G. S. 2003. Phylogeny and taxonomy of *Cladosporium*-like hyphomycetes, including *Davidiella* gen. nov., the teleomorph of *Cladosporium* s. str. Mycol. Prog., 2: 3-18.

Chou, H.-H. and Wu, W.-S. 2002. Phylogenetic analysis of internal transcribed spacer regions of the genus *Alternaria*, and the significance of filament-beaked conidia. Mycol. Res., 106: 164-169.

Covert, S. F. 1998. Supernumerary chromosomes in filamentous fungi. Curr. Genet., 33: 311-319.

Crous, P. W., Braun, U., Wingfield, M. J., Wood, A. R., Shin, H. D., Summerell, B. A., Alfenas, A. C., Cumagun, C. J. R. and Groenewald, J. Z. 2009. Phylogeny and taxonomy of obscure genera of microfungi. Persoonia, 22: 139-161.

David, J. C., Coles, K., Fisher, J. and Moss, S. T. 2000. A new species of *Embellisia* from soil with high levels of heavy metals. Mycoscience, 41: 533-537.

Elliott, J. A. 1917. Taxonomic characters of the genera *Alternaria* and *Macrosporium*. Amer. J. Bot., 4: 439-478.

Ellis, M. B. 1976. More Dematiaceous Hyphomycetes. CABI Publishing, Wallingford, UK.

Gannibal, Ph. B. and Yli-Mattila, T. 2007. Morphological and UP-PCR analyses and design of a PCR assay for differentiation of *Alternaria infectoria* species-group. Mikol. Fitopatol., 41: 313-322.

Groves, J. W. and Skolko, A. J. 1944. Notes on seed-born fungi II. *Alternaria*. Can. J. Res., C22: 217-234.

Hong, S. G., Liu, D. and Pryor, B. M. 2005a. Restriction mapping of the IGS region in *Alternaria* spp. reveals variable and conserved domains. Mycol. Res., 109: 87-95.

Hong, S. G., Robert, A. C., Lawrence, C. B. and Pryor, B. M. 2005b. Alt a 1 allergen homologs from *Alternaria* and related taxa: analysis of phylogenetic content and secondary structure. Fungal Genet. Biol., 42: 119-129.

de Hoog, G. S. and Horré, R. 2002. Molecular taxonomy of the *Alternaria* and *Ulocladium* species from humans and their identification in the routine laboratory. Mycoses, 45: 259-276.

Huson, D. H., Dezulian, T., Klöopper, T. and Stee, M. A. 2004. Phylogenetic Super-networks from Partial Trees, pp. 388-399. In: 4th Workshop on Algorithms in BioInformatics in Lecture Notes in Bioinformatics (LNBI). Vol. 3240. I. Jonassen and J. Kim (eds.). Springer-Verlag, Berlin–Heidelberg.

Inderbitzin, P., Shoemaker, R. A., O'Neill, N. R., Turgeon, B. G. and Berbee, M. L. 2006. Systematics and mating systems of two fungal pathogens of opium poppy: the heterothallic *Crivellia papaveracea* with *Brachycladium penicillatum* sexual state and a homothallic species with a *Brachycladium papaveris* asexual state. Can. J. Bot., 84: 1304-1326.

Jasalavich, C. A., Morales, V. M., Pelcher, L. E. and Séguen-Swartz, G. 1995. Comparison of nuclear ribosomal DNA sequence from *Alternaria* species pathogenic to crucifers. Mycol. Res., 99: 604-614.

Johnson, D. A., Simmons, E. G., Miller, J. S. and Stewart, E. L. 2002. Taxonomy and pathology of *Macrospora/Nimbya* on some North American bulrushes (*Scirpus* spp.). Mycotaxon, 84: 413-428.

Johnson, L. J., Johnson, R. D., Akamatsu, H., Salamiah, A., Otani, H., Kohmoto, K. and Kodama, M. 2001. Spontaneous loss of a conditionally dispensable chromosome from the *Alternaria alternata* apple pathotype leads to loss of toxin production and pathogenicity. Curr. Genet., 40: 65-72.

Joly, P. 1964. Le genre *Alternaria*. P. Lechevallier, Paris.

Kang, J.-C., Crous, P. W., Mchau, G. R. A., Serdani, M. and Song, S.-M. 2002. Phylogenetic analysis of *Alternaria* spp. associated with apple core rot and citrus black rot in South Africa. Mycol. Res., 106: 1151-1162.

Khaldi, N., Collemare, J., Lebrun, M.-H. and Wolfe, K. H. 2008. Evidence for horizontal transfer of a secondary metabolite gene cluster between fungi. Genome Biol., 9: R18.

Kwaśna, H. and Kosiak, B. 2003. *Lewia avenicola* sp. nov. and its *Alternaria* anamorph from oat grain, with a key to the species of *Lewia*. Mycol. Res., 107: 371-376.

Leach, C. M. 1970. Effect of temperature on conidium characteristics of *Ulocladium chartarum* and *Stemphylium floridanum*. Mycologia, 62: 1071-1076.

Mayr, E., Linsley, E. G. and Usinger, R. L. 1953. Methods and Principles of Systematic Zoology. McGraw-Hill, New York.

Misaghi, I. J., Grogan, R. G., Duniway, J. M. and Kimble, K. A. 1978. Influence of environment and culture media on spore morphology of *Alternaria alternata*. Phytopathology, 68: 29-34.

Neergaard, P. 1945. Danish species of Alternaria and Stemphylium. Taxonomy, parasitism, economical significance. E. Munksgaard, Copenhagen.

Nees von Esenbeck, C. G. 1816-1817. Das System der Pilze und Schwämme. Stahelschen Buchhandlung. Wurzburg. xxxviii + 329 p., Pl. I-XXVII (1986); Pl. XXVIII-XLIV (1817).

Nishimura, S. and Kohmoto, K. 1983. Host-specific toxins and chemical structures from *Alternaria* species. Ann. Rev. Phytopathol., 21: 87-116.

O'Donnell, K., Ward, T. J., Geiser, D. M., Kistler, H. C. and Aoki, T. 2004. Genealogical concordance between the mating type locus and seven other nuclear genes supports formal recognition of nine phylogenetically distinct species within the *Fusarium graminearum* clade. Fungal Genet. Biol., 41: 600-623.

Park, M. S., Romanoski, C. E. and Pryor, B. M. 2008. A re-examination of the phylogenetic relationship between the causal agents of carrot black rot, *Alternaria radicina* and *A. carotiincultae*. Mycologia, 100: 511-527.

Pavlinov, I. Ya. 2003. It is normal for the classification approaches to be diverse. Zh. Obshch. Biol., 64: 275-291 (in Russian).

Peever, T. L., Carpenter-Boggs, L., Timmer, L. W., Carris, L. M. and Bhatia, A. 2005. Citrus black rot is caused by phylogenetically distinct lineages of *Alternaria alternata*. Phytopathology, 95: 512-518.

Peever, T. L., Ibañez, A., Akimitsu, K. and Timmer, L. W. 2002. Worldwide phylogeography of the citrus brown spot pathogen, *Alternaria alternata*. Phytopathology, 92: 794-802.

Peever, T. L., Olsen, L., Ibanez, A. and Timmer, L. W. 2000. Genetic differentiation and host specificity among populations of *Alternaria* spp. causing brown spot of grapefruit and tangerine × grapefruit hybrids in Florida. Phytopathology, 90: 407-414.

Peever, T. L., Su, G., Carpenter-Boggs, L. and Timmer, L. W. 2004. Molecular systematics of citrus-associated *Alternaria* species. Mycologia, 96: 119-134.

Pryor, B. M. and Bigelow, D. M. 2003. Molecular characterization of *Embellisia* and *Nimbia* species and their relationship to *Alternaria, Ulocladium* and *Stemphylium*. Mycologia, 95: 1141-1154.

Pryor, B. M. and Gilbertson, R. L. 2000. Molecular phylogenetic relationship amongst *Alternaria* species and related fungi based upon analysis of nuclear ITS and mt SSU rDNA sequences. Mycol. Res., 104: 1312-1321.

Pryor, B. M. and Gilbertson, R. L. 2002. Relationships and taxonomic status of *Alternaria radicina, A. carotiincultae*, and *A. petroselini* based upon morphological, biochemical, and molecular characteristics. Mycologia, 94: 49-61.

Pryor, B. M., Creamer, R., Shoemaker, R. A., McLain-Romero, J. and Hambleton, S. 2009. *Undifilum*, a new genus for endophytic *Embellisia oxytropis* and parasitic *Helminthosporium bornmuelleri* on legumes. Botany, 87: 178-194.

Roberts, R. G. 2007. Two new species of *Alternaria* from pear fruit. Mycotaxon, 100: 159-167.

Serdani, M., Kang, J.-Ch., Andersen, B. and Crous, P. W. 2002. Characterisation of *Alternaria* species-groups associated with core rot of apples in South Africa. Mycol. Res., 106: 561-569.

Simmons, E. G. 1967. Typification of *Alternaria*, *Stemphylium*, and *Ulocladium*. Mycologia, 59: 67-92.

Simmons, E. G. 1971. *Helminthosporium allii* as type of a new genus. Mycologia, 63: 380-386.

Simmons, E. G. 1983. An aggregation of *Embellisia* species. Mycotaxon, 17: 216-241.

Simmons, E. G. 1986. *Alternaria* themes and variations (22-26). Mycotaxon, 25: 287-308.

Simmons, E. G. 1989. *Macrospora* Fuckel (Pleosporaceae) and related anamorphs. Sydowia, 41: 314-329.

Simmons, E. G. 1990a. *Alternaria* themes and variations (27-53). Mycotaxon, 37: 79-119.

Simmons, E. G. 1990b. *Embellisia* and related teleomorphs. Mycotaxon, 38: 251-265.

Simmons, E. G. 1992. *Alternaria* taxonomy: current status, viewpoint, challenge, pp. 1-36. In: Alternaria. Biology, Plant Diseases and Metabolites. J. Chełkowski and A. Visconti (eds.). Elsevier, Amsterdam.

Simmons, E. G. 1995. *Alternaria* themes and variations (112-144). Mycotaxon, 55: 55-163.

Simmons, E. G. 1997. *Alternaria* themes and variations (151-223). Mycotaxon, 65: 1-91.

Simmons, E. G. 1998. Multiplex conidium morphology in species of *Ulocladium atrum* group. Can. J. Bot., 76: 1533-1539.

Simmons, E. G. 2000. *Alternaria* themes and variations (244-286) Species on Solanaceae. Mycotaxon, 75: 1-115.

Simmons, E. G. 2002. *Alternaria* themes and variations (287-304). Mycotaxon, 82: 1-40.

Simmons, E. G. 2004. Novel dematiaceous hyphomycetes. Stud. Mycol., 50: 109-118.

Simmons, E. G. 2007. *Alternaria*. An Identification Manual. CBS, Utrecht.

Singh, U. P., Singh, S. K. and Sarma, B. K. 2000. Time-dependent sporulation, conidial size and germ tube formation in *Alternaria tenuissima* (Kunze ex Pers.) Wiltshire on different media. J. Phytopathol., 148: 413-416.

Sobers, E. K. and Doupnik, B. 1973. Stabilization of conidium morphology in cultures of *Alternaria longipes*. Phytopathology, 63: 191-192.

Taylor, J. W., Jacobson, D. J., Kroken, S., Kasuga, T., Geiser, D. M., Hibbett, D. S. and Fisher, M. C. 2000. Phylogenetic Species Recognition and Species Concepts in Fungi. Fungal Genet. Biol., 31: 21-32.

Tweedy, B. G. and Powell, D. 1963. The taxonomy of *Alternaria* and species of this genus reported on apples. Bot. Rev., 29: 405-412.

Wang, Y., Bruno, L. C. and Zhang, Z.-G. 2008. Two new species of *Ulocladium* from Southwest China. Mycologia, 100: 455-459.

Wang, Y., Pei, Y.-F., Zhang, K. and Zhang, X.-G. 2009. Molecular and morphological description of a new species of *Ulocladium* from Southern China. Mycol. Prog., 8: 207-214.

Wang, Y., Pei, Y.-F., O'Neill, N. R. and Zhang, X.-G. 2010. *Ulocladium cantlous* sp. nov. isolated from northwestern China: its morphology and molecular phylogenetic position. Mycologia, 102: 374-383.

Wiltshire, S. P. 1933. The foundation species of *Alternaria* and *Macrosporium*. Transact. British Mycol. Soc., 18: 135-160.

Zhang, X.-G. and Zhang, T.-Y. 2002. Studies on the genus *Ulocladium* Preuss from China. Mycosystema, 21: 25-26.

Zhao G. Z. and Zhang, T.-Yu. 2005. Notes on dictyosporous hyphomycetes from China VII. The genus *Nimbya*. Fungal Divers., 19: 201-215.

Chapter 12

Biomarker Discovery for Rapid Diagnosis of Invasive Candidiasis through Protein Profiling: A Review

Jalpa P. Tewari[1,*], **Sanjiv Tewari**[2] **and William G. Gardner**[3]

[1]Department of Agricultural, Food, and Nutritional Science
University of Alberta, Edmonton, Alberta, Canada T6G 2P5
E-mail: JP.Tewari@ualberta.ca

[2]Akron General Medical Center, Akron, Ohio U.S.A. 44307
E-mail: stewari@agmc.org

[3]Northeastern Ohio Universities College of Medicine
and Pharmacy, Rootstown, Ohio, U.S.A. 44272
E-mail: wgardner@neoucom.edu

ABSTRACT

Candida spp. are widespread innocuous commensal yeasts on human skin and mucosal membranes. They are also an emerging group of pathogens and cause candidiasis once the normal microflora of their abode is disrupted or the host immune defenses are compromised. Invasive candidiasis results in considerable morbidity and mortality in human adults and neonates requiring sensitive, rapid, and cost-effective diagnosis for optimum management. Initial treatment of the disease is based on clinical features which are not pathognomonic. Delayed targeted treatment may result in increased patient mortality. Also, *Candida* spp. differ in their sensitivity to various antifungal therapeutic interventions requiring rapid identification of the disease-causing agents in the clinical material at species level. Such a rapid species-level diagnosis is not possible at the moment. This chapter presents a critical review of proteomics research so far published

[*]*Corresponding author*

Note: Most of this article was prepared at the Department of Microbiology, Immunology, and Biochemistry, Northeastern Ohio Universities College of Medicine and Pharmacy, Rootstown, Ohio, U.S.A. 44272.

in relation to invasive candidiasis especially relating to various virulence factors of *Candida* spp. The information presented may help evolve potential strategies relating to development of proteomic biomarkers for sensitive, rapid, species-multiplex, and cost-effective diagnosis of invasive candidiasis caused by various species of *Candida*. The non-proteomic strategies are only mentioned briefly.

INTRODUCTION

Candida spp. are the most important cause of opportunistic invasive mycosis in human adults and neonates worldwide resulting in considerable morbidity and mortality (Hajjeh et al., 2004; Zaoutis et al., 2005; Pfaller et al., 2006; Pfaller and Diekema, 2007). These fungi are widespread as innocuous commensal yeasts on skin and mucosal membranes (oral cavity, gastrointestinal tract, and vagina), and cause diseases once the normal microflora is disrupted or the host immune defenses are compromised (Naglik et al., 2004). Anywhere from 1.0-80.6% humans are asymptomatic carriers of *Candida albicans* (Robin) Berkhout depending on the population surveyed (Niimi et al., 1999; Dolan et al., 2004). Primary or secondary infections by *Candida* spp. are termed as candidiasis and can assume numerous forms. They can range from relatively benign mucocutaneous infections such as oropharyngeal candidiasis, cutaneous candidiasis, or vulvovaginal candidiasis to more serious conditions such as hepatosplenic candidiasis, ocular candidiasis, or other deep organ involvements collectively referred to as disseminated candidiasis. (http://www.mycology.adelaide.edu.au/Mycoses/Cutaneous/Candidiasis/). The infections are termed invasive candidiasis once the pathogen breaches the epithelial barrier and grows into the underlying tissues (Jones, 1990). In such cases, the pathogen may be cultured from blood samples, cultured or demonstrated histologically in tissue biopsy samples, or recovered from body fluids such as cerebrospinal and joint fluids which are normally sterile (Jones, 1990). Many of the above are life-threatening complications especially in immune suppressed individuals and require immediate therapeutic intervention.

Candida is a yeast genus and includes about 154 species, out of which about 22 species are pathogenic or associated with humans (Hoog et al., 2000; Pfaller and Diekema, 2007; http://www.doctorfungus.org/thefungi/Candida_spp.htm). The pathogenic group consists of about nine medically significant species, although five species [*C. albicans*, *C. glabrata* (Anderson) S.A. Meyer & Yarrow, *C. parapsilosis* (Ashford) Langeron & Talice, *C. tropicalis* (Castellani) Berkhout, and *C. krusei* (Castellani) Berkhout] cause more than 90% of invasive infections due to *Candida* spp. (http://www.emedicine.com/med/topic264.htm; Pfaller and Diekema, 2007). Invasive candidiasis is reported to result in patient mortality rates of about 31 to 41% (Klingspor and Jalal, 2006) and in bone marrow transplantation recipients with *Candida* infections it can reach up to 73% (Pitarch et al., 2004). *Candida albicans* predominates among the various pathogenic species of this genus. However, some other species, such as *C. glabrata* and *C. krusei* have increased in incidence presumably coincident with high usage of fluconazole to treat human immunodeficiency virus (HIV)-infected patients during the early 1990s (Hajjeh et al., 2004; Pfaller et al., 2006). Both species exhibit tolerance to antifungal drugs. *Candida glabrata* is inherently less susceptible to fluconazole and amphotericin B, while *C. krusei* has intrinsic resistance to fluconazole, and has less susceptibility to amphotericin B and flucytosine (Pfaller et al., 2006). The other three most common pathogenic species (*C. albicans*, *C. parapsilosis*, and *C. tropicalis*) continue to be generally susceptible to polyenes, flucytosine, azoles, and echinocandins (Pfaller et al., 2006). *Candida lusitaniae* van Uden & do Carmo-Sousa and *C. guilliermondii* (Castell.) Langer & Guerra are also of clinical significance as they are frequently resistant to amphotericin B and *C. dubliniensis* is recovered primarily from HIV positive patients (http://www.emedicine.com/med/topic264.htm). Some species are specialists in causing certain types of diseases or colonizing certain ecological niches. For example, gastrointestinal tract infections and bloodborne dissemination usually occur through *C. albicans*, *C. tropicalis*, and *C. krusei* (Jones, 1990). *Candida parapsilosis* specializes in colonizing skin, intravascular devices, and equipment used to administer parenteral fluids (Jones 1990). *Candida* spp. are an emerging group of pathogens and no doubt more will be added to the medically significant species-list in future. The aforesaid highlights the need for a rapid and precise species level identification of *Candida* spp. infections for timely diagnosis and therapeutic intervention.

DIAGNOSIS OF INVASIVE CANDIDIASIS — PRESENT STATUS

The clinical diagnosis of invasive candidiasis is often difficult. Prompt diagnosis is essential because the mortality rate of untreated invasive candidiasis may exceed 70% whereas initiating treatment within 48 hours may reduce mortality by nearly half (Fraser et al., 1992; Nolla-Sallas et al., 1997). The decision to initiate treatment with potentially toxic antifungal agents is usually made based on the clinical setting, the presence of risk factors for invasive candidiasis, clinical signs and symptoms, cultures, radiographic findings, and histopathology of the involved tissue (Alexander and Pfaller, 2006; Perlroth et al., 2007; http://www.doctorfungus.org/mycoses/human/Candida/InvasiveOverview.htm). Some clinical risk factors are bone marrow and solid organ transplantation, immunocompromised patients, post-surgical patients especially abdominal surgery or bowel perforation, high-dose or long term corticosteroid therapy, cytotoxic drug therapy, broad-spectrum antimicrobial therapy, hemodialysis, central vascular catheters and other implanted devices, parenteral nutrition, mechanical ventilation, severe mucositis in neutropenic patients, HIV, neonatal age, and prolonged hospitalization (Blumberg et al., 2001; Sobel and Rex, 2001). The clinical signs of invasive candidiasis are nonspecific but may include fever or hypothermia, prolonged hypotension unresponsive to volume replacement, lack of response to antimicrobial therapy for suspected bacterial infection, and multisystem organ dysfunction or disseminated intravascular coagulopathy (DIC). No clinical features of this disease are pathognomonic (García-Ruiz et al., 1997; Moudni et al., 1998; Klingspor and Jalal, 2006). A retrospective multicenter study also could not identify reliable clinical predictors of non-*albicans* candidemia (Shorr et al., 2007). Multiple small, target-like, focal lesions in the liver or spleen demonstrated by ultrasound, computed tomography (CT) or magnetic resonance imaging are tentative indicators of hepatosplenic candidiasis but are apparent only after the patient has recovered from chemotherapy-induced neutropenia (Girmenia et al., 2004).

Histological and microbiological techniques have been used for a long time for diagnosing invasive candidiasis but they have poor sensitivity and specificity (García-Ruiz et al., 1997; Erjavec and Verweij, 2002; Girmenia et al., 2004; Klingspor and Jalal, 2006). *Candida* spp. are often commensal on skin, and mucous membranes of the mouth, gastrointestinal tract, and vagina. Hence, their presence alone on these surfaces as observed by microscopy or culturing techniques may not necessarily indicate a disease condition. Nevertheless, body fluids, smears, scrapings, and biopsy materials are examined microscopically for the presence of yeast cells with or without germ tubes, hyphae, and pseudohyphae (Sobel, 1997; O'Shaughnessy et al., 2003; Whiteway and Bachewich, 2007; http://www.merckmedicus.com/pp/us/hcp/diseasemodules/fungal/diagnosis.jsp). Biopsy of deep tissues for histological examination is often not feasible (Moudni et al., 1998). Blood cultures are positive in only 50-60% of cases of disseminated candidiasis (http://www.emedicine.com/med/topic264.htm) and diagnosis is generally established at a late stage or even at autopsy (Klingspor and Jalal, 2006). In another study, less than one-half patients with hepatosplenic candidiasis were reported to yield positive blood cultures (Chen, 2000). While lysis-centrifugation is considered to be the most sensitive technique, all culture techniques are time consuming with mean times for detection of positive cultures ranging from 1.8 to 7.9 days (Henry et al., 1983; Jones, 1990; Klingspor and Jalal, 2006). Germ tubes are produced by *C. albicans*, *C. dubliniensis* Sullivan et al., *C. stellatoidea* (Jones Martin & Durham) Langeron & Guerra, and rarely also by *C. tropicalis* (Tierno and Milstoc, 1977; Sudbery et al., 2004; Pincus et al., 2007; http://www.emedicine.com/med/topic264.htm). Hence, the germ tube test is not species-specific. In addition, a culture has to be prepared before conducting the germ tube test. Many growth media have been developed for distinguishing *C. albicans* from *C. dubliniensis* and other non-*albicans* species (Pinjon et al., 1998; Al Mosaid et al., 2001; Kim et al., 2002; Khan et al., 2004 a, b,; Girish Kumar and Menon, 2005; Sheth et al., 2005; Girish Kumar et al., 2006; Loreto et al., 2006). Rapid differentiation (<3 hour) of *C. albicans* from non-*albicans* species by germ tube production directly from BacTAlert blood culture bottles has been described recently but requires evaluation for diagnostic accuracy (Terlecka et al., 2006). Also, test samples containing other germ tube-forming species of *Candida* were not included in this study for validation of the technique. Phenotypic identification of *Candida* spp. based on hydrolysis of chromogenic substrates (Cooke et al., 2002; Pincus et al., 2007; Sendid et al., 2007) also requires prior preparation of cultures. Finally, empiric antifungal therapy may hamper isolation of causal fungi (Moudni et al., 1998) which is required for many diagnostic techniques.

Non-culture techniques, such as the *Candida* mannoprotein and (1->3)-ß-D-glucan assays, and the polymerase chain reaction (PCR), either lack sensitivity and specificity or require further validation (Coignard et al., 2004; Girmenia et al., 2004; White et al., 2004, 2005; Allan et al., 2005; Meyer et al., 2005; Klingspor and Jalal, 2006; Carvalho et al., 2007; Pincus et al., 2007; Schabereiter-Gurtner et al., 2007). For example, the (1->3)-ß-D-glucan assay is a sensitive and specific technique for diagnosing invasive fungal infections including candidiasis caused by several species of *Candida* (Odabasi et al., 2004). However, it does not distinguish between candidiasis and other diseases such as fusariosis, trichosporonosis, and aspergillosis since (1->3)-ß-D-glucan is a component of the cell walls of a wide variety of fungi (Odabasi et al., 2004; Ostrosky-Zeichner et al., 2005). Both *Mucor* and *Rhizopus* produce little or no (1->3)-ß-D-glucan and patients infected with these fungi gave negative results with this assay in a multicenter clinical evaluation (Ostrosky-Zeichner et al., 2005). Hence, the (1->3)-ß-D-glucan assay should have some value in distinguishing between zygomycosis caused by *Mucor* and *Rhizopus* and the aforesaid group of diseases. Page and Kurtzman (2005) developed direct hybridization and allele-specific primer extension DNA-based assays using flow cytometry for rapid identification of *Candida* spp. Diagnostic DNA microarray-based systems are also emerging techniques (Pincus et al., 2007). Some of the aforesaid techniques are also complicated, require highly trained manpower, sophisticated instrumentation, and are expensive.

Several antibody-based techniques have also been developed but they have limitations since the antibodies may be present in patients with *Candida*-colonization in their epithelial membranes or superficial tissues and also because antibody production may be confounded in immunocompromised patients (García-Ruiz et al., 1997; Quindós et al., 2004). However, there is some evidence that antibodies to *Candida* can be detected in highly immunodeficient patients, such as neutropenic bone marrow transplant recipients or liver transplant recipients (Quindós et al., 2004). *Candida albicans* germ tube antibodies (CAGTA) are also detectable in patients with invasive candidiasis caused by *C. tropicalis*, *C. parapsilosis*, *C. glabrata*, *C. dubliniensis*, *C. guilliermondii*, and *C. krusei*, although the titers are lower than in patients with candidiasis caused by *C. albicans* (Quindós et al., 2004). Hence, CAGTA are not species specific.

The Bactigard™ Candida is a rapid technique that monitors the activities of two enzymes, L-proline aminopeptidase and beta-galactosaminidase (Carrillo-Muñoz et al., 2002). However, it requires previous preparation of cultures, can identify only *C. albicans* which unlike other yeasts possesses both these enzymes, and does not discriminate between this species and *C. dubliniensis*. Several studies have addressed the use of *Candida* enolase antigen and antibody for diagnosis of invasive candidiasis (Walsh et al., 1991; Mitsutake et al., 1994, 1996; van Deventer et al., 1994; Laín et al., 2007). However, these approaches have limitations as enolase is produced by several species of *Candida* and also by another human yeast pathogen, *Saccharomyces cerevisiae* Meyen ex Hansen (Mitsutake et al., 1994; van Deventer et al., 1994; Tewari, Babu, Tewari, and Gardner unpublished data, http://genolist.pateur.fr/CandidaDB). Several methods of identifying *Candida* spp. (such as the API20C Aux from bioMérleux, Inc) are based on conventional assimilation tests of substrates (Schuffenecker et al., 1993; Pincus et al., 2007). These methods are cumbersome and time-consuming. D-arabinitol is a circulating metabolite of *Candida* spp. in human body fluids and this along with D-arabinitol/creatinine ratios have also been studied as markers for invasive candidiasis (Christensson et al., 1999; Yeo et al., 2006). However, several species of *Candida* are known to produce D-arabinitol, hence, the technique is not species specific. Also, in a serodiagnostic study, D-arabinitol test was least sensitive compared to (1->3)-ß-D-glucan, mannan, and *Candida* antigen by Cand-Tec tests (Kohno et al., 1993).

This chapter presents a critical review of proteomics research so far published in relation to invasive candidiasis especially relating to various virulence factors. The information presented may help evolve potential strategies relating to the development of proteomic biomarkers for diagnosis of invasive candidiasis. The non-proteomic strategies are mentioned only briefly.

PROTEOME PROFILING OF *CANDIDA* SPP.

Clinical chemistry research in its quest for disease biomarkers has focused on tissue specimens, and natural or clinical body fluids such as plasma, urine, tears, lymph, seminal plasma, milk, saliva, spinal fluid, lavages, and solubilized biopsy preparations (Castagna et al., 2005; Marko-Varga et al., 2005). Many studies on proteinaceous disease biomarkers have been carried out on human serum which is thought to contain most, if not all, human proteins (Castagna et al., 2005). Many holistic studies, mostly on *C. albicans*, have been carried out to identify proteomic biomarkers of systemic candidiasis. These biomarkers, besides use in disease diagnostics, could potentially help in clinical follow-up, following disease progression, and antifungal drug and vaccine designs (Pitarch et al., 2003, 2004). Progress in proteome profiling of *Candida* spp. and other systems have closely followed developments in protein separation technologies such as 2-dimensional polyacrylamide gel electrophoresis (2-D PAGE) and protein identification through mass spectrometry. Early work in this area is summarized by Niimi et al. (1999), Pitarch et al. (1999), and Hernández et al. (2004). One disadvantage with 2-D PAGE is that it favors identification of high abundance but not low abundance proteins (Rupp, 2004). Recent modifications in analytical techniques including removal of six most abundant proteins in human serum (albumin, IgG, IgA, transferrin, haptoglobin, and antitrypsin) which constitute over 85% of its proteome has resulted in considerable increase in protein detection in serum (Chromy et al., 2004). In another study, again on serum, employing immunoaffinity chromatography to remove the most abundant proteins, 2-D electrophoresis, and spot identification by MALDI and liquid chromatography-MS/MS (LC-MS/MS), Pieper at al. (2003) resolved nearly 3700 spots and identified 325 distinct proteins in the human serum proteome. Relatively fewer proteins have so far been resolved from the proteomes of *Candida* spp. (expected to encode about 7000 proteins, Niimi et al., 1999) and from sera of patients with candidiasis, and there is a need to apply the more modern techniques in such studies.

Pitarch et al. (1999) analyzed the proteomes of cytoplasmic extracts, protoplast lysates, and secretions of protoplasts of *C. albicans* and identified 18 immunoreactive proteins using 2-D PAGE and Western blotting with sera from patients with systemic candidiasis. Enolase was one of the two most immunoreactive proteins found in this study. In a further study from this group aimed at characterizing markers of systemic candidiasis, four immunogenic proteins (aconitase, pyruvate kinase, phosphoglycerate mutase, and methionine synthase) were identified using 2-D PAGE, Western blotting, and nanoelectrospray tandem mass spectrometry (Pardo et al., 2000). These studies were further extended into murine systemic candidiasis models where more than 31 immunoreactive proteins were detected by 2-D PAGE, immunoblotting with murine immune sera, and N-terminal amino acid sequencing (Pitarch et al., 2001). Pitarch et al. (2004) detected about 85 proteins in *C. albicans* protoplast lysates which were immunoreactive with systemic candidiasis patients' sera. This study used a combination of techniques such as 2-D PAGE, Western blotting, matrix-assisted laser desorption/ionization-time of flight-mass spectrometry (MALDI-TOF-MS), and nano-electrospray ionization-ion trap (ESI-IT) MS. The results suggested association of the production of antibodies to *C. albicans* phosphoglycerate kinase and alcohol dehydrogenase with differentiation of the human immune response. Also, especially the rise of high anti-enolase antibody concentration was related to recovery from systemic candidiasis in patients suggesting a possible marker for predicting disease outcome. They also noticed low levels of circulating antibodies directed against some *C. albicans* glycolytic and fermentative enzymes in some subjects with no evidence of candidiasis indicating exposure of these subjects to these antigens during harmless colonization of the pathogen. Bruneau et al. (2003) monitored proteome changes in cell extracts of *C. albicans* as a result of treatment with echinocandins and triazoles by 2-D PAGE, and MAIDI-TOF MS and/or nano-ESI MS/MS. This study showed comparable effects at the proteome level upon treatment with the antifungal compound with the same mode of action. Yin et al. (2004) identified 316 different proteins from disrupted cells of *C. albicans* by high-resolution 2D- PAGE and MALDI-TOF-MS. In the same study, 215 different proteins were identified from the relatively benign yeast *S. cerevisiae*. There were subtle differences between these two yeasts presumably reflecting their different life styles. However, it could also be that the major differences between the proteomes of these two yeasts were not revealed in this study. A number of studies have addressed the proteomics of filamentation as it is a well-known virulence factor in *C. albicans*. Hernández et al. (2004) analyzed the proteome of the cytoplasmic extract from hyphae of *C. albicans* and identified 66 proteins by using 2D-PAGE, MALDI-TOF-MS, or MALDI-TOF/TOF providing a benchmark for future investigations. Ebanks et al. (2006) carried out analysis of cell wall and associated proteins of *C. albicans* differentially regulated during transition from the yeast to hyphal phase and reported changes during this morphogenic switch. Insenser et al. (2006) analyzed the proteins of lipid rafts of *C. albicans* which have higher amounts of saturated fatty acids and sterols than other membranes, and are detergent-resistant. SDS (sodium dodecyl sulfate)-PAGE and MALDI-TOF/TOF MS yielded 29 proteins which were of potential diagnostic and prognostic significance. Also, antibodies to six of these proteins were reported earlier in patients with systemic candidiasis (Pitarch et al., 2004; Insenser et al., 2006). Yeater et al. (2007) used microarrays to identify changes in gene expression during biofilm development of *C. albicans* on two model substrates (denture and catheter) and identified some substrate-specific gene expression changes. Data analysis also identified expression of specific subsets of genes at different developmental phases.

DIAGNOSIS OF INVASIVE CANDIDIASIS — FUTURE PROSPECTS

It is clear that the numerous diagnostic tests for invasive candidiasis developed so far leave much to be desired with respect to the overall fungal disease- and *Candida* species-specific clues that can be used at the practical level. Virulence factors of a pathogen play key roles during pathogenesis and are likely to be unique strategies developed by the pathogen based on which diagnostic tests can be tailored. However, caution should be exercised in selecting these crucial strategies as many virulence factors in another yeast-pathogen, *Cryptococcus neoformans* (Sanfelice) Vuill., are thought to be "dual use", also important in survival of the organism in the environment (Casadevall et al., 2003). Like many other pathogenic organisms, virulence in *Candida* spp. is multicomponent and consists of adhesion, hyphal formation, phenotypic switching, and extracellular hydrolytic enzyme (lipolytic and proteolytic) production (Naglik et al., 2003b, 2004; Kadosh and Johnson, 2005). Each of these virulence components are regulated by many proteomic factors which could be potential candidates for developing biomarkers.

Adherence phase

Adhesion of *Candida* spp. in the infection court is essential for keeping it positioned for becoming invasive and it is considered to be a major virulence factor for this pathogen (Henriques et al., 2006). *Candida albicans* and other non-*Candida albicans Candida* (NCAC) species differ with respect to their affinity for adhering to various epithelial and endothelial cells, and artificial devices (Henriques et al., 2006). However, adhesion of the pathogen will be required both during the commensal (or colonization) and invasive phases, and any useful biomarker for invasive candidiasis during this stage must select specific strategies employed for the latter. Indeed, the yeast and hyphal/pseudohyphal forms differ in their adherence properties (Calderone and Braun, 1991; Kadosh and Johnson, 2005) and offer such diagnostic possibilities. The latter form of *C. albicans* have cell surface antigens that bind fibrinogen, laminin, and the complement (C3) conversion product C3d, while the former phase does not (Calderone and Braun 1991). However, it is generally not known if such antigens are synthesized *de novo* in the hyphal form or they merely move to the surface from a subsurface location upon change in form (Calderone and Braun, 1991). The hyphal form of *C. albicans* also has HWP1 (hyphal wall protein 1) which participates in cross-links with proteins on the mammalian mucosa, is synthesized *de novo* targeted to the hyphal surface, and is absent during the yeast growth (Staab et al., 1996, 2004).

Agglutinin-like sequence (Als) glycoproteins

The *ALS* (agglutinin-like sequence) gene family of *C. albicans* encodes eight large cell-surface glycoproteins out of which five (Als1p, Als2p, Als3p, Als4p, and Als9p) are known to have adhesive properties (Fu et al., 2002; Sheppard et al., 2004; Zhao et al., 2005, 2006, 2007; Hoyer et al., 2007). Products of these genes (adhesins) are known to mediate binding of germ tubes and hyphae to templates such as the endothelial cells (Als1p, Als2p, and Als4p) and model catheters (Als3p) (Fu et al., 2002; Zhao et al., 2005, 2006). The *ALS3* promotor has been shown to contain two activation regions essential for activation during hyphal development and for increase of the amplitude of this activation (Argimón et al., 2007). In another study, various Als proteins have been shown to possess distinct adherence profiles to diverse host substrates (Sheppard et al., 2004) underscoring the need of an array of antigens in *C. albicans* for binding to varied surfaces for facilitating infection. Green et al. (2006) also reported expression of a variety of *C. albicans* ALS genes in the hyposalivatory rat model of oral candidiasis and in clinical specimens from HIV-positive patients with oral candidiasis. However, in a recent study, it was found that *C. albicans* yeast forms deleted for *ALS5*, *ALS6*, and *ALS7* genes were more adherent to human vascular endothelial cell monolayers and buccal epithelial cells than the relevant control strain suggesting a potential anti-adhesive role for these genes (Zhao et al., 2007). In spite of this effect, this gene-deletion did not affect the degree of tissue destruction by the mutant compared to control strains. For the sake of clarification, it may be mentioned that *ALS8* was shown to be the same locus as *ALS3* (Zhao et al., 2004).

Candida albicans also produces a β-glucanase mannoprotein (Camp65p) which besides being essential for filamentation is also required for adherence to plastic (Sandini et al., 2007).

Filamentation phase

Candida spp. are unique pathogens as they undergo phenotypic switching from yeast (innocuous superficial colonization phase) to hyphal/pseudohyphal (virulent invasive phase) phase during pathogenesis (Saville et al., 2003; Sudbery et al., 2004; Zheng et al., 2004; Kadosh and Johnson, 2005). A number of genes have been implicated in the filamentation process. Signal transduction through adenylyl cyclase has been shown to be essential for dimorphic differentiation of *C. albicans* (Rocha et al., 2001). Using a DNA microarray approach, Sohn et al. (2003) showed that the gene *EFG1* plays a regulatory role in the induction and repression of cell wall genes both in yeast and hyphal forms of *C. albicans*. They also identified *YWP1* and *HWP2* that are specifically transcribed in the yeast and hyphal phases, respectively. Ramage et al. (2002) had also shown that Efg1p is crucial for filamentation in *C. albicans* (see below). The *C. albicans* adhesin, Als1p, has been shown to be a downstream effector of the *EFG1* filamentation pathway mediating adherence to it and to vascular endothelial cells, and contributing to virulence in a murine hematogenously disseminated model of candidiasis (Fu et al., 2002). Recently, a *C. albicans* strain was engineered in which the role of phenotypic switching and the individual impact of yeast and hyphal forms during various stages of the infection process could be studied through external manipulation in a mouse model (Saville et al., 2003). Studies using this strain showed that the hyphal form was required for mortality resulting from deep-seated infection and yeast cells played a role in the early infection process by disseminating to the target organs (Saville et al., 2003). Any useful diagnostic marker must selectively identify the invasive phase or specific events leading to it. Using whole-genome DNA microarray analysis, Kadosh and Johnson (2005) showed that 61 genes were significantly induced during yeast to hyphal/pseudohyphal transformation upon exposure to serum and 37°C. They also showed that about one-half of these genes were transcriptionally repressed in the yeast phase by three transcriptional repressors (Rfgl, Nrgl, and Tupl). A selected panel of genes induced during the hyphal/pseudohyphal phase could be useful molecular markers for it. Several biomolecules which play roles in this transformation have been identified (Stephenson, 2001; Zheng et al., 2004; Alvarez and Konopka, 2007; Shi et al., 2007) and development of diagnostic markers based on them will be useful.

Secreted aspartyl proteinases (Saps)

Out of four classes of proteinases (serine, cysteine, and aspartyl proteinases, and metalloproteinases), *C. albicans* produces only one such class of enzyme, i.e. the secreted aspartyl proteinases (Saps) (Naglik et al., 2003a, 2004; Robson et al., 2005). The Saps were originally known as *Candida* aspartyl proteases (Caps; Henriques et al., 2006). Numerous kinds of Saps are widely distributed in many organisms including vertebrates, invertebrates, and fungi (Bracq et al., 1997; Jean et al., 2001; Naglik et al., 2003a; Tacco et al., 2005). However, various *SAP* genes are expressed and regulated differentially, and several lines of evidence suggest that this family of gene products is involved in pathogenesis (Colina et al., 1996). The Saps of *C. albicans* are circulating antigens important in virulence and invasive hyphal growth (De Bernardis et al., 1999b; Na and Song, 1999; Naglik et al., 2003a, 2004). *Candida albicans* Saps are also thought to have roles in extracellular proteolytic activity, adhesion, penetration and invasion, and in evading immune responses (Morrison et al., 2003; Naglik et al., 2003a, 2004) Ten *SAP* genes (gene products Sapl to Sap 10 proteins; 35 to 50 kDa in size) have been identified in *C. albicans* and a distinct pathogenic role for Sap2 protein was shown in *Candida* vaginitis (De Bernardis et al., 1999a; Naglik et al., 2004; Tosun et al., 2005). Naglik et al. (2003b) found location-specific preferential expression of *SAP1*, *SAP3*, and *SAP9* genes in vaginal, rather than oral infections. They also found correlation of specific *SAP* gene expression with active (symptomatic) disease but not with asymptomatic (commensal stage) carriage (Naglik et al., 2003b; Hube, 2004). This observation is important in terms of identifying specific biomarkers for the invasive phase. *Candida tropicalis*, *C. parapsilosis*, and *C. dubliniensis* have about 4, 2, and 9 *SAP* genes, respectively (Naglik et al., 2003a).This offers the possibility of differential diagnosis of these species based on these proteomic biomarkers. Host signals which may activate individual *SAP* genes in *C. albicans* are largely unknown (Tosun et al., 2005). Staib et al. (1999) reported expression of *SAP*2 gene after fungal dissemination into deep tissue in a mouse model. In another study, in *in vitro* tests, only human immunodeficiency virus type-1 (HIV-1) but not *Lactobacillus* spp., *Gardnerella vaginalis* (Gardner & Dukes) Greenwood & Pickett, and human herpes simplex virus type-2 (HSV-2) were able to induce any Sap in *C. albicans* (Tosun et al., 2005). Expression of *SAP*1-3 has been reported during initial stages of epithelial colonization by *C. albicans* (Naglik et al., 2004). Some other functional parameters also correlate with regulation of *SAP* genes, e.g. greater levels of *SAP5*, *SAP6*, and *SAP9* mRNA transcripts were detected in biofilm than during the planktonic phase (Naglik et al., 2004). Hence, Saps and anti-Sap antibodies should be explored further for development as biomarkers for invasive candidiasis and indeed several studies have done so (Morrison et al., 1993, 2003; Na and Song, 1999). In parallel with some SAP proteins (see above), the *C. albicans ALS2, 3, 6, 7,* and *9* genes were highly expressed and *ALS4* and *5* genes repressed in models of vaginal candidiasis (Verstrepen and Klis, 2006). Also, in oral specimens, strong expression of *ALS1, 2, 3, 4, 5,* and *9*, and repression of *ALS6* and *7* was observed (Verstrepen and Klis, 2006). Protein products of such genes, individually or panels thereof, could form useful biomarkers for site-specific candidiasis.

Phospholipases

Four types of phospholipases (A, B, D, and D) are known to occur in *C. albicans* and some phospholipase (*PL*) genes (*PLB1*, *PLB2*, and *PLD1*) have been strongly implicated in virulence and invasive growth of this pathogen (Ghannoum, 2000; Mukherjee et al., 2001; Naglik et al., 2003b; Dolan et al., 2004; Kumamoto and Vines, 2005; Robson et al., 2005). Also, phospholipase B is the major secreted phospholipase in *C. albicans* (Naglik et al., 2003b; Robson et al., 2005). Mukherjee et al. (2001) showed that Plb1 facilitated penetration of the gastrointestinal tract and haematogenous dissemination by *C. albicans*. They also showed restoration of virulence *in vivo* in a *C. albicans* mutant deleted for *PLB1* following reintroduction of this gene. These studies provided firm evidence of the role of *PLB1* gene product in virulence. Naglik et al. (2003b) reported that *PLB1* expression correlated with active oral candidiasis and not with vaginal disease in human subjects. Compared to *C. albicans*, the non-albicans *Candida* spp. produce significantly lesser amounts of phospholipases (Ghannoum, 2000; Gokce et al., 2007) correlating with their degrees of virulence.

It is clear that the two groups of hydrolytic enzymes (proteinases and phospholipases) play major roles in invasive candidiasis. The protein sequence of PLB1 from *C. albicans* differs from the other known fungal PLBs (Ghannoum, 2000) offering potential strategies for development of diagnostic biomarkers. In addition, the sera of candidiasis patients may also contain antibodies to the secreted hydrolytic enzymes, again offering tools for diagnostic biomarkers.

Antibodies

Usefulness of antibodies in diagnosis of invasive diseases caused by *Candida* spp. is not clear at the moment. Antibodies to cell wall components may be present in healthy individuals and patients with *Candida*-colonization in their epithelial membranes or superficial tissue infection, and antibody production may be compromised in immunosuppressed patients (Jones, 1990; García-Ruiz et al., 1997; Moudni et al., 1998; Quindós et al., 2004). Nevertheless, many reports in the literature laud the usefulness of antibodies in diagnosis of invasive *Candida*-infections. These reports include monitoring antibodies to antigens such as *C. albicans* germ tube (CAGT; García-Ruiz et al., 1997), enolase (Mitsutake et al., 1994, 1996; van Deventer et al., 1994; Laín et al., 2007), Sap (Morrison et al., 1993; Na and Song, 1999; Naglik et al., 2004), Als (Spellberg et al., 2006), and metallopeptidase (Moudni et al., 1998). In a recent study, Pitarch et al. (2006) found that serum anti-glucan 1, 3-β-glucosidase (Bg12p) antibodies are a novel diagnostic biomarker for systemic candidiasis.

Other possible strategies

Recently, a gene, *IRS4* (increased rDNA silencing 4), which encodes an immunogenic protein that probably is a member of the mitogen-activated kinase cell integrity pathway, has been implicated to play a multifactorial role in hyphal formation and virulence of *C. albicans* after initial stages of disseminated but not oropharyngeal candidiasis in murine models (Badrane et al., 2005; Filler, 2006). However, it was ascertained that the gene is expressed during oropharyngeal candidiasis in HIV-infected patients and is implicated in adherence to human epithelial cells *in vitro* (Badrane et al., 2005). Also, it has been shown that the mitogen-activated protein (MAP) kinase *HOG1* (high-osmolarity glycerol) plays a major role in *C. albicans* virulence (Alonso-Monge et al., 1999, 2003; Fernández-Arenas et al., 2004). The protein, Hog1p is essential for oxidative stress and hyperosmolarity responses, and chlamydospore formation (Alonso-Monge et al., 1999, 2003; Fernández-Arenas et al., 2004). Additionally, it has been suggested that metallopeptidase of *C. albicans* may be involved in degradation of subendothelial extracellular matrix components and in the process of dissemination of this pathogen (Rodier et al., 1999; Imbert et al., 2002).

In an interesting study, the genomes of *C albicans* and *C. dubliniensis* were compared by co-hybridizing *C. albicans* microarrays with fluorescently labeled *C. albicans* and *C. dubliniensis* genomic DNA (Moran et al., 2004). A major portion of *C. dubliniensis* DNA sequences (95.6%) showed a significant degree of nucleotide sequence homology (>60%) with *C. albicans* DNA sequences. However, the remaining 4.4% sequences representing 247 genes indicated a significant degree of divergence (<60%) or absence in *C. dubliniensis*. In addition, sequence divergence or absence of some commonly accepted virulence factors of *C. albicans* such as the gene encoding for Hwp1p, *SAP5*, several *ALS* genes, and some others was noticed. These results are highly meaningful in terms of disease diagnostics and mechanism of pathogenesis as

C. dubliniensis is appreciably less virulent than *C. albicans* and also differs in epidemiology.

PROTEOMICS OF THE BIOFILM PHASE

The majority of infections leading to invasive candidiasis are associated with biofilm formation on indwelling medical devices, such as stents, shunts, prostheses, implants, endotracheal tubes, pace-makers, and various types of catheters (Kojic and Darouiche, 2004; Ramage et al., 2005). It has been shown that, in *C. albicans*, physical contact results in activation of the mitogen-activated kinase Mkc1p which is required for invasive hyphal growth and normal biofilm development (Kumamoto, 2005). Twenty to 70% episodes of candidiasis are associated with device-associated *Candida*-infections and lead to approximately 30% mortality among patients (Kojic and Darouiche, 2004; Kumamoto and Vines, 2005; Nett et al., 2007). Epidermal surfaces, and oral, gastrointestinal, anorectal, and vaginal mucosal membranes also serve as templates for biofilm formation (Kojic and Darouiche, 2004; Kumamoto and Vines, 2005).

Biofilms develop through adhesion of planktonic cells to the surfaces of medical devices and mucosal surfaces and further differentiation into complex three-dimensional structures which are reservoirs for infection, and have increased resistance to antifungal drugs, environmental stresses, and host immune defense mechanisms (Kojic and Darouiche, 2004; Ramage et al., 2005; Thomas et al., 2006; Li et al., 2007). Once adhesion has occurred, a complex structure is attained through cell growth, morphogenic switching between yeast and filamentous states, and quorum sensing (Li et al., 2007).

Removal of the involved device is a prerequisite for diagnosis and effective treatment of candidiasis; hence many devices may be removed unnecessarily as only 20 to 70% patients have catheter-related bloodstream infections (Nett et al., 2007). No laboratory methods exist for rapid detection of biofilm infections *in vivo* (Nett et al., 2007). Some presently available methods may indicate device-related infections but most of them are time-consuming as they generally require preparation of cultures (Safdar et al., 2005; Nett et al., 2007). Most of these methods may merely indicate the presence of *C. albicans* in association with the device and may not necessarily rule out disseminated candidiasis. In a recent interesting study on *C. albicans*, concentrations of the secreted polysaccharide ß-1,3 glucan under biofilm conditions were 4-10-fold greater *in vitro* compared to supernatants from planktonic cultures and were 10-fold greater *in vivo* in serum of rats with an intravascular catheter biofilm infection compared to disseminated candidiasis (Nett et al., 2007). Based on this information, upregulation of ß-1,3 glucan biosynthesis was proposed to be a marker for *Candida* biofilm. However, in this investigation, *C. glabrata* and *C. parapsilosis* also showed similar but varied magnitude of effects (Nett et al., 2007); hence this marker will likely not produce *Candida* species-specific information. Additionally, ß-1,3 glucan is also a component of the cell walls of a wide variety pathogenic fungi (see above under Diagnosis of Invasive Candidiasis – Present Status). Hence, potential of confusion of biofilm-initiated candidiasis infection with other fungal diseases also exists, although it is likely that high ß-1,3 glucan concentration in the sera of patients with the former condition may be able to distinguish it from the rest.

Cell adhesion is an essential component of biofilm development and the *EAP1* (elF4E-associated protein1) gene encoding an adhesin, glycosylphosphatidylinositol (GPI)-anchored glucan-cross-linked cell wall protein (Eap1p), has been shown to mediate this process both *in vitro* and *in vivo* on polystyrene and rat central venous catheter, and epithelial cells, respectively (Li and Palecek, 2003; Li et al., 2007). *ALS* genes have been shown to be differentially regulated between planktonic- and biofilm-grown cells of *C. albicans* (Chandra et al., 2001). Additionally, expression of several *ALS* genes (*ALS1*, *ALS2*, *ALS3*, *ALS4*, *ALS5*, and *ALS9*) was consistently observed through RT-PCR analysis of total RNA extracted from *C. albicans*-inoculated reconstituted human buccal epithelium (RHE) model of mucocutaneous candidiasis which showed destruction of the epithelial layer and formation of a biofilm (Green et al., 2004; Mukherjee et al., 2005). The adhesin protein, Als3p, which is associated with germ tube and hypha surfaces (see above under Diagnosis of Invasive Candidiasis – Future Prospects), has also been shown to mediate biofilm formation on silicone elastomer surfaces (Zhao et al., 2006). Hence, proteins encoded by the *ALS* gene family have dual functions in being both "social" (intraspecific interactions, and agglutination of colonies and biofilms to substrata for the purpose of survival and nutrient acquisition) and "antisocial" (binding to host organisms for pathogenic purposes) (Dranginis et al., 2007). Expression of the genes of sulfur metabolism is initiated after 30 minutes of adherence of *C. albicans* planktonic cells to polystyrene and upregulation of some of these genes continues in mature biofilms (Murillo et al., 2005; Blankenship and Mitchell, 2006). The most upregulated gene after attachment with known function (encoding ATP sulfurylase) is *MET3* (~23-fold after 30 minutes, Murillo et al., 2005). Additionally, the biofilms reveal upregulation of numerous amino acid biosynthetic genes (García-Sánchez et al., 2004; Blankenship and Mitchell, 2006). Murillo et al. (2005) also reported upregulation of a subset of genes homologous to the Ribi (for ribosome biogenesis) regulon of *S. cerevisiae*. These broad gene expression alterations are thought to be functionally significant in terms of biofilm development (Blankenship and Mitchell, 2006).

Filamentation is required for biofilm formation. Using wild type and mutants of *C. albicans* defective in genes for filamentation, it has been shown that Efg1 (elongation factor G) is a positive regulator of hyphal formation and is pivotal for biofilm development (Ramage et al., 2002; Blankenship and Mitchell, 2006).

As mentioned above (see under Diagnosis of Invasive Candidiasis – Future Prospects), some *SAP* genes are regulated by biofilm formation in *C. albicans* (Naglik et al., 2004). Genes encoding drug efflux pumps in *C. albicans* (*CDR1* and *CDR1*; *Candida* drug resistance; ATP binding cassette transporters) and *C. dubliniensis* (major facilitators encoded by the *MDR* genes) are upregulated during biofilm formation (Jabra-Rizk et al., 2004). Additionally, expressions of 24 proteins including alcohol dehydrogenase (Adh1p) are downregulated in early-phase *C. albicans* biofilms compared to planktonic cells (Mukherjee et al., 2006). This enzyme catalyzes the production of ethanol from acetaldehyde. Planktonic levels of Adh1p are higher than in the biofilms suggesting that it might be an inhibitor of biofilm formation (Blankenship and Mitchell, 2006; Mukherjee et al., 2006). In a study on cell surface-associated proteins of *C. albicans*, Thomas et al. (2006) reported upregulation of seven [Hsp70, pyruvate decarboxylase, inositil 1 phosphate synthase, enolase (two isoforms), O-acetylhomoserine O-acetylseriene sulfhydrylase, immunogenic alcohol dehydrogenase, and inosine-5′ monophosphate dehydrogenase] and downregulation of two (alcohol dehydrogenase and malate dehydrogenase) proteins during biofilm growth compared to the planktonic phase. Interestingly, out of 67 proteins identified in spots present in biofilm and planktonic extracts, none was unique to either of the two phases.

Candida glabrata differs from *C. albicans* in several significant aspects, for example hyphal formation and secreted proteinases (such as the SAPs), which are important virulence attributes in *C. albicans*, are not known to play any role in *C. glabrata* (Kaur et al., 2005). However, phenotypic switching has been recently described in this species (Li et al., 2007). *Candida glabrata* has many *EPA* (epithelial adhesin)-related genes which encode GPI-anchored cell wall lectins and mediate adherence to epithelial cells (Kaur et al., 2005; Filler, 2006). The *EPA* genes are not known in *C. albicans*. Such differences between the two species are interesting and may have value towards development of differential biomarkers. In *C. glabrata*, four genes (*RIF1*, *SIR4*, *EPA6*, *YAK1*) are involved in biofilm formation on inert surfaces and Epa6p is the main adhesin involved in this process (Iraqui et al., 2005). The aforesaid has special significance as *C. glabrata* infections are often associated with indwelling catheters (Kaur et al., 2005). Another gene, *EPA1*, has been shown to encode an adhesin that binds *C. glabrata* to human epithelial cells in culture (Cormack et al., 1999). A recent study has shown diversity in formation and morphology of biofilms among five species of *Candida* (Parahitiyawa et al., 2006). This underscores the need for detailed individual species-based studies on various *Candida* spp.

CONCLUDING REMARKS

Clearly, development of simple, rapid, sensitive, specific, and cost effective species-multiplex technology is required for diagnosing invasive candidiasis at an early stage from human samples. Appreciable morbidity and mortality result due to late initiation of targeted therapy (García-Ruiz et al., 1997). Proteomics is a powerful tool capable of resolving a large number of proteins and compared to genomics is recognized as being a better tool for disease diagnostics as "proteins are often expressed at levels and forms that cannot be predicted from DNA or mRNA analysis" (Srivastava and Srivastava, 2005; Kim et al., 2008). Most biomarker studies carried out so far on various systems have focused on high-molecular weight and high-abundance proteins and there is now a need to investigate proteins at the lower ends of these parameters (Chromy et al., 2004; Hortin, 2006; Hsieh et al., 2006). These and some other objectives are being presently achieved through use of more refined techniques for protein separation and identification (Adkins et al., 2002; Chromy et al., 2004; Echan et al., 2005; Hortin, 2006; Hsieh et al., 2006; Lee et al., 2006; Carberry and Doyle, 2007; Whiteaker et al., 2007a, b). DNA and protein microarray technologies can study many components simultaneously and are expanding the multiplexing capabilities of clinical diagnostics research (MacBeath, 2002; Templin et al., 2002; Leinberger et al., 2005; Garaizar et al., 2006). A few such studies deal with *Candida* spp. as well (Leinberger et al., 2005; Garaizar et al., 2006).

Discovery of clinical proteomic biomarkers is an emerging field of study in many biomedical areas (Hanash, 2003; Granger et al., 2004; Chandra et al., 2005; Garbis et al., 2005; Gillette et al., 2005; Marko-Varga et al., 2005; Vitzthum et al., 2005; Wilkins et al., 2006; Whiteaker et al., 2007b). However, the literature is markedly deficient with respect to the diagnostics of mycoses caused by opportunistic pathogens. Proteomic studies of *Candida* spp. are also in infancy but will no doubt benefit from databases now available (Arnaud et al., 2005, 2007; Braun et al., 2005; d'Enfert et al., 2005; Moran et al., 2006; Nantel, 2006; Rossignol et al., 2007).

In summary, for optimizing successful management of invasive mycoses caused by *Candida* spp., a system of precise, sensitive, and rapid diagnosis is required. Identification to species levels will be required as some species have to be treated differently. Such a system should also be simple, cost effective, and not requiring sophisticated instrumentation and technician training.

ACKNOWLEDGMENTS

Preparation of this review was supported through grants from the Akron General Development Foundation, Akron, Ohio.

REFERENCES

Adkins, J. N., Varnum, S. M., Auberry, K. J., Moore, R. J., Angell, N. H., Smith, R. D., Springer, D. L. and Pounds, J. G. 2002. Toward a human blood serum proteome. Analysis by multidimensional separation coupled with mass spectrometry. Mol. Cell. Proteomics, 1: 947-955.

Alexander, B. D. and Pfaller, M. A. 2006. Contemporary tools for the diagnosis and management of invasive mycoses. Clin. Infect. Dis., 43: S15-S27.

Allan, E. K., Jordanides, N. E., McLintock, L. A., Copland, M., Devaney, M., Stewart, K., Parker, A. N., Johnson, P. R. E., Holyoake, T. L. and Jones, B. L. 2005. Poor performance of galactomannan and mannan sandwich enzyme-linked immunosorbent assays in the diagnosis of invasive fungal infection. Brit. J. Haematol., 128: 578-579.

Al Mosaid, A., Sullivan, D., Salkin, I. F., Shanley, D. and Coleman, D. C. 2001. Differentiation of *Candida dubliniensis* from *Candida albicans* on staib agar and caffeic acid-ferric citrate agar. J. Clin. Microbiol., 39: 323-327.

Alonso-Monge, R., Navarro-García, F., Molero, G., Diez-Orejas, R., Gustin, M., Pla, J., Sánchez, M. and Nombella, C. 1999. Role of the mitogen-activated protein kinase Hog1p in morphogenesis and virulence of *Candida albicans.* J. Bact., 181: 3058-3068.

Alonso-Monge, R., Navarro-García, F., Román, E., Negredo, A. I., Eisman, B., Nombela, C. and Pla, J. 2003. The Hog1 mitogen-activated protein kinase is essential in the oxidative stress response and chlamydospore formation in *Candida albicans.* Eukaryt. Cell, 2: 351-361.

Alvarez, F. J. and Konopka, J. B. 2007. Identification of an N-acetylglucosamine transporter that mediates hyphal induction in *Candida albicans*[D]. Mol. Biol. Cell, 18: 965-975.

Argimón, S., Wishart, J. A., Leng, R., Macaskill, S., Mavor, A., Alexandris, T., Nicholls, S., Knight, A. W., Enjalbert, B., Walmsley, R., Odds, F. C., Gow, N. A. R. and Brown, A. J. P. 2007. Developmental regulation of an adhesin gene during cellular morphogenesis in the fungal pathogen *Candida albicans.* Eukaryot. Cell, 6: 682-692.

Arnaud, M. B., Costanzo, M. C., Skrzypek, M. S., Binkley, G., Lane, C., Miyasato, S. R. and Sherlock, G. 2005. The Candida Genome Database (CGD), a community resource for *Candida albicans* gene and protein information. Nucleic Acids Res., 33: D358-D363.

Arnaud, M. B., Costanzo, M. C., Skrzypek, M. S., Shah, P., Binkley, G., Lane, C., Miyasato, S. R. and Sherlock, G. 2007. Sequence resources at the Candida Genome database. Nucleic Acids Res., 35: D452-D456.

Badrane, H., Cheng, S., Nguyen, M. H., Jia, H. Y., Zhang, Z., Weisner, N. and Clancy, C. J. 2005. *Candida albicans IRS4* contributes to hyphal formation and virulence after the initial stages of disseminated candidiasis. Microbiology, 151: 2923-2931.

Blankenship, J. R. and Mitchell, A. P. 2006. How to build a biofilm: a fungal perspective. Curr. Opin. Microbiol., 9: 588-594.

Blumberg, H. M., Jarvis, W. R., Soucie, J. M., Edwards, J. E., Patterson, J. E., Pfaller, M. A., Rangel-Frausto, M. S., Rinaldi, M. G., Saiman, L., Wiblin, R. T., Wenzel, R. P. and the NEMIS Study Group. 2001. Risk factors for candidal bloodstream infections in surgical intensive care unit patients: the NEMIS prospective multicenter study. Clin. Infect. Dis., 33: 177-186.

Bracq, E., Levieux, A. and Levieux, D. 1997. Purification and immunochemical quantitation of *Penicillium roqueforti* acid aspartyl proteinase. J. Dairy Res., 64: 105-113.

Braun, B. R., van het Hoog, M., d'Enfert, C., Martchenko, M., Dungan, J., Kuo, A., Inglis, D. O., Uhl, M. A., Hogues, H., Berriman, M., Lorenz, M., Levitin, A., Oberholzer, U., Bachewich, C., Harcus, D., Marcil, A., Dignard, D., Iouk, T., Zito, R., Frangeul, L., Tekaia, F., Rutherford, K., Wang, E., Munro, C. A., Bates, S., Gow, N. A., Hoyer, L. L., Köhler, G., Morschhäuser, J., Newport, G., Znaidi, S., Raymond, M., Turcotte, B., Sherlock, G., Costanzo, M., Ihmels, J., Berman, J., Sanglard, D., Agabian, N., Mitchell, A. P., Johnson, A. D., Whiteway, M. and Nantel, A. 2005. A human-curated annotation of the *Candida albicans* genome. PLoS Genet., 1: 0036-0057.

Bruneau, J-M., Maillet, I., Tagot, E., Legrand, R., Supatto, F., Fudali, C., Le Caer, J-P., Labas, V., Lecaque, D. and Hodgson, J. 2003. Drug induced proteome changes in *Candida albicans*: Comparison of the effect of ß(1,3) glucan synthase inhibitors and two trizoles, fluconazole and itraconazole. Proteomics, 3: 325-336.

Calderone, R. A. and Braun, P. C. 1991. Adherence and receptor relationships of *Candida albicans*. Microbiol. Revs., 55: 1-20.

Carberry, S. and Doyle, S. 2007. Proteomic studies in biomedically and industrially relevant fungi. Cytotechnology, 53: 95-100.

Carrillo-Muñoz, A. J., Quindóz, G., Cárdenes, C. D., Alonso-Vargas, R., Brió, S., Arévalo, P., Pemán, J., Estivill, D. and Pontón, J. 2002. Performance of Bacticard™ Candida compared with the germ tube test for the presumptive identification of *Candida albicans*. Mycoses, 46: 467-470.

Carvalho, A., Cosia-de-Oliveira, S., Martins, M. L., Pina-Vaz, C., Rodgrigues, A. G., Ludovico, P. and Rodrigues, F. 2007. Multiplex PCR identification of eight clinically relevant *Candida* species. Med. Mycol., 45: 619-627.

Casadevall, A., Steenbergen, J. N. and Nosanchuk, J. D. 2003. 'Ready made' virulence and 'dual use' virulence factors in pathogenic environmental fungi – the *Cryptococcus neoformans* paradigm. Curr. Opin. Microbiol., 6: 332-337.

Castagna, A., Cecconi, D., Sennels, L., Rappsilber, J., Guerrier, L., Fortis, F., Boschetti, E., Lomas, L. and Righetti, P. G. 2005. Exploring the hidden human urinary proteome via ligand library beads. J. Proteome Res., 4: 1917-1930.

Chandra, J., Kuhn, D. M., Mukherjee, P. K., Hoyer, L. L., McCormick, T. and Ghannoum, M. A. 2001. Biofilm formation by the fungal pathogen *Candida albicans*: Development, architecture, and drug resistance. J. Bact., 183: 5385-5394.

Chandra, S., Muir, C., Silva, M. and Carr, S. 2005. Imaging biomarkers in drug development: An overview of opportunities and open issues. J. Proteome Res., 4: 1134-1137.

Chen, S. 2000. Invasive fungal infection – what's new in diagnosis. Inoculum, 9: 1-3.

Christensson, B., Sigmundsdottir, G. and Larsson, L. 1999. D-arabinitol – a marker for invasive candidiasis. Med. Mycol., 37: 391-396.

Chromy, B. A., Gonzales, A. D., Perkins, J., Choi, M. W., Corzett, M. H., Chang, B. C., Corzett, C. H. and McCutchen-Maloney, S. L. 2004. Proteomic analysis of human serum by two-dimensional differential gel electrophoresis after depletion of high-abundant proteins. J. Proteome Res., 3: 1120-1127.

Coignard, C., Hurst, S. F., Benjamin, L. E., Brandt, M. E., Warnock, D. W. and Morrison, C. J. 2004. Resolution of discrepant results for *Candida* species identification by using DNA probes. J. Clin. Microbiol., 42: 858-861.

Colina, A-R., Aumont, F., Deslauriers, N., Belhumeur, P. and de Repentigny, L. 1996. Evidence for degradation of gastrointestinal mucin by *Candida albicans* secretory aspartyl proteinase. Infect. Immun., 64: 4514-4519.

Cooke, V. M., Miles, J., Price, R. G., Midgley, G., Khamri, W. and Richardson, A. C. 2002. New chromogenic agar medium for the identification of *Candida* species. Appl. Environ. Biol., 68: 3622-3627.

Cormack, B. P., Ghori, N. and Falkow, S. 1999. An adhesion of the yeast pathogen *Candida glabrata* mediating adherence to human epithelial calls. Science, 285: 578-582.

De Bernardis, F., Arancia, S., Morelli, L., Hube, B., Sanglard, D., Schafer, W. and Cassone A. 1999a. Evidence that members of the secretory aspartyl proteinase gene family, in particular SAP2, are virulence factors for *Candida vaginitis*. J. Infect. Dis., 179: 201-208.

De Bernardis, F., Mondello, F., Scaravelli, G., Pachì, A., Girolamo, A., Agatensi, L. and Cassone, A. 1999b. High aspartyl proteinase production and vaginitis in human immunodeficiency virus-infected women. J. Clin. Microbiol., 37: 1376-1380.

d'Enfert, C., Goyard, S., Rodriguez-Arnaveilhe, S., Frangeul, L., Jones, L., Tekaia, F., Bader, O., Albrecht, A., Castillo, L., Dominguez, A., Ernst, J. F., Fradin, C., Gaillardin, C., Garcia-Sanchez, S., de Groot, P., Hube, B., Klis, F. M., Krishnamurthy, S., Kunze, D., Lopez, M-C., Mavor, A., Martin, N., Moszer, I., Onésime, D., Perez Martin, J., Sentandreu, R., Valentin, E. and Brown, A. J. P. 2005. CandidaDB: a genome database for *Candida albicans* pathogenomics. Nucleic Acids Res., 33: D353-D357.

Dolan, J. W., Bell, A. C., Hube, B., Schaller, M., Warner, T. F. and Balish, E. 2004. *Candida albicans* PLD1 activity is required for full virulence. Med. Mycol., 42: 439-447.

Dranginis, A. N., Rauceo, J. M., Coronado, J. E. and Lipke, P. N. 2007. A biochemical guide to yeast adhesins: Glycoproteins for social and antisocial occasions. Microbiol. Mol. Biol. Revs., 71: 282-294.

Ebanks, R. O., Chisholm, K., McKinnon, S., Whiteway, M. and Pinto, D. M. 2006. Proteomic analysis of *Candida albicans* yeast and hyphal cell wall and associated proteins. Proteomics, 6: 2147-2156.

Echan, L. A., Tang, H-Y., Ali-Khan, N., Lee, K. and Speicher, D. W. 2005. Depletion of multiple high-abundance proteins improves protein profiling capacities of human serum and plasma. Proteomics, 5: 3298-3303.

Erjavec, Z. and Verweij, P. E. 2002. Recent progress in the diagnosis of fungal infections in the immunocompromised host. Drug Res. Updates, 5: 3-10.

Fernández-Arenas, E., Molero, G., Nombela, C., Diez-Orejas, R. and Gil, C. 2004. Contribution of the antibodies response induced by a low virulent *Candida albicans* strain in protection against systemic candidiasis. Proteomics, 4: 1204-1215.

Filler, S. G. 2006. *Candida*-host cell receptor-ligand interactions. Curr. Opin. Microbiol., 9: 333-339.

Fraser, V. J., Jones, M., Dunkel, J., Storfer, S., Medoff, G. and Dunagan, W. C. 1992. Candidemia in a tertiary care hospital: epidemiology, risk factors, and predictors of mortality. Clin. Infect. Dis., 15: 414-421.

Fu, Y., Ibrahim, A. S., Sheppard, D. C., Chen, Y-C., French, S. W., Cutler, J. E., Filler, S. G. and Edwards, J. E., Jr. 2002. *Candida albicans* Als1p: an adhesion that is a downstream effector of the *EFG1* filamentation pathway. Mol. Microbiol., 44: 61-72.

Garaizar, J., Brena, S., Bikandi, J., Rementeria, A. and Pontón, J. 2006. Use of DNA microarray technology and gene expression profiles to investigate the pathogenesis, cell biology, antifungal susceptibility and diagnosis of *Candida albicans*. FEMS Yeast Res., 6: 987-998.

Garbis, S., Lubec, G. and Fountoulakis, M. 2005. Limitations of current proteomic technologies. J. Chromatogr. A., 1077: 1-18.

García-Ruiz, J. C., Arilla, M. D. C., Regúlez, P., Quindós, G., Álvarez, A. and Pontón, J. 1997. Detection of antibodies to *Candida albicans* germ tubes for diagnosis and therapeutic monitoring of invasive candidiasis in patients with hematologic malignancies. J. Clin. Microbiol., 35: 3284-3287.

García-Sánchez, S., Aubert, S., Iraqui, I., Janbon, G., Ghigo, J-M. and d'Enfert, C. 2004. *Candida albicans* biofilms: a developmental state associated with specific and stable gene expression patterns. Eukaryot. Cell, 3: 536-545.

Ghannoum, M. A. 2000. Potential role of phospholipases in virulence and fungal pathogenesis. Clin. Microbiol. Revs., 13: 122-143.

Gillette, M. A., Mani, D. R. and Carr, S. A. 2005. Place of pattern in proteomic biomarker discovery. J. Proteome Res., 4: 1143-1154.

Girish Kumar, C. P. and Menon, T. 2005. Tobacco agar: a new medium for chlamydosporulation in *Candida albicans* and *Candida dubliniensis*. Med. Mycol., 43: 473-475.

Girish Kumar, C. P., Menon, T., Prabu, D. and Nandakumar, B. 2006. Chlamydosporulation of *Candida albicans* and *Candida dubliniensis* on mustard agar. Mycoses, 50: 71-73.

Girmenia, C., Martino, P., De Bernardis, F., Boccanera, M. and Cassone, A. 2004. Lack of circulating *Candida* mannoprotein in patients with focal hepatosplenic candidiasis. J. Med. Microbiol., 53: 103-106.

Gokce, G., Cerikcioglu, N. and Yagci, A. 2007. Acid proteinase, phospholipase, and biofilm production of *Candida* species isolated from blood cultures. Mycopathologia, 164: 265-269.

Granger, C. B., Eyk, J. E. V., Mockrin, S. C. and Anderson, N. L.; on behalf of the Working Group Members. 2004. National Heart, Lung, and Blood Institute Clinical Proteomics Working Group Report. Circulation, 109: 1697-1703.

Green, C. B., Cheng, G., Chandra, J., Mukherjee, P., Ghannoum, M. A. and Hoyer, L. L. 2004. RT-PCR detection of *Candida albicans* ALS gene expression in the reconstituted human epithelium (RHE) model of oral candidiasis and in model biofilms. Microbiology, 150: 267-275.

Green, C. B., Marretta, S. M., Cheng, G., Faddoul, F. F., Ehrhart, E. J. and Hoyer, L. L. 2006. RT-PCR analysis of *Candida albicans* ALS gene expression in a hyposalivatory rat model of oral candidiasis and in HIV-postive human patients. Med. Mycol., 44: 103-111.

Hajjeh, R. A., Sofair, A. N., Harrison, L. H., Lyon, G. M., Arthington-Skaggs, B. A., Mirza, S. A., Phelan, M., Morgan, J., Lee-Yang, W., Ciblak, M. A., Benjamin, L. E., Sanza, L. T., Huie, S., Yeo, S. F., Brandt, M. E. and Warnock, D. W. 2004. Incidence of bloodstream infections due to *Candida* species and *in vitro* susceptibilities of isolates collected from 1998 to 2000 in a population-based active surveillance program. J. Clin. Microbiol., 42: 1519-1527.

Hanash, S. 2003. Disease proteomics. Nature, London, 422: 226-232.

Henriques, M., Azeredo, J. and Oliveira, R. 2006. *Candida* species adhesion to oral epithelium: factors involved and experimental methodology. Crit. Revs. Microbiol., 32: 217-226.

Henry, N. K., McLimans, C. A., Wright, A. J., Thompson, R. L., Wilson, W. R. and Washington II, J. A. 1983. Microbiological and clinical evaluation of the isolator lysis-centrifugation blood culture tube. J. Clin. Microbiol., 17: 864-869.

Hernández, R., Nombela, C., Diez-Orejas, R. and Gil, C. 2004. Two-dimensional reference map of *Candida albicans* hyphal forms. Proteomics, 4: 374-382.

Hoog, G. S. de, Guarro, J., Gené, J. and Figueras, M. J. 2000. Atlas of Clinical Mycology. 2nd edn. Centraalbureau voor Schimmelcultures, Utrecht, The Netherlands.

Hortin, G. L. 2006. The MALDI-TOF mass spectrometric view of the plasma proteome and peptidome. Clin. Chem., 52: 1223-1237.

Hoyer, L. L., Green, C. B., Oh, S-H. and Zhao, X. 2007. Discovering the secrets of the *Candida albicans* agglutinin-like sequence (ALS) gene family – a sticky pursuit. Med. Mycol., 1-15; DOI: 10.1080/13693780701435317.

Hsieh, S-Y., Chen, R-K., Pan, Y-H. and Lee, H-L. 2006. Systematical evaluation of the effects of sample collection procedures on low-molecular weight serum/plasma proteome profiling. Proteomics, 6: 3189-3198.

Hube, B. 2004. From commensal to pathogen: Stage- and tissue-specific gene expression of *Candida albicans*. Curr. Opin. Microbiol., 7: 336-341.

Imbert, C., Kauffmann-Lacroix, C., Daniault, G., Jacquemin, J. L. and Rodier, M. H. 2002. Effect of matrix metalloprotease inhibitors on the 95 kDa metallopeptidase of *Candida albicans*. J. Antimicrob. Chemother., 49: 1007-1010.

Insenser, M., Nombela, C., Molero, G. and Gil, C. 2006. Proteomic analysis of detergent-resistant membranes from *Candida albicans*. Proteomics, 6: S74-S81.

Iraqui, I., Garcia-Sanchez, S., Aubert, S., Dromer, F., Ghigo, J-M., d'Enfert, C. and Janbon, G. 2005. The Yak1p kinase controls expression of adhesins and biofilm formation in *Candida glabrata* in a Sir4p-dependent pathway. Mol. Microbiol., 55: 1259-1271.

Jabra-Rizk, M. A., Falkler, W. A. and Meiller, T. F. 2004. Fungal biofilms and drug resistance. Emerg. Infect. Dis., 10: 14-19.

Jean, L., Long, M., Young, J., Pery, P. and Tomley, F. 2001. Aspartyl proteinase genes from apicomplexan parasites: evidence for evolution of the gene structure. Trends Parasitol., 17: 491-498.

Jones, J. M. 1990. Laboratory diagnosis of invasive candidiasis. Clin. Microbiol. Rev., 3: 32-45.

Kadosh, D. and Johnson, A. D. 2005. Induction of the *Candida albicans* filamentous growth program by relief of transcriptional repression: a genome-wide analysis. Mol. Biol. Cell, 16: 2903-2912.

Kaur, R., Domergue, R., Zuponcic, M. L. and Cormack, B. P. 2005. A yeast by any other name: *Candida glabrata* and its interaction with the host. Curr. Opin. Microbiol., 8: 378-384.

Khan, Z. U., Ahmad, S., Mokaddas, E. and Chandy, R. 2004a. Simplified sunflower (*Helianthus annus*) seed agar for *Candida dubliniensis* from *Candida albicans*. Clin. Microbiol. Infect., 19: 590-592.

Khan, Z. U., Ahmad, S., Mokaddas, E. and Chandy, R. 2004b. Tobacco agar, a new medium for differentiating *Candida dubliniensis* from *Candida albicans*. Clin. J. Microbiol., 44: 4796-4798.

Kim, D., Shin, W-S., Lee, K-H., Kim, K., Park, J. Y. and Koh, C-M. 2002. Rapid differentiation of *Candida albicans* from other *Candida* species using its unique germ tube formation at 39°C. Yeast, 19: 957-962.

Kim, Y., Nandakumar, M. P. and Marten, M. R. 2008. The state of proteome profiling in the fungal genus *Aspergillus*. Brief. Funct. Genomics Proteomics, 7: 87-94.

Klingspor, L. and Jalal, S. 2006. Molecular detection and identification of *Candida* and *Aspergillus* spp. from clinical samples using real-time PCR. Clin. Microbiol. Infect., 12: 745-753.

Kohno, S., Mitsutake, K., Maesaki, S., Yasuoka, A., Miyazaki, T., Kaku, M., Koga, H. and Hara, K. 1993. An evaluation of serodiagnostic tests in patients with candidemia: beta-glucan, mannan, *Candida* antigen by Cand-Tec and D-arabinitol. Microbiol. Immunol., 37: 207-212.

Kojic, E. M. and Darouiche, R. O. 2004. *Candida* infections of medical devices. Clin. Microbiol. Revs., 17: 255-267.

Kumamoto, C. A. 2005. A contact-activated kinase signals *Candida albicans* invasive growth and biofilm development. Proc. Natl. Acad. Sci., U.S.A., 102: 5576-5581.

Kumamoto, C. A. and Vines, M. D. 2005. Alternative *Candida albicans* lifestyles: Growth on surfaces. Annu. Rev. Microbiol., 59: 113-133.

Laín, A., Moragues, M. D., García-Ruiz, J. C., Mendoza, J., Camacho, A., del Palacio, A. and Pontón, J. 2007. Evaluation of a novel enzyme-linked immunosorbent assay to detect immunoglobulin G antibody to enolase for serodiagnosis of invasive candidiasis. Clin. Vacc. Immunol., 14: 318-319.

Lee, H-J., Lee, E-Y., Kwon, M-S. and Paik, Y-K. 2006. Biomarker discovery from the plasma proteome using multidimensional fractionation proteomics. Curr. Opion. Chem Biol., 10: 42-49.

Leinberger, D. M., Schumacher, U., Autenrieth, I. B. and Bachmann, T. T. 2005. Development of a DNA microarray for detection and identification of fungal pathogens involved in invasive mycoses. J. Clin. Microbiol., 43: 4943-4953.

Li, F. and Palecek, S. P. 2003. *EAP1*, a *Candida albicans* gene involved in binding human epithelial cells. Eukaryot. Cell, 2: 1266-1273.

Li, F., Svorovsky, M. J., Karlsson, A. J., Wagner, J. P., Marchillo, K., Oshel, P., Andes, D. and Palecek, S. P. 2007. EAP1p, an adhesin that mediates *Candida albicans* biofilm formation *in vitro* and *in vivo*. Eukaryot. Cell, 6: 931-939.

Li, L., Redding, S. and Dongari-Bagtzoglou, A. 2006. *Candida glabrata*, an emerging oral opportunistic pathogen. J. Dent. Res., 86: 204-215.

Loreto, E. S., Bolzan, A. R., Linares, C. E., Boff, E., Santurio, J. M. and Alves, S. H. 2006. Evaluation of 5 new media containing extracts of seeds applied to *Candida dubliniensis* screening. Diagn. Microbiol. Infect. Dis., 55: 191-193.

MacBeath, G. 2002. Protein microarrays and proteomics. Nature Gen., Suppl., 32: 526-532.

Marko-Varga, G., Lindberg, H., Löfdahl, C-G., Jönsson, P., Hansson, L., Dahlbäck, M., Lindquist, E., Johansson, L., Foster, M. and Fehniger, T. E. 2005. Discovery of biomarker candidates within disease by protein profiling: principles and concepts. J. Proteome Res., 4: 1200-1212.

Meyer, D. W., Latouche, G. N., Daniel, H-M., Thanos, M., Mitchell, T. G., Yarrow, D., Schönian, G. and Sorrell, T. C. 2005 (1997). Identification of pathogenic yeasts of the imperfect genus *Candida* by polymerase chain reaction fingerprinting. Electrophoresis, 18: 1548-1559.

Mitsutake, K., Kohono, S., Miyazaki, T., Mizazaki, H., Maesaki, S. and Koga, H. 1994. Detection of *Candida* enolase antibody in patients with candidiasis. J. Clin. Lab. Anal., 8: 207-210.

Mitsutake, K., Miyazaki, T., Tashiro, T., Yamamoto, Y., Kakeya, H., Otsubo, T., Kawamura, M. S., Hossain, A., Noda, T., Hirakata, Y. and Kohno, S. 1996. Enolase antigen, mannan, antigen, Cand-Tec antigen, and ß-glucan in patients with candidemia. J. Clin. Microbiol., 34: 1918-1921.

Moran, G. P. and Jabra-Rizk, M. A. 2006. 8th ASM conference on *Candida* and candidiasis: Molecular tools provide insights into host-pathogen interactions. Mycopathologia, 162: 17-24.

Moran, G., Stokes, C., Thewes, S., Hube, B., Coleman, D. C. and Sullivan, D. 2004. Comparative genomics using *Candida albicans* DNA microarrays reveals absence and divergence of virulence-associated genes in *C. dubliniensis*. Microbiology, 150: 3363-3382.

Morrison, C. J., Hurst, S. F. and Reiss, E. 2003. Competitive binding inhibition enzyme-linked immunosorbent assay that uses the secreted aspartyl proteinase of *Candida albicans* as an antigenic marker for diagnosis of disseminated candidiasis. Clin. Diagn. Lab. Immunol., 10: 835-848.

Morrison, C. J., Hurst, S. F., Bragg, S. L., Kuykendall, R. J., Diaz, H., Pohl, J. and Reiss, E. 1993. Heterogeneity of the purified extracellular aspartyl proteinase from *Candida albicans*: characterization with monoclonal antibodies and N-terminal amino acid sequence analysis. Infect. Immun., 61: 2030-2036.

Moudni, B. E., Rodier, M-H., Daniault, G. and Jacquemin, J. L., 1998. Improved immunodiagnosis of human candidiasis by an enzyme-linked immunosorbent assay using a *Candida albicans* 52-kilodalton metallopeptidase. Clin. Diagn. Lab. Immunol., 5: 823-825.

Mukherjee, P. K., Mohamed, S., Chandra, J., Kuhn, D., Liu, S., Antar, O. S., Munyon, R., Mitchell, A. P., Andes, D., Chance, M. R., Rouabhia, M. and Ghannoum, M. A. 2006. Alcohol dehydrogenase restricts the ability of the pathogen *Candida albicans* to form a biofilm on catheter surfaces through an ethanol-based mechanism. Infect. Immun., 74: 3804-3816.

Mukherjee, P. K., Seshan, K. R., Leidich, S. D., Chandra, J., Cole, G. T. and Ghannoum, M. A. 2001. Reintroduction of the *PLB1* gene into *Candida albicans* restores virulence *in vivo*. Microbiology, 147: 2585-2597.

Mukherjee, P. K., Zhou, G., Munyon, R. and Ghannoum, M. A. 2005. *Candida* biofilm: a well designed protected environment. Med. Mycol., 43: 191-208.

Murillo, L. A., Newport, G., Lan, C-Y., Habelitz, S., Dungan, J. and Agabian, N. M. 2005. Genome-wide transcription profiling of the early phase of biofilm formation by *Candida albicans*. Eukaryot. Cell, 4: 1562-1573.

Na, B-K. and Song, C-Y. 1999. Use of monoclonal antibody in diagnosis of candidiasis caused by *Candida albicans*: detection of circulating aspartyl proteinase antigen. Clin. Diagn. Lab. Immunol., 6: 924-929.

Naglik, J., Albrecht, A., Bader, O. and Hube, B. 2004. *Candida albicans* proteinases and host/pathogen interactions. Cell. Microbiol., 6: 915-926.

Naglik, J. R., Challacombe, S. J. and Hube, B. 2003a. *Candida albicans* secreted aspartyl proteinases in virulence and pathogenesis. Microbiol. Mol. Biol. Revs., 67: 400-428.

Naglik, J. R., Rodgers, C. A., Shirlaw, P. J., Dobbie, J. L., Fernandes-Naglik, L. L., Greenspan, D., Agabian, N. and Challacombe, S. J. 2003b. Differential expression of *Candida albicans* secreted aspartyl proteinase and phopholipase B genes in humans correlates with active oral and vaginal infections. J. Infect. Dis., 188: 469-479.

Nantel, A. 2006. The long hard road to a completed *Candida albicans* genome. Fungal Gen. Biol., 43: 311-315.

Nett, J., Lincoln, L., Marchillo, K. and Andes, D. 2007. ß-1,3 glucan as a test for central venous catheter biofilm infection. J. Infect. Dis., 195: 1705-1712.

Niimi, M., Cannon, R. D. and Monk, B. C. 1999. *Candida albicans* pathogenicity: a proteomic perspective. Electrophoresis, 20: 2299-2308.

Nolla-Salas, J., Sitges-Serra, A., León-Gil, C., Martínez-González, J., León-Regidor, M. A., Ibáñez-Lucía, P., Torres-Rodríguez, J. M. and Study Group of Fungal Infection in the ICU. 1997. Candidemia in non-neutropenic critically ill patients: analysis of prognostic factors and assessment of systemic antifungal therapy. Intensive Care Med., 23: 23-30.

Odabasi, Z., Mattiuzzi, G., Estey, E., Kantarjian, H., Saeki, F., Ridge, R. J., Ketchum, P. A., Finkelman, M.A., Rex, J. H. and Ostrosky-Zeichner, L. 2004. ß-D-glucan as a diagnostic adjunct for invasive fungal infections: validation, cutoff, development, and performance in patients with acute myelogenous leukemia and myelodysplastic syndrome. Clin. Infect. Dis., 39: 199-205.

O'Shaughnessy, E. M., Shea, Y. M. and Witebsky, F. G. 2003. Laboratory diagnosis of invasive mycoses. Infect. Dis. Clin. N. Am., 17: 135-158.

Ostrosky-Zeichner, L., Alexander, B. D., Kett, D. H., Vasquez, J., Pappas, P. G., Saeki, F., Ketchum, P. A., Wingard, J., Schiff, R., Tamura, H., Finkelman, M. A. and Rex, J. H. 2005. Multicenter clinical evaluation of the (1->3) ß-D-glucan assay as an aid to diagnosis of fungal infections in humans. Clin. Infect. Dis., 41: 654-659.

Page, B. T. and Kurtzman, C. P. 2005. Rapid identification of *Candida* species and other clinically important yeast species by flow cytometry. J. Clin. Microbiol., 43: 4507-4514.

Parahitiyawa, N. B., Samaranayake, Y. H., Samaranayake, L. P., Ye, J., Tsang, P. W. K., Cheug, B. P. K., Yau, J. Y. Y. and Yeung, S. K. W. 2006. Interspecies variation in *Candida* biofilm formation studied using the Calgary biofilm device. APMIS, 114: 298-306.

Pardo, M., Ward, M., Pitarch, A., Sánchez, M., Nombela, C., Blackstock, W. and Gil, C. 2000. Cross-species identification of novel *Candida albicans* immunogenic proteins by combination of two-dimensional polyacrylamide gel electrophoresis and mass spectrometry. Electrophoresis, 21: 2651-2659.

Perlroth, J., Choi, B. and Spellberg, B. 2007. Nosocomial fungal infections: epidemiology, diagnosis, and treatment. Med. Mycol., 45: 321-346.

Pfaller, M. A. and Diekema, D. J. 2007. Epidemiology of invasive candidiasis: a persistent public health problem. Clin. Microbiol. Revs., 20: 133-163.

Pfaller, M. A., Pappas, P. G. and Wingard, J. R. 2006. Invasive fungal pathogens: current epidemiological trends. Clin. Infect. Dis., 43: S3-S14.

Pieper, R., Gatlin, C. L., Makusky, A. J., Russo, P. S., Schatz, C. R., Miller, S. S., Su, Q., MacGrath, A. M., Estock, M. A., Parmar, P. P., Zhao, M., Huang, S-T., Zhou, J., Wang, F., Esquer-Blasco, R., Anderson, N. L., Taylor, J. and Taylor, S. 2003. The human serum proteome: display of nearly 3700 chromatographically separated protein spots on two-dimensional electrophoresis gels and identification of 325 distinct proteins. Proteomics, 3: 1345-1364.

Pincus, D. H., Orenga, S. and Chatellier, S. 2007. Yeast identification – past, present, and future methods. Med. Mycol., 45: 97-121.

Pinjon, E., Sullivan, D., Salkin, I., Shanley, D. and Coleman, D. 1998. Simple, inexpensive, reliable method for differentiation of *Candida dubliniensis* from *Candida albicans*. J. Clin. Microbiol., 36: 2093-2095.

Pitarch, A., Abian, J., Carrascal, M., Sánchez, M., Nombela, C. and Gil, C. 2004. Proteomics-based identification of novel *Candida albicans* antigens for diagnosis of systemic candidiasis in patients with underlying hematological malignancies. Proteomics, 4: 3084-3106.

Pitarch, A., Diez-Orejas, R., Molero, G., Pardo, M., Sánchez, M., Gil, C. and Nimbela, C. 2001. Analysis of the serologic response to systemic *Candida albicans* infection in a murine model. Proteomics, 1: 550-559.

Pitarch, A., Jiménez, A., Nombela, C. and Gil, C. 2006. Decoding serological response to *Candida* cell wall immunome into novel diagnostic, prognostic, and therapeutic candidates for systemic candidiasis by proteomic and bioinformatic analyses. Mol. Cell. Proteomics, 5: 79-96.

Pitarch, A., Pardo, M., Jiménez, A., Pla, J., Gil, C., Sánchez, M. and Nombela, C. 1999. Two-dimensional gel electrophoresis as analytical tool for identifying *Candida albicans* immunogenic proteins. Electrophoresis, 20: 1001-1010.

Pitarch, A., Sánchez, M., Nombela, C. and Gil, C. 2003. Analysis of the *Candida albicans* proteome. I. Strategies and applications. J. Chromatogr. B, 787: 101-128.

Quindós, G., Moragues, M. D. and Pontón, J. 2004. Is there a role for antibody testing in the diagnosis of invasive candidiasis? Rev. Iberoam. Micol., 21: 10-14.

Ramage, G., Saville, S. P., Thomas, D. P. and López-Ribot, J. L. 2005. *Candida* biofilms: an update. Eukaryot. Cell, 4: 633-638.

Ramage, G., VandeWalle, K., López-Ribot, J. L. and Wickes, B. L. 2002. The filamentation pathway controlled by the Efg1 regulator protein is required for normal biofilm formation and development in *Candida albicans*. FEMS Microbiol. Lett., 214: 95-100.

Robson, G. D., Huang, J., Wortman, J. and Archer, D. B. 2005. A preliminary analysis of the process of protein secretion and the diversity of putative secreted hydrolases encoded in *Aspergillus fumigatus*: insights from the genome. Med. Mycol. Suppl. 1, 43: S41-S47.

Rocha, C. R. C., Schröppel, K., Harcus, D., Marcil, A., Dignard, D., Taylor, B. N., Thomas, D. Y., Whiteway, M. and Leberer, E. 2001. Signaling through adenylyl cyclase is essential for hyphal growth and virulence in the pathogenic fungus *Candida albicans*. Mol. Biol., 12: 3631-3643.

Rodier, M-H., Moudni, B. E., Kauffmann-Lacroix, C., Daniault, G. and Jacquemin, J-L. 1999. A *Candida albicans* metallopeptidase degrades constitutive proteins of extracellular matrix. FEMS Microbiol. Lett., 177: 205-210.

Rossignol, T., Lechat, P., Cuomo, C., Zeng, Q., Moszer, I. and d'Enfert, C. 2007. CandidaDB: a multi-genome database for *Candida* species and related Saccharomycotina. Nucleic Acids Res., 1-5. doi:10.1093/nar/gkm1010.

Rupp, S. 2004. Proteomics on its way to study host-pathogen interaction in *Candida albicans*. Curr. Opin. Microbiol., 7: 330-335.

Safdar, N., Fine, J. P. and Maki, D. F. 2005. Meta-analysis: Methods for diagnosing intravascular device-related bloodstream infection. Ann. Intern. Med., 142: 451-466.

Sandini, S., Valle, R. L., De Bernardis, F., Macrì, C. and Cassone, A. 2007. The 65 kDa mannoprotein gene of *Candida albicans* encodes a putative β-glucanase adhesin required for hyphal morphogenesis and experimental pathogenicity. Cell. Microbiol., 9: 1223-1238.

Saville, S. P., Lazzell, A. L., Monteagudo, C. and Lopez-Ribot, J. L. 2003. Engineered control of cell morphology in vivo reveals distinct roles for yeast and filamentous forms of *Candida albicans* during infection. Eukaryot. Cell, 2: 1053-1060.

Schabereiter-Gurtner, C., Selitsch, B., Rotter, M. L., Hirschl, A. M. and Willinger, B. 2007. Development of novel real-time PCR assays for detection and differentiation of eleven medically important *Aspergillus* and *Candida* species in clinical specimens. J. Clin. Microbiol., 45: 906-914.

Schuffenecker, I., Freydiére, A., de Montclos, A. and Gille, Y. 1993. Evaluation of four commercial systems for identification of medically important yeasts. Eur. J. Clin. Microbiol. Infect. Dis., 12: 255-260.

Sendid, B., François, N., Standaert, A., Dehecq, E., Zerimech, F., Camus, D. and Poulain, D. 2007. Prospective evaluation of the new chromogenic medium Candi*Select* 4 for differentiation and presumptive identification of the major pathogenic *Candida* species. J. Med. Microbiol., 56: 495-499.

Sheppard, D. C., Yeaman, M. R., Welch, W. H., Phan, Q. T., Fu, Y., Ibrahim, A. S., Filler, S. G., Zhang, M., Waring, A. J. and Edwards, J. E., Jr. 2004. Functional and structural diversity in the Als protein family of *Candida albicans*. J. Biol. Chem., 279: 30480-30489.

Sheth, C. C., Johnson, E., Baker, M. E., Haynes, K. and Mühlschlegel, F. A. 2005. Phenotypic identification of *Candida albicans* by growth on chocolate agar. Med. Mycol., 43: 735-738.

Shi, Q-M., Wang, Y-M., Zheng, X-D., Lee, R.T.H. and Wang, Y. 2007. Critical role of DNA checkpoints in mediating genotoxic-stress-induced filamentous growth in *Candida albicans*. Mol. Biol. Cell, 18: 815-826.

Shorr, A. F., Lazarus, D. R., Sherner, J. H., Jackson, W. L., Morrel, M., Fraser, V. J. and Kollef, M. H. 2007. Do clinical features allow for accurate prediction of fungal pathogenesis in bloodstream infections? Potential implications of the increasing prevalence of non-albicans candidemia. Crit. Care Med., 35: 1077-1083.

Sobel, J. D. 1997. Vaginitis. New Engl. J. Med., 337: 1896-1903.

Sobel, J. D. and Rex, J. H. 2001. Invasive candidiasis: turning risk into practical prevention policy? Clin. Infect. Dis., 33: 187-190.

Sohn, K., Urban, C., Brunner, H. and Rupp, S. 2003. *EFG1* is a major regulator of cell wall dynamics in *Candida albicans* as revealed by DNA microarrays. Mol. Microbiol., 47: 89-102.

Spellberg, B. J., Ibrahim, A. S., Avanesian, V., Fu, Y., Myers, C., Phan, Q. T., Filler, S. G., Yeaman, M. R. and Edwards, J. E., Jr. 2006. Efficacy of the anti-*Candida* rAls3p-N or rAls1p-N vaccines against disseminated and mucosal candidiasis. J. Infect. Dis., 194: 256-260.

Srivastava, S. and Srivastava, R-G. 2005. Proteomics in the forefront of cancer biomarker discovery. J. Proteome Res., 4: 1098-1103.

Staab, J. F., Bahn, Y-S., Tai, C-H., Cook, P. F. and Sundstrom, P. 2004. Expression of transglutaminase substrate activity on *Candida albicans* germ tubes through a coiled, disulfide-bonded N-terminal domain of Hwp1 requires C-terminal glycosylphosphatidylinositol modification. J. Biol. Chem., 279: 40737-40747.

Staab, J. F., Ferrer, C. A. and Sundstrom, P. 1996. Developmental expression of a tandemly repeated, proline- and glutamine-rich amino acid motif on hyphal surfaces of *Candida albicans*. J. Biol. Chem., 271: 6298-6305.

Staib, P., Kretschmar, M., Nichterlein, T., Köhler, G., Michel, S., Hof, H., Hacker, J. and Morschhäuser, J. 1999. Host-induced, stage specific virulence gene activation in *Candida albicans* during infection. Mol. Microbiol., 32: 533-546.

Stephenson, J. 2001. Can a common medical practice transform *Candida* infections from benign to deadly? JAMA, 286: 2531-2532.

Sudbery, P., Gow, N. and Berman, J. 2004. The distinct morphologic states of *Candida albicans*. Trends Microbiol., 12: 317-324.

Tacco, B. A. C. A., Costa, M., Parente, J. A., Felipe, M. S. S., Castro, N. S., Pereira, M. and Soares, C. M. A. 2005. Characterization of an aspartyl proteinase of *Paracoccidioides brasiliensis*. Rev. Inst. Med. Trop. Sao Paulo, 1 p. http://findarticles.com/p/articles/mi_qa3855/is_200510/ai_n15744028

Templin, M. F., Stoll, D., Schrenk, M., Traub, P. C., Vohringer, C. F. and Joos, T. O. 2002. Protein microarray technology. Trends Biotechnol., 20: 160-166.

Terlecka, J. A., du Cros, P. A., Morrissey, C. O. and Spelman, D. 2006. Rapid differentiation of *Candida albicans* from non-*albicans* species by germ tube test directly from BacTAlert blood culture bottles. Mycoses, 50: 48-51.

Thomas, D. P., Bachmann, S. P. and Lopez-Ribot, J. L. 2006. Proteomics for the analysis of the *Candida albicans* biofilm lifestyle. Proteomics, 6: 5795-5804.

Tierno, P. M., Jr. and Milstoc, M. 1977. Germ tube positive *Candida tropicalis*. Am. J. Clin. Pathol., 68: 294-295.

Tosun, I., Aydin, F., Kaklikkaya, N. and Erturk, M. 2005. Induction of secretory aspartyl proteinase of *Candida albicans* by HIV-1 but not HSV-2 or some other microorganisms associated with vaginal environment. Mycopathologia, 159: 213-218.

van Deventer, A. J. M., van Vliet, H. J. A., Hop, W. C. J. and Goessens, W. H. F. 1994. Diagnostic value of anti-*Candida* enolase antibodies. J. Clin. Microbiol., 32: 17-23.

Verstrepen, K. J. and Klis, F. M. 2006. Flocculation, adhesion and biofilm formation in yeasts. Mol. Microbiol., 60: 5-15.

Vitzthum, F., Behrens, F., Anderson, N. L. and Shaw, J. H. 2005. Proteomics: From basic research to diagnostic application. A review of requirements and needs. J. Proteome Res., 4: 1086-1097.

Walsh, T. J., Hathorn, J. W., Sobel, J. D., Merz, W. G., Sanchez, V., Maret, S. M., Buckley, H. R., Pfaller, M. A., Schaufele, R., Sliva, C., et al. 1991. Detection of circulating candida enolase by immunoassay in patients with cancer and invasive candidiasis. New Engl. J. Med., 324: 1026-1031.

White, P. L., Archer, A. E. and Barnes, R. A. 2005. Comparison of non-culture-based methods for detection of systemic fungal infections, with emphasis on invasive *Candida* infections. J. Clin. Microbiol., 43: 2181-2187.

White, P. L., Wiliams, D. W., Kuriyama, T., Samad, S. A., Lewis, M. A. O. and Barnes, R. A. 2004. Detection of *Candida* in concentrated oral rinse cultures by real-time PCR. J. Clin. Microbiol., 42: 2101-2107.

Whiteaker, J. R., Zhang, H., Eng, J. K., Fang, R., Piening, B. D., Feng, L-C., Lorentzen, T. D., Schoenherr, R. M., Keane, J. F., Holzman, T., Fitzgibbon, M., Lin, C., Zhang, H., Cooke, K., Liu, T., Camp II, D. G., Anderson, L., Watts, J., Smith, R. D., McIntosh, M. W. and Paulovich, A. G. 2007a. Head-to-head comparison of serum fractionation techniques. J. Proteome Res., 6: 828-836.

Whiteaker, J. R., Zhang, H., Zhao, L., Wang, P., Kelly-Spratt, K. S., Ivey, R. G., Piening, B. D., Feng, L-C., Kasarda, E., Gurley, E., Eng, J. K., Chodosh, L. A., Kemp, C. J., McIntosh, M. W. and Paulovich, A. G. 2007b. Integrated pipeline for mass spectrometry-based discovery and confirmation of biomarkers demonstrated in a mouse model of breast cancer. J. Proteome Res., 6: 3962-3975.

Whiteway, M. and Bachewich, C. 2007. Morphogenesis in *Candida albicans*. Annu. Rev. Microbiol., 61: 529-553.

Wilkins, M. R., Appel, R. D., Van Eyk, J. E., Chung, M. C. M., Görg, A., Hecker, M., Huber, L. A., Langen, H., Link, A. J., Paik, Y-K., Patterson, S. D., Pennington, S. R., Rabilloud, T., Simpson, R. J., Weiss, W. and Dunn, M. J. 2006. Guidelines for the next 10 years of proteomics. Proteomics, 6: 4-8.

Yeater, K. M., Chandra, J., Cheng, G., Mukherjee, P. K., Zhao, X., Rodriguez-Zas, S. L., Kwast, K. E., Ghannoum, M. A. and Hoyer, L. L. 2007. Temporal analysis of *Candida albicans* gene expression during biofilm development. Microbiology, 153: 2373-2385.

Yeo, S. F., Huie, S., Sofair, A. N., Campbell, S., Durante, A. and Wong, B. 2006. Measurement of serum D-arabinitol/creatinine ratios for initial diagnosis and for predicting outcome in an unselected, population-based sample of patients with *Candida* fungemia. J. Clin. Microbiol., 44: 3894-3899.

Yin, Z., Stead, D., Selway, L., Walker, J., Riba-Garcia, I., McInerney, T., Gaskell, S., Oliver, S. G., Cash, P. and Brown, A. J. P. 2004. Proteomic response to amino acid starvation in *Candida albicans* and *Saccharomyces cerevisiae*. Proteomics, 4: 2425-2436.

Zaoutis, T. E., Argon, J., Chu, J., Berlin, J. A., Walsh, T. J. and Feudtner, C. 2005. The epidemiology and attributable outcomes of candidemia in children hospitalized in the United States: a propensity analysis. Clin. Infect. Dis., 41: 1232-1239.

Zhao, X., Oh, S-H., Cheng, G., Green, C. B., Nuessen, J. A., Yeater, K., Leng, R. P., Brown, A. J. P. and Hoyer, L. L. 2004. *ALS3* and *ALS8* represent a single locus that encodes a *Candida albicans* adhesion; functional comparison between Als3p and Als1p. Microbiology, 150: 2415-2428.

Zhao, X., Daniels, K. J., Oh, S-H., Green, C. B., Yeater, K. M., Soll, D. R. and Hoyer, L. L. 2006. *Candida albicans* Als3p is required for wild-type biofilm formation on silicone elastomer surfaces. Microbiology, 152: 2287-2299.

Zhao, X., Oh, S-H., Yeater, K. M. and Hoyer, L. L. 2005. Analysis of the *Candida albicans* Als2p and Als4p adhesins suggests the potential for compensatory function within the Als family. Microbiology, 151: 1619-1630.

Zhao, X., Oh, S-H. and Hoyer, L. L. 2007. Deletion of *ALS5*, *ALS6* or *ALS7* increases adhesion of *Candida albicans* to human vascular endothelial and buccal epithelial cells. Med. Mycol., 45: 429-434.

Zheng, X., Wang, Y. and Wang, Y. 2004. Hgc1, a novel hypha-specific G1 cyclin-related protein regulates *Candida albicans* hyphal morphogenesis. EMBO J., 23: 1845-1856.

Chapter 13

Comparative Methods in Fungal Evolutionary Biology: Theory, Examples and Potential Applications

László G. Nagy*, **Tamás Papp and Csaba Vágvölgyi**

Department of Microbiology, Faculty of Science and Informatics
University of Szeged, Közép fasor 52, Szeged, H-6726, Hungary
E-mail: mucor1959@gmail.com

ABSTRACT

Phylogenetic comparative approaches represent a very powerful group of methods to infer, recover or analyze patterns of evolution. They are often applied to reconstruct ancestral protein or nucleic acid sequences and ancestral states of morphological or physiological traits on a given phylogeny through which the species or sequences are related to each other. It is now possible to identify ancestral environments and geographic distribution or correlation between characters. Due to a surge in development of comparative methods in the last few years, new statistics, various Bayesian approaches and hypothesis testing frameworks have been proposed, which offer several advantages over traditional approaches. This chapter reviews some of the methods and the corresponding basic theory, as well as fungal examples and potential fields where comparative analyses can be used. Of the fungal examples, inference of the ancestral fruiting body types in the Agaricomycotina, the discovery that lichenization emerged very early in the Ascomycota, and that most ascomycete lineages are derived from lichen symbiotic ancestors have been dealt with, besides several other interesting questions have also been addressed. Today, the complexity of hypotheses, that can be tested, is surprisingly high. This tempts the fungal systematists and evolutionary biologists to test their ideas about evolution and adaptation.

**Corresponding author*: E-mail: cortinarius2000@yahoo.co.uk

INTRODUCTION

The evolution of various life-history traits has become the focus of many taxonomic, systematic and evolutionary studies recently. Since evolutionary biologists intend to study patterns of adaptation, radiations or character change, the evolutionary dynamics of distinctive characters is of prime interest. From the systematic point of view, one of the most serious problems systematists can face while attempting to characterize a group of organisms (either defined by some classification or inferred from molecular data) is the evolutionary history of the character that ought to be used for defining the group. This is particularly true for fungi, where fossilization is very scanty. However, given some morphological characters, a phylogeny that is inferred from molecular data, it is possible to reconstruct the evolutionary history of the character *in silico*.

This chapter presents a brief introduction to the basics of the theory, nature of comparative methods, and details how these can be used to address basic evolutionary questions. These questions may range from how many times a character evolved in a particular group (Character Mapping), what state does the character take in one of the common ancestors of a clade or a taxon (Ancestral State Reconstruction), to what is the model that best describes the evolutionary process of the trait given our data, or have any of our traits co-evolved on the phylogeny. It should be stressed at this point, that although our consideration and examples here are restricted to morphological characters, the theory and properties of methods are applicable to molecular data as well, e.g. reconstructing ancestral protein or nucleic acid sequences, as well as for examining whether the sequences evolved in a correlated fashion (Harvey and Pagel, 1991).

Systematists often seek morphological, biochemical, or physiological characters to circumscribe a group of organisms they would like to consider as a taxon (species, family, etc.). However it is often difficult to infer the phylogeny, i.e. the bifurcating tree through which the species are related, from morphological characters. *Studying the evolution of the group*, based on morphological characters, requires the researcher to present a large number of characters and a model of evolution for each morphological character, which describes the processes that generated the contemporary distribution of character states in our tree (i.e. in the terminal branches thereof, in practice taxa or sequences). However, there are several problems associated with using morphological characters to infer phylogeny. The most serious of these is the reliability of morphological characters that is often unknown due to, for example, potential convergent evolutionary processes, hidden homologies or coding bias. *Studying the evolution of morphological characters* in a phylogenetic framework has, however, a great potential for recovering patterns of evolution of the character in a statistically justified framework. This requires the researcher to have a molecular phylogeny of the taxa and a corresponding matrix of character states of the trait(s) of interest. Given today's high throughput sequencing facilities, the easy access to standard molecular biological reagents and techniques, it is more or less straightforward to obtain molecular data (sequences) of several genes from a set of taxa. The resulting molecular data sets can then be analyzed by over a dozen methods to reconstruct the phylogeny of the group. Detailed treatment of these methods would be beyond the scope of the present chapter, hence only a few relevant concepts are given.

If we reconstruct phylogenetic trees by methods such as Maximum Parsimony (MP), Maximum Likelihood (ML) or Bayesian inference, the analyses will almost surely result in a set of more or less equally good trees (topologies and branch lengths), which we refer hereafter as the *phylogeny*. The expression phylogeny includes branch length estimates based on molecular data. This will be important when we apply these branch lengths to our morphological data. It is also well known that all reconstructions of the phylogeny are associated with some error, because phylogeny inference is merely a statistical estimation procedure. This error can affect a number of things during phylogenetic reconstruction, but from our point of view the error in the estimates of topologies and branch lengths are the most important. This error will affect any type of analysis performed downstream to reconstruction of the trees, and will have to be accounted for. Another issue which has to be defined is to how to interpret the credibility of clades on trees. There are various types of measures to characterize robustness of the clades on trees, of which the best known are bootstrap proportions and posterior probabilities. The former usually applies for MP and ML trees, while the latter usually characterizes Bayesian trees, although it is also possible to apply the bootstrap analysis to Bayesian inference and *vice versa* (Douady et al., 2003). If one attempts to reconstruct an ancestral character state for a clade, one will have to take into account the probability that the clade is correctly inferred, i.e. the statistical support for that clade has to be considered.

CODING CHARACTERS AS DISCRETE STATES

The characters one studies in a phylogenetic framework can either be discrete (=qualitative) or continuous (=quantitative). Since most of the characters in mycology are discrete traits, in this chapter we will focus on this type of characters. It should be noted, however, that continuously varying traits can also be discretized (for details see Felsenstein, 2004), and that a wealth of methods to analyze continuous characters are also available.

Discrete coding means states of the character take discrete values, such as 0 or 1 in case of a binary, or 0, 1, 2, ... *n* in case of a multistate trait. Familiar binary characters may include presence/absence of a character, whether a species is pathogenic or not, and a gene is expressed or not. It is the binary coding regime that is most often used in comparative studies, whereas multistate traits are seldom used and entail a more difficult analytical challenge to the researcher (Hibbett, 2004). The difficulties of estimating parameters of models of multistate characters compared to binary ones are well exemplified by a study of fruiting body evolution of Agaricomycetes (Hibbett, 2004). In this study, different types of fruiting bodies were coded either as binary or as multistate, and the statistical power and outcome of two different coding strategies were explored (Hibbett, 2004). Congruent with general views, it turned out that multistate coding is statistically less powerful as it requires estimation of many more parameters (of the evolutionary model) than binary coding does. In that study, ancestral state reconstructions based on multistate coding were inconclusive, whereas binary coding provided significant support for certain ancestral states. Binary coding, on the other hand, is often unable to accommodate/capture all aspects of the natural variation, such as that of fruiting bodies in the Agaricomycetes. As an alternative to multistate coding, Hibbett (2004) incorporated five different fruiting body types recognized in the study in several binary matrices. Note that in such a case, four of the five functionally equivalent states of recognized fruiting body types should be merged in one state of the binary matrix, whereas in multistate coding they are recognized as individual character states. Therefore, binary models may be biologically unrealistic when applied to characters that may be more naturally coded in a multistate manner. Further, as exemplified by the mentioned study, results obtained from multistate coding can in some cases contradict those obtained from binary coding. Therefore, coding choice should be justified very strongly in evolutionary studies.

To mitigate the potential errors of estimating multistate models Hibbett (2004) employed a minimal model analysis, which successfully reduced the number of parameters of the evolutionary model from 20 to 4. This seeks the model that has the minimum number of parameters which adequately describe the underlying evolutionary process, by reducing the number of parameters (i.e. merging two rates into one rate class) in the model step by step and examining if the resulting model is significantly worse than the null model. This logic is similar to the hierarchical likelihood ratio tests used to select among members of the GTR family of models of gene-sequence evolution and implemented in the program ModelTest (Posada and Crandall, 1998). Because the study mentioned above uses a partly hierarchical LRT strategy, it shares some important limitations with the hierarchical likelihood ratio tests of ModelTest, for instance, the result being dependent on the hierarchy and significance threshold used, or the inability to compare multiple models simultaneously. Later in this chapter we will show that Reversible-Jump MCMC analyses could be used as a surrogate to develop minimal models of evolution, while evading the restraints of hierarchical tests.

PARSIMONY MAPPING OF CHARACTERS

The traditional approach to map characters onto a tree takes advantage of the parsimony principle. Under the parsimony criterion, the hypothesis that requires the minimum number of evolutionary events is looked for. Thus, parsimony mapping of characters seeks the reconstruction of the characters requiring the minimum number of changes in the character states along a tree. This is accomplished by a down-pass and an up-pass procedure on the tree. During the down-pass step, states of the terminal taxa are used to calculate ancestral states at the subsequent nodes of the tree as follows: first, a clade of a tree that contains two taxa A and B (Fig. 13.1) is considered. We have information about a binary character (e.g., presence or absence of a trait) which we would like to map on the tree. These two taxa have a neighboring node that represents their most recent common ancestor (MRCA). If both terminal taxa have state 0 (or 1), then we infer their MRCA having had state 0 (or 1) as well, unambiguously. If one of them has a state 0, the other has 1, then we infer an ambiguous state as the ancestral condition of our character. Going further up the tree to the next node, we carry this information about the MRCA of taxon A and B. The ancestral state of this node is inferred by the following rule: if both subsequent nodes have been assigned the same state (either ambiguously or unambiguously), this state is inferred for that node, while if they have different states, we assign an ambiguous state to this node. This way, we can pass the tree down up to its root node which is the ancestor of all the taxa on our tree. When this is completed, we start from the root and go upwards on the tree, inferring a state for each interior node applying the above rules. Hence, we will have two states inferred for each interior node (except the root) which we extrapolate according the rules we applied formerly to one state assignment. This means that if we inferred a 0 during the down-pass step and a 1 during the up-pass step, the final assignment will be ambiguous (0/1). If one has on the other hand a state 0 and a 0/1, then 0 is inferred. Of course, if both passes gave the same states to a node, that state will be inferred unambiguously in the final step.

There are cases when multiple most parsimonious reconstructions exist for the node(s) of interest. In these cases, one might interpret the result as ambiguous, or attempt to resolve the ambiguity. There are two well-known ways for resolving ambiguously reconstructed nodes. The first, termed Accelerated Transformations (ACCTRAN), minimizes the number of changes on all subtrees, i.e. places character changes closer to the root, emphasizing reversals and minimizing convergence (Fig. 13.1). On the other hand, the Delayed Transformations approach (DELTRAN) places changes closer to the tip of the tree which suggests convergent evolution. On a theoretical basis, it was suggested that losses of complex traits may be more probable than parallel gains (Britto, 2003; Gaubert et al., 2005; Agnarsson and Miller, 2008), which in practice led to preference of ACCTRAN optimizations over DELTRAN. However it was

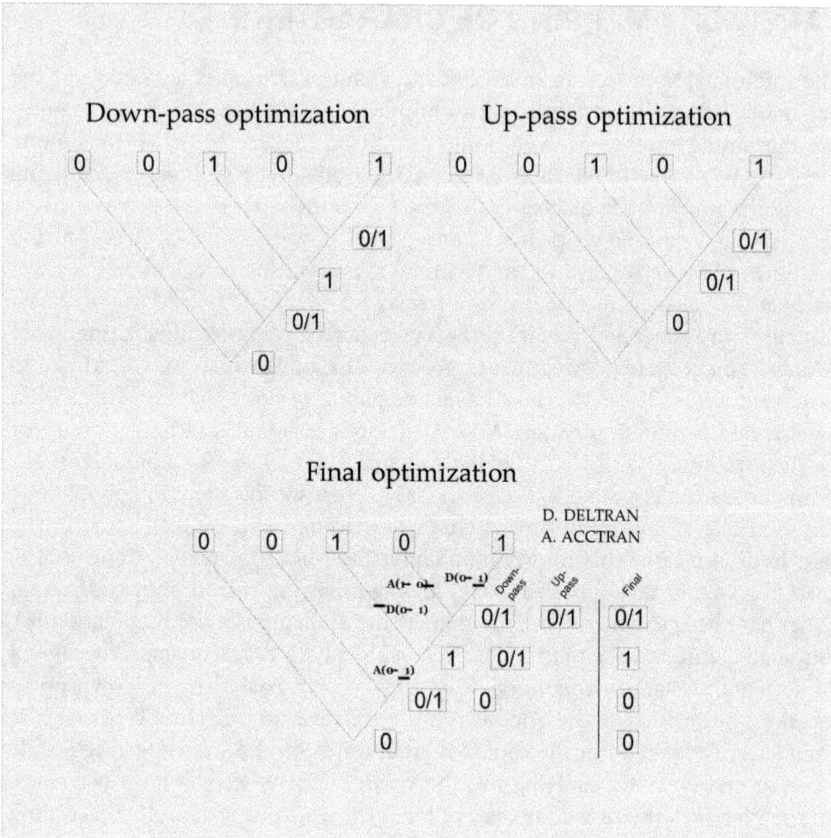

Fig. 13.1 Parsimony optimization of ancestral states on trees. Three optimization steps are depicted as explained in the text. ACCTRAN and DELTRAN optimizations of an ambiguously inferred ancestral state assignment (resulting from multiple most parsimonious reconstructions) are shown. For explanation, please refer to the text.

recently argued that this extrapolation from theoretical reasoning to a practice preference of ACCTRAN is based on an incorrect association and each case should be considered separately (Agnarsson and Miller, 2008).

It is noteworthy that in parsimony analyses, differential weighting of characters may give better, or at least different results on trees, although choosing a method for weighting can be problematic. While, in phylogeny reconstruction evaluation of various weighting schemes is easily accomplished by observing tree lengths, in comparative studies with one or a few characters it can be difficult (Hall, 2006).

The parsimony approach obviously has a number of limitations. The most serious one is that parsimony is unable to infer more than one change along a branch of the tree. This results in incorrect reconstructions if the rates of evolution are high, i.e. more than one change is likely on a branch (Pagel, 1999 a, b). Similarly, if there are some long branches on the tree, there is a higher probability that a change will occur along that branch which parsimony completely ignores. Another problem with parsimony is that it is not possible to estimate the confidence in our results. Hence, we cannot compare directly the most parsimonious reconstruction with less parsimonious reconstructions, although there are cases where the true evolutionary history is not the most parsimonious solution. Probability distributions have been developed to be able to compare reconstructions inferred from parsimony (Maddison, 1995), but these have not been used widely.

Parsimony mapping has been applied to many of the central questions in fungal evolutionary biology, among others, to the evolution of fruiting body morphologies within Agaricomycetes (Hibbett and Binder, 2002; Hibbett, 2007; Nagy et al., 2009). While more examples will be given later in this chapter, one potential application is mentioned here. It is possible to examine the extent of homoplasy exhibited by certain characters by parsimony, and to draw conclusions about the utility of these characters in classification (Nagy et al., 2009). The logic behind this is that homoplasious characters on a given phylogeny are less reliable as diagnostic characters for defining taxa within the group assuming that the phylogeny is a good description of the evolutionary history. In a study by Nagy et al. (2009), characters traditionally used to define *Parasola* species (Psathyrellaceae, Agaricales) and the genus itself, were parsimony mapped onto ML, MP and Bayesian trees. The results implied that certain characters (e.g., deliquescence, presence of pseudoparaphyses and plicate pileus) were gained only once within the genus *Parasola*, hence they are suitable to define the group in a phylogenetic context.

MODEL BASED APPROACHES

Besides parsimony, probabilistic methods have been developed for calculation of ancestral states in phylogenies. These require a stochastic model of evolution that describes how the character evolves along a branch of a tree. For discrete characters, continuous-time Markov models are used. They are equivalent to models of the GTR family used to reconstruct phylogenies, assuming that characters evolve in a Markovian fashion, i.e. changing from one state to another is dependent only upon the current state, and that change along branches of a tree is independent from changes on other branches (Pagel, 1994).

The simplest case is to consider a discrete character with two states, 0 and 1 (Fig. 13.2). For this example, two substitution rates exist, one that denotes the rate of change from state 0 to state 1 (q_{01}) and the other from 1 to 0 (q_{10}). The model allows the states to revert or evolve multiple occasions over longer periods of time. Given that the branches of the tree are defined as expected number of substitutions per site, we can calculate transition probabilities (P_{01} and P_{10}) for states along branches. How this can be done is beyond the scope of this chapter, but can be found in other papers and reviews (Pagel, 1994; Huelsenbeck and Bollback, 2001; Pagel et al., 2004; Gascuel, 2005). For instance, $P_{01}(t)$ denotes the probability that state 0 changes to state 1 over a branch of length t and $P_{11}(t)$ means the probability that branch t has state 1 at the beginning and at the end point, as well. A property of this is that there is higher probability of change of a character along longer branches than on shorter ones. This is easily understandable, as lengths of branches are usually somehow related to time passed between two speciation events.

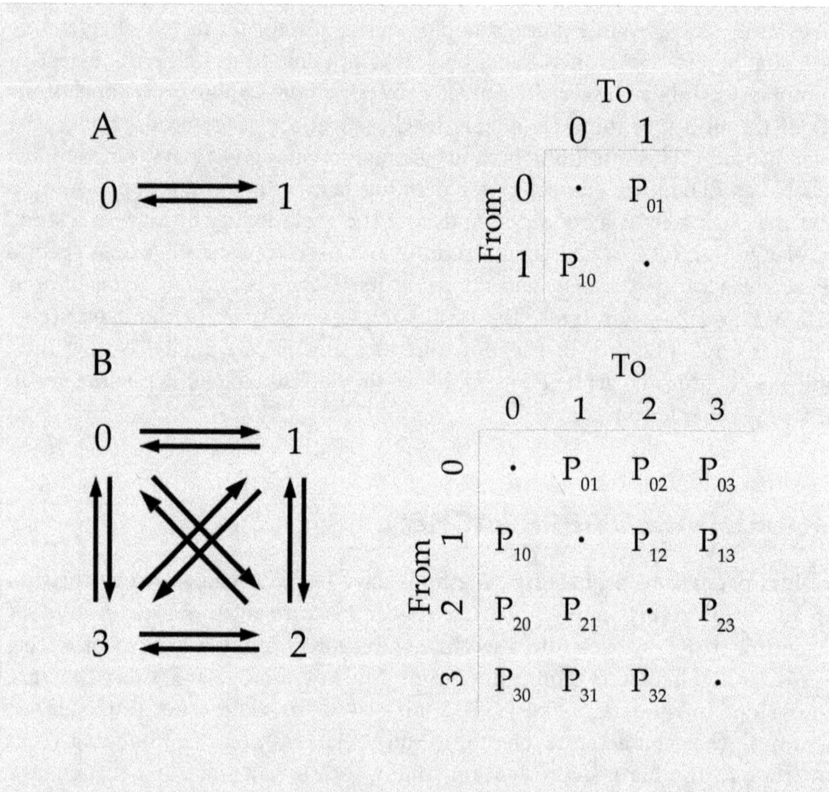

Fig. 13.2 Models and transition probabilities for binary and multistate characters. Note that for binary characters *(A)* there are two, whereas for multistate characters *(B)* there are 12 different transition probabilities to be estimated from the data.

ANCESTRAL STATE RECONSTRUCTION BY MAXIMUM LIKELIHOOD

Maximum Likelihood estimation of ancestral states implies calculating conditional probabilities of each state at the nodes of interest and estimating the parameters of the Markov model. To calculate likelihood values, let us consider a tree with three taxa (A, B, and C). We shall calculate the probabilities of ancestral states at the ancestor of the species A and B, called Node 1, which then denotes a subtree with two species, both in state 1. The conditional probabilities can be written for the two possible states at Node 1 as the product of the transition probabilities for the branches, given the model of evolution used:

$$\Pr(0|M,T) = P_{01}(t_1)\,P_{01}(t_2)$$
$$\Pr(1|M,T) = P_{11}(t_1)\,P_{11}(t_2)$$

Pr(0|M,T) is, in this case, the conditional probability of the subtree designated by Node 1, as it has state 0. The likelihood of subtree (A, B) can be written as the sum of the two conditional probabilities, given the model of evolution:

$$\mathrm{L(A,B)} = P_{01}(t_1)\,P_{01}(t_2) + P_{11}(t_1)\,P_{11}(t_2)$$

The conditional probabilities at the root node [the ancestor of the subtree (A, B, C)] are as follows:

$$\Pr(0|M,T) = P_{00}(t_4)\,P_{00}(t_3)\,P_{01}(t_1)\,P_{01}(t_2) + P_{00}(t_4)\,P_{01}(t_3)\,P_{11}(t_1)\,P_{11}(t_2)$$
$$\Pr(1|M,T) = P_{10}(t_4)\,P_{10}(t_3)\,P_{01}(t_1)\,P_{01}(t_2) + P_{10}(t_4)\,P_{11}(t_3)\,P_{11}(t_1)\,P_{11}(t_2)$$

and the likelihood is:

$$\begin{aligned} L\,(A,B,C) &= P_{00}(t_4)\,P_{00}(t_3)\,P_{01}(t_1)\,P_{01}(t_2) + P_{00}(t_4)\,P_{01}(t_3)\,P_{11}(t_1)\,P_{11}(t_2) \\ &\quad + P_{10}(t_4)\,P_{10}(t_3)\,P_{01}(t_1)\,P_{01}(t_2) + P_{10}(t_4)\,P_{11}(t_3)\,P_{11}(t_1)\,P_{11}(t_2) \end{aligned}$$

Note that in the above example model parameters were considered as known, although generally they are unknown parameters. In most cases, they are unknown parameters to estimate, and the Maximum Likelihood solution will be those values of the model parameters that maximize the above equation. If the goal is to test alternative hypotheses, the node of interest should be fixed at one of the states, the likelihood calculated, with model parameters maximized and compared to the likelihood obtained when fixing the node in the other state. This way, likelihood of each hypothesis will be conditioned on the hypothesis being true, allowing comparison.

There are two basic approaches to reconstruct ancestral character states by Maximum Likelihood. The first, termed "local" by Pagel (1999a), calculates the likelihoods of each state at a given node, estimating model parameters for each state, as described above. On the other hand, "global" methods (e.g., Yang et al., 1995; Koshi and Goldstein, 1996; Maddison and Maddison, 2007) calculate the overall likelihood and partition it to find the proportion associated with each of the character states, without fixing the node in any of the states. Model parameters are maximized only once in the global method, hence, they are not conditional on any of the states being true. Therefore, the global method is not suitable to test alternative hypotheses about ancestral states (Pagel, 1999 a, b). Instead, the global method was used to reconstruct ancestral protein and nucleic acid sequences (Yang et al., 1995; Koshi and Goldstein, 1996). Of the available programs, Mesquite uses the global method (Maddison and Maddison, 2007). Miadlikowska and Lutzoni (2004) used this method to reconstruct ancestral symbiotic states in the lichenized fungal order Peltigerales, across a set of trees sampled from MrBayes. As a measure of significance, proportion of trees on which ancestral states were reconstructed with "significant" support over alternative states. A reconstructed ancestral state on a tree was considered significant if the difference between the proportional log likelihoods contributed by each state to the overall likelihood was ≥2.00, as is the default setting in Mesquite. However, Pagel (1999 a, b) argued that this kind of hypothesis test is invalid because certain parameters of the hypothesis (estimates of model parameters) are not maximized under the assumption that the hypothesis is true. In other words, the hypotheses are not ML estimates. Further, reporting results only on those trees that provided significant reconstructions, fail to incorporate uncertainty in those trees on which ancestral states have not been reconstructed with proper support. A similar approach was applied to the evolution of codon usage and the number of isoprene molecules in the Co-Q system in the order Saccharomycetales (Diezmann et al., 2004) and to the evolution of polyspory in euascomycetes (Reeb et al., 2004).

There have been some concerns about the advantages of using *branch length information* in ancestral state estimation. Incorporating this information in Maximum Likelihood (and Bayesian) ancestral state estimation proved to have serious effects on the results as compared to the case when branches are considered to be of equal length (Cunningham, 1999). Advocates of Maximum Likelihood correctly emphasized that neglecting the information carried by branch lengths, as in Maximum Parsimony, potentially introduces unnecessarily high degrees of ambiguity in the results (Cunningham et al., 1998, Cunningham, 1999; Pagel, 1999 a, b; Huelsenbeck and Bollbeck, 2001; Ronquist, 2004). There are, however, two major sources of potential error of using branch lengths in studies of morphological evolution that rely on molecular phylogenies (Cunningham, 1999). First, it should be kept in mind that the branch lengths used by us refer to the molecular characters, not the morphological. The speed of evolution can, in certain cases, differ significantly between molecular and morphological characters, for example in adaptive radiations, when large amounts of morphological variability can accumulate over a short period of time, whereas the rate of molecular evolution is not affected (Schluter et al., 1997; Cunningham, 1999). On phylogenies, adaptive radiations can be noted by the emergence of numerous lineages that are connected by short branches. The other case when use of branch lengths may result in incorrect reconstructions, originate from an erroneous choice of molecular data used for phylogeny reconstruction. If the substitution process of DNA sequences is saturated, the terminal branches of the phylogeny will be too long, while length of internal branches may be systematically underestimated (Cunningham, 1999; Felsenstein, 2004). This may lead to the inference of character changes on longer branches. This is due to a property of the model, i.e. the assumption that the rate of evolution is constant over the whole phylogeny.

HYPOTHESIS TESTING IN MAXIMUM LIKELIHOOD

The Maximum Likelihood estimate means choosing the hypothesis with the largest likelihood, no matter how much larger it is to the second largest value. This holds no more information than favoring the most parsimonious reconstruction over somewhat less parsimonious reconstructions. Therefore, we need to measure the confidence in our hypothesis, such as inferred ancestral states or evolutionary models.

It is crucial in comparative studies to know if reconstructed ancestral states and evolutionary models are supported by the data significantly better than alternative states or models. Therefore, statistical tests should be conducted to be able to choose among competing hypotheses unequivocally.

In phylogenetics, Likelihood Ratio Tests (*LRT*) are well-known and widely used to choose among nested models of evolution (Huelsenbeck and Crandall, 1997; Posada and Crandall, 1998; Huelsenbeck et al., 2004; Posada and Buckley, 2004; Sullivan and Joyce, 2005). The *LRT* is defined as twice the difference in the likelihoods of the models to be compared:

$$LRT = 2(L_1 - L_2)$$

where the *LRT* statistic is asymptotically χ^2-distributed, with degrees of freedom equal to the number of parameters that vary between the two models (Posada and Crandall, 1998; Pagel, 1999 a, b; Posada and Buckley, 2004). This stands only for nested models, i.e. when one of the models is a special case of the other. In comparative studies this quite often happens to be untrue, and in these cases the *LRT* statistic cannot be applied. For instance, two models of a binary trait, where in one model both rates are equal, while one of the rates in the other is zero. In this simple case, neither model can be written as the special case of the other. For non-nested models, Pagel (1999 a, b, advocating Edwards, 1972) proposed a difference of two log likelihood units to be considered a "rule-of-thumb" criterion of significant support for the better fitting model (Pagel, 1999a, b; Hibbett, 2004).

BAYESIAN INFERENCE OF ANCESTRAL STATES AND MODELS OF MORPHOLOGICAL EVOLUTION

It is evident from the above description that Maximum Likelihood estimation has a number of serious limitations. One of the major drawbacks of Maximum Likelihood estimation is that it uses point estimates for parameters such as rates of the evolutionary model, branch lengths of the tree, and tree topology. Many studies exemplified the role of these parameters in comparative analyses and the rate of error that is caused by neglecting uncertainty of branch length estimates (Huelsenbeck and Bollback, 2001; Pagel et al., 2004) topology (Lutzoni et al., 2001; Pagel and Lutzoni, 2002; Huelsenbeck et al., 2004; Pagel, 2004; Ronquist, 2004; Bollbeck, 2006; Pagel and Meade, 2006; Ekman et al., 2008) or other parameters (Mooers and Schluter, 1999).

Methods based on Bayesian statistics offer a convenient way to accommodate many types of uncertainty in comparative studies. They take advantage of Bayes' theorem:

$$\mathrm{P}\ (\mathrm{H}|\mathrm{D}) = \frac{P(H)\ P(D|H)}{P(D)}$$

where $P(H|D)$ is the posterior probability (belief) of hypothesis H, $P(H)$ is the prior probability of the hypothesis, $P(D|H)$ is the likelihood of the hypothesis and $P(D)$ is the probability of the data. The denominator of this equation includes the probability of our data considering all possible tree topologies, branch length combinations and model parameters, which can be thought of as a multidimensional integral. Because the number of tree topologies is very high, even for moderate numbers of taxa, normally it is impossible to compute the denominator analytically (Mau and Newton, 1997; Yang and Rannala, 1997; Larget and Simon, 1999; Huelsenbeck et al., 2001).

The prior probability in Bayes' formula is a mathematical quantification of one's prior belief about the problem. As the prior belief can vary from person to person, it may be a source of subjectivity in comparative studies. In order to avoid this subjectivity, generally, uniform (uninformative) prior distributions are used in phylogenetics. If a uniform prior is used, the differences in the posterior probabilities will depend on how the data can transform the prior distribution into posterior distribution (Fig. 13.3), i.e. how strong the signal is in the data (Huelsenbeck et al., 2001, 2002). In phylogeny reconstruction, for instance, this poses no problem because there are hundreds or thousands of sites in an alignment, so the posterior distribution will be dominated by the evidences entailed by our data. In comparative studies researchers have very few data, usually only one or two characters. In such cases, we may expect too much from our data when uniform (=uninformative) priors are applied, so prior choice becomes a capital issue (Huelsenbeck et al., 2003; Pagel et al., 2004; Pagel and Meade, 2006, 2007; Ekman et al., 2008). Therefore, we need to restrict the prior distribution so that the posterior distribution will be close to the true underlying distribution. Several strategies were proposed to restrict prior distributions, including empirical Maximum Likelihood based priors (Pagel et al., 2004), various distributions (gamma, exponential, beta) and hyperpriors (Pagel et al., 2004; Bollback, 2006; Pagel and Meade, 2006, 2007; Bollbeck et al., 2007). The latter removes at least some of the subjectivity associated with prior choice by seeding exponential, gamma- or beta- distributed priors from a uniform distribution (Pagel and Meade, 2007).

The denominator of Bayes' formula is integral over all possible tree topologies, branch lengths and model parameters, and normally it is very difficult to compute analytically, even for finite sized data sets. Therefore, the posterior probability is approximated by Markov Chain Monte Carlo (MCMC) algorithms (Huelsenbeck and Bollbeck, 2001; Altekar et al., 2004; Pagel et al., 2004; Lunter et al., 2005).

During a Bayesian MCMC run, the algorithm wanders in the space of tree topologies, branch lengths, and parameters of the evolutionary model(s). In each generation, a new value is proposed for one of the mentioned parameters and the posterior probability of the tree with that parameter value is calculated. If there is an increase in the posterior probability, the new parameter is accepted, if there is not, then the new parameter is accepted with some probability, dependent on the ratio of the null and the alternative posterior probabilities. This way, by running the MCMC long enough, the posterior probability of trees will increase so that it converges to the stationary distribution of parameters (Fig. 13.4). Acceptance of lower posterior probability states ensures that the algorithm can come out of locally optimal regions of the state space and can reach global optimum. After reaching stationarity, trees, branch lengths and model parameters are sampled at regular intervals from the posterior distribution, resulting in a statistically valid sample. Individual elements of this sample are autocorrelated, but the autocorrelation can be mitigated by saving trees reasonably rare, as a general practice every 100th to 1000th generation of the MCMC algorithm. This sample will contain plausibly probable tree topologies, branch lengths and model parameters, which will be representative of the underlying posterior probability distribution.

The above description implies simultaneous inference of tree topologies, branch lengths, and parameters of the evolutionary models (i.e. describing our molecular and morphological data, the given phylogeny is reconstructed from molecular sequences). It has, however, a number of drawbacks as in many cases it is likely that morphological data would make the Markov chains

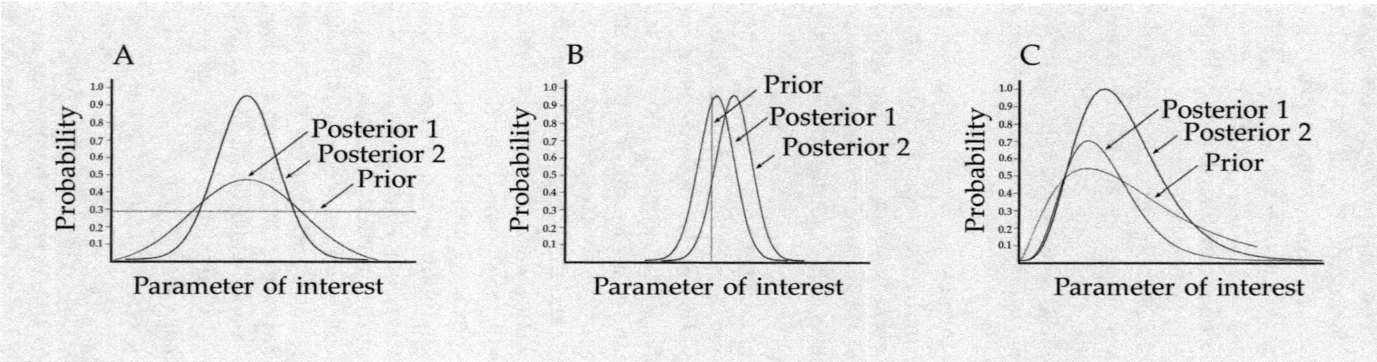

Fig. 13.3 Prior and posterior probability distributions in Bayesian analyses. A: uniform (flat) prior distribution and corresponding posteriors when the "signal" in the data is weak (posterior 1) or strong (posterior 2). Note that strong signal in the data is able to transform the prior into a more definite posterior distribution than a weak signal. This emphasizes the crucial role of priors in comparative analyses, where the signal is normally weak (due to the small number of characters). B: Empirical Maximum Likelihood prior often used in empirical Bayesian analyses, and corresponding posteriors (posterior 1: weak signal, posterior 2: strong signal). C: gamma distributed prior and corresponding posteriors when the signal is weak (posterior 1) or strong (posterior 2). In part A and B posterior distributions are modeled by normal distributions, whereas in part C they are depicted as gamma distributions. It should be noted that in real-life data sets the posteriors can often make up arbitrary distributions.

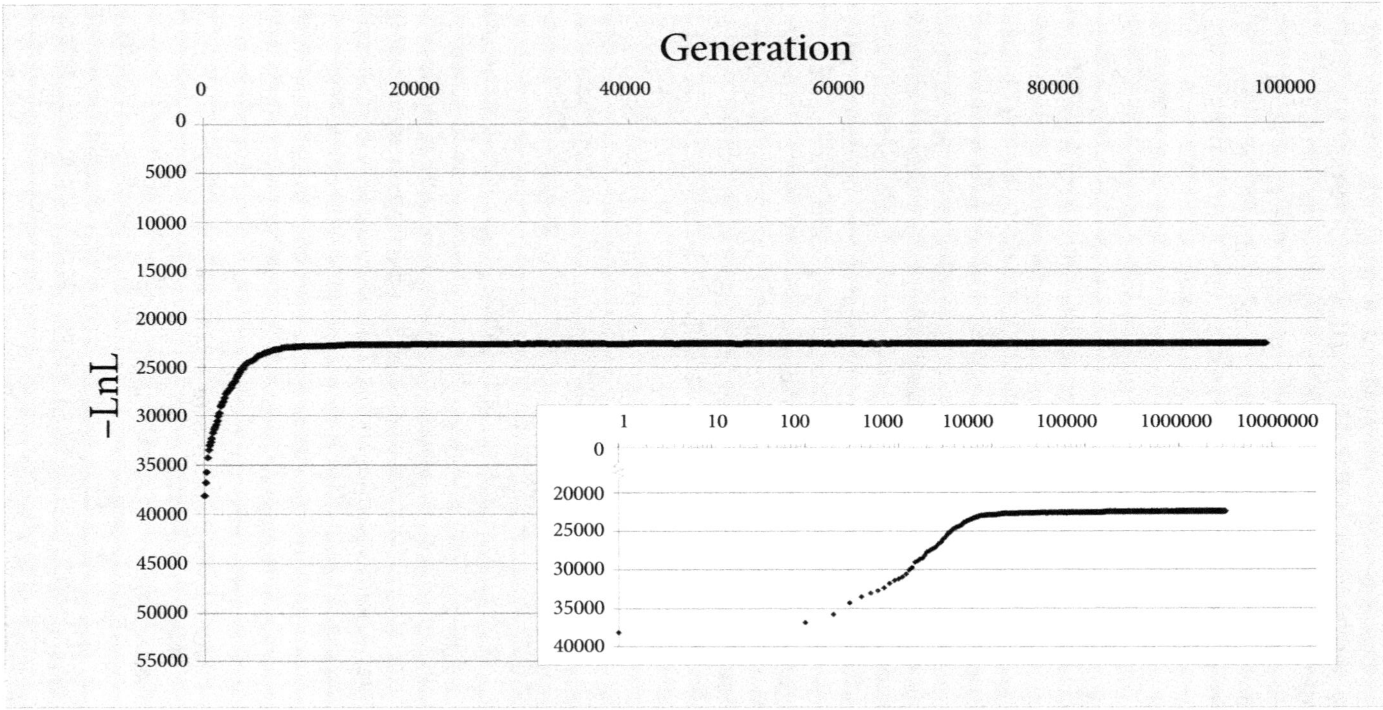

Fig. 13.4 Convergence of a Markov Chain in a Bayesian MCMC analysis (data from Nagy et al. 2009). The first 10^5 generations of the analysis are depicted. Convergence to optimal likelihood values takes place in as short as 10^4 generations in this rather complex data set, which shows the efficiency of MCMC algorithms. Inset figure shows the same analysis depicted on a logarithmic scale.

converge to different optima than sequence data would, which may result in poor quality posterior distributions. In practice, tree inference is separated from reconstructing ancestral states and inference of morphology models (Huelsenbeck et al., 2003; Pagel et al., 2004). Reconstruction of ancestral protein or nucleotide sequences may be more easily combined with tree inference, as is allowed for instance by the program MrBayes (Ronquist and Huelsenbeck, 2003; Lunter et al., 2005; Redelings and Suchard, 2005).

ACCOUNTING FOR PHYLOGENETIC UNCERTAINTY

It is well known that no tree used in comparative studies is devoid of potential errors (uncertainties). It is because our inferences of the trees are based on a statistical estimation procedure from finite sized samples (usually molecular data). Therefore, if we use a single (most parsimonious or Maximum Likelihood) tree, the analyses will be influenced by the errors the tree possesses. In other words, results of comparative analyses will be conditioned on the correctness of tree (Huelsenbeck et al., 2001; Lutzoni et al., 2001; Pagel and Lutzoni, 2002; Pagel et al., 2004). As we cannot make this assumption, this source of error, often referred to as phylogenetic uncertainty (Ronquist, 2004), should be integrated by some statistical approach. This can be most easily accomplished by using more than one tree in the analyses. At first, authors used either a set of equally parsimonious trees, or a MP and a ML tree, or other haphazard tree samples (e.g. Gargas et al., 1995; Hibbett and Binder, 2002, Hibbett, 2004; Nagy et al., 2009; Schoch et al., 2009). These are, however, not derived from a statistically justified sampling procedure and, therefore cannot be considered a correct way of integrating phylogenetic uncertainty.

Bayesian Markov Chain Monte Carlo (MCMC) methods offer a good alternative to generate valid samples of trees. In phylogenetic tree inference practice usually more than one Markov Chain is run in order to increase the efficiency of the MCMC analysis with respect to finding best regions of the state space (for more detailed reviews of Bayesian phylogeny inference see, e.g. Huelsenbeck et al., 2001, 2002; Altekar et al., 2004; Felsenstein, 2004). Our sample will contain plausibly probable tree topologies, branch lengths and model parameters, which will be representative of the underlying posterior probability distribution.

By using a MCMC sample of trees to map morphological characters onto them, we can effectively incorporate uncertainty in tree topology and branch lengths (i.e., phylogenetic uncertainty) into comparative studies (Huelsenbeck et al., 2000, 2001, 2003; Pagel et al., 2004; Ronquist, 2004; Pagel and Meade, 2006). Such MCMC tree samples have been used to account for phylogenetic uncertainty in two different ways. The earlier approach takes Maximum Likelihood as the optimality criterion for inferring morphology models and ancestral states and examines all trees one by one in the sample (Lutzoni et al., 2001; Pagel and Lutzoni 2002). This is often called, the *empirical Bayesian estimation*. In this approach, instead of using flat priors, at least one of the prior probabilities, e.g. tree topology, branch lengths or model parameters are fixed at their Maximum Likelihood values. In this case, the posterior distribution is biased towards our empirically pre-specified values which may be correct or incorrect. Pagel et al. (2004) found that constraining the prior distributions to their ML values returned very narrow posterior distributions. On the contrary, in a *fully Bayesian approach* (often called hierarchical Bayesian estimation), none of the parameters take *a priori* selected values, and a random tree proposal mechanism is used in the MCMC analysis to consider all trees in the sample. In other words, in a fully Bayesian approach, for each generation of the Markov chain new values of the morphology model parameters, and a new tree (out of the MCMC tree sample inferred formerly from sequence data) are proposed.

Reconstructing ancestral states for a node while taking phylogenetic uncertainty into account has to consider if the node of interest is not necessarily present on all of the trees. This problem can be overcome in two ways. The first is to reconstruct the ancestral states only on those trees that contain the node of interest. This also has the drawback of neglecting all trees that do not share that node, therefore, confidence in the results will be overestimated. An alternative and more conservative way is called the *most recent ancestor approach* (MRCA, Fig. 13.5) (Pagel and Meade, 2006, for examples see Binder and Hibbett, 2006; Moore and Donoghue, 2007; Xiang and Thomas, 2008; Nagy et al., 2010). By using this method, the ancestor of the clade that contains all species of interest plus optionally other species is reconstructed. This means, if a group of species is not monophyletic on all trees, i.e. the posterior probability of the clade containing those species is lower than 1.00, then on trees not containing the group uniquely, the nearest node that groups the desired species together is reconstructed. Therefore, those trees that do not contain the node of interest will also contribute to the outcome giving more confidence to the researcher in the results (Fig. 13.5). Binder and Hibbett (2006) when inferring ancestral states of fruiting body types and nutritional modes in the mushroom order Boletales used the MRCA approach to integrate uncertainty posed by trees that do not contain the desired node.

HYPOTHESIS TESTING

Although the posterior probability represents our "posterior belief" in the inferred ancestral states and model parameters, it is often required to decide if the maximum posterior probability (MAP) hypothesis fits the data significantly better than other competing hypotheses. In a Bayesian perspective, the Likelihood ratio test or the AIC (discussed under Maximum Likelihood) cannot

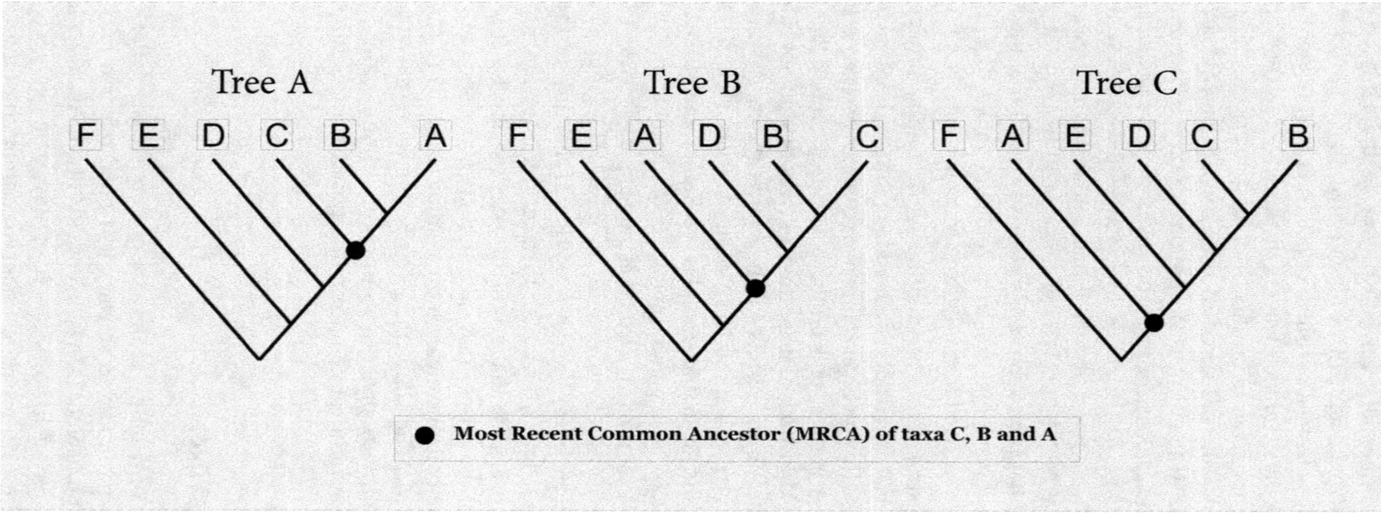

Fig 13.5 The Most Recent Common Ancestor (MRCA) approach of Pagel et al. (2004). The method is used to effectively incorporate phylogenetic uncertainty if the aim was to infer the ancestral state for the taxa A, B and C. Due to the ambiguous position of taxon A (i.e. an uncertain phylogeny), there are trees where taxa A, B and C form a clade of its own (Tree 1), but on Tree 2 and Tree 3 other taxa are nested in the clade. When reconstructing the ancestral condition for the most recent common ancestor of A, B and C, the MRCA approach selects the node (denoted by a black dot) which minimally contains taxa A, B and C, and reconstructs the ancestral state for that node. Therefore it selects different nodes for the three different trees depicted.

be used to test hypotheses because they assume that likelihoods are near or at their maximum values, a requirement that is not necessarily met in Bayesian MCMC samples. Instead, Bayes factors are used to estimate how strongly our data support our choice of hypothesis. The Bayes Factor are Bayesian analogues of the LRT, it expresses the relative increase in the posterior probabilities when the better hypothesis is preferred. It can be written as the ratio of the marginal likelihoods of the two competing hypotheses:

$$BF = \frac{P(D|H_i)}{P(D|H_j)}$$

For larger data sets it is difficult to analytically calculate the marginal likelihoods (Pagel and Meade, 2006). In practice, $P(D|H)$ is approximated by the harmonic mean of the likelihoods of the hypotheses sampled during the MCMC run. Although the harmonic mean of the likelihoods can be unstable, they are assumed to converge to $P(D|H)$ when the Markov chains are allowed to run long enough (Kass and Raftery 1995; Raftery 1996; Pagel and Meade, 2006).

As is evident from the above definition, a BF>1 means that hypothesis *i* is strongly supported by the data than hypothesis *j*. As a rule of thumb, it was suggested that a Bayes Factor between 3 and 12 represents "positive" evidence favoring hypothesis *i*, between 12 and 150 as strong evidence, and a value over 150 as very strong evidence. In phylogenetics likelihood values are often reported on a logarithmic scale, for which the corresponding log Bayes Factor values are: 2-5, 5-10, and over 10.

If model selection is of primary importance it is possible to achieve it in Bayesian framework estimating the fit of a model using Reversible Jump MCMC (RJ-MCMC) algorithms. A RJ-MCMC analysis allows exploration of very complex model spaces where the possible models differ in the number of parameters. As a simple example, let us consider the selection of the best fit model for a binary trait. For a binary model with two rates (q_{01} and q_{10}), we can think of five alternative models, i.e. [$q_{01}=q_{10}$], [$q_{01}<q_{10}$], [$q_{01}>q_{10}$], [$q_{01}=0$, q_{10}] and [q_{01}, $q_{10}=0$]. Of these, three models have only onc parameter to be estimated ([$q_{01}=q_{10}$] [$q_{01}=0$, q_{10}] and [q_{01}, $q_{10}=0$]), while the others have two parameters. Using traditional statistics (such as the LRT) it is impossible to compare all of them to each other. The RJ-MCMC chain enables increasing and decreasing the number of rate classes in the models, therefore, it can "jump" between models with varying dimensions during the analysis depending on a proposal mechanism, similar to that of conventional tree inference MCMC algorithms. Because the chain visits models in proportion to their posterior probability, the sample drawn from such analyses will directly measure the extent to which better fitting models can be preferred over worse fitting ones. The RJ-MCMC approach has been applied to choose among models of gene-sequence evolution (Suchard et al., 2002; Huelsenbeck et al., 2004). In comparative evolutionary biology it has been shown to be particularly efficient for testing for correlated evolution of two discrete traits, because the number of possible models in this case is very large (Pagel and Meade, 2006).

In general, RJ-MCMC algorithms could be applied effectively in any case where the number of models is large. For instance, model testing for multistate characters can be prohibitive or tedious, considering that even for a four-state character (such as DNA sequences), there are over a hundred models which should be tested pairwise, if finding the best-fit model is of primary importance. In such cases, the RJ-MCMC approach offers a convenient way of identifying the model that best describes the data and requires estimation of minimum number of parameters (Suchard et al., 2002; Huelsenbeck et al., 2004; Pagel and Meade, 2006).

STOCHASTIC MAPPING

Recently, another method has been proposed that allows inference of histories of character state changes along phylogenies (Nielsen, 2002; Huelsenbeck et al., 2003). It uses the same continuous time Markov models as Maximum Likelihood and Bayesian approaches. Stochastic mapping of characters is performed in three steps: first, ancestral states at all nodes of the phylogeny are calculated. Second, a combination of ancestral states at internal nodes is drawn randomly and, third, being aware of the character states at both ends of each branch, specific substitution histories are simulated on the branches. If a large number of such histories are simulated this way, it will represent a sample from the posterior distribution of mappings which can be used to draw conclusions about the evolution of a trait. Further, this sample can be used to test hypotheses about various evolutionary questions, like correlated evolution, asymmetries in substitution rates, etc. (refer to Huelsenbeck et al., 2003; Bollbeck, 2006 for further details).

In conjunction with stochastic character mapping, Huelsenbeck et al. (2003) proposed a framework for hypothesis testing, termed posterior predictive distributions. It represents a Bayesian analogue of parametric bootstrapping which is a much better known approach in phylogenetics and statistics. As in the parametric bootstrap, data sets are simulated from the posterior probability distribution sampled from the original data via MCMC. Each data set is then treated in the same way as was the original data set and the results are compared to the results obtained from the original data by some test statistic. The authors of the method applied posterior predictive distributions to test for correlation between characters by calculating the fraction of time the characters are associated on the tree from a sample of mappings. The logic behind this is that co-varying characters are assumed to spend roughly the same amount of time in the associated state, while in non-co-evolving traits the time, the characters are associated, is different.

EXAMPLES OF APPLICATIONS OF COMPARATIVE METHODS TO FUNGAL EVOLUTIONARY BIOLOGY

Since a decade, the surge in the development of phylogenetic comparative approaches has inspired several mycologists to apply these methods to questions about fungal evolution. Parsimony mapping and reconstruction methods, were the first available in popular software (such as MacClade), have now become routine techniques to examine how a trait evolved along a phylogenetic tree. One example of parsimony mapping is the inference of the number of transformations from non-gilled to gilled fruitbodies (≈agaricoid, see below) in the Agaricomycotina (Hibbett et al., 1997). Although sampling of taxa was rather restricted, polyphyletic origin of agaricoid fruiting bodies has been demonstrated with agaricoid mushroom species having arisen at least six times.

With increasing notice of drawbacks of parsimony methods, and the invent of probabilistic inference (first, Maximum Likelihood), several researchers switched over to inferring ancestral states under probabilistic models (Hibbett et al., 2000; Hibbett and Donoghue, 2001; Lutzoni et al., 2001; Hibbett and Binder, 2002; Hibbett, 2004; Miadlikowska and Lutzoni, 2004; Binder and Hibbett, 2006). Bayesian framework was first applied to fungi by Lutzoni et al. (2001) using a set of trees sampled from Bayesian MCMC methods for phylogeny reconstruction which is often referred to as the empirical Bayesian inference because some of the parameters are fixed at their Maximum Likelihood values (Huelsenbeck and Bollback, 2001). Fully Bayesian statistics have been developed more recently (Nielsen, 2002; Huelsenbeck et al., 2003; Pagel et al., 2004) and, hence only very few fungal examples are available.

In fungi, comparative methods were mainly applied to "macrofungi", i.e. mainly larger Ascomycota and Basidiomycota (Hibbett et al., 2000; Hibbett and Donoghue, 2001, Humpert et al., 2001; Lutzoni et al., 2001; Hibbett and Binder, 2002; Hibbett, 2004; Aanen and Eggleton, 2005; Hosaka et al., 2006; Binder and Hibbett, 2006; Hibbett, 2007; Ekman et al., 2008) but there are examples from other fungal groups also (Diezmann et al., 2004; James et al., 2006). The above instances on macrofungi pertain mainly to the fruiting body evolution and the evolution of nutritional modes of mushrooms, although lichenization has also been studied (Gargas et al., 1995; Lutzoni et al., 2001; Miadlikowska and Lutzoni, 2004; Blanco et al., 2005; Schmitt et al., 2005; Gueidan et al., 2007). In the following pages an attempt has been made to summarize the progress in these fields along with notes on methodological issues.

EVOLUTION OF NUTRITIONAL MODES IN THE BASIDIOMYCETES

Basidiomycete fungi are known to have a wealth of different lifestyles, including wood- and leaf-litter decaying saprotrophs, ectomycorrhizal species forming mutualistic associations with plants, parasites of plants, animals and fungi as well as ant-associated taxa. Hence, it is not surprising that the process that gave rise to this astonishing diversity and the question of what lifestyle the ancestors of certain groups occupied. This became a central question of fungal evolutionary biology because there is increasing interest in research in horticultural and conservational applications of ectomycorrhizae. Also as knowledge of the evolution of ectomycorrhizal association is of primary importance. Whether mycorrhizal association represents a monophyletic grade or not, was among the earliest questions addressed in fungal evolution. The trees inferred by Hibbett et al. (1997), Moncalvo et al. (2000) and Binder and Hibbett (2002) have evidenced non-monophyly of the ectomycorrhizal life form. Early papers reported that the mycorrhizal mode of lifestyle is widespread among genera of mushroom-forming fungi as well as of Ascomycetes, and that multiple gains and reversals that happened across the Agaricomycetes in the course of evolution (Hibbett et al., 2000). Hibbett et al. (2000) used parsimony optimization and Maximum Likelihood ancestral state reconstruction on equally most parsimonious trees inferred from three genes. Ancestors of the Agaricales, Boletales and the Thelephorales were inferred as having been mycorrhizal, whereas the ancestor of Agaricomycetes (=Homobasidiomycetes) was reconstructed as non-mycorrhizal. Multiple switches from and to mycorrhizal associations were inferred which entails instability of nutritional modes in mushroom-forming fungi. On a smaller scale, but with considerably denser sampling of taxa, this question was further investigated in the Boletales (Binder and Hibbett, 2006). The authors used a Bayesian sample of trees to factor out phylogenetic uncertainty and applied the MRCA approach to include trees not having the particular node. Eight nodes were reconstructed corresponding to the suborders and families Tapinellineae, Coniophorineae, Serpulaceae, Hygrophoropsidaceae, Suillineae, Sclerodermatineae, Boletineae, as well as the ancestor of the Boletales. In contradiction with the former results (Hibbett et al., 2000), the ancestor of Boletales was inferred as a brown rot species with high probability, although alternative hypotheses of mycoparasitism of mycorrhizae have not been tested. Similarly, brown rot was inferred as the ancestral state for the Tapinellineae, Coniophorineae, Serpulaceae and Hygrophoropsidaceae. This gives rise to the hypothesis that mycorrhizal lifestyle was gained on the branch leading to the common ancestor of Suillinae, Paxillinae and Boletinae.

The contradiction with former results underlines the importance of taxon sampling for comparative studies. Generally speaking, the preferred method is to include as many species as possible, in order to represent true distribution of the particular trait as closely as possible (Salisbury and Kim, 2001), but priority should always be given to the representation of all lineages in the group. Counter examples do exist, as shown by Li et al. (2008). Although their conclusions are largely consistent with a former study examining the effects of taxon sampling density (Salisbury and Kim, 2001), i.e. reconstruction accuracy first decreases, then increases with addition of taxa to the phylogeny, but it should be kept in mind that tree shape and structure can have profound effects on accuracy. Special care should be given to cases where the tree is imbalanced, because, in these cases a single basal species on a short branch may provide more important signals, than using hundreds of taxa on otherwise well-sampled parts of the tree. However, they concluded that the general trend is that increased taxon sampling increases reconstruction accuracy (Li et al., 2008).

Besides Basidiomycota, it would be interesting to know the ancestral nutritional strategy of even earlier ancestors of fungi. The question has been addressed rather recently (James et al., 2006). The authors reported ambiguous reconstructions on most nodes of the phylogeny, although more details and the test of significance used are not given. The only ancestor which was reported to be reconstructed unequivocally is the MRCA of Basidiomycota which was suggested to be parasitic. Unfortunately, no inferred state is reported for the common ancestor of the informal taxon "Symbiomycota" which is supposed to contain all mycorrhiza forming fungal taxa.

A strikingly unique life strategy in Agaricomycetes is presented by termite and ant associated fungi (Rouland-Lefevre et al., 2002; Froslev et al., 2003; Aanen and Eggleton, 2005). Employing a Maximum Likelihood and a hierarchical Bayesian framework, Aanen and Eggleton (2005) inferred the ancestral geographic range of the fungus-termite association. Phylogenetic uncertainty was taken into account by performing the analyses on 28,000 trees sampled during a Bayesian MCMC analysis. Although these taxa predominantly inhabit savanna regions today, the results strongly supported that fungus-growing termites emerged in tropical rain forests of Africa, and multiple dispersals to the savanna occurred.

RESOLVING THE HISTORY OF DIVERSITY OF FRUITING BODY MORPHOLOGIES IN BASIDIOMYCOTA

Fruiting body evolution is a crucial question as it has the potential to resolve a number of issues about classification, phylogeny and, hence conservation. Making exact inferences about fruiting body evolution is complicated by the fact that the fossil record of fungal fruiting bodies is very poor, represented only by a few examples (Poinar and Singer, 1990; Hibbett et al., 1997). Therefore, computational methods are the only tools to recover evolutionary patterns of fruiting body development. Since scientists started to work with mushrooms, the origin of the perplexing diversity of fruiting body morphologies has been a central question in research. The invention of comparative phylogenetic methods enabled much deeper insights into the evolution of fruiting bodies than it was possible before.

The origin of fruiting body formation has been addressed only recently in the Ascomycota. Two independent gains of fruiting body formation have been proposed, one in the Pezizomycotina, and the other in the Neolectomycetes (Schoch et al., 2009). Within the former, the organization of reproductive structures shows great diversity across lineages. These include a range of fairly simple forms, such as cleistothecia, or perithecia, but also more complex forms with more structurally specialized tissues, such as apothecia or stromata. Another, probably biologically more realistic grouping classifies fruiting bodies as open (apothecia), partially enclosed (perithecia and stromata thereof) and enclosed (cleisthothecia). Exposed (open) fruiting bodies have been claimed as ancestral in the ascomycetes, with several independent origins of both the perithecia (e.g. Dothideomycetes, Sordariomycetes) and cleistothecia (e.g. Eurotiomycetes, Erysiphales). This is consistent with one of the oldest hypotheses of fruiting body evolution by John-Axel Nannfeldt, dating back to 1932.

Different types of basidiomycete fruiting bodies include the well-known agaricoid type, i.e. those of the best known agarics with cap, stipe and lamellae, boletoid (with tubes instead of lamellae), cantharelloid, phalloid (that of stinkhorns), clavarioid (coral fungi), gasteroid (puffballs) or resupinate. The resupinate type, being widespread in all of the larger clades of Basidiomycota, consists of a membranous, cortical "layer" on mainly woody substrates. It is often considered a rather simple form of fruiting bodies, being composed of some layers of hyphae attached to the wood (subiculum), basidia, and often certain types of spacer cells (cystidia). As judged from their simple structure, resupinate fruiting bodies have either been interpreted as reduced forms, or ancestral morphologies. These hypotheses have recently been tested by Maximum Parsimony and Maximum Likelihood approaches (Hibbett and Binder, 2002; Hibbett, 2004; Binder and Hibbett, 2006). The results suggested that the ancestral condition of fruiting body morphologies in the Agaricomycetes (=Homobasidiomycetes) and a number of other larger clades (e.g. Polyporales, Boletales) was resupinate. Further, Hibbett (2004) inferred that the best-fit model to fruiting body evolution on their sample implies an asymmetry in the rates from resupinate to non-resupinate with significant support [$2\Delta\ln(L_0-L_1)>2.00$]. This is partially in contradiction with results of parsimony mapping which suggested that the number of transformations inferred on an equally weighted MP tree was roughly the same in both directions (Hibbett and Binder, 2002). The authors suggested that this difference may be due to the property of parsimony that uses an implicit model of evolution that assumes the two rates to be equal (Hibbett and Binder, 2002). On the other hand, because the rate of change from non-resupinate to resupinate fruiting bodies is larger than zero, and parsimony analyses suggested several reversals (on average 20-30, dependent upon the tree used). It seems that resupinate fruiting bodies can represent a derived condition too, emerging by reduction of more complex forms (e.g., agaricoid). Looking at trees from several studies, where resupinate taxa are nested in large agaricoid clades (e.g., Moncalvo et al., 2002; Binder and Hibbett, 2006; Larsson, 2007), often makes one anticipate such trends, although before making such conclusions it should be kept in mind that on these trees resupinate taxa may represent undersampled sister groups (Crisp and Cook, 2004). Therefore, ancestral states cannot be predicted this way, without statistical testing. Similarly, it is intuitively likely that cyphelloid taxa which are simple, pendant cup-like fungi, represent a grade of simplification, being derived from agaricoid ancestors, but this has never been tested statistically.

Besides resupinate fruiting bodies, gasteroid, and puffball-like fungi (which have their spore-bearing surface enclosed inside the fruiting body) also comprise a great challenge to researchers. Hypotheses about the development of gasteroid forms were published in the early 20th century (Lohwag, 1925). Early molecular phylogenies suggested that gasteroid fungi do not form a monophyletic group (Bruns et al., 1989; Hibbett et al., 1997; Moncalvo et al., 2002; Binder and Bresinsky, 2002; Binder and Hibbett, 2002), providing evidence for placement of these taxa at various levels in various clades of Agaricomycetes. Examples include single-species or small clades of gasteroid fungi nested within large agaricoid clades (e.g. *Thaxterogaster, Quadrispora, Protoglossum, Hymenogaster* or *Setchelliogaster* in the Cortinariaceae), as well as large, well-established monophyletic groups forming big clades (e.g. Lycoperdaceae or clades of the Phallales). These findings prompted a number of studies to further evaluate the position of specific groups of gastromycetous taxa (Miller et al. 2001; Peintner et al., 2001; Miller and Buyck 2002; Vellinga, 2004; Geml et al., 2005; Hosaka et al., 2006). All of these studies point towards the hypothesis that gasteroid fungi evolved from agaricoid ancestors (gastromycetation) (Thiers 1984; Bruns et al., 1989). As it has been suggested that gastromycetation is an unidirectional process (Thiers, 1984), Hibbett (2004) addressed the irreversibility of gastromycetation by Maximum Likelihood analyses. Unrestricted models of evolution suggested that the rate of change from gasteroid to non-gasteroid fruiting bodies is greater than zero, allowing for reversals of gastromycetation, but this model did not fit the data significantly better than the restricted model in which this rate was constrained to zero. It is likely that this result can largely be attributed to the topology of the gomphoid-phalloid clade. Considering the special anatomy of gasteroid gomphoid-phalloid fungi, it would not be surprising if future studies would show that these differ markedly from "normal" gasteroid fruiting bodies. Further, it may be possible in the future, that studies of gastromycetation, the gomphoid-phalloid clade would be considered separately.

Although many aspects about phylogenetic distribution and reversibility of gastromycetation have been revealed recently, it seems that this is a very complex process requiring further attention. For instance, the causes and the physiological changes accompanying gastromycetation are poorly understood. It seems that there are several species that can serve as a model to study gastromycetation in more detail (Desjardin, 2003; Miller, 2004).

Besides gasteroid species, another spectacular type of fruiting bodies is the coprinoid fruiting body found in some dark-spored groups, such as the Psathyrellaceae, Agaricaceae or Bolbitiaceae. These fungi have short-lived, ephemeral fruiting bodies, which produce chitinases and glucanases to digest themselves (hence the name "inky caps") into a black ink upon maturation (Kües, 2000). For a long time, most species have been known under the generic name *Coprinus*, which included all deliquescent (autodigesting) taxa with black spore print. A significant number of the species prefer rather exposed habitats, such as dung, vegetable refuse, or logs. Interestingly, the number of coprinoid species occupying forests and wet habitats is lower than in closely related lineages of *Psathyrella* species. Taxonomy and classification of this group have been in a major flux with several issues remaining to be resolved. The old *Coprinus sensu lato* genus has been split into four smaller monophyletic genera (Redhead et al., 2001). Molecular phylogenies suggest that these smaller genera represent extensive parallelism or convergence by independently gaining deliquescence (Padamsee et al., 2008; Nagy et al., 2009). Bayesian and Maximum Likelihood reconstruction of ancestral fruiting body types at the major nodes of Psathyrellaceae revealed that there have been 4-6 independent acquisitions of deliquescence in the Psathyrellaceae (Nagy et al., 2010). Losses of deliquescence proved unlikely, as all models sampled during Bayesian RJ-MCMC runs, as well as model testing in ML suggested that the rate of fruiting body evolution is in the reverse direction (i.e. loss of autodigestion) approaches zero. That is, once deliquescence is gained, it is probably never lost, which suggests some major adaptive benefit for deliquescent mushrooms over their non-deliquescent relatives. Anatomy of basidia, hymenial cystidia, basidioles and pileus structure were found to be strongly correlated with changes in fruiting body types. Both hymenial cystida and basidioles were found to have undergone significant increase in volume. A compound model of correlated evolution (Pagel and Meade, 2006) suggested that these four traits and fruiting body types evolved in a correlated fashion. In addition, the evolutionary pathway tests designed by Nagy et al. (2010) inferred very short waiting times between successive changes in the correlated characters. Hence, it is likely that characters associated with fruiting body evolution emerged as a result of an adaptive burst within a very short time interval in the Psathyrellaceae. The name coprinoidization have been proposed for the complex procedure of the emergence of deliquescent fruiting bodies. The authors have also tried to decipher the timing and cause of changes from non-deliquescent to deliquescent fruiting bodies in the Psathyrellaceae, which would allow for speculations about the adaptive benefit of deliquescence. Molecular clock analyses rejected simultaneous gains of deliquescence in the four main lineages (individual acquisitions ranging from 226 to 72 million years ago), thereby excluding the possibility of major geologic event as triggers of changes in fruiting body types. Instead, the authors suggested that fruiting body evolution in the Psathyrellaceae (and probably in general) may be driven by accidental events such as dispersal to new niches or the development of key innovations. Fruiting bodies of coprinoid fungi are widely known for having accelerated ontogeny, chiefly resulting from intense cell inflation and elongation synchronized with spore maturation (Kües, 2000). We hypothesize that the increased cell size have actually been the prerequisite for the fast ontogeny of these mushrooms, which in combination with the observed habitat preferences of coprinoid fungi suggests that it may primarily be desiccation which deliquescent fruiting bodies effectively overcome. This is consistent with the general observation that stipe elongation and cap expansion takes place during the humid night/early morning hours and spore discharge happens in the morning. When spores have been released from the gills, the entire fruiting body loses its turgor and starts to deliquesce and/or collapse. Therefore, smaller species are usually hardly seen in the field, except early in the morning.

Although the above hypothesis fits well with the ecological and ontogenetic observations, it does not explain the autolysis and chitinase production itself. Spore dispersal might then be achieved through the black inky droplets to which the deliquescing pileus is transformed. However, it is well-known that spores inhibit each other's germination in masses by exuding fungistatic substances, so spore dispersal in the form of droplets is unlikely (Bristov and Lockwood, 1975).

EVOLUTION OF LICHEN SYMBIOSIS

Lichenization is one of the most challenging and throughly studied physiological phenomena in fungi. Lichen forming fungi can be found both in the Basidiomycota and the Ascomycota, the latter contributing the vast majority of species. Lichen forming fungi evidently do not form a monophyletic group (Grube and Hawksworth, 2007). Even within Ascomycota, lichen association is not a monophyletic grade of development (Gargas et al., 1995; Lutzoni et al., 2001; Schoch et al., 2009). It is uncertain if it should be regarded as poly- or paraphyletic. Therefore, a basic question that requires clarification is how many times and when the lichen symbiosis evolved and was lost during the evolution. Based on one of the first molecular phylogenies of fungi, Gargas et al. (1995) performed parsimony mapping on a MP tree reconstructed from a very restricted sample of lichen-forming and non-lichen-forming fungal basidio- and ascomycete taxa. The inferred tree evidenced non-monophyly of lichen-forming fungi, and made them conclude that at least two gains of lichen formation in the Ascomycota and three in the Basidiomycota. Later, however, contradicting results were published by Lutzoni et al. (2001), suggesting that lichen formation has been ancestral in the Ascomycota and that losses of lichen formation happened multiple times during evolution, although most basal nodes were not reconstructed as having been lichen-forming with significant support. The results implied that some widespread and well-known groups, such as the Eurotiales (e.g. *Aspergillus* spp.) evolved from lichenized ancestors. The Ascomycetes, is the most species-rich group of fungi, with several controversial lineages, which necessitates careful sampling of taxa for such studies. Recently, the question of lichenization has been re-evaluated on the basis of a molecular phylogeny comprising more than 400 ascomycete taxa. Ancestral state reconstructions performed on a bootstrap tree sample suggested that lichenized lineages evolved multiple times from within non-lichenized ancestors. At least five independent origins of lichenization have been inferred. However, the ancestor of the Ascomycota is unequivocally reconstructed as non-lichenized, which excludes the possibility of lichenization as the ancestral condition (Lutzoni et al., 2001).

CONCLUSIONS AND FUTURE PERSPECTIVES

The ability to draw significant conclusions about evolutionary events and processes of fungi has long been hampered by the lack of proper fossil record and statistical approaches. Today, however, a wealth of computational methods, including parsimony, Maximum Likelihood and Bayesian methods, are available that can help clarify such questions. Although uncertainties often remain, these methods have been successfully applied in a number of cases where molecular phylogenies were available, such as fruiting body evolution of higher fungi, evolution of nutritional modes, lichenization, and evolution of different morphological traits and so on. The potential applications of these computational approaches and other tools are practically endless. Not mentioned in this chapter are examples and methods for reconstructing ancestral gene or protein sequences and studying the evolution of continuous characters or correlation between discrete or continuous traits.

Several studies showed that Bayesian methods are superior to both maximum parsimony and Maximum Likelihood in many respects, a fact that could be exploited by revisiting former data sets where results were not supported enough by the data. For instance, inferring models of fruiting body evolution proved difficult for multistate analyses, an example where Reversible-Jump MCMC methods should perform better for identifying the best-fit model. Since statistical and computational methods are continuously getting improved, the number of questions that need to be answered or tested is also increasing continuously. This gives a great chance to mycologists to find answers to complex questions using available comparative phylogenetic methods.

ACKNOWLEDGEMENTS

The authors are grateful to Gábor M. Kovács for critical comments for the improvement of the manuscript. The project was supported by a bilateral grant between HSRF (OTKA-NN75255) and DFG (No. Vo 772/9-1).

REFERENCES

Aanen, D. K. and Eggleton, P. 2005. Fungus-growing termites originated in African rain forest. Curr. Biol., 15: 851-855.

Agnarsson, I. and Miller, J. A. 2008. Is ACCTRAN better than DELTRAN. Cladistics, 24: 1-7.

Altekar, G., Dwarkadas, S., Huelsenbeck, J. P. and Ronquist, F. 2004. Parallel Metropolis Coupled Markov Chain Monte Carlo for Bayesian Phylogenetic Inference. Bioinformatics, 20: 407-412.

Binder, M. and Bresinsky, A. 2002. Derivation of a polymorphic lineage of gasteromycetes from boletoid ancestors. Mycologia, 94: 85-98.

Binder, M. and Hibbett, D. H. 2006. Molecular systematics and biological diversification of Boletales. Mycologia, 98: 971-981.

Binder, M. and Hibbett, D. S. 2002. Higher-level phylogenetic relationships of homobasidiomycetes (mushroom-forming fungi) inferred from four rDNA regions. Mol. Phyl. Evol., 22: 76-90.

Blanco, O., Crespo, A., Divakar, P. K., Elix, J. A. and Lumbsch, H. T. 2005. Molecular phylogeny of parmotremoid lichens (Ascomycota, Parmeliaceae). Mycologia, 97: 150-159.

Bodensteiner, P., Binder, M., Moncalvo, J-M., Agerer, R. and Hibbett, D. S. 2004. Phylogenetic relationships of cyphelloid homobasidiomycetes. Mol. Phyl. Evol., 33: 501-515.

Bollback, J. P. 2006. SIMMAP: Stochastic character mapping of discrete traits on phylogenies. BMC Bioinformatics, 7: 88.

Bollback, J. P., Gardner, P. P. and Nielsen, R. 2007. Estimating the history of mutations on a phylogeny. In: Ancestral Sequence Reconstruction. D. Liberles (ed.). Oxford University Press, UK.

Bristov, B. L. and Lockwood, J. L. 1975. Soil Fungistasis: Role of Spore Exudates in the Inhibition of Nutrient-independent Propagules. J. Gen. Microbiol., 90: 140-146.

Britto, M. R. 2003. Phylogeny of the subfamily Corydoradinae Hoedeman, 1952 (Siluriformes: Callichthyidae), with a definition of its genera. Proc. Acad. Nat. Sci. Philadelphia, 153: 119-154.

Bruns, T. D., Fogel, R., White, T. J. and Palmer, J. D. 1989. Accelerated evolution of a false-truffle from a mushroom ancestor. Nature, 339: 140-142.

Crisp, M. D. and Cook, L. G. 2004. Do early branching lineages signify ancestral traits? Trends in Ecol. Evol., 20: 122-128.

Cunningham, C. W. 1999. Some limitations of ancestral character-state reconstruction when testing evolutionary hypotheses. Syst. Biol., 48: 665-674.

Cunningham, C. W., Omland, K. E. and Oackley, T. H. 1998. Reconstructing ancestral character states: a critical reappraisal. Trends in Ecol. Evol., 13: 361-366.

Desjardin, D. E. 2003. A unique ballistosporic hypogeous sequestrate *Lactarius* from California. Mycologia, 95: 148-155.

Diezmann, S., Cox, C. J., Schönian, G., Vilgalys, R. and Mitchell, T. G. 2004. Phylogeny and Evolution of Medical Species of Candida and Related Taxa: A Multigenic Analysis. J. Clin. Microbiol., 42: 5624-5635.

Douady, C. J., Delsuc, F., Boucher, Y., Doolittle, W. F. and Douzery, E. J. P. 2003. Comparison of Bayesian and Maximum Likelihood bootstrap measures of phylogenetic reliability. Mol. Biol. Evol., 20: 248-254.

Edwards, A. W. F. 1972. Likelihood. Cambridge University Press, Cambridge, U.K.

Ekman, S., Andersen, H. L. and Wedin, M. 2008. The limitations of ancestral state reconstruction and the evolution of the ascus in the Lecanorales (Lichenized Ascomycota). Syst. Biol., 57: 141-156.

Felsenstein, J. 2004. Inferring Phylogenies, Sinauer Assoc., pp. xx + 664.

Frøslev, T. G., Aanen, D. K., Læssoe, T. and Rosendahl, S. 2003. Phylogenetic relationships of *Termitomyces* and related taxa. Mycol. Res., 107: 1277-1286.

Gargas, A., DePriest, T. P., Grube, M. and Tehler, A. 1995. Multiple origins of lichen symbiosis in fungi suggested by SSU rDNA phylogeny. Science, 268: 1492-1495.

Gascuel, O. 2005. Mathematics of Evolution and Phylogeny. Oxford University Press, Oxford, UK.

Gaubert, P., Wozencraft, W. C., Cordeiro-Estrela, P. and Veron, G. 2005. Mosaics of convergences and noise in morphological phylogenies: what's in a viverrid-like carnivoran? Syst. Biol., 54: 865-894.

Geml, J., Davis, D. D. and Geiser, D. M. 2005. Phylogenetic analyses reveal deeply divergent lineages in the genus *Sphaerobolus* (Phallales, Basidiomycota). Mol. Phylogenet. Evol., 35: 313-322.

Grube, M. and Hawksworth, D. L. 2007. Trouble with lichen: the re-evaluation and re-interpretation of thallus form and fruit body types in the molecular era. Mycol. Res., 111: 1116-1132.

Gueidan, C., Roux, C. and Lutzoni, F. 2007. Using a multigene phylogenetic analysis to assess generic delineation and character evolution in Verrucariaceae (Verrucariales, Ascomycota). Mycol. Res., 111: 1145-1168.

Hall, G. B. 2006. Simple and accurate estimation of ancestral protein sequences. PNAS, 103: 5431-5436.

Harvey, P. H. and Pagel, M. 1991. The Comparative Method in Evolutionary Biology (Ecology and Evolution). Oxford University Press, U.K.

Hibbett, D. S. 2004. Trends in morphological evolution in Homobasidiomycetes inferred using Maximum Likelihood: a comparison of binary and multistate approaches. Syst. Biol., 53: 889-903.

Hibbett, D. S. 2006. A phylogenetic overview of the Agaricomycotina. Mycologia, 98: 917-925.

Hibbett, D. S. 2007. After the gold rush, or before the flood? Evolutionary morphology of mushroom-forming fungi (Agaricomycetes) in the early 21st century. Mycol. Res., 111: 1001-1018.

Hibbett, D. S. and Binder, M. 2002. Evolution of complex fruiting-body morphologies in homobasidiomycetes. Proc. R. Soc. Lond. B, 269: 1963-1969.

Hibbett, D. S. and Donoghue, M. J. 2001. Analysis of character correlations among wood decay mechanisms, mating systems, and substrate ranges in Homobasidiomycetes. Syst. Biol., 50: 215-242.

Hibbett, D. S., Gilbert, L. B. and Donoghue, M. J. 2000. Evolutionary instability of ectomycorrhizal symbioses in basidiomycetes. Nature, 407: 506-508.

Hibbett, D. S., Pine, E. M., Langer, E., Langer, G. and Donoghue, M. J. 1997. Evolution of gilled mushrooms and puffballs inferred from ribosomal DNA sequences. PNAS, 94: 12002-12006.

Hosaka, K., Bates, S. T., Beever, R. E., Castellano, M. A., Colgan, W., Domínguez, L. S., Nouhra, E. R., Geml, J., Giachini, A.J., Kenney, S. R., Simpson, N. B., Spatafora, J. W. and Trappe, J. M. 2006. Molecular phylogenetics of the gomphoid-phalloid fungi with an establishment of the new subclass Phallomycetidae and two new orders. Mycologia, 98: 949-959.

Huelsenbeck, J. H. and Bollback, J. P. 2001. Empirical and hierarchical Bayesian estimation of ancestral states. Syst. Biol., 50: 351-366.

Huelsenbeck, J. P. and Crandall, K. A. 1997. Phylogeny estimation and hypothesis testing using Maximum Likelihood. Annu. Rev. Ecol. Syst., 28: 437-466.

Huelsenbeck, J. P., Larget, B. and Alfaro, M. 2004. Bayesian phylogenetic model selection using reversible jump Markov chain Monte Carlo. Mol. Biol. Evol., 21: 1123-1133.

Huelsenbeck, J. P., Larget, B., Miller, R. E. and Ronquist, F. 2002. Potential applications and pitfalls of Bayesian inference of phylogeny. Syst. Biol., 51: 673-688.

Huelsenbeck, J. H., Nielsen, R. and Bollback, J. P. 2003. Stochastic mapping of morphological characters. Syst. Biol., 52: 131-158.

Huelsenbeck, J. P., Rannala, B. and Masly, J. P. 2000. Accommodating phylogenetic uncertainty in evolutionary studies. Science, 288: 2349-2350.

Huelsenbeck, J. P., Ronquist, F., Nielsen, R. and Bollback, J.P. 2001. Bayesian inference of phylogeny and its impact on evolutionary biology. Science, 294: 2310-2314.

Humpert, A. J., Muench, E. L., Giachini, A. J., Castellano, M. A. and Spatafora, J. W. 2001. Molecular phylogenetics of Ramaria and related genera: evidence from nuclear large subunit and mitochondrial small subunit rDNA sequences. Mycologia, 93: 465-477.

James, T. Y., Kauff, F., Schoch, C. L., Matheny, P. B., Hofstetter, V., Cox, C. J., Celio, G., Gueidan, C., Fraker, E., Miadlikowska, J., Lumbsch, T. H., Rauhut, A., Reeb, A., Arnold, A. E., Amtoft, A., Stajich, J. E., Hosaka, K., Sung, G. H., Johnson, D., O'Rourke, B., Crockett, M., Binder, M., Curtis, J. M., Slot, J. C., Wang, Z., Wilson, A. Z., Schüßler, A., Longcore, J. E., O'Donnell, K., Mozley-Standridge, S., Porter, D., Letcher, P. M., Powell, M. J., Taylor, J. V., White, M. M., Griffith, G. W., Davies, D. R., Humber, R. A., Morton, J. B., Sugiyama, J., Rossman, A. Y., Rogers, J. D., Pfister, D. H., Hewitt, D., Hansen, D., Hambleton, S., Shoemaker, R. A., Kohlmeyer, A., Volkmann-Kohlmeyer, B., Spotts, R. A., Serdani, M., Crous, P. W., Hughes, K. W., Matsuura,

K., Langer, E., Langer, G., Untereiner, W. A., Lücking, R., Büdel, B., Geiser, D. M., Aptroot, A., Diederich, P., Schmitt, I., Schultz, M., Yahr, R., Hibbett, D. S., Lutzoni, F., McLaughlin, D. J., Spatafora, J. W. and Vilgalys, R. 2006. Reconstructing the early evolution of fungi using a six-gene phylogeny. Nature, 443: 818-822.

Kass, R. E. and Raftery, A. E. 1995. Bayes factors. J. Amer. Stat. Assoc., 90: 773-795.

Koshi, J. M. and Goldstein, R. A. 1996. Probabilistic reconstruction of ancestral protein sequences. J. Mol. Evol., 42: 313-320.

Kües, U. 2000. Life history and developmental processes in the basidiomycete *Coprinus cinereus*. Microb. Mol. Biol. Rev., 64: 316-353.

Larget, B. and Simon, D. 1999. Markov chain Monte Carlo algorithms for the Bayesian analysis of phylogenetic trees. Mol. Biol. Evol., 16: 750-759.

Larsson, K-H. 2007. Re-thinking the classification of corticioid fungi. Mycol. Res., 111: 1040-1063.

Lewis, P. O. 2001. A likelihood approach to estimating phylogeny from discrete morphological character data. Syst. Biol., 50: 913-925.

Li, G., Steel, M. and Zhang, L. 2008. More Taxa Are Not Necessarily Better for the Reconstruction of Ancestral Character States. Syst. Biol., 57: 647-653.

Lohwag, H. 1925. Zur Entwicklungsgeschichte und Morphologie der Gastromyceten. Ein Beitrag zur Systematik der Basidiomyceten. Beih. Bot. Centralbl. Abt. II, 42: 177-334.

Lunter, G., Miklós, I., Drummond, A., Jensen, J. L. and Hein, J. 2005. Bayesian coestimation of phylogeny and sequence alignment. BMC Bioinformatics, 6: 83.

Lutzoni, F., Pagel, M. and Reeb, V. 2001. Major fungal lineages are derived from lichen symbiotic ancestors. Nature, 411: 937-940.

Maddison, W. P. 1995. Calculating the probability distributions of ancestral states reconstructed by parsimony on phylogenetic trees. Syst. Biol., 44: 474-481.

Maddison, W. R. and Maddison, D. R. 2007. Mesquite: a modular system for evolutionary analysis. Version 2.5. http://mesquiteproject.org; http://mesquiteproject.org

Matheny, P. B., Curtis, J. M., Hofstetter, V., Aime, M. C., Moncalvo, J. M., Ge, Z. W., Slot, J. C., Ammirati, J. F., Baroni, T. J., Bougher, N. L., Hughes, K. W., Lodge, D. J., Kerrigan, R. W., Seidl, M. T., Aanen, D. K., DeNitis, M., Daniele, G. M., Desjardin, D. E., Kropp, B. R., Norvell, L. L., Parker, A., Vellinga, E. C., Vilgalys, R. and Hibbett, D. S. 2006. Major clades of Agaricales: a multilocus phylogenetic overview. Mycologia, 98: 982-995.

Mau, B. and Newton, M. 1997. Phylogenetic inference for binary data on dendrograms using Markov chain Monte Carlo. J. Comp. Graph. Stat., 6: 122-131.

Miadlikowska, J. and Lutzoni, F. 2004. Phylogenetic classification of peltigeralean fungi (Peltigerales, Ascomycota) based on ribosomal small and large subunits. Am. J. Bot., 91: 449-464.

Miller, S. L. 2004. Hypogeous fungi from the southeastern United States 3. The genus *Macowanites*. Mycotaxon, 89: 283-288.

Miller, S. L. and Buyck, B. 2002. Molecular phylogeny of the genus Russula in Europe with a comparison of modern infrageneric classifications. Mycol. Res., 106: 259-276.

Miller, S. L., McClean, T. M., Walker, J. and Buyck, B. 2001. A molecular phylogeny of the Russulaceae including agaricoid, gasteroid and pleurotoid taxa. Mycologia, 93: 344-354.

Moncalvo, J-M., Lutzoni, F., Rehner, S. A., Johnson, J. and Vilgalys, R. 2000. Phylogenetic Relationships of Agaric Fungi Based on Nuclear Large Subunit Ribosomal DNA Sequences. Syst. Biol., 49: 278-305.

Moncalvo, J-M., Vilgalys, R., Redhead, S. A., Johnson, J. E., James, T. Y., Aime, M. C., Hofstetter, V., Verduin, S. J. W., Larsson, E., Baroni, T. J., Thorn, G., Jacobsson, S., Clémencon, H. and Miller, O. K. 2002. One hundred and seventeen clades of euagarics. Mol. Phyl. Evol., 23: 357-400.

Moore, G. W. and Donoghue, M. J. 2007. Correlates of diversification in the plant clade Dipsacales: geographic movement and evolutionary innovations. Am. Nat., 170: S28-S55

Mooers, A. O. and Schluter, D. 1999. Reconstructing ancestor states with Maximum Likelihood: support for one- and two-rate models. Syst. Biol., 48: 623-633.

Nagy, G. L., Walther, G., Házi, J., Vágvölgyi, Cs. and Papp, T. 2010. Understanding evolutionary processes of fungal fruiting bodies: Correlated evolution and divergence times in the Psathyrellaceae. Syst. Biol. (in press).

Nagy, L., Kocsubé, S., Papp, T. and Vágvölgyi, Cs. 2009. Phylogeny and character evolution of the coprinoid mushroom genus *Parasola* as inferred from LSU and ITS data. Persoonia, 22: 28-37.

Nielsen, R. 2002. Mapping Mutations on Phylogenies. Syst. Biol., 51(5): 729-739.

Omland, K. E. 1999. The assumptions and challenges of ancestral state reconstructions. Syst. Biol., 48: 604-611.

Padamsee, M., Matheny, P. B., Dentinger, B. T. and McLaughlin, D. J. 2008. The mushroom family Psathyrellaceae: Evidence for large-scale polyphyly of the genus *Psathyrella*. Mol. Phylogenet. Evol., 46: 415-429.

Pagel, M. 1994. Detecting correlated evolution on phylogenies: A general method for the comparative sanalysis of discrete characters. Proc. Roy. Soc. Lond. B, 255: 37-45.

Pagel, M. 1999a. The Maximum Likelihood approach to reconstructing ancestral character states of discrete characters on phylogenies. Syst. Biol., 48: 612-622.

Pagel, M. 1999b. Inferring the historical patterns of biological evolution. Nature, 401: 877-884.

Pagel, M. and Lutzoni, F. 2002. Accounting for phylogenetic uncertainty in comparative studies of evolution and adaptation, pp. 148-161. In: Biological Evolution and Statistical Physics. M. Lassig and A. Valleriani (eds.). Springer, Berlin.

Pagel, M. and Meade, A. 2006. Bayesian analysis of correlated evolution of discrete characters by reversible-jump Markov chain Monte Carlo. Am. Nat., 167: 808-825.

Pagel, M. and Meade, A. 2007. Bayestraits 1.0 Manual. University of Reading, UK.

Pagel, M., Meade, A. and Barker, D. 2004. Bayesian estimation of ancestral character states on phylogenies. Syst. Biol., 53: 673-684.

Peintner, U., Bougher, N. L., Castellano, M. A., Moncalvo, J-M., Moser, M. M., Trappe, J. M. and Vilgalys, R. 2001. Multiple origins of sequestrate fungi related to *Cortinarius* (Cortinariaceae). Am. J. Bot., 88: 2168-2179.

Poinar, Jr. G. O. and Singer, R. 1990. Upper Eocene Gilled Mushroom from the Dominican Republic. Science, 248: 1090-1101.

Posada, D. and Buckley, T. R. 2004. Model selection and model averaging in phylogenetics: advantages of the AIC and Bayesian approaches over likelihood ratio tests. Syst. Biol., 53: 793-808.

Posada, D. and Crandall, K. A. 1998. Modeltest: testing the model of DNA substitution. Bioinformatics, 14(9): 817-818.

Raftery, A. E. 1996. Hypothesis testing and model selection, pp. 163-188. In: Markov Chain Monte Carlo in Practice. W. R. Gilks, S. Richardson and D. J. Spiegelhalter (eds.). Chapman & Hall, London.

Redelings, B. D. and Suchard, M. A. 2005. Joint Bayesian Estimation of Alignment and Phylogeny. Syst. Biol., 54: 401-418.

Redhead, S. A., Vilgalys, R., Moncalvo, J. M., Johnson, J. and Hopple, J. S. 2001. *Coprinus* Persoon and the disposition of *Coprinus* species sensu lato. Taxon, 50: 203-241.

Reeb, V., Lutzoni, F. and Roux, C. 2004. Contribution of RPB2 to multilocus phylogenetic studies of the euascomycetes (Pezizomycotina, Fungi) with special emphasis on the lichen forming Acarosporaceae and evolution of polyspory. Mol. Phylogenet. Evol., 32: 1036-1060.

Ronquist, F. 2004. Bayesian inference of character evolution. Trends Ecol. Evol.,19: 475-481.

Ronquist, F. and Huelsenbeck, J. P. 2003. MRBAYES 3: Bayesian phylogenetic inference under mixed models. Bioinformatics, 19: 1572-1574.

Rouland-Lefevre, C., Diouf, M. N., Brauman, A. and Neyra, M. 2002. Phylogenetic Relationships in *Termitomyces* (Family Agaricaceae) Based on the Nucleotide Sequence of ITS: A First Approach to Elucidate the Evolutionary History of the Symbiosis between Fungus-growing Termites and Their Fungi. Mol. Phyl. Evol., 22: 423-429.

Salisbury, B. A. and Kim, J. 2001. Ancestral state estimation and taxon sampling density. Syst. Biol., 50: 557-564.

Schluter, D., Price, T., Mooers, A. O. and Ludwig, D. 1997. Likelihood of Ancestor states in adaptive radiation. Evolution, 51: 1699-1711.

Schmitt, I., Mueller, G. and Lumbsch, H. T. 2005. Ascoma morphology is homoplaseous and phylogenetically misleading in some pyrenocarpous lichens. Mycologia, 97: 362-374.

Schoch, C. L., Sung, G. H., López-Giráldez, F., Townsend, J. P., Miadlikowska, J., Hofstetter, V., Robbertse, B., Matheny, P. B., Kauff, F., Wang, Z., Gueidan, C., Andrie, R. M., Trippe, K., Ciufetti, L. M., Wynns, A., Fraker, E., Hodkinson, B. P., Bonito, G., Groenewald, J. Z., Arzanlou, M., de Hoog, G. S., Crous, P. W., Hewitt, D., Pfister, D. H., Peterson, K., Gryzenhout, M., Wingfield, M. J., Aptroot, A., Suh, S. O., Blackwell, M., Hillis, D. M., Griffith, G. W., Castlebury, L. A., Rossman, A. Y., Lumbsch, H. T., Lücking, R., Büdel, B., Rauhut, A., Diederich, P., Ertz, D., Geiser, D. M., Hosaka, K., Inderbitzin, P., Kohlmeyer, J., Volkmann-Kohlmeyer, B., Mostert, L., O'Donnell, K., Sipman, H., Rogers, J. D., Shoemaker, R. A., Sugiyama, J., Summerbell, R. C., Untereiner, W., Johnston, P. R., Stenroos, S., Zuccaro, A., Dyer, P. S., Crittenden, P. D., Cole, M. S., Hansen, K., Trappe, J. M., Yahr, R., Lutzoni, F. and Spatafora, J. W. 2009. The Ascomycota tree of life: a phylum-wide phylogeny clarifies the origin and evolution of fundamental reproductive and ecological traits. Syst. Biol., 58(2): 224-239.

Sullivan, J. and Joyce, P. 2005. Model selection in phylogenetics. Annu. Rev. Ecol. Evol. Syst., 36: 445-466.

Suchard, M. A., Weiss, R. E. and Sinsheimer, J. S. 2002. Bayesian Selection of Continuous-Time Markov Chain Evolutionary Models. Mol. Biol. Evol., 18: 1001-1013.

Thiers, H. D. 1984. The secotioid syndrome. Mycologia, 76: 1-8.

Vellinga, E. C. 2004. Genera in the family Agaricaceae - Evidence from nrITS and nrLSU sequences. Mycol. Res., 108: 354-377.

Xiang, Q. Y. and Thomas, D. T. 2008. Tracking character evolution and biogeographic history through time in Cornaceae—Does choice of methods matter? J. Syst. Evol., 46(3): 349-374.

Yang, Z., Kumar, Nei, M. 1995. A new method of inference of ancestral nucleotide and amino acid sequences. Genetics, 141: 1641-1650.

Yang, Z. and Rannala, B. 1997. Bayesian phylogenetic inference using DNA sequences: a Markov chain Monte Carlo method. Mol. Biol. Evol., 14: 717-724.

Index

C

G

H

I

Z

About the Editors

J.K. Misra, D.Sc., has 40 years of teaching and research experience. He has been involved in research projects as Principal Investigator funded by various agencies, and in guiding mycological investigations. Dr. Misra worked in the Philippines, the UK and the USA under various fellowships. He is a widely traveled researcher who has published more than 70 research papers in various scientific journals and edited several books including one that has been published by the International Rice Research Institute, Manila, The Philippines. He is a Fellow of the Indian Botanical Society (FBS) and the International Society for Conservation of Natural Resources (FNRS).

Jalpa P. Tewari, M.Sc., Ph.D. is a Professor Emeritus of Plant Pathology and Member of Alberta Veterinary Research Institute (AVRI) Council at the Department of Agricultural, Food, and Nutritional Science (AFNS), University of Alberta, Edmonton, Alberta, Canada. He has more than 50 years experience of working in multidisciplinary areas such as the Pathology of Crop and Urban Forestry Plants, Mycology (Plant, Human, and Veterinary), Fungal and Biomedical Cell Biology, Composting of Municipal Waste, and International Development Research. Dr. Tewari has worked as a tenured faculty member at the Departments of Botany and AFNS, Universities of Lucknow in India and Alberta in Canada, respectively, Acting Chair of the Department of Plant Science at the University of Alberta, Research Professor at the Department of Microbiology, Immunology, and Biochemistry, Northeastern Ohio Universities College of Medicine (NEOUCOM), Rootstown, Ohio, U.S.A., and as a Visiting Professor at institutions in P. R. China, France, Germany, India, and Spain. He has supervised numerous Ph.D. and Masters' graduate students and post-doctoral fellows from around the world, and has two patents and more than 450 publications to his credit. He is a recipient of an Award for Outstanding Research by the Canadian Phytopathological Society. Dr. Tewari is an Emeritus Member of the Canadian Phytopathological Society, American Phytopathological Society, and Mycological Society of America. He is also a Life Member of the Indian Phytopathological Society, Society for Plant Biochemistry and Biotechnology, and The Indian Science Congress Association.

S.K. Deshmukh has a Ph.D. degree from the University of Saugar and is currently working at the Piramal Life Sciences Limited, Goregaon, Mumbai as Assistant Director- Natural Products. He is a mycologist, and in-charge of the microbial culture collection and fungal screening. He has more than two decades of experience in industrial research that includes development and maintenance of the culture collection of Piramal Life Sciences Limited and fungal screening applicable to both pharmaceutical and agricultural sections. He has been trained in the Laboratory of Hoechst Pharmaceutical Ltd., Frankfurt and also at DSM and GBF Braunsweig, Germany. He has to his credit 52 publications and 3 patents, and has written chapters in twelve books published in India and abroad. He has edited four books on various aspects of fungi and is a member of several scientific societies. He is a Fellow of the Mycological Society of India (FMSI), and has been on the executive of the Mycological Society of India (2002-2006), Member of the editorial board of Kavaka (2007-2009), and Vice President Mycological Society of India (2009- 2010). He delivered the prestigious Shome Memorial Lecture at the 35th Annual Meeting of Mycological Society of India (2009). He is a reviewer of journals like Mycoses, Journal of Complementary and Integrative Medicine, Mycopathologia, Medical Mycology, Indian Journal of Microbiology, and Kavaka.

Color Plate Section

CHAPTER-10

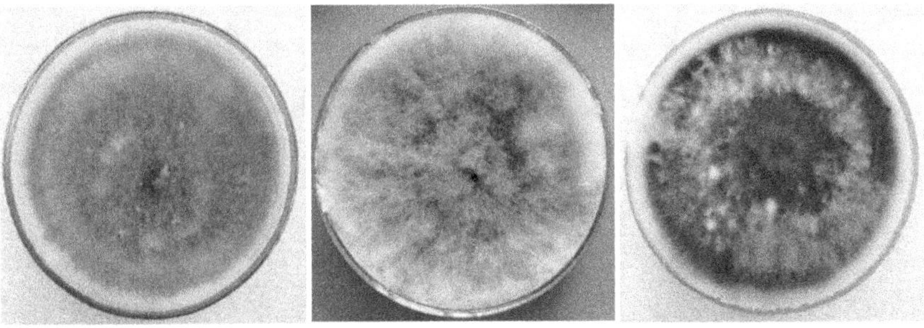

Fig. 10.1 *Fusarium* species (PSA, 2 weeks in dark): *F. graminearum, F. cerealis, F. culmorum*

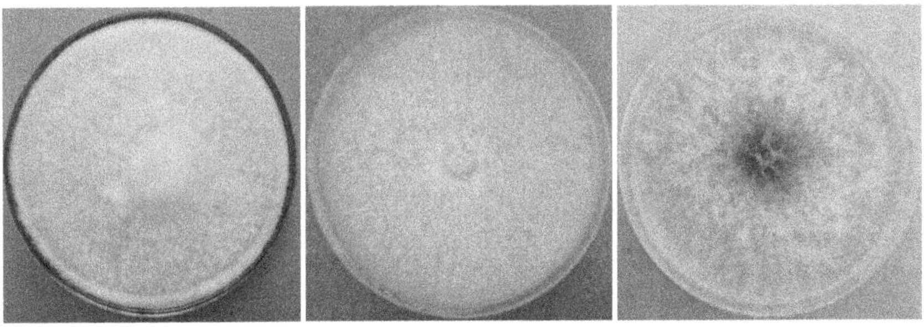

Fig. 10.3 *Fusarium* species (PSA, 2 weeks in dark): *F. poae, F. langsethiae, F. sporotrichioides*

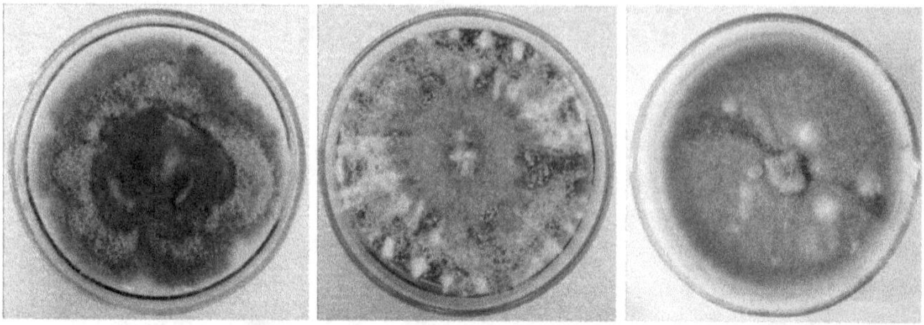

Fig. 10.6 *Fusarium* species (PSA, 2 weeks in dark): *F. tricinctum*, *F. avenaceum*, *F. arthrosporioides*.

For Product Safety Concerns and Information please contact our EU
representative GPSR@taylorandfrancis.com
Taylor & Francis Verlag GmbH, Kaufingerstraße 24, 80331 München, Germany

www.ingramcontent.com/pod-product-compliance
Lightning Source LLC
LaVergne TN
LVHW022314150926
848178LV00003B/62

* 9 7 8 1 5 7 8 0 8 7 2 3 5 *